Solving Large-scale Problems in Mechanics

Concentric Rinds. 1953 Escher, Maurits Cornelius, 1898–1972 Wood engraving, 240 × 240mm (9.5" × 9.5") © 1953 Escher Foundation, Baarn, Holland. All rights reserved.

Solving Large-scale Problems in Mechanics

The Development and Application of Computational Solution Methods

Edited by

Manolis Papadrakakis
National Technical University of Athens, Greece

JOHN WILEY & SONS
Chichester · New York · Brisbane · Toronto · Singapore

Copyright © 1993 by John Wiley & Sons Ltd,
  Baffins Lane, Chichester,
  West Sussex PO19 1UD, England

All rights reserved.

No part of this book may be reproduced by any means,
or transmitted, or translated into a machine language
without the written permission of the publisher.

Other Wiley Editorial Offices

John Wiley & Sons, Inc., 605 Third Avenue,
New York, NY 10158-0012, USA

Jacaranda Wiley Ltd, G.P.O. Box 859, Brisbane,
Queensland 4001, Australia

John Wiley & Sons (Canada) Ltd, 22 Worcester Road,
Rexdale, Ontario M9W 1L1, Canada

John Wiley & Sons (SEA) Pte Ltd, 37 Jalan Pemimpin #05-04
Block B, Union Industrial Building, Singapore 2057

Library of Congress Cataloging-in-Publication Data

Solving large-scale problems in mechanics : the development and
  application of computational solution methods / edited by Manolis
  Papadrakakis.
    p. cm.
  Includes bibliographical references and index.
  ISBN 0 471 93809 2
    1. Structural analysis (Engineering)—Data processing.
  2. Computer-aided engineering.   3. Finite element method—Data
  processing.   I. Papadrakakis, Manolis.
  TA647.S58   1993
  624.1′7—dc20   92-43970
                                                                  CIP

British Library Cataloguing in Publication Data

A catalogue record for this book is available from the British Library

ISBN 0 471 93809 2

Typeset in 10/12pt Palatino by Lasertext Ltd, Stretford, Manchester
Printed and bound in Great Britain by Bookcraft (Bath) Ltd

τό δέ ζητούμενον αλωτόν
Seek, and ye shall find

(Sophocles, Oedipus Rex, v. 110)

Contents

Preface		xi
List of contributors		xv

1 Solving large-scale linear problems in solid and structural mechanics   1
M. Papadrakakis

1.1	Introduction	1
1.2	Iterative equation solving	2
1.3	Global preconditioners	7
1.4	Element-based preconditioners	10
1.5	Domain decomposition-based preconditioners	21
1.6	Performance of the methods under finite precision arithmetic	26
1.7	Concluding remarks	30
	Acknowledgements	32
	Bibliography	32

2 Solving large linearized systems in mechanics   39
B. Nour-Omid

2.1	Introduction	39
2.2	Lanczos algorithm	43
2.3	Conjugate gradient algorithm	45
2.4	Loss of orthogonality	47
2.5	Element-by-element preconditioning	49
2.6	Substructure by substructure preconditioning	51
2.7	Numerical examples	54
2.8	Conclusions	62
	Acknowledgements	62
	Bibliography	62

3 Adaptive iterative solvers in finite elements   65
J. Mandel

3.1	Introduction	65
3.2	Development of robust iterative solvers	66
3.3	Adaptive solver for the p-version FEM in three dimensions	70
3.4	Computational results	76
3.5	Discussion and directions for future research	81

		3.6	Conclusion	86
			Acknowledgements	86
			Bibliography	86

4 Solution of large boundary element equations — 89
J. H. Kane and K. G. Prasad

4.1	Essentials of multi-zone boundary element analysis	89
4.2	Equation solving in BEA	97
4.3	Numerical results	107
4.4	Conclusions	119
	Acknowledgements	121
	Bibliography	121

5 Solution of large eigenvalue problems — 125
G. Gambolati

5.1	Introduction	125
5.2	Eigenproblems in engineering	128
5.3	Optimization of Rayleigh quotients by accelerated conjugate gradients	131
5.4	Preconditioning	135
5.5	Lanczos method	136
5.6	Numerical results	140
5.7	Conclusion	150
	Acknowledgements	151
	Bibliography	151

6 Lanczos eigensolution method for high-performance computers — 157
S. W. Bostic

6.1	Introduction	157
6.2	Application to structural problems	158
6.3	Lanczos method	162
6.4	Computational analysis	164
6.5	Implementation of the Lanczos method	168
6.6	Applications	172
6.7	Summary	179
	Bibliography	180

7 Solving large-scale non-linear problems in solid and structural mechanics — 183
M. Papadrakakis

7.1	Introduction	183
7.2	Explicit methods	184
7.3	Implicit methods	186
7.4	Tracing equilibrium paths	194
7.5	The line search	204
7.6	Numerical examples	208
7.7	Concluding remarks	210
	Acknowledgements	218
	Bibliography	218

8 Mode superposition method — 225
P. Léger

- 8.1 Introduction — 225
- 8.2 Dynamic response analysis by vector superposition methods — 226
- 8.3 Load-dependent transformation vectors — 231
- 8.4 Application examples — 234
- 8.5 Mode superposition methods for non-linear dynamic analysis — 242
- 8.6 Conclusions — 253
- Bibliography — 254

9 Recent developments in time integration methods for structural and interaction system dynamics — 259
K. C. Park

- 9.1 Introduction — 259
- 9.2 Solution techniques for multibody dynamics — 260
- 9.3 Algorithms for control-structure interaction simulation — 269
- 9.4 Solution methods for coupled thermal-structural analysis — 277
- 9.5 Application examples — 282
- 9.6 Closing remarks — 297
- Acknowledgements — 298
- Bibliography — 298

10 Coupled field problems — solution techniques for sequential and parallel processing — 301
I. St. Doltsinis

- 10.1 Introduction — 301
- 10.2 Formulation of the coupled field problem — 303
- 10.3 Solution of the coupled problem taken as a whole — 312
- 10.4 The coupled problem divided into physical subdomains — 315
- 10.5 Parallel processing of coupled field problems — 326
- 10.6 Finite element model — 336
- 10.7 Applications — 341
- Bibliography — 353

11 Direct time-integration methods: stabilized space–time finite element formulation of incompressible flows — 357
S. Mittal and T. E. Tezduyar

- 11.1 Introduction — 357
- 11.2 The governing equations of unsteady incompressible flows — 360
- 11.3 The stabilized space–time finite element formulation — a direct time-integration method — 361
- 11.4 The grouped element-by-element (GEBE) iteration method and vectorization/parallelization concepts — 364
- 11.5 The clustered element-by-element (CEBE) iteration method — 369
- 11.6 Applications: unsteady flows past a circular cylinder — 370
- 11.7 Concluding remarks — 386
- Acknowledgements — 387
- Bibliography — 387

12 Methods for optimization of large-scale systems   391
J. S. Arora

    12.1 Introduction   391
    12.2 Design optimization model   391
    12.3 Basic concepts related to numerical algorithms   394
    12.4 Linearization of the problem   399
    12.5 Methods based on linear approximations   399
    12.6 Sequential quadratic programming: quasi-Newton methods   410
    12.7 Numerical implementation aspects   417
    12.8 Applications of optimization techniques   418
    12.9 Practical design optimization: structural design with finite elements   418
    12.10 Optimization of large systems   422
    12.11 Concluding remarks   424
    Bibliography   425

13 Automatic generation of finite element models   431
M. S. Shephard

    13.1 Introduction   431
    13.2 Definition of a valid finite element mesh   432
    13.3 Key components/issues in automatic mesh generation   437
    13.4 Octree mesh generators   442
    13.5 Delaunay, advancing front (paving), and medial (symmetric) axis transformation mesh generators   447
    13.6 Integration of automatic mesh generation with geometric modelling   451
    13.7 Efficient parallel solution of automatically generated adaptive meshes   453
    Acknowledgements   458
    Bibliography   458

Author index   461

Subject index   467

Preface

The solution of 'large-scale' problems in computational mechanics poses its own particular challenge: the efficient use of available resources. Unless powerful techniques are used, extensive computer resources are required; but on the other hand, the bigger and faster computers become, the larger are the problems scientists and engineers want to solve. The mathematical models become ever more sophisticated, and the computer simulations even more complex and extensive. The pressure on resources remains with us always; and so there is an urgent need to develop and test powerful solution techniques capable of exploiting the potential of modern computers and of accomplishing solutions in acceptable computing time.

This volume is born of that need. It comprises a number of self-contained chapters contributed by leading researchers from the United States and Europe, dealing with recent advances in the solution of 'large-scale' problems in mechanics. The techniques explored here are, in fact, applicable to any problem in the field where available computing power is liable to be stretched to its limit; additionally, much emphasis is given to computational procedures suitable for computing systems with vector and parallel architectures. Each chapter proceeds logically first with theory, then with algorithmic-computational analysis, followed by application to real problems. The result is a comprehensive state-of-the-art treatment of theory and practice — the latter illustrated by extensive numerical examples — which should serve as an essential reference book on this subject.

The first three chapters are concerned with the solution of linear problems resulting from the application of the finite element method. Chapter 1 reviews solution procedures for linear finite element analysis in solid and structural mechanics. Particular emphasis is put on the preconditioned conjugate gradient method. Preconditioning techniques based on global, element-by-element and domain-decomposition concepts are discussed and evaluated in a number of test examples, as well as the performance of the methods under finite precision arithmetic. Chapter 2 explores the relationship between Lanczos and conjugate gradient procedures giving the theoretical framework of each method. It discusses the influence of the loss of orthogonality in the performance of the methods and the use of partial reorthogonalization for improving the robustness of the Lanczos algorithm. Substructure-by-substructure preconditioning is introduced as an extension of the element-by-element preconditioning. In Chapter 3 is developed a flexible and self-adaptive combination of direct and iterative methods which is controlled by an *a priori* estimator of the condition number. The method is illustrated for linear systems arising from the *p*-version finite element method for three-dimensional elasticity. Theoretical results and estimates of convergence are combined with computational results on real-world test problems with distorted elements.

Chapter 4 discusses recent advances in the direct and iterative solution of the algebraic equations resulting from the multi-zone boundary element analysis. Static condensation is shown to reduce dramatically the computational burden associated with matrix fill-in. Algorithms based on the unsymmetric conjugate gradient and generalized minimum residual methods are examined and their performances reported. Preconditioned block iterative unsymmetric equation solution schemes are shown generally to out-perform direct approaches for single and multi-zone models.

Chapters 5 and 6 review methods for the solution of the algebraic eigenvalue problem. In Chapter 5, the accelerated conjugate gradient method for the evaluation of the leftmost part of the eigenspectrum is described via the optimization of the Rayleigh quotient, along with a discussion on effective preconditioners. The Lanczos method is also presented in a sparse finite element context, and it is shown that the conjugate gradient method plays an important role within the Lanczos approach as well. A deflation-conjugate gradient optimization scheme is compared with a combined conjugate gradient–Lanczos method. Chapter 6 discusses the implementation of the Lanczos method for solving the eigenvalue problem. It mainly concentrates on techniques that optimize the solution process by exploiting the vector and parallel capabilities of high performance computers. Issues related to dedicated versus batch mode, direct solvers and storage schemes, efficiency and performance of reorthogonalization, are examined and tested for complex aerospace structures.

Chapter 7 surveys a family of solution procedures for non-linear finite element analysis based on quasi-Newton methods with reduced storage. Depending on the number of updates and the updating formula, a number of solution schemes may be constructed such as the conjugate and secant-Newton methods, the conventional and modified Newton–Raphson methods, as well as a variety of quasi-Newton updates. The non-linear iterative scheme is combined with the preconditioned truncated Lanczos method for the solution of the linearized problem inside the non-linear iteration.

Chapters 8–11 deal with dynamic and coupled-field problems. Chapter 8 presents recent developments and applications of eigen- and load-dependent vector bases for the solution of structural dynamic problems. Classical modal summation techniques, and solution techniques using load-dependent vectors for various forms of loading, are presented in the first section. The effectiveness of mode-superposition methods is then studied in the framework of elasto-plastic seismic analysis of MDOF systems. Chapter 9 surveys three recent developments in time integration methods for dynamic systems analysis: simulation of flexible multi-body dynamics problems; control-structure interaction problems; and coupled thermal-structure transient analysis. These developments have been brought about by the need for realistic modelling and prediction of complex systems, such as high-speed ground transportation vehicles and space satellites, the increased demand for simultaneous design and synthesis of coupled-field problems and the availability of parallel computers. Chapter 10 outlines solution techniques for coupled-field problems. Consideration is given to solution approaches of the coupled problem taken as a whole or divided into physical subdomains. Special attention is given to the implementation of the iterative coupling algorithm on parallel computers with distributed memory. The chapter concludes by addressing the subject of the finite element discretization in connection with mesh generation, adaptive modification, and decomposition for parallel computations. Chapter 11 reviews the stabilized space–time finite element formulation of incompressible flows, including those involving moving boundaries and interfaces. Also reviewed are the efficient iteration techniques employed to solve the equation systems resulting from the space–time finite element discretization of these flow problems. The high vectorizable and parallelizable generalized minimal residual iteration algorithm with the clustered element-by-element preconditioners is given special attention.

Chapter 12 describes the most modern algorithms in structural optimization based on linearization and quadratic programming. Special attention is given to a sequential quadratic programming algorithm that generates and uses second-order information. This is the most recent method of optimization that has proven to be very reliable in solving highly non-linear problems. The numerical difficulties in solving large complex problems are presented as well as a discussion on robustness versus efficiency of optimization methods.

Chapter 13 addresses issues associated with reliable automatic generation of finite element meshes for general three-dimensional geometries and presents the concept of geometric triangulation, which provides a definition of a valid finite element mesh, as well as octree-based meshing procedures. The last part of the chapter considers the influence the types of meshes, generated by automatic mesh generators, have on the efficiency of adaptive solution procedures on parallel computers.

Finally, I would like to express my appreciation to the contributing authors for entering into this venture and for their spirit of collaboration during the preparation of this book.

M. Papadrakakis

List of Contributors

J. S. Arora
Optimal Design Laboratory
Departments of Civil & Environmental Engineering, and
   Mechanical Engineering
College of Engineering
University of Iowa
Iowa City, IA 52242, USA

S. W. Bostic
National Aeronautics and Space Administration
Structural Mechanics Division
Computational Mechanics Branch
Langley Research Center
Hampton, VA 23681-0001, USA

I. St. Doltsinis
Institute for Computer Applications
University of Stuttgart
Pfaffenwaldring 27,
D-7000 Stuttgart 80, Germany
(Address for correspondence:
Sandstrasse 34
7405 Dettenhausen
Germany)

G. Gambolati
Dipartmento di Metodi e Modelli Matematici per le
   Scienze Applicate
Università degli Studi di Padova,
35 131 Padova, Italy

J. H. Kane
Mechanical and Aeronautical Engineering Department
Clarkson University
Potsdam, NY 13699, USA

P. Léger
Department of Civil Engineering
Ecole Polytechnique
University of Montréal Campus
P.O. Box 6079 Station A
Montréal, QC H3C 3A7, Canada

J. Mandel Computational Mathematics Group
Department of Mathematics
University of Colorado at Denver
Denver, CO 80217-3364, USA

S. Mittal Department of Aerospace Engineering and Mechanics,
Army High Performance Computing Research Center
and Minnesota Supercomputing Institute
University of Minnesota,
1200 Washington Avenue South
Minneapolis, MN 55455, USA

B. Nour-Omid Scopus Technology Inc.
1900 Powell St., Suite 900
Emeryville, CA 94608, USA

M. Papadrakakis Institute of Structural Analysis and Aseismic Research
National Technical University of Athens
Zografou Campus,
157 73 Athens, Greece

K. C. Park Department of Aerospace Engineering Sciences and
Center for Space Structures and Controls
University of Colorado at Boulder
Boulder, CO 80309-0429, USA

K. G. Prasad Mechanical and Aeronautical Engineering Department
Clarkson University
Potsdam, NY 13699, USA

M. S. Shephard Scientific Computation Research Center
Schools of Engineering and Science
Rensselaer Polytechnic Institute
Troy, NY 12180-3590, USA

T. E. Tezduyar Department of Aerospace Engineering and
Mechanics, Army High Performance Computing
Research Center and Minnesota Supercomputing
Institute
University of Minnesota
1200 Washington Avenue South
Minneapolis, MN 55455, USA

1
Solving Large-scale Linear Problems in Solid and Structural Mechanics

M. Papadrakakis

National Technical University of Athens, Greece

1.1 INTRODUCTION

The application of the finite element (FE) method in large-scale problems of solid and structural mechanics and in particular in three-dimensional (3D) applications generally leads to sparse matrices of high order in which more than ninety per cent of the elements under the skyline are zero. The solution of such problems based on a direct method requires huge storage demands which is often the limiting factor to the size of problems that can be solved for a given computer configuration. The costs of interest for equation solving are CPU execution time, I/O time for secondary storage and the amount of storage. These costs are related to the size of the bandwidth of the coefficient matrix, as well as to the performance properties of the particular computer system. For large 3D problems the storage required by the factorization matrix of a direct solver necessitates the use of secondary storage resources, and the overall time for the solution not only depends upon the processor speed but also upon the data transfer between secondary memory and main memory. Thus, the execution time of the direct solution technique cannot be estimated for any given problem.

For direct methods it is advisable that the elements and nodes be numbered judiciously to reduce the skyline for both storage and efficiency. In some cases this is difficult to accomplish without performing some preprocessing to renumber the

---

Solving Large-scale Problems in Mechanics, Edited by M. Papadrakakis
© 1993 John Wiley & Sons Ltd

nodes or the elements or both. Moreover, once the mesh in some part of the domain is changed or modified, as in adaptive refinement procedures, the elements and/or nodes must be renumbered for the optimum performance of direct methods. However, in large problems, even the optimum band or front-width may still be prohibitively large, requiring too much core storage and computation time for a given computer.

Among the most desirable advantages of iterative methods for solving large-scale linear FE equations is the low storage requirements. The coefficient matrix is preserved in its original form and is treated as a linear operator for computing matrix–vector products only. Thus, no excessive storage is required for the fill-in elements encountered in the factorization. The unfavourable asymptotic cost growth for 3D applications that direct methods display has inspired the study and development of more efficient iterative linear equation solvers. There are a number of reasons for the increasing popularity of iterative methods: (i) the desire to solve larger and larger problems, especially in three-dimensions; (ii) the development of highly effective preconditioners which can enormously improve the speed and robustness of the iterative procedures; (iii) the ability to solve relatively large-scale problems in mini- and microcomputers; and (iv) they naturally lend themselves to vector and parallel programming.

Iterative methods for solving linear equations also play an important role in other applications of implicit 3D FE analysis. The solution of non-linear problems, in the framework of tracing equilibrium paths through Newton-like iterations [1], requires repeated solution of the linearized equations, which is the most expensive part of the non-linear procedure. In the solution of the eigenvalue problem the advantages of iterative methods are well documented [2,3]. Also implicit time-integration schemes, whether for linear or non-linear simulation, provide another promising application of iterative methods with the additional advantage that the mass matrix helps to regularize the eigenvalue spectrum of the algebraic system.

This chapter reviews solution procedures for linear FE analysis in solid and structural mechanics. Particular emphasis is given to iterative methods based on the conjugate gradient method and the Lanczos method and the application of various preconditioners. Preconditioning techniques based on global, element-by-element and domain decomposition concepts are discussed and evaluated in a number of test examples.

1.2 ITERATIVE EQUATION SOLVING

Linear equation solving represents the major cost of implicit 3D stress analysis. With moderately sized problems of a few tens of thousands degrees of freedom (d.o.f.), the traditional direct solver uses over 90% of total CPU time, while for implicit non-linear applications, it still accounts for 70 to 80% of the total CPU. It is therefore obvious that reducing the cost of linear equation solving will have a significant impact on the total cost of the analysis whether it is linear or non-linear. It has already been stated in Section 1.1 that iterative linear methods are gaining in popularity in large-scale FE analysis. Among the most efficient iterative solvers are the conjugate gradient method and the Lanczos method.

Both methods have a number of attractive properties when compared with a large class of iterative algorithms. In contrast to successive over-relaxation and some of the methods belonging to the family of the three-term recursive expression [4,5], they do not require an estimation of parameters and they take advantage of the distribution of the eigenvalues of the iteration operator. Overall, these methods are optimal in that for linear problems they reduce the particular error norm more than does any other iterative algorithm for the same number of iterations.

1.2.1 The conjugate gradient (CG) method

The CG method was first presented by Hestenes and Stiefel [6] in 1952 but it was not until 1970 that the method was presented as an iterative method for the solution of large sparse systems of linear equations [7]. Solving the linear system of equations

$$Ax = b \tag{1.1}$$

where A is an $N \times N$ symmetric and positive definite matrix, can be viewed as minimizing the associated function $\Phi(x) = 1/2 x^T A x - x^T b$ over all $x \in \mathbb{R}^N$. The essence of the CG method consists in obtaining a new vector x_{i+1} from x_i along a direction d_i at a distance α_i: $x_{i+1} = x_i + \alpha_i d_i$. The direction vectors d_i are mutually conjugate and satisfy the relationships: $d_i^T A d_j = 0$ for $i \neq j$ and $d_{i+1} = r_i + \beta_i d_i$ with $r_i = b - Ax_i$. The step length parameter α_i is fixed by the condition $\partial \Phi(x_i + \alpha_i d_i)/\partial \alpha_i = 0$, while the parameter β_i is determined by the A-conjugate property of the search direction d_i.

An important factor in the success of any iterative method is the preconditioning technique employed. This typically consists of replacing the original linear system by, for example, the equivalent system

$$B^{-1} A x = B^{-1} b \tag{1.2}$$

in which the CG algorithm may be applied by replacing the standard Euclidean inner product by the inner product $(x, y)_B = (Bx, y)$. Different expressions for the characteristic parameters of the method exist. Although all of the versions are mathematically equivalent their computer implementation is not. The following preconditioned conjugate gradient (PCG) version, which is applied to the preconditioned system (1.2) is considered more efficient with respect to computational labour, accuracy and storage requirements [7]:

Initialize:
$$x_0 = 0, r_0 = b, z_0 = B^{-1} r_0, d_0 = z_0$$
for $i = 0, 1, \ldots$
$$\alpha_i = \frac{r_i^T z_i}{d_i^T K d_i}$$
$$x_{i+1} = x_i + \alpha_i d_i$$
$$r_{i+1} = r_i - \alpha_i A d_i$$

Convergence check: (1.3)

if $\|r_{i+1}\| < \varepsilon \|r_0\|$ stop

$z_{i+1} = B^{-1} r_{i+1}$

$\beta_i = \dfrac{r_{i+1}^T z_{i+1}}{r_i^T z_i}$

$d_{i+1} = z_{i+1} + \beta_i d_i$

end for

1.2.2 The Lanczos method

The Lanczos method was introduced in 1950 as a method for solving the symmetric eigenvalue problem and subsequently for the solution of linear systems [8]. The Lanczos algorithm in exact arithmetic may be described as the process of constructing a set of orthonormal vectors $(q_1, q_2, \ldots, q_i)$ by applying Gram–Schmidt orthogonalization to the set of vectors $(r_0, Ar_0, \ldots, A^{i-1} r_0)$ which form the Krylov subspace. The characteristic equation of the method is the three-term recursion

$$\beta_{i+1} q_{i+1} = \hat{q}_{i+1} = A q_i - \alpha_i q_i - \beta_i q_{i-1} \qquad (1.4)$$

where $\alpha_i = q_i^T A q_i$ and $\beta_i = \|q_i\|$. This special choice of the base vectors for the subspace has the property that the projection of A onto this subspace is a tridiagonal matrix T_i:

$$T_i = Q_i^T A Q_i = \begin{bmatrix} \alpha_1 & \beta_2 & & \\ \beta_2 & \alpha_2 & \ddots & \\ & \ddots & \ddots & \beta_i \\ & & \beta_i & \alpha_i \end{bmatrix} \qquad (1.5)$$

where $Q_i = [q_1 q_2 \ldots q_i]$. The key relationships between the quantities computed by the Lanczos method can be summarized in three equations:

$$Q_i^T Q_i = I \qquad (1.6)$$

$$AQ_i - Q_i T_i = \hat{q}_{i+1} e_i^T \qquad (1.7)$$

$$Q_i^T r_i = 0 \qquad (1.8)$$

where e_i is the ith column of the $i \times i$ identity matrix I_i. The approximation to the solution may be obtained from the solution y_i of the Krylov subspace by

$$x_i = Q_i y_i \qquad (1.9)$$

where y_i is the solution of the weak form of $Ax = b$

$$T_i y_i = Q_i^T b \tag{1.10}$$

with $r_0 = b$. The residual norm can be computed during the iteration process without the need to perform explicitly the operation $r_i = b - Ax_i$.

With the ability to monitor the norm of the residual without explicit calculation of the residual vector, the Lanczos method need not accumulate the current approximation x_{i+1}, at each step. The iterative procedure may therefore proceed until convergence by storing the coefficients of T_i in main memory and the basis vectors in secondary storage. The reduced system of equation (1.10) may then be solved and the final solution is obtained from equation (1.9) by recalling the basis vectors.

The preconditioned Lanczos algorithm is better formulated by considering an LU decomposition of the preconditioning matrix B applied to the transformed system $L^{-1}AU^{-1}\bar{x} = \bar{b}$ with $\bar{x} = Ux$, $\bar{b} = L^{-1}b$. Based on an implementation suggested by Paige and Saunders [9] which transforms the Lanczos process to an iterative algorithm similar to CG, the accumulated version of the algorithm in its preconditioned form may be described as follows [10]:

Initialize
$$\bar{x}_0 = q_0 = 0$$
$$\bar{b} = L^{-1}b, \beta_1 = \|\bar{r}_0\| = \|\bar{b}\|$$
$$q_1 = \frac{\bar{b}}{\beta_1}, u_1 = L^{-1}AU^{-1}q_1$$
$$\alpha_1 = q_1^T u_1$$
$$d_1 = \alpha_1, \zeta_1 = \beta_1/d_1$$
$$c_1 = q_1$$
$$\bar{x}_1 = \bar{x}_0 + \zeta_1 c_1$$

for $i = 1, 2, \ldots$
$$\hat{q}_{i+1} = u_i - \alpha_i q_i - \beta_i q_{i-1}$$
$$\beta_{i+1} = \|\hat{q}_{i+1}\|$$

Convergence check
$$\|\hat{r}_i\| = \beta_{i+1}|\zeta_i| \tag{1.11}$$
$$\text{if } \|\hat{r}_i\| < \varepsilon\|\hat{r}_0\| \text{ then } x_i = U^{-1}\bar{x}_i$$
stop

$$q_{i+1} = \hat{q}_{i+1}/\beta_{i+1}$$
$$u_{i+1} = L^{-1}AU^{-1}q_{i+1}$$
$$\alpha_{i+1} = q_{i+1}^T u_{i+1}^i$$

Compute factorized elements of T_{i+1}
$$l_{i+1} = \beta_{i+1}/d_i$$
$$d_{i+1} = \alpha_{i+1} - \beta_{i+1}l_{i+1}$$

Update the solution

$$\zeta_{i+1} = -\zeta_i \beta_{i+1}/d_{i+1}$$
$$c_{i+1} = q_{i+1} - l_{i+1} c_i$$
$$\bar{x}_{i+1} = \bar{x}_i + \zeta_{i+1} c_{i+1}$$

end for

The above handling of the preconditioner is referred by Shakib [11] as 'wrap-around' preconditioning and is better suited with reorthogonalization strategies.

The main feature of the Lanczos method is that it gradually transforms the original space specified by the coefficient matrix to another space governed by a tridiagonal matrix and a number of orthonormal vectors. This equivalence between the two spaces offers some advantages over CG as a consequence of the transfer of some properties from one space to the other. The Lanczos method is considered a very powerful method for the solution of the eigenvalue problem [2,3]. It can also handle more effectively indefinite coefficient matrices[1] and a sequence of right-hand sides in linear problems. In the latter case it was found that an accumulated Lanczos implementation proposed in Reference [10], may improve the computing time by a factor of four for every ten right-hand sides considered.

The Lanczos method is directly related to the CG method. In fact, the diagonal and subdiagonal elements, (α_i^{LAN}, α_i^{CG}, β_i^{CG}) of the tridiagonal matrix are formed with respect to the parameters α_i^{CG}, β_i^{CG} of the CG method according to

$$\alpha_i^{LAN} = \frac{1}{\alpha_i^{CG}} + \frac{\beta_{i-1}^{CG}}{\alpha_{i-1}^{CG}}, \qquad \beta_{i+1}^{LAN} = -\frac{\sqrt{\beta_i^{CG}}}{\alpha_i^{CG}}. \tag{1.12}$$

In addition to the storage of A and B, the CG and Lanczos algorithms require four and five N-length vectors respectively.

With exact arithmetic both algorithms will produce identical approximations at every iteration. In a finite arithmetic environment the search/basis vectors do not maintain the conjugacy/orthogonality, and may in fact become linearly dependent. The loss of conjugacy/orthogonality depends on the conditioning of the coefficient matrix and the round-off unit of the computer. Thus, the convergence properties of the method may be severely affected in ill-conditioned problems and in a computer environment with low precision arithmetic. A remedy against this phenomenon is to apply some sort of reorthogonalization procedure. An effective but very expensive process is to reorthogonalize each new basis vector against all previous Lanczos vectors. More elaborate but computationally more efficient techniques have been proposed by Parlett and Scott [12] and Simon [13] based on the principle of semi-reorthogonalization. These techniques, although they do have beneficiary effects on the rate of convergence of the methods [14,15], are of questionable practical interest for large-scale computations due to the tremendous amount of the I/O operations involved. In fact, for a range of problems considered by Papadrakakis and Smerou [10], it was found more advantageous to allocate storage for a more effective preconditioner rather than retaining a set of basis vectors (see also Ferencz [16]).

1.2.3 *Computational strategies*

A number of special strategies can be used to increase the efficiency of the linear iterative solver. These strategies are based on the selection and implementation of proper preconditioning matrices combined with different ways of handling the matrix equations. The resulting hybrid techniques are more effective than the exclusive use of either a direct or an iterative approach. In order to be effective, a preconditioner has to combine the following properties: (i) to reduce the ellipticity of the coefficient matrix as much as possible; (ii) to be easily computed and handled by the computer; (iii) to be sparse in order to reduce storage demands and to keep matrix–vector operations at low cost; and (iv) to be susceptible to vectorization and/or parallelization. Since some of these requirements are conflicting, a dynamic balance has to be kept between these properties according to the type of problem and the available computing resources. Herein, three preconditioning and matrix handling strategies are discussed: global, element-by-element, and block and domain decomposition implementations.

1.3 GLOBAL PRECONDITIONERS

1.3.1 *SSOR-based preconditioners*

With the term global preconditioners we refer to preconditioning strategies which are applied directly to the coefficient matrix A. An easily applied preconditioner of this type is derived from the combination of CG with the symmetry successive over-relaxation method (SSOR) [17–20]. The preconditioner takes the form of the SSOR characteristic matrix

$$B = (D + \omega C)D^{-1}(D + \omega C^{\mathrm{T}}) \tag{1.13}$$

where the scalar factor $1/[\omega(2 - \omega)]$ has been dropped since it has no influence on the condition number of $B^{-1}A$. $A = D + C + C^{\mathrm{T}}$, D is diagonal and ω is a relaxation parameter usually taken in the vicinity of one.

The SSOR preconditioner does not require any extra storage outside A, is quite insensitive to the choice of ω and can be used with Eisenstat [21] implementation to decrease the computational cost per iteration. A modification to the SSOR preconditioner is proposed by Papadrakakis [22,23] by discarding those elements of C which influence to a lesser extent the ellipticity of A. Thus,

$$B = (D + \omega C_p)D^{-1}(D + \omega C_p^{\mathrm{T}}) \tag{1.14}$$

The criterion which specifies the elements of A to be retained in C_p is defined by a user-controlled parameter ψ. All elements of A satisfying $\alpha_{ij}^2 < \psi \alpha_{ii} \alpha_{jj}$ are considered small with diminishing effect on the condition number of A and are not included in B.

1.3.2 Incomplete factorization-based preconditioners

An alternative preconditioning method is based on an incomplete factorization of A. The reason for performing an incomplete factorization is to obtain a reasonably accurate factorization of A without generating too many fill-ins. Such an approach leads to the factorization $LDL^T = A - E$, where E is an error matrix which does not have to be formed. For this class of methods, E is defined either by the prescribed positions of the elements to be rejected [24–28] or by the computed positions of 'small' elements in L which do not satisfy a specified magnitude criterion and therefore are discarded [24,28–35]. The rejection of certain elements inside the preconditioning matrix often leads to an unstable factorization process, with small diagonal elements in B, or to an indefinite B. To avoid this phenomenon, several techniques have been proposed by either modifying A before factorization, making it more diagonal dominant [25,27], or by correcting the diagonal elements during the factorization [24,27,29–32,35,36].

Experience in FE applications has shown that the modification proposed by Jennings and co-workers [29–32] gives the more robust and efficient incomplete factorization preconditioner [32–34]. This incomplete factorization procedure can be described using a row-by-row approach and referring to L^T rather than L, as follows:

for each row $i = 1, 2, \ldots, N$
  for each column $j = i + 1, \ldots, N$

$$a_{ij}^* = a_{ij} - \sum_{k=1}^{i-1} l_{ki} d_k l_{kj}$$

$$\bar{a}_{ii} = a_{ii} + \sum_{k=1}^{i-1} e_{ii}^{(k)} + \sum_{k=i+1}^{j-1} e_{ii}^{(k)} - \sum_{k=1}^{i-1} l_{ki} d_k l_{ki}$$

$$\bar{a}_{jj} = a_{jj} + \sum_{k=1}^{i-1} e_{jj}^{(k)}$$

  check for rejection of a_{ij}^* (1.15)
  if yes: $s = (\bar{a}_{ii}/\bar{a}_{jj})^{1/2}$

$$e_{ii}^{(j)} = s |a_{ij}^*|$$

$$e_{jj}^{(i)} = s^{-1} |a_{ij}^*|$$

$$a_{ij}^* = 0$$

  next j
  $d_i = \bar{a}_{ii}, \ l_{ii} = 1$
  for each column $j = i + 1, \ldots, N$
    $l_{ij} = a_{ij}^*/d_i$
  next j
next i

In the incomplete factorization by 'position', a_{ij}^* is not retained if it does not fit into the sparsity pattern adopted which is usually forced to coincide with that of A. In the incomplete factorization by 'magnitude' the criterion to decide whether a_{ij}^* is large or small is defined by $(a_{ij}^*)^2 < \psi \bar{a}_{ii} \bar{a}_{jj}$, where ψ again is a user-controlled parameter. The choice of $\psi = 0$ corresponds to the complete factorization, while in the case of $\psi = 1$ it corresponds to a form of diagonal scaling. The computer implementation of the incomplete factorization by magnitude, as was given in a Fortran listing in Reference [31], is a little cumbersome due to the complicated addressings and the high demands for auxiliary storage during the factorization. Improved computational versions of the incomplete factorization by magnitude are proposed in References [33, 34].

The incomplete factorization by position is more suitable where the equations are of regular form, such as for a set of finite difference equations specified over a regular grid, or when they are derived by an adaptive p-version FE procedure [37]. However, when the sparsity pattern of the equations is irregular, or some individual elements give much stronger coupling between the variables than others, then incomplete factorization by magnitude performs better. Furthermore, in highly ill-conditioned problems, and in those cases where some additional storage is available, the incomplete factorization by magnitude is preferable. Applications of incomplete factorization preconditioners in different FE problems may be found in References [22, 29–34, 37–41].

1.3.3 *Numerical results*

The performance of the global preconditioners presented in this section is investigated in two test examples. The first is a clamped square plate, discretized in its quarter with a mesh of 50 × 50 finite elements, having 7400 d.o.f. with half-bandwidth 159. The second example is the space frame dome of Figure 1.1 with 3126 d.o.f. and half-bandwidth 318. Both structures are subject to a central point load. The tests were performed on a SUN Sparc station. Figure 1.2 shows the total computer storage required for handling a direct method of solution, with the skyline storage scheme and Cholesky factorization of Reference [42], and PCG with different preconditioners. OIC(ψ) corresponds to the Optimized Incomplete Cholesky preconditioner of algorithm (1.15), as optimized in Reference [34], with the rejection parameter ψ. OICP stands for the Optimized Incomplete Cholesky by Position in which the prescribed positions of the elements to be retained coincide with the sparsity pattern of the coefficient matrix, PPR(ψ) is the modified SSOR preconditioner of equation (1.14) and DCG corresponds to the diagonally scaled CG. In estimating the computer storage it is assumed that integers are stored as INTEGER*2 or INTEGER*4, according to their maximum values, and the floating point variables as REAL*8, using double precision arithmetic.

The number of iterations and the total CPU time for the two examples are depicted in Figures 1.3 and 1.4, respectively (convergence tolerance $\varepsilon = 10^{-6}$). These figures clearly indicate a superiority in both computing time and storage of PCG methods

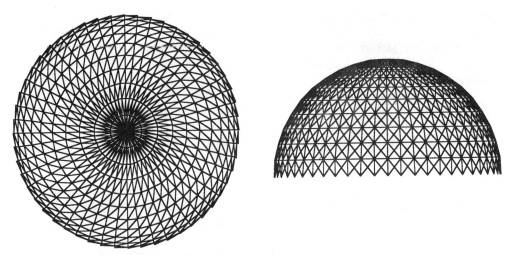

Figure 1.1 Space frame dome (3126 d.o.f., half bandwidth = 318).

over direct methods. An overwhelming advantage of PCG is observed in the well-conditioned 3D example, where the diagonally scaled CG was found much faster than the direct method. The presented results give the opportunity to assess the efficiency of preconditioners in problems with different degrees of ill-conditioning. In the ill-conditioned plate example, the performance of OIC(ψ) is much better than OICP, SSOR, PPR and DCG at the expense of storing more elements in B. The optimum values, in terms of computing time, of the rejection parameters ψ are problem dependent. However, in a number of tests performed they were quite often found to be in the range $(10^{-6} - 10^{-5})$ for OIC and $(10^{-2} - 10^{-1})$ for PPR.

Figure 1.5 presents some tests with the optimized incomplete Cholesky factorization in which different techniques have been applied to preserve the stability of the factorization process. OIC(ψ) and OICP implement the diagonal correction proposed by Jennings and included in algorithm (1.15). The symbol 'DM(α)' corresponds to a Diagonal Modification before factorization by adding αa_{ii} to the diagonal elements of the original matrix A [25,27,40,41]. The optimum scalar parameter α for each case is shown in parenthesis. The symbol 'NDC' means that No Diagonal Correction is applied either before or during the factorization. The results indicate the superiority of Jennings' correction. More details on the performance of global preconditioners may be found in Reference [34].

1.4 ELEMENT-BASED PRECONDITIONERS

Element-by-element (EBE) methods were motivated by the intrinsic characteristic of FE implementations in which global data structure can be stored and maintained at the local (element) level. Fox and Stanton [43] and Fried [44] were among the first to point out that the assembly of the stiffness matrix is not essential and that the matrix-vector product required for the CG iterations can be obtained by performing operations at the element level. Hughes *et al.* [45] and Nour-Omid and Parlett [46] were the first to address the problem of preconditioning for the solution of differential

1.4 ELEMENT-BASED PRECONDITIONERS

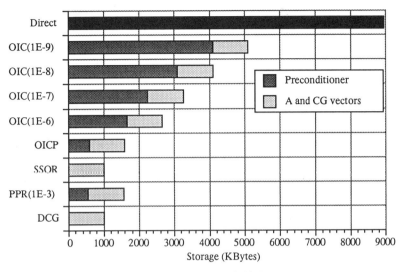

(a) Plate example (50x50 elements)

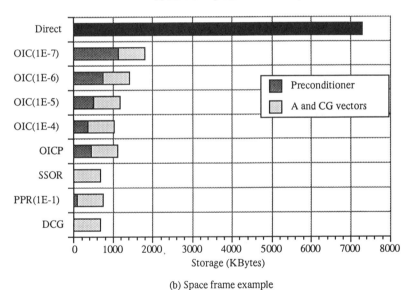

(b) Space frame example

Figure 1.2 Total storage requirements of the skyline direct method and PCG with different global preconditioners.

equations based on EBE implementations [47,48]. The primary purpose of the EBE applications was to keep storage requirements to a minimum, but the advent of vector and parallel computing has given to this approach a new interest especially in large-scale 3D problems [23,49–56].

The assembling of A can be expressed as

$$A = \sum_{j=1}^{p} A_j \qquad (1.16)$$

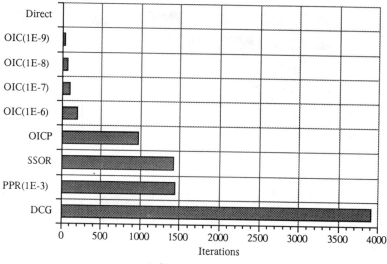

(a) Plate example (50x50 elements)

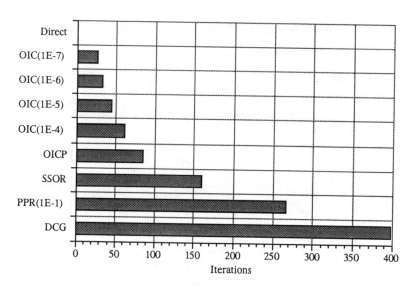

(b) Space frame example

Figure 1.3 Iteration performance of different global preconditioners.

in which $\underset{\sim}{A}_j$ is the so-called macro-element matrix given by $P_j^T A_j P_j$ where P_j is the global connectivity matrix of element j which maps the global to element d.o.f. The global nodal displacements can be related to the unassembled global nodal displacements by the transformation $x_e = Px$, where x_e is defined from the element nodal displacements in global d.o.f. as: $x_e = [x_{(1)} x_{(2)} \ldots x_{(p)}]^T$ and $P = [P_1 P_2 \ldots P_p]^T$. Similarly, the unassembled global force vector can be defined as: $b_e = [b_1 b_2 \ldots b_p]^T$ and $b = P^T b_e$.

1.4 ELEMENT-BASED PRECONDITIONERS

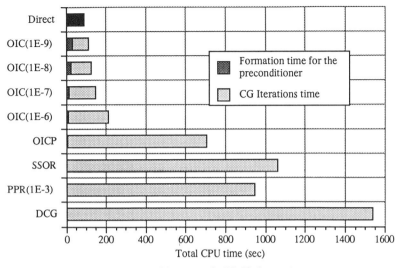

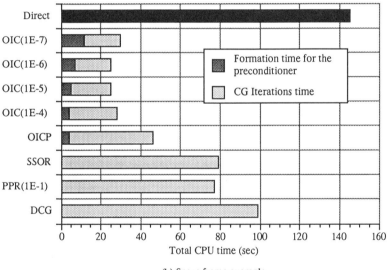

Figure 1.4 Total CPU performance of the skyline direct method and PCG with different global preconditioners.

1.4.1 Product form EBE preconditioners

The most widely used EBE preconditioners are based on approximate factorization techniques of element data structures [49]. The so-called Crout EBE preconditioner is defined as follows:

$$B = D^{1/2} \left(\prod_{j=1}^{p} P_j^T \tilde{L}_j P_j \right) \left(\prod_{j=1}^{p} P_j^T \tilde{D}_j P_j \right) \left(\prod_{j=p}^{1} P_j^T \tilde{L}_j P_j \right) D^{1/2} \qquad (1.17)$$

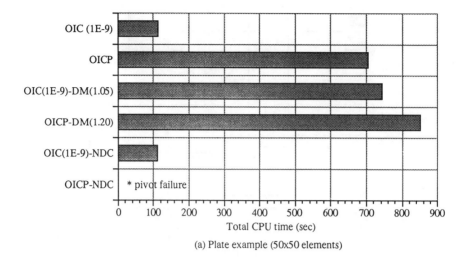

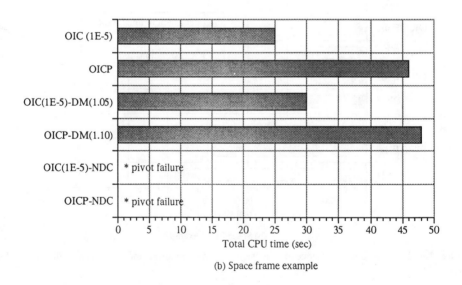

Figure 1.5 Performance of different incomplete factorization schemes.

where $\tilde{L}_j, \tilde{D}_j, \tilde{L}_j^T$ are the Crout factors of the regularized matrix $\tilde{A}_{j-1} + (\bar{A}_j - \bar{D}_j)$ with $\bar{D}_j = \text{diag}(\bar{A}_j)$ and $\bar{A}_j$ is the scaled element matrix

$$\bar{A}_j = P_j(D^{-1/2}A_j D^{-1/2})P_j^T \tag{1.18}$$

where D is the assembled global diagonal matrix $D = \text{diag}(A)$. A drawback of the Crout preconditioning is the need to store twice as much data as for the simple CG.

An alternative preconditioner called symmetrized Gauss–Seidel EBE preconditioner

$$B = D^{1/2} \left(\prod_{j=1}^{p} P_j^T [I + L_j^s] P_j \right) \left(\prod_{j=p}^{1} P_j^T [I + (L_j^s)^T] P_j \right) D^{1/2} \qquad (1.19)$$

with $\bar{A}_j = L_j^s + D_j^s + (L_j^s)^T$, uses the same element data as required to compute the matrix product Ad_i, thereby reducing storage requirements to 50% compared to Crout–EBE.

1.4.2 An additive EBE preconditioner

An EBE preconditioner based on polynomial preconditioning techniques is proposed by Papadrakakis *et al* [57]. The inverse of $\bar{A} = D^{-1/2} A D^{-1/2}$ is expressed as

$$\bar{A}^{-1} = (I - V)^{-1} \qquad (1.20)$$

with

$$V = D^{-1/2}(D - A)D^{-1/2}. \qquad (1.21)$$

For $|V| < 1$ the inverse of $I - V$ can be written by the polynomial expansion

$$(I - V)^{-1} = \eta_0 I + \eta_1 V + \eta_2 V^2 + \cdots + \eta_k V^k \qquad (1.22)$$

where η_k are properly selected coefficients for accelerated convergence. Following the same definitions for the element quantities of V as for A with $V_j = D^{-1/2}(D_j - A_j)D^{-1/2}$ the inverse expression (1.22) may be written as

$$(I - V)^{-1} = \eta_0 I + \eta_1 \sum_{j=1}^{p} V_j + \eta_2 \left(\sum_{j=1}^{p} V_j \right)^2 + \cdots + \eta_k \left(\sum_{j=1}^{p} V_j \right)^k. \qquad (1.23)$$

An approximation to equation (1.23) is obtained by replacing $(\Sigma\ \underline{V}_j)^k$ by $\Sigma\ (\underline{V}_j)^k$ which is computationally more efficient in terms of operations and communication between elements. Thus, equation (1.23) becomes

$$(I - V)^{-1} \approx \eta_0 I + \eta_1 \sum_{j=1}^{p} V_j + \eta_2 \sum_{j=1}^{p} V_j^2 + \cdots + \eta_k \sum_{j=1}^{p} V_j^k \qquad (1.24)$$

and the preconditioning matrix may now be written as

$$B^{-1} = M = D^{-1/2}(I - V)^{-1}D^{-1/2}$$

or

$$M = \eta_0 D^{-1} + \sum_{j=1}^{p} \underset{\sim}{R}_j. \tag{1.25}$$

The inverse preconditioning matrix is formed from element quantities

$$\underset{\sim}{M} = \sum_{j=1}^{p} \underset{\sim}{M}_j \tag{1.26}$$

where $\underset{\sim}{M}_j \approx \eta_0 \underset{\sim}{D}_j + \underset{\sim}{R}_j$ and $\underset{\sim}{D}_j$ is the part of D which corresponds to the element d.o.f.

The EBE-CG algorithm may be stated as follows:

Initialize:
$x_0 = 0, r_0 = b$
$\{r_e\}_0 = Pr_0 \equiv [r_{(1)} r_{(2)} \ldots r_{(p)}]_0^T$
$\{z_j\}_0 = M_j r_{(j)}, j = 1, \ldots, p$
$\{z_e\}_0 = [z_1 z_2 \ldots z_p]_0^T$
$z_0 = P^T \{z_e\}_0$
$d_0 = z_0$
for $i = 0, 1, \ldots$
$\quad \{u_j\}_i = A_j \{d_{(j)}\}_i, j = 1, \ldots, p$
$\quad \{u_e\}_i = [u_1 u_2 \ldots u_p]_i^T$
$\quad u_i = P^T \{u_e\}_i$ $\qquad\qquad\qquad\qquad\qquad$ (1.27)
$\quad \alpha_i = \dfrac{r_i^T z_i}{d_i^T u_i}$
$\quad x_{i+1} = x_i + \alpha_i d_i$
$\quad r_{i+1} = r_i - \alpha_i u_i$
$\quad$ Convergence check:
$\qquad$ if $\|r_{i+1}\| \leq \varepsilon \|r_0\|$ stop
$\quad \{r_e\}_{i+1} = Pr_{i+1}$
$\quad \{z_j\}_{i+1} = M_j \{r_{(j)}\}_{i+1}, j = 1, \ldots, p$
$\quad \{z_e\}_{i+1} = [z_1 z_2 \ldots z_p]_{i+1}^T$
$\quad z_{i+1} = P^T \{z_e\}_{i+1}$
$\quad \beta_i = \dfrac{r_{i+1}^T z_{i+1}}{r_i^T z_i}$

$$d_{i+1} = z_{i+1} + \beta_i d_i$$
end for

The element contributions may also be considered to be contributions by groups of elements or superelements depending on the partition of the structure into a number of substructures. This is equivalent to an overlapping block partitioning in which each block corresponds to the assembled stiffness matrix of each substructure.

The handling of R_j in equation (1.25) is an important factor for the efficiency of the preconditioner. In order to avoid operations with the components of $\mathbf{R}_j$ in each PCG iteration, this matrix is formed only once and stored before the iterative process begins. In an effort to further reduce the operations involved in the multiplication $M_j\{r_{(j)}\}_i$ a rejection by magnitude criterion is adopted for the off-diagonal elements of M: $(R_{kl})_j^2 \leq \psi D_k^{-1} D_l^{-1}$ which controls the number of elements to be retained in M through the parameter ψ ($0 \leq \psi \leq 1$). An important feature of this preconditioner is its local additive nature, which is inherently parallelizable without requiring complicated manipulation or any colouring schemes.

1.4.3 Polynomial preconditioning

Polynomial preconditioning consists of choosing a polynomial $s(\bar{A})$ to replace $\bar{A}^{-1}$. The simplest polynomial expansion of equation (1.22) is given by the Neumann series [58] with $\eta_0 = \cdots = \eta_k = 1$: $s(\bar{A}) = I + V + V^2 + \cdots$, . More efficient polynomials can be obtained by making the spectrum of the preconditioned matrix as close as possible to that of the identity. One way to achieve this is to minimize the residuals $\|I - s(\bar{A})\bar{A}\|$, where $\|.\|$ represents the L_2-norm.

According to Johnson et al. [59] and Saad [60] this can be achieved by defining the best polynomial which minimizes

$$\max_{\lambda \varepsilon [a,b]} |1 - \lambda s(\lambda)| \tag{1.28}$$

where $[a, b]$ is some interval containing the eigenvalues of $\bar{A}$, with $0 < a < b$. The Chebyshev polynomials or alternatively a least squares minimization procedure may be implemented to solve this problem. The latter approach takes the form: Find s that minimizes $\|1 - \lambda s(\lambda)\|_w$, where $\|.\|_w$ is the L_2-norm on $[a, b]$ with respect to some weight function $w(\lambda)$. It has been observed in Reference [58] that least squares polynomials tend to perform better than those based on the Chebyshev ones, in that they lead to a better overall clustering of the spectrum.

Following the derivation of Saad [60] the parameters a and b are computed as the Gershgorin estimates of $\lambda_{\min}$ and $\lambda_{\max}$. However, since the Gershgorin lower bound 'a' may be non-positive even when A is positive definite, the least squares polynomials are thus defined in the interval $[0, b]$. The computed first two polynomials take the form

$$S_1(\lambda) = -\frac{5}{4}b + \lambda \tag{1.29}$$

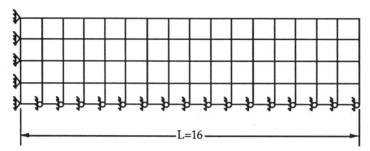

Figure 1.6 Kinematic boundary conditions for the plane strain cantilever beam.

$$S_2(\lambda) = \frac{7}{8}b^2 - \frac{7}{4}b\lambda + \lambda^2 \qquad (1.30)$$

with $\eta_0 = -5b/4$, $\eta_1 = 1$ and $\eta_0 = 7b^2/8$, $\eta_1 = -7b/4$, $\eta_2 = 1$, the coefficients of λ in equations (1.29) and (1.30), respectively.

The main attraction of polynomial preconditioning is that the only operations involving matrices are products with vectors. As a consequence of this, the performance of this preconditioning strategy in vector and parallel machines is facilitated and portability is achieved by optimizing matrix by vector multiplications only [61]. The performance of polynomial preconditioning is enhanced on parallel machines with a large number of processors. On the other hand, global incomplete factorization preconditioning exhibits poor performance during forward and backward substitutions in vector and parallel machines. To alleviate this inefficiency, approximations to the inverse factors are obtained using polynomial expansions [62]. More general type polynomial preconditioners may be formed when combined with k steps of a relaxation type iterative method [63–65]. Application of polynomial preconditioning in structural analysis problems has been recently presented in References [57, 66]

1.4.4 Numerical results

In order to investigate the behaviour and efficiency of the additive EBE preconditioner, a series of test examples has been performed on a cantilever beam subjected to a transverse end load and a simply supported plate under a central point load. The kinematic boundary conditions of half the beam example are illustrated in Figure 1.6 for a mesh with 4 × 16 plain strain elements; however, all numerical results reported are for a mesh containing 16 × 43 elements (1122 d.o.f.). The plate example is discretized with a mesh of 16 × 16 elements (855 d.o.f.). We will index the results by the element aspect ratio (ratio of the largest dimension to the smallest) ignoring discretization errors. The length of both structures is held constant but the other dimension is varied to alter the condition number of the stiffness matrix. The ill-conditioning is increased with large element aspect ratios.

1.4 ELEMENT-BASED PRECONDITIONERS

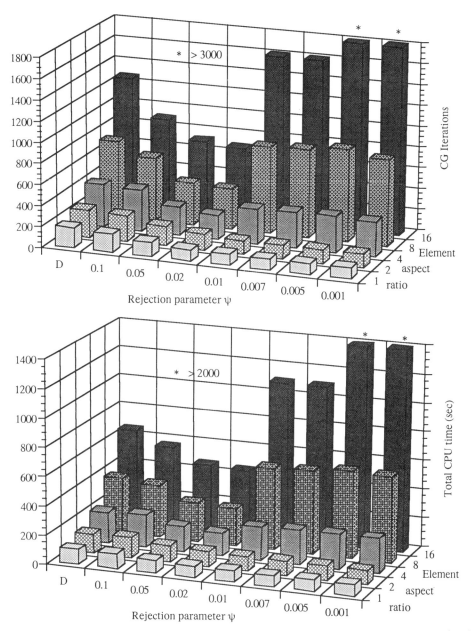

Figure 1.7 Performance of diagonal and EBE(ψ) preconditioners versus element aspect ratio for the plane strain cantilever beam (16 × 32 elements) with 32 superelements.

The results depicted in Figures 1.7 and 1.8 provide a comparison of the EBE(ψ) preconditioner with the diagonally scaled CG ('D') (convergence tolerance $\varepsilon = 10^{-5}$). Different values of the rejection parameter ψ which controls the number of retained elements in R_j of equation (1.25) are used for the EBE preconditioner. The EBE formulation of algorithm (1.27) is implemented with 32 superelements for the beam

20 SOLVING LINEAR PROBLEMS IN SOLID AND STRUCTURAL MECHANICS

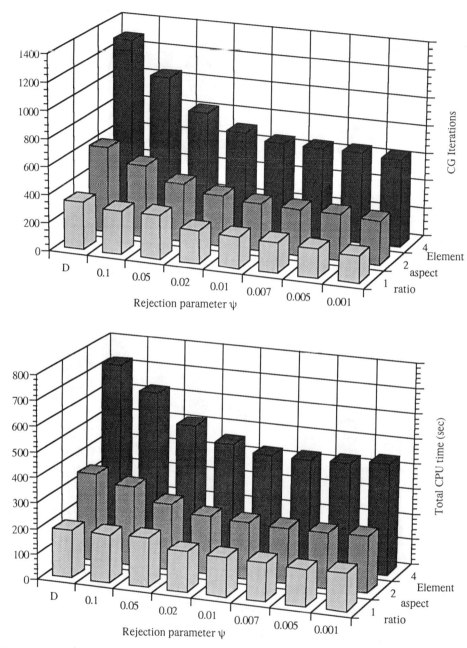

Figure 1.8 Performance of diagonal and EBE(ψ) preconditioners versus element aspect ratio for the plate problem (16 × 16 elements) with 8 superelements.

and 8 superelements for the plate. Each superelement contains 68 and 135 d.o.f. respectively. A second-order polynomial approximation is used with coefficients obtained from equation (1.30). The results clearly indicate that the superiority of the

EBE preconditioner is improved with an increase of the ellipticity of the problem. It is also anticipated that in vector and parallel computers the cost advantage of this EBE preconditioner over DCG will be increased. Further insight into the performance of this additive EBE scheme is provided in References [57, 67].

1.5 DOMAIN DECOMPOSITION-BASED PRECONDITIONERS

The partition of structural problems into sub-problems or substructures has served as a practical computational technique since the early 1960s [68] in the aerospace industry primarily as a technique to facilitate the solution of large problems. Its use was also motivated by manufacturing considerations and computational savings. A revived interest in this approach has been observed with the widespread use of microcomputers and the increasing availability of multiprocessor computers. It is due to this strong emergence of this new generation of computers that the method of substructures has attracted growing popularity over the last few years. The numerical analysis community used the substructuring concept as a way of breaking up the solution of elliptic problems into problems on smaller domains and provided a new understanding and a theoretical foundation for the use of iterative methods in this process.

Let us consider, for example, an FE domain D subdivided into D_j non-overlapping subdomains. The mesh nodes which are common to the subdomain interfaces define a global interface noted by D_I and are not part of any D_j. Numbering first the nodal point unknowns within D_I results in an arrow pattern for the coefficient matrix. To simplify our notation, however, we restrict our discussion to the case when a region is partitioned into two substructures 1 and 2 with dimensions N_1, N_2, and an interior interface 3 with dimension N_0. The extension of the implementation to cases with a larger number of substructures is immediate. The corresponding coefficient matrices are $A_{11}^{(1)}, A_{11}^{(2)}, A_{00}$, and the coupling between the substructure and the interface $A_{(10)}^{(1)}$ and $A_{10}^{(2)}$. This defines the following splitting position of the global equations (1.1)

$$\begin{bmatrix} A_{11} & A_{10} \\ A_{01} & A_{00} \end{bmatrix} \begin{bmatrix} x_1 \\ x_0 \end{bmatrix} = \begin{bmatrix} b_1 \\ b_0 \end{bmatrix} \qquad (1.31)$$

or

$$\begin{bmatrix} A_{11}^{(1)} & 0 & A_{10}^{(1)} \\ 0 & A_{11}^{(2)} & A_{10}^{(2)} \\ A_{01}^{(1)} & A_{01}^{(2)} & A_{00} \end{bmatrix} \begin{bmatrix} x_1^{(1)} \\ x_1^{(2)} \\ x_0 \end{bmatrix} = \begin{bmatrix} b_1^{(1)} \\ b_1^{(2)} \\ b_0 \end{bmatrix} \qquad (1.32)$$

in which A_{00} is assembled from two parts coming from elements belonging to the two different substructures $A_{00} = A_{00}^{(1)} + A_{00}^{(2)}$.

Domain decomposition techniques can also be identified at the algebraic level by partitioning the coefficient matrix in an appropriate way and grouping together the associated degrees of freedom.

1.5.1 Partitioned matrix implementation

Consider now the exact blockwise symbolic factorization of the partitioned A:

$$A = \begin{bmatrix} A_{11}^{(1)} & & \\ 0 & A_{11}^{(2)} & \\ A_{01}^{(1)} & A_{01}^{(2)} & S \end{bmatrix} \begin{bmatrix} I & 0 & A_{11}^{(1)-1} A_{10}^{(1)} \\ & I & A_{11}^{(2)-1} A_{10}^{(2)} \\ & & I \end{bmatrix} \tag{1.33}$$

where

$$S = S^{(1)} + S^{(2)} = A_{00}^{(1)} - A_{01}^{(1)} A_{11}^{(1)-1} A_{10}^{(1)} + A_{00}^{(2)} - A_{01}^{(2)} A_{11}^{(2)-1} A_{10}^{(2)} \tag{1.34}$$

is the capacitance matrix or the Schur complement of A_{00} in the matrix A. One way to derive a preconditioner for A is to approximate the exact factorization of equation (1.33) under three different options [69]: (i) the approximation $Y_{11}^{(i)} \approx A_{11}^{(i)-1}$, used in forming an approximation to the Schur complement $M = A_{00} - A_{01}^{(1)} Y_{11}^{(1)} A_{10}^{(1)} - A_{01}^{(2)} Y_{11}^{(2)} A_{10}^{(2)}$; (ii) the approximation $Z_{11}^{(i)} \approx A_{11}^{(i)-1}$, that will be used to solve linear subsystems during the implementation of the preconditioner; (iii) further approximations of the approximate Schur complement may be applied. We therefore arrive at the following incomplete factorization which is a preconditioner for the system (1.32):

$$B = \begin{bmatrix} & & \\ & & \\ A_{01}^{(1)} & A_{01}^{(2)} & \end{bmatrix} \begin{bmatrix} Z_{11}^{(1)-1} & & \\ 0 & Z_{11}^{(2)-1} & \\ A_{01}^{(1)} & A_{01}^{(2)} & M \end{bmatrix} \begin{bmatrix} I & 0 & Z_{11}^{(1)} A_{10}^{(1)} \\ & I & Z_{11}^{(2)} A_{10}^{(2)} \\ & & I \end{bmatrix} \tag{1.35}$$

Early work on this type of partitioning was associated mainly with block diagonal partitioning techniques [19,30]. The idea of a block-type factorization preconditioner was first suggested by Underwood [70] and later further elaborated by Concus et al. [71] and Axelsson et al. [72]. A complete bibliography regarding this approach may be found in Reference [73], while detailed analyses of block preconditioners have been presented in References [69, 74, 75].

1.5.2 Schur complement matrix implementation

An alternative representation of the solution of equation (1.31) with the blockwise factorization of equation (1.33) is derived by eliminating the interior unknowns and solving the Schur complement system

1.5 DOMAIN DECOMPOSITION-BASED PRECONDITIONERS

$$Sx_0 = \tilde{b}_0 \tag{1.36}$$

where

$$\tilde{b}_0 = b_0 - A_{01}^{(1)} A_{11}^{(1)-1} b_1 - A_{01}^{(2)} A_{11}^{(2)-1} b_2. \tag{1.37}$$

The right-hand side of equation (1.37) can be obtained by solving two subdomain problems and performing the multiplication with the sparse matrices $A_{01}^{(1)}$ and $A_{01}^{(2)}$.

The matrix S is generally dense, and therefore a direct method for solving system (1.36) could be prohibitively expensive if N_0 is large. The usual implementation of solving the Schur complement matrix equations is to apply PCG methods. In applying the PCG method one needs to evaluate the matrix–vector product Sd_i, for the direction vector d_i, which requires a solution for each subdomain. It also requires the solution of the system $Mz_i = r_i$, where M is the preconditioning matrix of the Schur complement. Keyes and Gropp [76] referred to methods which depart from equation (1.31) as partitioned matrix methods and from equation (1.36) as Schur complement matrix methods. The implementation of the PCG algorithm in the domain decomposition formulation of equation (1.31) based on the Schur complement equation (1.36) may be realized as follows:

Initialize:
$\{x_0\}_0$
$\{x_1\}_0 = -A_{11}^{-1}(A_{10}\{x_0\}_0 - b_1)$
$\{r_0\}_0 = b_0 - (A_{01}\{x_1\}_0 + A_{00}\{x_0\}_0)$
$\{z_0\}_0 = M^{-1}\{r_0\}_0$
$\{d_0\}_0 = \{z_0\}_0$
for $i = 0, 1, \ldots$
$\quad \{d_1\}_i = -A_{11}^{-1} A_{10} \{d_0\}_i$ $\tag{1.38}$
$\quad \{u_0\}_i = A_{00}\{d_0\}_i + A_{01}\{d_1\}_i$
$\quad \{x_0\}_{i+1} = \{x_0\}_i + \alpha_i \{d_0\}_i$
$\quad \{r_0\}_{i+1} = \{r_0\}_i - \alpha_i \{u_0\}_i$
$\quad$Convergence check:
$\quad \quad$ if $\|\{r_0\}_{i+1}\| \leq \varepsilon \|\{r_0\}_0\|$ stop
$\quad \{z_0\}_{i+1} = M^{-1}\{r_0\}_{i+1}$
$\quad \{d_0\}_{i+1} = \{z_0\}_{i+1} + \beta_i \{d_0\}_i$
end for

The parameters α_i, β_i are computed as in the algorithm (1.3) in which the inner products operate on dimension N_0. Once the interface variables x_0 are determined, the internal variables x_1 can be solved from

$$x_1 = -A_{11}^{-1}(A_{10} x_0 - b_1) \tag{1.39}$$

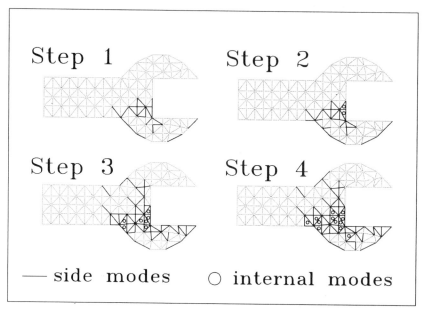

Figure 1.9 Wrench problem. Meshes of adaptive refinement.

or, alternatively can be updated at each iterative step as

$$\{x_1\}_{i+1} = \{x_1\}_i + \alpha_i \{d_1\}_i. \tag{1.40}$$

Drija [77,78], Golub and Mayers [79] and Bjorstad and Widlund [80] were among the first to introduce preconditioning techniques for the Schur complement matrix. Later many other techniques were proposed along this approach [81–89]. The reader is also referred to recent surveys [61,76] and to the collection of papers which can be found in the proceedings [90,91]. Other effective preconditioning methods based on the domain decomposition principle are connected with the use of hierarchical basis functions (p-version finite elements) [37,92–100] and the related approach of the multigrid method [101,102].

1.5.3 Numerical results

Domain decomposition techniques are applied for the solution of equations resulting from an adaptive (p-version) finite element formulation. Following an error estimation rule, the discretization error is reduced by providing some additional higher-order modes on a hierarchical basis. Figure 1.9 depicts a test example of a wrench with its initial discretization and the additional modes (side and internal) for four consecutive steps of the adaptive procedure. The wrench has initially 227 d.o.f.

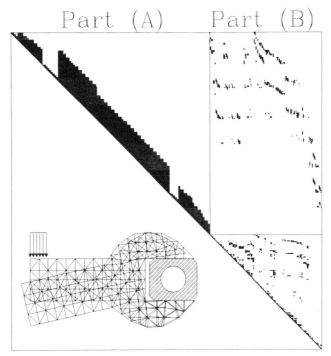

Figure 1.10 Non-zero terms of the stiffness matrix after step 4 and the deformed shape of the wrench example.

Figure 1.10 shows the sparsity pattern of the stiffness matrix for four adaptive steps with the initial matrix (A) being fully factorized. The stiffness matrix becomes very sparse with very large bandwidth since the additional modes are coupled with the initial nodal degrees of freedom. The number of adaptive steps and the corresponding energy factor achieved are illustrated in Figure 1.11. The results indicate that after four adaptive steps the energy of the system becomes more than 90% of the total energy corresponding to all additional second- and third-order side modes and internal modes of the system.

The performance of different solution techniques applied to the adaptive finite element procedure with four additional steps (227 + 397 d.o.f.), and to the full problem with all additional modes (227 + 1743 d.o.f.), is illustrated in Figure 1.12 (convergence tolerance $\varepsilon = 10^{-4}$). Domain decomposition techniques are applied at the algebraic level by performing iterations either on part (A) or (B) of the coefficient matrix. Thus, DD(A)-ICP means that iterations are performed on the Schur complement of part (A) of the matrix using incomplete Cholesky factorization by position (ICP) for the solutions involving part (B) of the matrix. DD(B)-ICP corresponds to the Schur complement of part (B) with an ICP of the initial block matrix (B) as preconditioner. In both cases the complete factorization of part (A) is used.

The abbreviation PCG-ICP corresponds to a global application of PCG with the complete factorized matrix of part (A) and the incomplete Cholesky factorization by

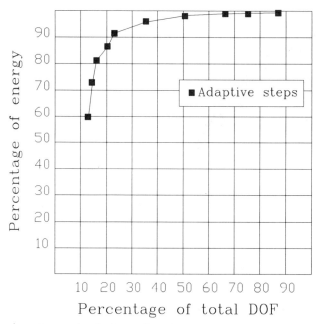

Figure 1.11 Energy levels during the adaptive steps for the wrench problem.

position of part (B) as preconditioner. The abbreviations 'SSOR' and 'D' mean that the SSOR and diagonal preconditioners are employed respectively. Further insight into the behaviour of PCG strategies in connection with adaptive finite element techniques and domain decomposition may be found in Reference [37].

1.6 PERFORMANCE OF THE METHODS UNDER FINITE PRECISION ARITHMETIC

In FE computations the formation and solution of equation (1.1) entail some or all of the following types of errors: (i) errors arising due to the lack of definition of the original element and loading data; (ii) round-off errors and numerical integration errors in the computation of the element matrices and element vectors; (iii) errors caused by the truncation of information associated with the lower eigenvalues of the global matrix in the assembly of A and the assembly of b; (iv) round-off errors occurring during the process of solving equation (1.1). It will be interesting to study how the ellipticity of A influences the behaviour of solution algorithms in the presence of truncation errors in the formation of A and round-off errors during the solution phase. As a result the equations represented in the computer due to these errors take the form

$$(A + \delta A)(x + \delta x) = b \tag{1.41}$$

in which δA contains the truncation error and the perturbation in order to account

1.6 PERFORMANCE OF THE METHODS UNDER FINITE PRECISION ARITHMETIC

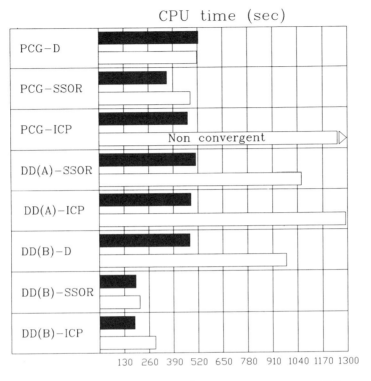

■ Adaptive refinement (Error in energy 5%)
□ Full problem

Figure 1.12 Performance of solution techniques for the wrench problem.

for the round-off errors and δx is the corresponding deviation due to both types of errors.

A method for solving equation (1.1) is numerically stable in a t-digit precision computer, and in the presence of round-off errors only, if the relative change in the solution vector satisfies

$$\frac{\|\delta x\|}{\|x\|} \leq c_1 10^{-t} \kappa(A) \tag{1.42}$$

where $\kappa(A)$ is the spectral condition number of A and c_1 is a constant which depends only on the size N of the matrix. In order to evaluate an estimate of the influence of initial truncation errors, it can be assumed that $\|\delta A\|/\|A\| < 10^{-t}$ and $\|\delta x\|/\|x\| < 10^{-s}$, and an estimate of the number of accurate digits obtained in the solution can be given by

$$s \geq t - \log[\kappa(A)]. \tag{1.43}$$

It is therefore clear that the influence of these errors on the solution of the system

Table 1.1 Condition number versus element aspect ratio of the plane strain cantilever beam (16 × 64 elements)

Element Aspect ratio	1	2	4	8	16	40
Condition Number	3.0×10^5	1.5×10^6	2.9×10^7	3.8×10^8	5.0×10^9	1.7×10^{11}

is dependent on the round-off error unit of the computer and the value of the condition number. Ill-conditioning may be the result of large differences in the material or stiffness properties of the elements, of irregular mesh geometry or fine mesh discretizations. In addition, structures, such as those which can experience large rigid-body movements associated with small strains suffer from ill-conditioning.

Wilkinson [103], followed by a number of investigators [104–107], has considered iterative improvement to eliminate the effect of round-off errors from direct methods, but the effectiveness of this scheme has materialized for well-conditioned problems only. Besides Rossanoff *et al.* [108] and Roy [109] have pointed out that round-off and truncation errors may be in opposite directions, and that removing round-off errors does not always result in an improved solution. Iterative refinement procedures aiming at reducing both truncation and round-off errors have been presented in References [110–113], while an improved approach, based on PCG, is presented by Papadrakakis and Bitoulas [114]. On the other hand different implementations of CG have been reported in the past [115, 116] in an effort to remove the influence of truncation and/or round-off errors, while the Lanczos method has implemented with partial reorthogonalization techniques in order to alleviate the effect of round-off errors [10, 14–16].

1.6.1 *Numerical results*

The aim of this section is to investigate the behaviour of PCG in the presence of truncation and round-off errors, and to shed some light on what can be expected in a 'real world' of computation. A series of test examples have been performed on the beam example of Figure 1.6 under a transverse end load. The numerical results reported in this case concern a mesh containing a 16 × 64 mesh with 2112 d.o.f. Table 1.1 depicts the variation of the condition number for the unpreconditioned matrix with respect to the element aspect ratio. The tests were performed on a SUN 3/50 workstation with single precision $\sim 10^{-7}$ and double precision $\sim 10^{-16}$.

Due to a variety of solution schemes evaluated, it is useful to adopt a shorthand notation based on the appellation.

(*prec, residual, precision*)

where: *prec* = 'IC', 'SSOR' or 'D', corresponding to the incomplete Cholesky factorization by magnitude (algorithm 1.15) [31,34], or SSOR preconditioner (equation (1.13)), or a diagonal scaling; *residual* = R or D, corresponding to the Recursive expression for the residual computation of algorithm (1.3), or the Defining formula

Table 1.2 Convergence paths and decimal places of accuracy of PCG algorithms for the plane strain cantilever beam (16 × 64 elements) with element aspect ratio = 4

Method	\multicolumn{6}{c	}{Iterations to achieve tolerance}	Total CPU (s)	Accuracy
	10^{-1}	10^{-2}	10^{-3}	10^{-4}	10^{-5}	10^{-6}		
IC(1E-3)-R-DP	151	157	164	168	173	177	337	9.9
IC(1E-3)-R-SP	163	210	279	349	443	462	751	1.7
IC(1E-3)-R-MP	151	160	193	252	259	285	538	6.9
IC(1E-3)-D-DP	151	157	164	168	173	177	457	9.9
IC(1E-3)-D-SP	—	—	—	—	—	—	**	—
IC(1E-3)-D-MP	169	—	—	—	—	—	*	4.9
SSOR-R-DP	209	216	221	227	232	236	411	9.9
SSOR-R-SP	228	298	372	506	548	637	1026	1.7
SSOR-R-MP	209	225	257	343	350	387	604	6.7
SSOR-D-DP	209	216	221	227	232	236	564	9.9
SSOR-D-SP	—	—	—	—	—	—	*	—
SSOR-D-MP	224	—	—	—	—	—	*	3.9
D-R-DP	686	705	722	739	756	771	786	9.9
D-R-SP	726	947	1139	1706	1808	1864	1680	1.7
D-R-MP	691	742	839	1055	1181	1575	1512	6.6
D-D-DP	686	706	724	741	757	773	1342	9.9
D-D-SP	—	—	—	—	—	—	*	—
D-D-MP	—	—	—	—	—	—	*	3.9(700)

*no convergence, **divergence

$r_i = Ax_i - b$; precision = DP, SP or MP, corresponding to Double, Single or Mixed Precision computation.

An initial insight into the level of effort, the efficiency and the accuracy achieved by different versions of PCG is provided by Table 1.2. The amount of truncation and round-off errors is measured by comparing the tip displacement with the solution obtained by a direct method with all calculations performed in double precision arithmetic. In the mixed-precision tests all calculations are performed in single precision, except for Kd_i of algorithm (1.3) for the recursive computation of residuals, and Kx_i for the defining expression of residuals, where double precision is used. Numbers in parentheses indicate the corresponding number of iterations to achieve the depicted accuracy. Figure 1.13 summarizes the performance of IC preconditioner with recursive evaluation of residuals for double-, single- and mixed-precision computations (convergence tolerance $\varepsilon = 10^{-6}$).

An insight into the effectiveness of the proposed mixed-precision technique is provided in Figure 1.14 where two different IC(ψ) preconditioners with the same storage requirements are compared. The amount of storage is controlled by the value of the rejection parameter ψ. In mixed-precision computations ψ is allowed to take smaller values. The results show that, with the mixed formulation proposed, the computational efficiency of PCG can be further increased without severely affecting the accuracy of the computed solution. For more tests and results the reader is referred to Reference [114].

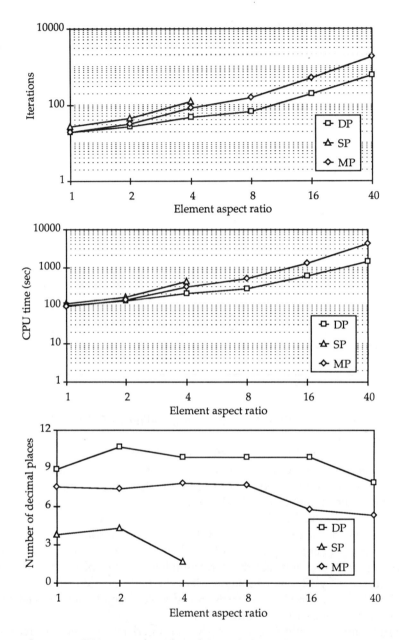

Figure 1.13 Iterations, CPU time and number of decimal places of IC-R ($\psi = 1\text{E-}6$) preconditioner for the plane strain cantilever beam (16 × 64 elements).

1.7 CONCLUDING REMARKS

The numerical results presented in this chapter confirm the superiority of the iterative methods over direct equation solvers in large-scale FE applications. PCG-like methods

1.7 CONCLUDING REMARKS

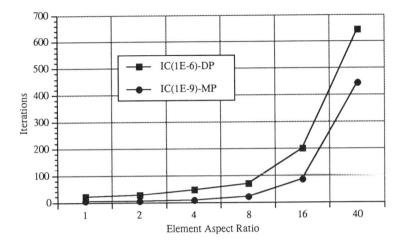

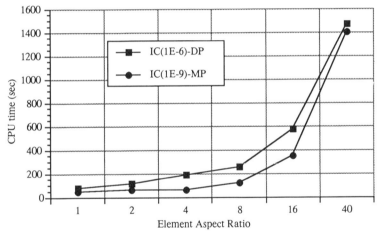

Figure 1.14 Performance of double and mixed precisions versus element aspect ratio for the plain strain cantilever beam (16 × 64 elements).

offer considerable flexibility in the selection of the proper preconditioning technique according to the type of problem and the available computing resources.

Stronger preconditioners, based on the incomplete Cholesky factorization by magnitude, provide a significant improvement of PCG performance in comparison to the SSOR preconditioners and diagonal scaling, at the expense of more storage requirements. This improvement is more pronounced for very ill-conditioned problems. The incomplete Cholesky factorization, based on an optimized computer implementation, avoids complicated addressings during the formation of the preconditioner and keeps the required storage to a minimum.

The element-based preconditioners offer certain advantages with respect to storage requirements and their implementation on vector and parallel computer architectures.

The presented additive element-by-element preconditioner appears to have some favourable characteristics. It is based on a multi-element group partitioning of the entire structure where formation of the coefficient matrix, as well as the iterative procedure, are performed on the superelement level. The resulting preconditioner has an additive form which is susceptible to parallelization without complicated manipulations. The preconditioned matrix–vector operations are performed at the superelement level in exactly the same manner as the stiffness matrix operation and communication between processors is required only after the completion of these multiplications.

Domain decomposition techniques, applied either on the physical domain or on the algebraic level, can substantially improve the performance of the solution procedure in conventional and, even more dramatically, in multiprocessor computer applications. The application of PCG techniques to the Schur complement matrix equations gives the opportunity to implement efficient global and element-by-element preconditioners and create very powerful tools for the solution of large-scale FE problems.

Finally, the proposed implementation of the incomplete Cholesky preconditioner with a mixed-precision formulation may even further improve the efficiency of preconditioned iterative solution methods.

ACKNOWLEDGEMENTS

The author is very grateful to G. Babilis, N. Bitoulas and K. Hatzikonstantinou for their cooperation in implementing the algorithms and obtaining the numerical results presented in this chapter.

BIBLIOGRAPHY

[1] Papadrakakis, M. Solving large-scale problems in non-linear solid and structural mechanics, This volume
[2] Bostic, S.W., Lanczos eigenvalues method for high performance computers, This volume.
[3] Gambolati, G., Solution of large-scale eigenvalue problems, This volume.
[4] Engeli, M., Ginsburg, T., Rutishauser, H. and Stiefel, E. (1959) *Refined Iterative Methods for Computation of the Solution and the Eigenvalues of Self-adjoint Boundary Value Problems*, Birkhauser Verlag, Basel/Stuttgart.
[5] Papadrakakis, M. (1982) A family of methods with three-term recursion formulae, *Int. J. Num. Meth. Engg*, **18**, 1785–1799.
[6] Hestenes, M.R. and Stiefel, E. (1952) Methods of conjugate gradients for solving linear systems, *J. Res. Nat. Bur. Stand.*, **49**, 409–436.
[7] Reid, J.K. (1971) On the method of conjugate gradients for the solution of large sparse systems of linear equations, in J.K. Reid (ed.), *Large Sparse Sets of Linear Equations*, Academic Press, New York, pp. 231–254.
[8] Lanczos, C. (1952) Solution of systems of linear equations by minimized iterations, *J. Res. Nat. Bur. Stand.*, **49**, 33–53.
[9] Paige, C.C. and Saunders, M.A. (1975) Solution of sparse indefinite systems of linear equations, *SIAM J. Numer. Anal.*, **12**, 617–629.
[10] Papadrakakis, M. and Smerou, S. (1990) A new implementation of the Lanczos method

in linear problems, *Int. J. Num. Meth. Engg*, **29**, 141–159.
[11] Shakib, F., *Finite Element Analysis of the Compressible Euler and Navier–Stokes Equations*, Ph.D. Thesis, Stanford University, Stanford CA, 1989.
[12] Parlett, B.N. and Scott, D.S. (1984) The Lanczos algorithm with selective orthogonalization, *Math. Comp.*, **42**, 115–142.
[13] Simon, H.D. (1984) The Lanczos algorithm with partial reorthogonalization, *Math. Comput.*, **42**, 115–142.
[14] Nour-Omid, B., Solving large linearized systems in mechanics, This volume.
[15] Coutinho, A.L.G.A., Alves, J.L.D., Landau, L., Lima, E.C.P. and Ebecken, U.F.F. (1987) On the application of an element-by-element Lanczos solver to large offshore structural engineering problems, *Comp. Struct.*, **27**, 27–37.
[16] Ferencz, R.M. (1989) *Element-by-element Preconditioning Techniques for Large-scale, Vectorized Finite Element Analysis in Nonlinear Solid and Structural Mechanics*, Ph.D. Dissertation, Department of Mechanical Engineering, Stanford University, USA.
[17] Evans, D.J. (1967) The use of pre-conditioning in iterative methods for solving linear equations with symmetric positive definite matrices, *J. Inst. Math. Applics.*, **4**, 295–314.
[18] Axelsson, O. (1976) A class of iterative methods for finite element equations, *Comput. Meth. Appl. Mech. Engg.*, **9**, 123–137.
[19] Concus, P., Golub, G.H. and O'Leary, D.P. (1976) A generalized conjugate gradient method for the numerical solution of elliptic partial differential equations, in J.R. Bunch and D.J. Rose (eds.), *Sparse Matrix Computations*, Academic Press, pp. 309–332.
[20] Young, D.M., Hayes, L. and Schleicher, E. (1975). The use of the accelerated SSOR method to solve large linear systems, Abstract 1975 *SIAM Fall Meeting*, San Francisco.
[21] Eisenstat, S.C. (1981) Efficient implementation of a class of preconditioned conjugate gradient methods, *SIAM J. Sci. Statist. Comput.*, **2**, 1–4.
[22] Papadrakakis, M. (1986) Accelerating vector iteration methods, *J. Appl. Mech.*, ASME, **53**, 291–297.
[23] Papadrakakis, M. and Dracopoulos, M.C. (1991) A global preconditioner for the element-by-element solution methods, *Comp. Meth. Appl. Mech. Engg.*, **88**, 275–286.
[24] Kershaw, D.S. (1978) The incomplete Cholesky-conjugate gradient method for the iterative solution of systems of linear equations, *J. Comput. Phys.*, **26**, 43–65.
[25] Manteuffel, T.A. (1980) An incomplete factorization technique for positive definite linear systems, *Math. Comput.*, **34**, 473–497.
[26] Meijerink, J.A. and van der Vorst, H.A. (1977) An iterative solution of linear systems of which the coefficient matrix is a symmetric M-matrix, *Math. Comput.*, **32**, 148–162.
[27] Gustafsson, A.I. (1978) A class of first order factorization methods, *BIT*, **18**, 142–156.
[28] Munksgaard, N. (1980) Solving sparse symmetric sets of linear equations by preconditioned conjugate gradients, *ACM Trans. Math. Software*, **6**, 206–219.
[29] Tuff, A.D. and Jennings, A. An iterative method for large systems of linear structural equations, *Int. J. Num. Meth. Engg.*, **7**, 175–183.
[30] Jennings, A. and Malik, G.M. (1978) The solution of sparse linear equations by the conjugate gradient method, *Int. J. Num. Meth. Engg*, **12**, 141–158.
[31] Ajiz, M.A. and Jennings, A. (1984) A robust incomplete Choleski conjugate gradient algorithm, *Int. J. Num. Meth. Engg*, **20**, 949–966.
[32] Jennings, A. (1983) Development of an ICCG alogirthm for large sparse systems, in D.J. Evans (ed.), *Preconditioning Methods, Analysis and Applications*, Gordon & Breach, New York, pp. 425–438.
[33] Papadrakakis, M. and Dracopoulos, M.C. (1991) Improving the efficiency of incomplete Choleski preconditionings, *Comm. Appl. Num. Meth.*, **7**, 603–612.
[34] Papadrakakis, M. and Bitoulas, N. (1991) *An Optimized Computer Implementation of Incomplete Factorization*, Report, Institute of Structural Analysis and Aseismic Research,

National Technical University of Athens.

[35] Axelsson, O. and Munksgaard, N. (1983) Analysis of incomplete factorizations with fixed storage allocation, in D.J. Evans (ed.), *Preconditioning Methods, Analysis and Applications*, Gordon & Breach, New York, pp. 219–241.

[36] Axelsson, O. and Lindskog, G. (1986) On the eigenvalue distribution of a class of preconditioning methods, *Num. Math.*, **48**, 479–498.

[37] Babilis, G. and Papadrakakis, M. (1992) *Adaptive Strategies and Solution Techniques for the p-version of the Finite Element Method* Report, Institute of Structural Analysis and Aseismic Research, National Technical University of Athens.

[38] Becovier, M. and Rosenthal, A. (1986) Using the conjugate gradient method with preconditioning for solving FEM approximations of elasticity problems, *Eng. Comp.*, **3**, 77–80.

[39] Gambolati, G., Pini, G. and Zilli, G. (1988) Numerical comparison of preconditionings for large sparse finite element problems, *Num. Meth. Part. Differ. Eqs.*, **4**, 139–157.

[40] Angeleri, F., Sonnad, V. and Bathe, K.J. (1989) Studies of finite element procedures — An evaluation of preconditioned iterative solvers, *Comp. Struct.*, **32**, 671–677.

[41] Tan, L.H. and Bathe, K.J. (1991) Studies of finite element procedures — The conjugate gradient and GMRES Methods in ADINA and ADINA-F, *Comp. Struct.*, **40**, 441–449.

[42] Bathe, K.J. and Wilson, E.L. (1976) *Numerical Methods in Finite Element Analysis*, Prentice-Hall, Englewood Cliffs, New Jersey.

[43] Fox, R.L. and Stanton, E.L. (1968) Developments in structural analysis by direct energy minimization, *AIAA J.*, **6**, 1036–1042.

[44] Fried, I. (1969) More on gradient iterative methods in finite element analysis, *AIAA J.*, **7**, 565–567.

[45] Hughes, T.J.R., Winget, J., Levit, I. and Tezduyar, T. (1983) New alternating direction procedures in finite element analysis based upon EBE approximate factorizations, in S.N. Atluri and N. Perrone (eds.), *Computer Methods for Nonlinear Solids and Structural Mechanics*, AMD-Vol. 4, ASME, New York, pp. 75–109.

[46] Nour-Omid, B. and Parlett, B.N. (1985) Element preconditioning using splitting techniques, *SIAM J. Sci. Stat. Comput.*, **6**, 761–770.

[47] Ortiz, M., Pinsky, P.M. and Taylor, R.L. (1986) Unconditionally stable element-by-element algorithm for dynamic problems, *Comp. Meth. Appl. Mech. Engg.*, **59**, 327–362.

[48] Hughes, T.J.R., Levit, I. and Winget, J.M. (1983) Implicit stable algorithms for heat conduction analysis, *J. Engg. Mech. Div.*, ASCE, **109**, 576–585.

[49] Hughes, T.J.R., Ferencz, R.M. and Hallquist, J.O. (1987) Large scale vectorized implicit calculations in solid mechanics on a Cray X-MP/48 utilizing EBE preconditioned conjugate gradients, *Comp. Meth. Appl. Mech. Engg.*, **61**, 215–248.

[50] Barragy, E. and Carey, G.F. (1988) A parallel element-by-element solution scheme, *Int. J. Num. Meth. Engg.*, **26**, 2367–2382.

[51] Nour-Omid, B., Parlett, B.N. and Raefsky, A. (1988) Comparison of Lanczos with conjugate gradient using element preconditioning, in R. Glowinski, G.H. Golub, G.A. Meurant and J. Periaux (eds.), *Domain Decomposition Methods for Partial Differential Equations*, SIAM, Philadelphia, pp. 250–260.

[52] Hughes, T.J.R. and Ferencz, R.M. (1988) Fully vectorized EBE preconditioners for nonlinear solid mechanics: Applications to large-scale three-dimensional continuum, shell and contact/impact problems, in R. Glowinski, G.H. Golub, G.A. Meurant and J. Periaux (eds.), *Domain Decomposition Methods for Partial Differential Equations*, SIAM, Philadelphia, pp. 261–280.

[53] Tezduyar, T.E., Liou, J., Nguyen, T. and Poole, S. (1989) Adaptive implicit-explicit and parallel element-by-element iteration schemes, in T.F. Chan, R. Glowinski, J. Periaux and O.B. Widlund (eds.), *Domain Decomposition Methods, Proc. Second International Symposium*, SIAM, Philadelphia, pp. 443–463.

[54] Axelsson, O., Carey, G. and Lindskog, G. (1989) On a class of preconditioned iterative methods on parallel computers, *Int. J. Num. Meth. Engg.*, **27**, 637–654.

[55] Hayes, L.J. (1989) Advances and trends in element-by-element techniques, in A.K. Noor and T.J. Oden (eds.), *State-of-the-Art Surveys on Computational Mechanics*, ASME, New York, pp. 219–236.

[56] Gustafsson, I. and Lindskog, G. (1986) A preconditioning technique based on element matrix factorizations, *Comp. Meth. Appl. Mech. Engg.*, **55**, 201–220.

[57] Papadrakakis, M., Bitoulas, N. and Hatzikonstantinou, K. (1992) An efficient superelement-by-superelement method for large finite element computations, *Computing Systems in Engg*, **2**, 535–540.

[58] Dubois, P.F., Greenbaum, A. and Rodrigue, G.H. (1979) Approximating the inverse of a matrix for use in iterative alogrithms on vector processors, *Computing*, **22**, 257–268.

[59] Johnson, O.G., Micheli, C.A. and Paul, G. (1983) Polynomial preconditioners for conjugate gradient calculations, *SIAM J. Numer. Anal.*, **20**, 362–376.

[60] Saad, Y. (1985) Practical use of polynomial preconditionings for the conjugate gradient method, *SIAM J. Sci. Stat. Comput.*, **6**, 865–881.

[61] Saad, Y. (1989). Krylov subspace methods on supercomputers, *SIAM J. Sci. Stat. Comput.*, **10**, 1200–1232.

[62] Axelsson, O. (1984) *A Survey of Vectorizable Preconditioning Methods for Large Scale Finite Element Problems*, Report CNA-190 University of Texas, Austin, Texas.

[63] Adams, L. (1985) m-step preconditioned conjugate gradient methods, *SIAM J. Sci. Stat. Comput.*, **6**, 452–463.

[64] Adams, L. and Ong, E.G. (1988/89) Additive polynomial preconditioners for parallel computers, *Parallel Computing*, **9**, 333–345.

[65] Eisenstat, S.C., Ortega, J.M. and Vaughan, C.T. (1990) Efficient polynomial preconditioning for the conjugate gradient method, *SIAM J. Sci. Stat. Comput.*, **11**, 859–872.

[66] Coutinho, A.L.G.A., Alves, J.L.D., Ebecken, N.F.F. and Troina, L.M. (1990) Conjugate gradient solution of finite element equations on the IBM 3090 vector computer utilizing polynomial preconditionings, *Comp. Meth. Appl. Mech. Engg*, **84**, 129–145.

[67] Papadrakakis, M., Bitzarakis, S. and Hatzikonstantinou, K. (1993) *Element-by-element and Domain Decomposition Preconditioners for Large-scale Problems*, Report, Institute of Structural Analysis & Aseismic Research, National Technical University of Athens.

[68] Przemieniecki, J.S. (1963) Matrix structural analysis of substructures, *AIAA J.*, **1**, 138–147.

[69] Eijkhout, V. (1988) A general formulation for incomplete blockwise factorizations, *Comm. Appl. Num. Meth.*, **4**, 161–164.

[70] Underwood, R.R. (1976) *An Approximate Factorization Procedure Based on the Block Cholesky Decomposition and its Use With the Conjugate Gradient Method*, Report NEDO-11386, General Electric Co., Nuclear Energy Div., San Jose, California.

[71] Concus, P., Golub, G.H. and Meurant, G. (1985) Block preconditioning for the conjugate gradient method, *SIAM J., Sci. Stat. Comput.*, **6**, 220–252.

[72] Axelsson, O., Brinkkemper, S. and Ilin, V.P. (1984) On some versions of incomplete block-matrix factorization iterative methods, *Lin. Alg. Appl.*, **58**, 3–15.

[73] Axelsson, O. (1985) A survey of preconditioned iterative methods for linear systems of algebraic equations, *BIT*, **25**, 166–167.

[74] Axelsson, O. and Polman, B. (1986) On approximate factorization methods for block matrices suitable for vector and parallel processors, *Lin. Alg. Appl.*, **77**, 3–26.

[75] Beauwens, R. and Ben Bouzid, M. (1987) On sparse block factorization iterative methods, *SIAM J. Num. Anal.*, **24**, 1066–1076.

[76] Keyes, D.E. and Gropp, W.D. (1987) A comparison of domain decomposition techniques for elliptic partial differential equations and their parallel implementation, *SIAM J. Sci. Stat. Comput.*, **8**, 5166–5201.

[77] Drija, M. (1982) A capacitance matrix method for Dirichlet problem on polygon region, *Num. Math.*, **39**, 51–64.

[78] Drija, M. (1984) A finite element-capacitance method for elliptic problems on regions partitioned into subregions, *Num. Math.*, **44**, 153–168.

[79] Golub, G.H. and Mayers, D. (1983) The use of pre-conditioning over irregular regions, *Lecture at Sixth Conf. on Computing Methods in Applied Sciences and Engineering*, Versailles, December.

[80] Björstad, P.E. and Widlund, O.B. (1984) Solving elliptic problems on regions partitioned into substructures, in G. Birkhoff and A. Schoenstadt (eds.), *Elliptic Problem Solvers* II, Academic Press, New York, pp. 245–256.

[81] Chan, T.F. and Resasco, D.C. (1985) *A Survey of Preconditioners for Domain Decomposition*, Tech. Report/DCS/RR-414, Yale University, New Haven, CT.

[82] Bramble, J.M., Pasciak, J.E. and Schatz, A.H. (1986) The construction of preconditioners for elliptic problems by substructuring I, *Math. Comput.*, **47**, 103–134.

[83] Chan, T.F. (1987) Analysis of preconditioners for domain decomposition, *SIAM J. Num. Anal.*, **24**, 382–390.

[84] Chan, T.F. and Goovaerts, D. (1990) A note on the efficiency of domain decomposed incomplete factorizations, *SIAM J. Sci. Stat. Comput.*, **11**, 794–803.

[85] Björstad, P.E. and Hvidsten, A. (1988) Iterative methods for substructured elasticity problems in structural analysis, in R. Glowinski, G.H. Golub, G.A. Meurant and J. Periaux (eds.), *Domain Decomposition Methods for Partial Differential Equations*, SIAM Philadelphia, pp. 301–312.

[86] Carter, W.T., Sham, T.L. and Law, K.H. (1984) A parallel finite element method and its prototype implementation on a hypercube, *Comp. Struct.*, **31**, 921–934.

[87] De Roeck, Y.H. (1989) *A Local Preconditioner in a Domain Decomposed Method*, CERFACS Report TR89/10, 31057 Toulouse CEDEX France.

[88] Nour-Omid, B., Raefsky, A. and Lyzenga, G. (1987) Solving finite element equations on concurrent computers, in A.K. Noor (ed.), *Parallel Computations and their Impact on Mechanics*, ASME, New York, pp. 209–227.

[89] Börgers, C. (1989) The Neumann–Dirichlet domain decomposition method with inexact solvers on the subdomains, *Num. Math.*, **55**, 123–136.

[90] Glowinski, R., Golub, G.H., Meurant, G.A. and Periaux, J. (eds.) (1988) Domain Decomposition Methods for Partial Differential Equations, *Proc. 1st Int. Symp.*, SIAM, Philadelphia.

[91] Chan, T.F., Glowinski, R., Periaux, J. and Widlund, O.B. (eds.) (1989) Domain decomposition methods, *Proc. 2nd Int. Symp.*, SIAM, Philadelphia.

[92] Crisfield, M.A. (1985) New solution procedures for linear and nonlinear finite element analysis, in J. Whiteman (ed.), *The Mathematics of Finite Elements and Applications* V, Academic Press, pp. 49–81.

[93] Axelsson, O. and Gustafsson, I. (1983) Preconditioning and two-level multigrid methods for arbitrary degree of approximation, *Math. Comput.*, **40**, 214–242.

[94] Adams, L.M. and Ong, E.G. (1987) A comparison of preconditioners for GMRES on parallel computers, in A.K. Noor (ed.), *Parallel Computations and their Impact on Mechanics*, ASME, New York, pp. 171–208.

[95] Smith, B.F. and Widlund, O. (1990) A domain decomposition algorithm using a hierarchical basis, *SIAM J. Sci. Stat. Comput.*, **6**, 121–122.

[96] Mandel, J. (1990) Two-level domain decomposition preconditioning for the p-version finite element method in three dimensions, *Int. J. Num. Meth. Engg.*, **29**, 1095–1108.

[97] Mandel, J. (1990) Iterative solvers by substructuring for the p-version finite element method, *Comp. Meth. Appl. Mech. Engg*, **80**, 117–128.

[98] Mandel, J. (1990) On block diagonal and Schur complement preconditioning, *Num.*

Math., **58**, 79–73.

[99] Axelsson, O. and Vassilevski, P.S. (1989) Algebraic multilevel preconditioning methods I, *Num. Math.*, **56**, 157–177.

[100] Yserentant, H. (1986) On the multi-level splitting of finite element spaces, *Num. Math.*, **49**, 379–412.

[101] Farhat, C. and Sobh, N. (1989) A coarse/fine preconditioner for very ill-conditioned finite element problems, *Int. J. Num. Meth. Engg*, **28**, 1715–1723.

[102] Parsons, I.D. and Hall, J.F. (1990) The multi-grid method in solid mechanics: Part I — Algorithm description and behaviour, Part II — Practical applications, *Int. J. Num. Meth. Engg*, **29**, 719–753.

[103] Wilkinson, J.H. (1963) *Rounding Errors in Algebraic Processes*, Prentice Hall, Englewood Cliffs, New Jersey.

[104] Oettli, W. and Prager, W. (1964) Compatibility of approximate solution of linear equations with given error bounds for coefficients and right-hand sides, *Num. Math.*, **6**, 405–409.

[105] Forsythe, G. and Moler, C.B. (1967) *Computer Solution of Linear Algebraic Systems*, Prentice Hall, New Jersey.

[106] Jankowski, M. and Wozniakowski, H. (1977) Iterative refinement implies numerical stability, *BIT*, **7**, 303–311.

[107] Skeel, R.D. (1980) Iterative refinement implies numerical stability for Gaussian elimination, *Math. Comput.*, **35**, 817–832.

[108] Rosanoff, R.A., Gloudeman, J.F. and Levy, S. (1968) Numerical condition of stiffness matrix formulation for frame structures, *Proc. Conf. Matrix Methods in Structural Mechanics*, AFFDL-TR-68-150, Wright-Patterson Air Force Base, Ohio.

[109] Roy, J.R. (1971) Numerical error in structural solutions, *J. Struct. Div.*, ASCE, **97**, 1039–1053.

[110] Roy, J.R. (1972) Numerical error in structural solutions, *J. Struct. Div.*, ASCE, **98**, 1663–1665.

[111] Argyris, J.H., Johnsen, Th.L., Rosanoff, R.A. and Roy, J.R. (1976) On numerical error in the finite element method, *Comp. Meth. Appl. Mech. Engg*, **7**, 261–282.

[112] Martin, C.W. and Harrold, A.J. (1976) Removal of truncation error in finite element analysis, in J.R. Whiteman (ed.), *The Mathematics of Finite Element and Applications* II, Academic Press, pp. 525–533.

[113] Arioli, M., Demmel, J.W. and Duff, I.S. (1989) Solving large linear systems with sparse backward error, *SIAM J. Matrix Anal. Appl.*, **10**, 165–190.

[114] Papadrakakis, M. and Bitoulas, N. (1993) Accuracy and effectiveness of preconditioned conjugate gradient algorithms for large and ill-conditioned problems, *Comp. Meth. Appl. Mech. Engg.* (to be published).

[115] Axelsson, O. and Gustafsson, I. (1980) A preconditioned conjugate gradient method for finite element equations, which is stable for rounding errors, in S.H. Lavington (ed.), *Information Processing 80*, North Holland, pp. 723–728.

[116] Wozniakowski, H. (1980) Round-off error analysis of a class of conjugate gradient algorithms, *Lin. Alg. Applics.*, **29**, 507–529.

M. Papadrakakis
Institute of Structural Analysis and Aseismic Research
National Technical University of Athens
Zografou Campus
15773 Athens
GREECE

2
Solving Large Linearized Systems in Mechanics

B. Nour-Omid

Scopus Technology, Inc., Emeryville, CA, USA

2.1 INTRODUCTION

In this chapter we consider the solution of symmetric linear systems of equations arising from applications of the finite element method using Krylov based algorithms (e.g. Lanczos and conjugate gradient procedures). The Lanczos algorithm [1] was first introduced in 1950 as a method for computing eigenvalues and the corresponding eigenvectors of a matrix. In 1952 Hestenes and Stiefel [2] introduced the method of conjugate gradients (CG) for solving linear systems of equations. In the same year, Lanczos showed that his algorithm, then called the method of minimized iteration [3], can also be used to obtain the solution of a linear system of equations. In fact, these methods are closely related in the sense that in exact arithmetic (when no round-off errors are present) they compute the same approximate solution at each step, a fact known to both Lanczos and Hestenes.

An important motivating factor for using Lanczos and CG methods is the theoretical result showing that in exact arithmetic both methods are able to compute the solution in less than n iterations, where n is the number of equations in the system. In fact the number of iterations is less than the total number of distinct eigenvalues in the system. In 1960, the CG method was first used to solve system of linear equations arising in structural mechanics [4]. In this paper, Lively showed that CG was not effective for solving the ill-conditioned systems that often arise in structural analysis. The popularity of the CG method vanished when it was found that for certain problems it required well over n steps to converge to the correct solution. CG was then abandoned and the more effective direct methods based on

---

Solving Large-scale Problems in Mechanics, Edited by M. Papadrakakis
© 1993 John Wiley & Sons Ltd

triangular factorization of the matrix were adopted by structural analysts as their method of choice.

Non-linear transient finite element problems may be characterized by the equations of dynamic equilibrium

$$\mathbf{M}\ddot{\mathbf{w}} + \mathbf{f}^{int}(\mathbf{w}) = \mathbf{f}^{ext}(t) \tag{2.1}$$

where $\mathbf{M}$ is the mass matrix, $\mathbf{f}^{int}$ is the vector of internal resisting forces due to the displacements $\mathbf{w}$, and $\mathbf{f}^{ext}$ is the time dependent external force vector [5]. The above system is generally solved by applying a step-by-step time integration procedure resulting in a system of non-linear algebraic equations. The solution to this system is obtained using a Newton–Raphson iteration or related schemes. At the heart of this iteration is a set of linear equations

$$\mathbf{A}\mathbf{x} = \mathbf{b} \tag{2.2}$$

where $\mathbf{x}$ is the correction to the approximate solution vector in the nonlinear iteration loop. $\mathbf{A}$ is symmetric, positive definite and sparse, and is related to the system in (2.1) through

$$\mathbf{A} = \mathbf{M} + \mathbf{K}\delta \tag{2.3}$$

where $\mathbf{K} = [(\partial \mathbf{f}^{int})/\partial \mathbf{w}]$ and δ is a scalar parameter that depends on the time step (e.g. $\delta = \frac{1}{2}\Delta t^2$). For problems with small bandwidth or problems which result in small fill-in in the triangular factorization of $\mathbf{A}$, direct methods are the fastest solvers [6]. When the bandwidth of $\mathbf{A}$ is large (e.g. large complex two- and three-dimensional problems), the solution of the linear system may be a formidable task and alternative procedures must be considered.

In 1971 the advantages of Lanczos and CG methods were recognized when attention was focused on linear systems with large sparse matrix coefficients, (see [7]). An important property of these methods is that, at each step, the solution (2.2) can be obtained without an explicit knowledge of the matrix. Only a means of computing the matrix vector product $\mathbf{A}\mathbf{v}$ for a given vector $\mathbf{v}$ is required. This is an elegant way of exploiting the sparsity structure of $\mathbf{A}$. Typically, in finite element analysis there are fewer than 100 non-zero terms in each row of $\mathbf{A}$. The number of non-zero terms in $\mathbf{A}$ is independent of the number of equations. It depends on the number of nodes per element, the number of parameters per node, and the number of elements attached to a node. The number of equations in this system depends on the complexity of the structure (i.e. the domain) and also on the amount of required detail in describing the solution.

Preconditioning is introduced to alleviate the difficulties associated with the slow convergence of Lanczos and CG methods. Instead of solving (2.2) one solves

$$\mathbf{A}\mathbf{B}^{-1}\mathbf{y} = \mathbf{b} \tag{2.4}$$

where $\mathbf{x} = \mathbf{B}^{-1}\mathbf{y}$. The matrix $\mathbf{B}$ is referred to as the preconditioning matrix. The advantages of preconditioning can also be realized by solving $\mathbf{B}^{-1}\mathbf{A}\mathbf{x} = \mathbf{B}^{-1}\mathbf{b}$. The number of iterations required to solve (2.4) depends on the condition number, κ, of its coefficient matrix. $\kappa(\mathbf{AB}^{-1})$ is the ratio of the largest eigenvalue of the eigenproblem $[\mathbf{A} - \lambda\mathbf{B}]\mathbf{z} = \mathbf{0}$ to the smallest (i.e. $\kappa = \|\mathbf{AB}^{-1}\| \|\mathbf{BA}^{-1}\|$). Theoretical considerations suggest that at the end of each of the first few iterations of both Lanczos and CG methods the residual norm is reduced by a factor of $\sqrt{[\kappa(\mathbf{AB}^{-1}) - 1]}/\sqrt{[\kappa(\mathbf{AB}^{-1}) + 1]}$. Note that when κ is unity a single iteration is sufficient to solve the equation. Of course κ is one only when $\mathbf{B} = \mathbf{A}$! However, this provides us with a guideline for choosing $\mathbf{B}$. $\mathbf{B}$ must be chosen such that one can easily compute the solution of a linear system of equations with $\mathbf{B}$ as its matrix coefficient while at the same time it is as close to $\mathbf{A}$ as possible. For non-trivial matrices $\mathbf{A}$, these are contradictory requirements which makes the problem of finding good preconditioners a challenging one. For a well chosen $\mathbf{B}$ only a few iterations is required to reduce the residual norm to the desired level. It is important to note that the condition number of $\mathbf{AB}^{-1}$ depends on the time step through δ in equation (2.3) as the following example demonstrates.

Example 1: Let

$$\mathbf{K} = \begin{bmatrix} 1+\zeta & -1 \\ -1 & 1+\zeta \end{bmatrix} \quad \text{and } \mathbf{M} = \begin{bmatrix} 1 & 0 \\ 0 & 1 \end{bmatrix}.$$

Then

$$\mathbf{A} = \begin{bmatrix} 1+(1+\zeta)\delta & -\delta \\ -\delta & 1+(1+\zeta)\delta \end{bmatrix}.$$

The eigenvalues of the preconditioned system satisfies the quadratic equation $\det[\mathbf{A} - \lambda\mathbf{B}] = 0$. There is no change in the condition number for this problem with diagonal preconditioning since the diagonal of $\mathbf{A}$ is a scalar multiple of the identity matrix. Then the condition number for this problem becomes

$$\kappa = \frac{1+(2+\zeta)\delta}{1+\zeta\delta}.$$

For a sufficiently small time step the condition number is close to unity. On the other hand as the time step increases the condition number approaches $1 + 2/\zeta$ which may be arbitrarily large for small ζ.

This example demonstrates that (a) the performance of iterative methods for solving linear systems of equations arising from transient finite element problems depends strongly on the time step, and (b) for a given finite element discretization static problems result in worse conditioned system of equations than transient problems. These facts should be considered when assessing the performance of iterative methods.

Here, we focus on systems which arise from the application of the finite element method to engineering problems whose sparsity structure may be characterized by

$$\mathbf{A} = \sum_e \mathbf{N}_e \mathbf{a}_e \mathbf{N}_e^T \qquad (2.5)$$

where $\mathbf{N}_e$ is long and thin Boolean connectivity matrix and $\mathbf{a}_e$ denotes the small stiffness matrix for element e. We can take advantage of this structure of $\mathbf{A}$ when using either CG or Lanczos methods to solve (2.2). In [8] and [9], it is pointed out that the matrix vector product

$$\mathbf{Au} = \sum_e (\mathbf{N}_e \mathbf{a}_e \mathbf{N}_e^T \mathbf{u}) \qquad (2.6)$$

can be computed without ever assembling $\mathbf{A}$. The evaluation of $\mathbf{Au}$ using (2.6) requires more arithmetic operations than that using an assembled $\mathbf{A}$ (assuming some compact structure where no zero entries of $\mathbf{A}$ are stored). Typically, the number of arithmetic operations would increase by about threefold. However, the use of parallel and vector computers produces only a modest increase in the elapsed time and in certain cases might even reduce it.

In 1983, Hughes et al. [10] proposed a time integration algorithm for the solution of heat conduction equations that uses an element-by-element (E × E) splitting. In the same year, Ortiz et al. [11] proposed a novel extension of the E × E procedure to the solution of dynamic equations. These E × E time integration algorithms are unconditionally stable, but they lack accuracy, which limits their use. Hughes et al. [12] reformulate the E × E procedure as an iterative solver to achieve the accuracy and stability of standard finite element algorithms. In [13] Nour-Omid and Parlett addressed the problem of preconditioning (2.2). The idea is to employ the methods for solving differential equations presented in [10, 11, 12] as preconditioners. The resulting preconditioners use the element representation of $\mathbf{A}$ in (2.6), and requires no globally assembled matrix. They are defined as the product of positive definite element matrices obtained by applying a diagonal shift to the positive semi-definite element stiffness matrices. Winget and Hughes [14] further developed the ideas of element preconditioners and constructed a variation that replaces the terms on the diagonals of the element matrices with the corresponding ones in the assembled matrix. This modification also results in positive definite element matrices. A product algorithm similar to that in [11] is then constructed using the Cholesky factorization of these modified element matrices. It is worth noting that these preconditioners, though computed E × E, are an approximate Cholesky factorization of $\mathbf{A}$ (see [15]). Our primary interest in E × E preconditioning is in keeping storage requirements down in the analysis of regular structures, but the advent of vector and parallel computing may make this approach a fast one as well, especially in three dimensions.

In Section 2.2 we briefly describe the Lanczos algorithm and present a derivation of the conjugate gradient method from the Lanczos algorithm in Section 2.3. We then turn to the problem of orthogonality loss that affects both methods. As a remedy, we consider the approach of partial reorthogonalization proposed by Simon [16]. The element preconditioners used in this study are described in Section 2.5, together with a discussion of some implementation issues. In Section 2.6 we describe the S × S preconditioners and its relation to E × E method. The results of numerical tests on two characteristically different problems are presented in Section 2.7.

2.2 LANCZOS ALGORITHM

When used as a method for solving linear systems, the Lanczos process starts from a given initial approximation to the solution, x_0. Associated with x_0, define the residual vector $r_0 = b - Ax_0$. Unless a good estimate to the desired solution is available, the best choice for x_0 is the zero vector. Then $r_0 = b$. Normalizing r_0 gives the first Lanczos vector, q_1. Implicitly, at the end of the first step the algorithm obtains a Galerkin approximation to the solution of (2.2) from the one dimensional space with q_1 as the base vector. Associated with this approximation is a new residual vector. The normalization of this residual results in the second Lanczos vector, q_2. Repeating the Galerkin process but using the two-dimensional space, span (q_1, q_2), followed by the normalization of the residual one obtains q_3.

At a typical step, j, the Lanczos algorithm computes a residual vector associated with a best approximation, x_j, (in a Galerkin sense) to x from the j dimensional space, span $[q_1, q_2, \ldots, q_j]$. This space is often referred to as the space of trial vectors. The next Lanczos vector, q_{j+1}, is the normalized residual $b - Ax_j$. This process is repeated until the norm of the current residual is small compared to the that of the starting residual. The Galerkin method chooses the approximate solution x_j by forcing the associated residual to be orthogonal to the space of trial (Lanczos) vectors. The Lanczos method computes neither x_j nor the associated residual. Instead it computes, at step j, a j-vector s_j that contains the weighting parameters for constructing the Galerkin solution.

Alternatively, Lanczos may be described as the Gram–Schmit orthogonalization process applied to the Krylov space, $[r_0, AB^{-1}r_0, (AB^{-1})^2r_0, \ldots, (AB^{-1})^{j-1}r_0]$, associated with equation (2.4). The orthogonalization is performed with respect to the B^{-1} inner product. The result of this orthogonalization is the set of Lanczos vectors $[q_1, q_2, \ldots, q_j]$. The vector q_{j+1} is obtained by orthonormalizing $(AB^{-1})^j r_0$ against the computed Lanczos vectors. The same vector q_{j+1} is obtained if $AB^{-1}q_j$ is used instead of $(AB^{-1})^j r_0$. It turns out that the components of $AB^{-1}q_j$ along the first $j - 2$ Lanczos vectors are zero and orthogonalization needs to be performed only against q_j and q_{j-1}. The result is a vector r_j in the same direction as the residual due to the Galerkin approximation described above.

The algorithm can then be rewritten as the three term relation

$$r_j = \beta_{j+1} q_{j+1} = AB^{-1} q_j - q_j \alpha_j - q_{j-1} \beta_j \qquad (2.7)$$

where $\alpha_j = q_j^T B^{-1} AB^{-1} q_j$ and r_j is normalized with respect to the inverse of the preconditioner to obtain q_{j+1} with normalizing factor $\beta_{j+1} = (r_j^T B^{-1} r_j)^{1/2}$.

The jth step of the Lanczos algorithm involves the calculation of α_j, β_{j+1}, and q_{j+1}, in that order. In addition to the storage needs for A and B, the algorithm requires storage for 5 vectors of length n; one for each of the vectors, q_{j-1}, q_j, r_j, $p_j = B^{-1} q_j$ and p_{j-1}. The total cost for one step of the algorithm involves one solve with the preconditioner B as the coefficient matrix, a multiplication of A by a vector, two inner products and four products of a scalar by a vector. A summary of the Lanczos algorithm is presented in Table 2.1.

After m Lanczos steps all the quantities obtained from equation (2.7) can be arranged in a global matrix form

$$[AB^{-1}][Q_m] - [Q_m][T_m] = [0 \quad r_m] = r_m e_m^T. \tag{2.8}$$

Here $e_m^T = (0, 0, \ldots, 0, 1)$, Q_m is an $n \times m$ matrix with columns q_i, $i = 1, 2, \ldots, m$, and T_m is the tridiagonal matrix

$$T_m = \begin{bmatrix} \alpha_1 & \beta_2 & & & & \\ \beta_2 & \alpha_2 & \beta_3 & & & \\ & \beta_3 & \cdot & \cdot & & \\ & & \cdot & \cdot & \cdot & \\ & & & \cdot & \cdot & \beta_m \\ & & & & \beta_m & \alpha_m \end{bmatrix}. \tag{2.9}$$

The orthogonality property of the Lanczos vectors, $Q_m^T B^{-1} Q_m = I_m$, where I_m is the $m \times m$ identity matrix, can be used in equation (2.8) to obtain

$$Q_m^T B^{-1} A B^{-1} Q_m = T_m. \tag{2.10}$$

A Galerkin approximation to y in (2.4) can be constructed by taking a linear combination of the Lanczos vectors. Accordingly,

$$y_m = Q_m s_m \tag{2.11}$$

where s_m satisfies the tridiagonal system of equations

$$T_m s_m = Q_m^T B^{-1} r_0 = \beta_1 e_1. \tag{2.12}$$

The last equality is obtained using the fact that the starting vector is $r_0 = \beta_1 q_1$. The vector e_1 is the first column of the identity matrix. Equation (2.12) is a weak form of (2.4) and is obtained by first substituting the approximation to y from (2.11) into equation (2.4) to obtain the residual

$$g_m = AB^{-1} Q_m s_m - b. \tag{2.13}$$

Orthogonalizing g_m against Q_m with respect to the B^{-1} inner product results in equation (2.12). The residual g_m is simply related to r_m through

$$g_m = r_m \sigma_m \tag{2.14}$$

where σ_m is the bottom element of s_m. The norm of this residual, $\rho_m = \|g_m\| = \beta_{m+1} |\sigma_m|$, can be used to monitor the convergence. Once ρ_m is sufficiently small the Lanczos algorithm is terminated and the solution is constructed using (2.11). The Lanczos vectors can be put on secondary storage as they are being generated. There are two main reasons for keeping the Lanczos vectors.

(a) They are used occasionally in subsequent steps to restore orthogonality (see the following section on Loss of Orthogonality for more details).

(b) They can be called and used to construct the solution to a new right hand side [17]. The algorithm in Table 2.1 is particularly well suited for multiple right hand sides.

Table 2.1 The Lanczos algorithm.

Given an approximate solution vector $\mathbf{x}_0$:

(1) Set

    (a) $\mathbf{r}_0 = \mathbf{b} - \mathbf{A}\mathbf{x}_0$,

    (b) $\mathbf{q}_0 = \mathbf{0}$,

    (c) Solve $\mathbf{B}\bar{\mathbf{p}}_1 = \mathbf{r}_0$.

    (d) $\beta_1 = (\bar{\mathbf{p}}_1^T \mathbf{r}_0)^{\frac{1}{2}}$,

    (e) $\mathbf{q}_1 = \dfrac{1}{\beta_1}\mathbf{r}_0$

    (f) $\mathbf{p}_1 = \dfrac{1}{\beta_1}\bar{\mathbf{p}}_1$

(2) for $j = 1, 2, \ldots$ repeat;

    (a) $\bar{\mathbf{r}}_j = \mathbf{A}\mathbf{p}_j$

    (b) $\hat{\mathbf{r}}_j = \bar{\mathbf{r}}_j - \mathbf{q}_{j-1}\beta_j$

    (c) $\alpha_j = \mathbf{q}_j^T \mathbf{B}^{-1}\hat{\mathbf{r}}_j = \mathbf{p}_j^T \hat{\mathbf{r}}_j$

    (d) $\mathbf{r}_j = \hat{\mathbf{r}}_j - \mathbf{q}_j\alpha_j$

    (e) Solve $\mathbf{B}\bar{\mathbf{p}}_j = \mathbf{r}_j$

    (f) $\beta_{j+1} = (\mathbf{r}_j^T \mathbf{B}^{-1}\mathbf{r}_j)^{\frac{1}{2}} = (\bar{\mathbf{p}}_j^T \mathbf{r}_j)^{\frac{1}{2}}$

    (g) if residual norm is small then terminate the loop

    (h) $\mathbf{q}_{j+1} = \dfrac{1}{\beta_{j+1}}\mathbf{r}_j$

    (i) $\mathbf{p}_{j+1} = \dfrac{1}{\beta_{j+1}}\bar{\mathbf{p}}_j$

(3) Solution $\mathbf{x} = \mathbf{x}_0 + \mathbf{B}^{-1}\mathbf{Q}_m\mathbf{s}_m$

2.3 CONJUGATE GRADIENT ALGORITHM

In this section we give a derivation of the conjugate gradient algorithm directly from the Lanczos process [18–20]. The conjugate gradient method can be viewed as a procedure that implicitly computes the triangular factorization of $\mathbf{T}_m$ through an update algorithm to combine the steps 2 and 3 of the Lanczos algorithm given in Table 2.1. Accordingly

$$\mathbf{T}_m = \mathbf{L}_m \mathbf{D}_m \mathbf{L}_m^T \qquad (2.15)$$

where $\mathbf{D}_m = \mathrm{diag}[\delta_1, \delta_2, \ldots, \delta_m]$ and

$$\mathbf{L}_m = \begin{bmatrix} 1 & & & & & \\ -\omega_1 & 1 & & & & \\ & -\omega_2 & 1 & & & \\ & & \cdot & & & \\ & & & \cdot & & \\ & & & & 1 & \\ & & & & -\omega_{m-1} & 1 \end{bmatrix} \qquad (2.16)$$

The components of $\mathbf{T}_m$, $\mathbf{L}_m$, and $\mathbf{D}_m$ are related through the following pair of equations:

$$\delta_k = \alpha_k - \omega_{k-1}^2 \delta_{k-1}$$
$$\omega_k = -\frac{\beta_{k-1}}{\delta_k}. \qquad (2.17)$$

These two equations completely define the algorithm for triangular factorization of $\mathbf{T}_m$. Next we define

$$\mathbf{Z}_m = \mathbf{B}^{-1} \mathbf{Q}_m \mathbf{L}_m^{-T}. \qquad (2.18)$$

An important property of $\mathbf{Z}_m$ is that its columns are orthogonal with respect to $\mathbf{A}$. To show this consider

$$\mathbf{Z}_m^T \mathbf{A} \mathbf{Z}_m = \mathbf{L}_m^{-1} \mathbf{Q}_m^T \mathbf{B}^{-1} \mathbf{A} \mathbf{B}^{-1} \mathbf{Q}_m \mathbf{L}_m^{-T}$$
$$= \mathbf{L}_m^{-1} \mathbf{T}_m \mathbf{L}_m^{-T}$$
$$= \mathbf{D}_m.$$

The columns of $\mathbf{Z}_m$ are said to be conjugate and the orthogonality condition of $\mathbf{Z}_m$ with respect to $\mathbf{A}$ is referred to as the conjugacy condition.

Multiplying both sides of equation (2.18) by $\mathbf{L}_m^T$ and using the bi-diagonal structure of $\mathbf{L}_m$ to equate the k-th column on either side of this equation, yields

$$\mathbf{z}_k - \mathbf{z}_{k-1} \omega_{k-1} = \mathbf{B}^{-1} \mathbf{q}_k. \qquad (2.19)$$

Defining $\mathbf{d}_k = \mathbf{B}^{-1} \mathbf{q}_k$ the above equation reduces to

$$\mathbf{z}_k = \mathbf{d}_k + \mathbf{z}_{k-1} \omega_{k-1}. \qquad (2.20)$$

Due to the conjugacy property of Z_m we are able to update the solution vector x_k by simply adding a component of z_k. Thus, using equation (2.4) we have

$$\begin{aligned} x_k &= B^{-1} y_k \\ &= B^{-1} Q_k T_k^{-1} \beta_1 e_1 \\ &= B^{-1} Q_k L_k^{-T} D_k^{-1} L_k^{-1} \beta_1 e_1 \\ &= Z_k D_k^{-1} L_k^{-1} \beta_1 e_1 \\ &= x_{k-1} + \gamma_k z_k \end{aligned}$$

where γ_k is the kth element of $D_m^{-1} L_m^{-1} \beta_1 e_1$. This way the residual vector can also be updated using the update relation

$$g_k = g_{k-1} - \gamma_k u_k \qquad (2.21)$$

and $u_k = A z_k$. γ_k is related to the component of D_m and L_m through

$$\gamma_k = \frac{\rho_k}{\delta_k} \qquad (2.22)$$

where ρ_k is related to ω_{k-1} through

$$\rho_k = \omega_{k-1} \rho_{k-1} \qquad (2.23)$$

with $\rho_1 = \beta_1$. The above equation is simply the forward reduction algorithm to compute $L_m^{-1} \beta_1 e_1$.

Thus the CG method directly computes the triangular factor of T_m by updating the factors of T_{m-1}. The result is the algorithm in Table 2.2. It is important to note that T_m is often indefinite when A is a symmetric indefinite matrix. In this case the conjugate gradient algorithm is not reliable since the triangular factorization of T_m may be numerically unstable. This instability occurs whenever δ_k is small. Note that $\delta_k = z_k^T u_k$ is the denominator of the right hand side of (2.2b) in Table 2.2.

The CG algorithm generates a sequence of approximations, x_k, to the solution x with a corresponding residual vector g_k. The termination criterion can be chosen based on these quantities. In addition to storage demands for A and B the algorithm requires storage for four vectors.

2.4 LOSS OF ORTHOGONALITY

In finite precision, each computation introduces a small error and therefore the computed quantities differ from their exact counterparts. Our objective here is to state the effect of round-off error on the Lanczos process. For this purpose we denote

Table 2.2 The conjugate gradient algorithm

Given an approximate solution x_0 then:
(1) Set
   (a) $g_0 = b - Ax_0$
   (b) $z_1 = g_0$
   (c) Solve $Bd_1 = g_0$
   (d) $\rho_1 = g_0^T d_1$
(2) for $k = 1, 2, \ldots$ repeat;
   (a) $u_k = Az_k$
   (b) $\gamma_k = \dfrac{\rho_k}{z_k^T u_k}$
   (c) $x_k = x_{k-1} + \gamma_k z_k$
   (d) $g_k = g_{k-1} - \gamma_k u_k$
   (e) Solve $Bd_{k+1} = g_k$.
   (f) $\rho_{k+1} = g_k^T d_{k+1}$
   (g) if $\rho_{k+1} <= tol \cdot \rho_0$ then terminate the loop.
   (h) $\omega_k = \dfrac{\rho_{k+1}}{\rho_k}$
   (i) $z_{k+1} = d_{k+1} + \omega_k z_k$

by ε the smallest number in the computer such that $1 + \varepsilon > 1$. It is known as the unit round-off error.

Although the tridiagonal relation, equation (2.8), is preserved to within round-off, the B^{-1} orthogonality property of the Lanczos vectors completely breaks down after a certain number of steps depending on ε and the distribution of the eigenvalues of $B^{-1}A$ [16, 19]. The Lanczos vectors not only lose their orthogonality, but may even become linearly dependent. This problem also affects the conjugate gradient method in the form of loss of conjugacy. A direct consequence of this loss of orthogonality is delay in convergence to the desired solution.

The loss of orthogonality can be viewed as the subsequent amplification of the errors introduced after each computation. We let Q_m denote the computed Lanczos vectors and define the matrix

$$H_m = Q_m^T B^{-1} Q_m. \tag{2.24}$$

In exact arithmetic H_m is the identity matrix. The off-diagonals of H_m depend on ε, the unit round-off error. Simon [16] found a recurrence relation that can be used to estimate the elements of a column of H_m from the elements of T_m and the elements in the previous columns of H_m. This recursion can be stated in vector form

$$\beta_{j+1} h_{j+1} \approx T_{j-1} h_j - \alpha_j h_j - \beta_j h_{j-1} \tag{2.25}$$

where $\mathbf{h}_{j-1}$, $\mathbf{h}_j$ and $\mathbf{h}_{j+1}$ are vectors of length $j-1$ containing the top $j-1$ elements of the $j-1, j$, and $(j+1)$th columns of $(\mathbf{H}_m - \mathbf{I}_m)$. Here, the bottom element of $\mathbf{h}_{j-1}$ is ε. The orthogonality state can be monitored by updating $\mathbf{h}_{j+1}$ in the course of the Lanczos algorithm.

A number of preventive measures can be taken to maintain a certain level of orthogonality. Lanczos was aware of the effects of round-off on the algorithm when he presented his work. He proposed that the newly computed vector, $\mathbf{q}_{j+1}$, be explicitly orthogonalized against all the preceding vectors at the end of each step. We will refer to this technique as the *full reorthogonalization* method. It enforces orthogonality to within round-off (i.e. $|\mathbf{q}_i^T \mathbf{B}^{-1} \mathbf{q}_j| < n\varepsilon,\ i \neq j$). In [16] Simon showed that the computed tridiagonal remains accurate to within round-off if the more relaxed orthogonality condition

$$|\mathbf{q}_i^T \mathbf{B}^{-1} \mathbf{q}_j| < \sqrt{(n\varepsilon)}, \qquad i \neq j \tag{2.26}$$

is enforced. We refer to this as the semi-orthongonality condition, and to procedures that adopt the weaker condition as selective orthogonalization methods. Simon proposed to update $\mathbf{h}_{j+1}$ using (2.26) and monitor the magnitude of its elements. Whenever any component of $\mathbf{h}_{j+1}$ is greater than $\sqrt{\varepsilon}$ then semi-orthogonality may be lost between $\mathbf{q}_{j+1}$ and some columns of $\mathbf{Q}_j$. At this step the appropriate Lanczos vectors are brought in from secondary store and $\mathbf{q}_{j+1}$ is orthogonalized against each of them. This operation must be carried out in two successive steps to avoid propagation of the errors.

A variant of Simon's scheme is to restore orthogonality of $\mathbf{q}_j$ and $\mathbf{q}_{j+1}$ at the same time. In this way no reorthogonalization of $\mathbf{q}_{j+2}$ will be necessary at the end of the next step. The number of operations for this scheme is the same as that of the scheme above, but vectors are retrieved only once and therefore the I/O overhead is halved.

The disadvantage of reorthogonalization is that additional storage is required to keep the Lanczos vector. If m steps are required to reduce the residual to the desired level, then storage for m vectors of length n is needed. The advantage is that reorthogonalization can significantly reduce the number of steps. This is demonstrated by the numerical examples.

2.5 ELEMENT-BY-ELEMENT PRECONDITIONING

The idea of using solution algorithms from discretized partial differential equations for constructing preconditioners is not new. As early as 1963 Wachspress proposed one such preconditioner based on the ADI method [21]. Recently, in [13] we proposed a number of preconditioners, based on the element-by-element representation of $\mathbf{A}$ for solving $\mathbf{M}\dot{\mathbf{x}} + \mathbf{A}\mathbf{x} = \mathbf{b}$. Here $\mathbf{M}$ is a diagonal matrix. The steady state solution of this equation is the same as that of (2.2).

The first preconditioner uses a Cholesky factorization of the element matrices shifted by a diagonal matrix. We denote the diagonally scaled $\mathbf{A}$ by

$$\bar{\mathbf{A}} = \mathbf{M}^{-1/2}\mathbf{A}\mathbf{M}^{-1/2} = \sum_{e=1}^{n_e} \mathbf{N}_e \bar{\mathbf{a}}_e \mathbf{N}_e^T \qquad (2.27)$$

where n_e is the total number of elements. Then the proposed preconditioner can be constructed in the following manner. First, the Cholesky factors $\bar{\mathbf{c}}_e \bar{\mathbf{c}}_e^T = \sigma \mathbf{I} + \bar{\mathbf{a}}_e$ is computed. The shift was applied to eliminate the singularity of $\bar{\mathbf{a}}_e$. A lower triangular matrix $\bar{\mathbf{C}}$ is formed as the product of the $\bar{\mathbf{c}}_e$. The matrix $\bar{\mathbf{C}}$ is an approximation to the Cholesky factorization of $\bar{\mathbf{A}}$. The resulting preconditioner is given by

$$\mathbf{B}^{-1} = \mathbf{M}^{1/2} \prod_{e=1}^{n_e} \mathbf{N}_e \bar{\mathbf{c}}_e^{-1} \mathbf{N}_e^T \prod_{e=n_e}^{1} \mathbf{N}_e \bar{\mathbf{c}}_e^{-T} \mathbf{N}_e^T \mathbf{M}^{1/2}. \qquad (2.28)$$

Note that the second product is carried out in the reverse order of the first. Numerical results indicated that a shift $\sigma = 1$ results in a preconditioner that is close to the optimum.

Writing $\bar{\mathbf{a}}_e = \bar{\mathbf{u}}_e + \bar{\mathbf{u}}_e^T + \bar{\mathbf{d}}_e$, where $\bar{\mathbf{d}}_e$ and $\bar{\mathbf{u}}_e$ denote the diagonal and strict upper triangular part of $\bar{\mathbf{a}}_e$, a second preconditioner was constructed as

$$\mathbf{B}^{-1} = \mathbf{M}^{1/2} \prod_{e=1}^{n_e} \mathbf{N}_e (\mathbf{I} + \bar{\mathbf{d}}_e + \bar{\mathbf{u}}_e^T)^{-1} \mathbf{N}_e^T \prod_{e=n_e}^{1} \mathbf{N}_e (\mathbf{I} + \bar{\mathbf{d}}_e + \bar{\mathbf{u}}_e)^{-1} \mathbf{N}_e^T \mathbf{M}^{1/2}. \qquad (2.29)$$

A comparison of these two preconditioners on small problems indicated that the Cholesky form is more effective.

E × E Cholesky: In[14] Winget and Hughes replaced the diagonal of $\bar{\mathbf{a}}_e$ with the identity matrix to form

$$\mathbf{B}^{-1} = \mathbf{M}^{1/2} \prod_{e=1}^{n_e} \mathbf{N}_e \tilde{\mathbf{c}}_e^{-1} \mathbf{N}_e^T \prod_{e=n_e}^{1} \mathbf{N}_e \tilde{\mathbf{c}}_e^{-T} \mathbf{N}_e^T \mathbf{M}^{1/2} \qquad (2.30)$$

where $\tilde{\mathbf{c}}_e \tilde{\mathbf{c}}_e^T = \mathbf{I} + \bar{\mathbf{u}}_e + \bar{\mathbf{u}}_e^T$. This avoids the artificial shifting of $\bar{\mathbf{a}}_e$.

E × E LU Split: Similarly, by dropping $\bar{\mathbf{d}}_e$ in (2.29), a new but simpler preconditioner

$$\mathbf{B} = \mathbf{M}^{1/2} \prod_{e=1}^{n_e} \mathbf{N}_e (\mathbf{I} + \bar{\mathbf{u}}_e^T) \mathbf{N}_e^T \prod_{e=n_e}^{1} \mathbf{N}_e (\mathbf{I} + \bar{\mathbf{u}}_e)^{-1} \mathbf{N}_e^T \mathbf{M}^{1/2} \qquad (2.31)$$

can be constructed that eliminates the need for forming Cholesky factorization of element matrices. The main advantage is the reduction in storage since $\mathbf{B}^{-1} \mathbf{v}$ can be evaluated for any $\mathbf{v}$ using the element representation of $\mathbf{A}$. In the next section we compare the two preconditioners defined in (2.30) and (2.31) using both Lanczos and CG methods.

2.5.1 Implementation

We view the computation of the matrix–vector product **Au** via equation (2.6) as a mechanism for saving storage in return for extra arithmetic work. The reduction in storage demand is due to the following:

1. In most practical finite element problems there is a considerable amount of repetition of a given element in the mesh structure.
2. The element matrices of a number of element types, such as beams, trusses, etc., are known explicitly and depend on only a few fundamental parameters.

The first observation allows us to create a data structure which keeps the element matrix of one element to represent a whole group of elements. The second observation results in a canonical form for each element type, and therefore only a few parameters need be stored to define each element matrix. Hence the storage requirements for all the distinct $\mathbf{a}_e$ is often significantly less than the number of words required to hold **A**, even when a sophisticated sparse storage scheme is used (see [6]). Furthermore, one can always compute the element matrices $\mathbf{a}_e$ each time the product **Au** is required.

The overhead for the reduction in storage is the increased number of operations. However, two comments are in order:

1. The cost of a multiply no longer dominates arithmetic evaluations.
2. Vector and parallel computers or other special purpose devices can execute $\sum_e (\mathbf{N}_e \mathbf{a}_e \mathbf{N}_e^T \mathbf{u})$ very efficiently.

When using the implicit form of **Au** it can be seen that different elements operate on different parts of the vector **v**. One can take advantage of this fact by performing some of the element matrix operations in parallel. The elements are simply arranged into p groups, where p is the number of available processors. Each processor then computes the contribution of the product of all the element matrices in its assigned group by the corresponding components of **v**. Finally, the contribution from each group is combined to obtain **Au**. The implicit product increases the cost of matrix operations by a factor of, say μ, where μ depends on the average number of elements connected to a node. Typically μ ranges between 1.5 and 3 [13], although one could design examples that result in large μ.

2.6 SUBSTRUCTURE BY SUBSTRUCTURE PRECONDITIONING

An obvious generalization of the E × E preconditioner described above is the substructure-by-substructure (S × S) preconditioner. Here the finite element mesh is partitioned into a number of substructures (also referred to as subdomains or superelements). Each substructure consists of a group of elements. Associated with each substructure one can define a stiffness matrix. The assembly of the substructure stiffness matrices results in the global matrix

$$A = \sum_S N_S A_S N_S^T \qquad (2.32)$$

where A_S denotes the stiffness matrix for a substructure. This is a generalization of the E × E preconditioner in the sense that for the special case when each substructure consists of a single element, equation (2.32) reduce to (2.5).

Following a similar approach to the development of E × E preconditioner we are required to perform some form of factorization with A_S. It is important to note that any preconditioner obtained from the factorization A_S depends on the ordering of the unknowns associated with the parameters in the given substructure. By adopting a special ordering where the interior nodes for each substructure are numbered first one can arrive at the hybrid scheme proposed in [23]. Then, the unknowns associated with each sub-structure can be partitioned as

$$x_S = \begin{Bmatrix} x_S^I \\ x_S^B \end{Bmatrix} \qquad (2.33)$$

where x_S^I denotes the unknowns in the *interior*, and x_S^B denotes the unknowns on the *boundary*. Here, all quantities associated with boundary and interior unknowns are denoted with superscripts B and I, respectively. The terms in equation (2.32) can be expressed as

$$A_S = \begin{bmatrix} A_S^{II} & A_S^{IB} \\ (A_S^{IB})^T & A_S^{BB} \end{bmatrix} \qquad (2.34)$$

and

$$N_S = [0, 0 \ldots I, \ldots 0, N_S^B]. \qquad (2.35)$$

The first part of N_S consists of zero blocks except for an identity block corresponding to the interior unknowns. The right hand side vector may also be partitioned in a similar manner to obtain

$$b_S = \begin{Bmatrix} b_S^I \\ b_S^B \end{Bmatrix}. \qquad (2.36)$$

With the above ordering the linear system in (2.2) takes the form

$$\begin{bmatrix} A_1^{II} & & & & A_1^{IB}(N_1^B)^T \\ & A_2^{II} & & & A_2^{IB}(N_2^B)^T \\ & & \ddots & & \vdots \\ & & & A_S^{II} & A_S^{IB}(N_S^B)^T \\ N_1^B(A_1^{IB})^T & N_2^B(A_2^{IB})^T & \ldots & N_S^B(A_S^{IB})^T & A^{BB} \end{bmatrix} \begin{Bmatrix} x_1^I \\ x_2^I \\ \vdots \\ x_S^I \\ x^B \end{Bmatrix} = \begin{Bmatrix} b_1^I \\ b_2^I \\ \vdots \\ b_S^I \\ b^B \end{Bmatrix} \qquad (2.37)$$

2.6 SUBSTRUCTURE BY SUBSTRUCTURE PRECONDITIONING

where $\mathbf{A}^{BB}$, $\mathbf{b}^B$, and $\mathbf{x}^B$ are related to quantities in equations (2.33), (2.34), (2.35) and (2.36) through

$$\mathbf{A}^{BB} = \sum_S \mathbf{N}_S^B \mathbf{A}_S^{BB} (\mathbf{N}_S^B)^T$$
$$\mathbf{b}^B = \sum_S \mathbf{N}_S^B \mathbf{b}_S^B \qquad (2.38)$$
$$\mathbf{x}^B = (\mathbf{N}_S^B)^T \mathbf{x}^B.$$

The block arrow structure of the coefficient matrix in (2.37) is due to the fact that the interior unknowns in one substructure interact with those in a second substructure only through the boundary unknowns $\mathbf{x}^B$. Eliminating the interior unknowns and using the definitions in (2.38) one obtains the linear system of equations for the boundary unknowns

$$\left[\sum_S \mathbf{N}_S^B \bar{\mathbf{A}}_S^{BB} (\mathbf{N}_S^B)^T \right] \mathbf{x}^B = \sum_S \mathbf{N}_S^B \bar{\mathbf{b}}_S^B \qquad (2.39)$$

where $\bar{\mathbf{b}}_S^B = \mathbf{b}_S^B - (\mathbf{A}_S^{IB})^T (\mathbf{A}_S^{II})^{-1} \mathbf{b}_S^I$ and $\bar{\mathbf{A}}_S^{BB} = \mathbf{A}_S^{BB} - (\mathbf{A}_S^{IB})^T (\mathbf{A}_S^{II})^{-1} \mathbf{A}_S^{IB}$ is the Schur complement of $\mathbf{A}_S^{II}$ in (2.34). The main advantage with this approach is that the matrix in (2.39) has a better condition number than the original system (see [23] for detail). It is important to note that these quantities can be computed independently of the other substructures and therefore completely in parallel.

The structure of the matrix coefficient in equation (2.39) is similar to the matrix in (2.5) in the sense that they are both constructed using an assembly process. The only difference is that in (2.39) one is dealing with larger matrices. This similarity may be used to construct preconditioners for equation (2.39) in much the same way as in Section 2.5. However, the difficulty with the product form for the preconditioners defined in Section 2.5 is the sequential nature of the algorithm. In general, the product form in equations (2.30) and (2.31) when applied to the S × S partition requires the processing of substructure one at a time. This can hinder parallelism when implemented on multi-processor computers. Although there are schemes designed to minimize the impact of the product form on parallel implementation (through graph colouring algorithms), when the number of substructures are close to the number of available processors these schemes are not effective.

An alternative preconditioner that is suitable for concurrent implementation can be constructed using the splitting algorithm proposed in [24–26] for transient finite element analysis. This scheme when applied to the problem in (2.1) results in an algorithm that has an additive form and thus lends itself to parallel implementation. Thus, the new preconditioner takes the form

$$\mathbf{B}^{-1} = \left[\sum_S \mathbf{N}_S^B (\mathbf{U}_S^{\bar{B}B})^{-T} \mathbf{N}_S^{BT} \right] \bar{\mathbf{D}} \left[\sum_S \mathbf{N}_S^{BT} (\mathbf{U}_S^{\bar{B}B})^{-1} \mathbf{N}_S^B \right] \qquad (2.40)$$

where $\bar{\mathbf{D}}$ are diagonal matrices and $\bar{\mathbf{U}}_S^{BB}$ is the lower triangle of $\bar{\mathbf{A}}_S^{BB}$. A good choice

54   SOLVING LARGE LINEARIZED SYSTEMS IN MECHANICS

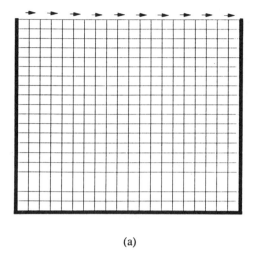

(a)

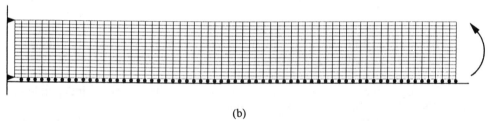

(b)

Figure 2.1 (a) 20 × 20 mesh for the incompressible fluid flow in a cavity; (b) 16 × 64 mesh for the beam in pure bending.

for $\bar{\mathbf{D}}$ is the diagonal of the coefficient matrix in (2.39) obtain by simply assembling the diagonals of $\bar{\mathbf{A}}_S^{BB}$.

The steps in evaluating $\mathbf{B}^{-1}\mathbf{v}$ for a given vector $\mathbf{v}$ are similar to those for computing the product of $\bar{\mathbf{A}}_S^{BB}$ and a vector $\mathbf{v}$. First, $\mathbf{v}$ is localised to each substructure through $\mathbf{N}_S^B\mathbf{v}$. This is followed by a step of forward reduction using the lower part of $\bar{\mathbf{A}}_S^{BB}$. Note that the forward reduction is performed for each substructure independently of the others. The result is then assembled to obtain a new vector, which is then multiplied by $\bar{\mathbf{D}}$. A second localization of the last result followed by a back-substitution and assembly completes the operation with the preconditioners.

2.7 NUMERICAL EXAMPLES

In this section we discuss the results obtained from the application of the Lanczos and CG methods to two different example problems. The first example is a cavity driven flow problem (Stokes flow). The incompressibility of the fluid is represented by local volumetric constraints. These constraints are enforced in each finite element using a penalty method. The penalty parameter represents the bulk modulus of the fluid. 400 elements are used to model this problem (see Figure 2.1(a)). The condition

2.7 NUMERICAL EXAMPLES

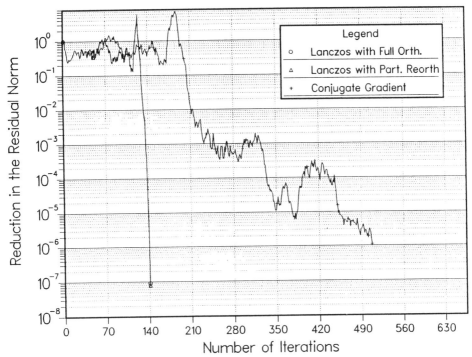

Figure 2.2 Effect of reorthogonalization on the number of iterations 4×16 beam with $T = 1.0$ and diagonal scaling.

number of **A** increases with the penalty parameter. We refer to this as material ill-conditioning.

The second example we used is a beam in pure bending. Taking advantage of symmetry, a quarter of the beam is modelled using plane stress elements (see Figure 2.1(b)). The beam was analysed for a range of different thicknesses while keeping the length constant. This way the element aspect ratio (ratio of the largest dimension to the smallest) can be varied. Again, the condition of **A** increases with the aspect ratio. We refer to this as geometric ill-conditioning. Three different levels of mesh refinement were used to study the effect of problem size on the algorithms; 4×16, 8×32 and 16×64.

To illustrate the advantages of semi-orthogonality we evaluate the solution for the 4×16 beam problem using the CG method, Lanczos method with full reorthogonalization, and Lanczos with Simon's scheme for maintaining semi-orthogonality. In Figure 2.2 we plot the residual norm against the iteration number for each of the methods. The results for the two implementations of the Lanczos method are indistinguishable. The residual norm in the CG method starts off the same as that in the Lanczos method, but it deviates quickly and takes four times as many steps to converge. The difference in the curves for the CG and Lanczos methods is due to loss of orthogonality in the CG method.

We use the 20×20 Stokes flow problem to make a direct comparison between three different preconditioners; diagonal scaling, the E × E Cholesky defined in (2.30),

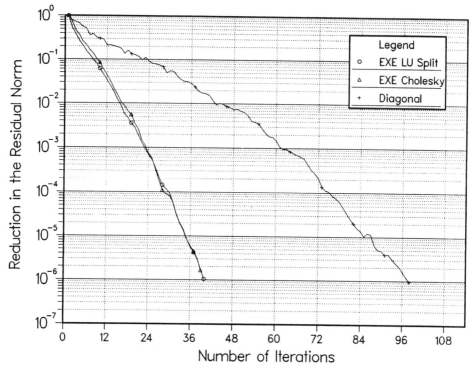

Figure 2.3 Effect of preconditioning on the number of Lanczos steps for 20 × 20 mesh with penalty parameter of 10.

Table 2.3 Result of tests using 20 × 20 mesh. $n = 722$

	Lanczos with partial reorthogonalization						Conjugate gradients		
	E × E LU split		E × E Cholesky		Diagonal		E × E LU split	E × E Cholesky	Diagonal
Penalty Param.	κ	# iter.	Reorth. cost*	# iter.	Reorth. cost*	# iter.	Reorth. cost*	# iterations	# iterations	# iterations
10^4	10^6	236	19 445	228	17 245	393	95 424	473	424	792
10^3	10^5	193	11 134	184	8 900	348	60 869	277	261	517
10^2	10^4	117	440	110	768	247	21 393	117	110	252
10^1	10^3	40	0	39	0	98	708	40	39	98
1	10^2	26	0	25	0	60	0	26	25	60

*Unit is one dot product and one SAXPY (vector plus a scalar times a vector)

and the E × E LU split defined in (2.31). We solve this problem for a range of different penalty parameters using both Lanczos and CG methods. A summary of the results is given in Table 2.3. Sample plots of the residual norm against the iteration number are illustrated in Figures 2.3 and 2.4. The number of iterations required to obtain the solution using the E × E LU split is marginally more than

2.7 NUMERICAL EXAMPLES

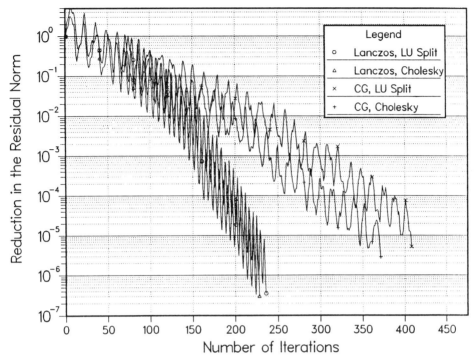

Figure 2.4 Comparison of different E × E preconditioners using 20 × 20 mesh with penalty parameter of 10^4.

that using E × E Cholesky. On average the cost of reorthogonalization for the E × E Cholesky was slightly less. On the other hand, the E × E Cholesky requires additional storage to keep the preconditioning matrix. It is interesting to note that the number of reorthogonalizations increases with the penalty parameter (condition number). This indicates that Simon's scheme performs reorthogonalizations whenever it is needed. The number of iterations given in Table 2.3 are plotted against the penalty parameter (see Figure 2.5). One can observe from this plot that maintaining orthogonality can result in reductions of factors of two in the number of iterations for this example.

We obtain a similar set of results for the beam example; see Table 2.4. Both diagonal and E × E LU split are used to precondition the problem. The solution is evaluated for three different levels of discretization using Lanczos and CG methods. Sample plots for the 8 × 32 mesh are illustrated in Figure 2.6 for thickness $T = 1.0$ and in Figure 2.7 for $T = 0.5$. When $T = 1.0$, CG method required three times as many steps to converge as the Lanczos method. More crucial is the fact that CG failed to converge after 6000 iterations when $T \leqslant 0.5$, even for E × E preconditioners. On the other hand Lanczos delivered the solution in less than 300 iterations using E × E preconditioners.

The 16 × 64 beam problem was also analysed using the CG method with S × S preconditioner using $T = 1.0$ and $T = 0.05$. The results for these runs are given in Table 2.5. The elimination of the interior unknowns in each substructure was carried

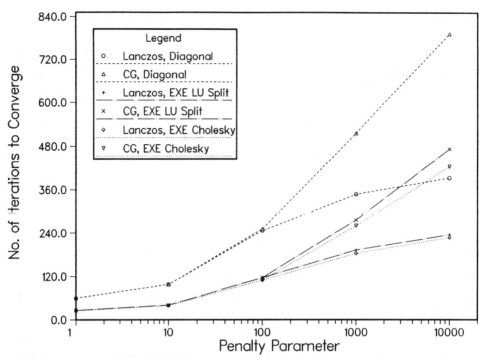

Figure 2.5 Effect of ill-conditioning due to element properties of Lanczos and CG methods, 20 × 20 mesh.

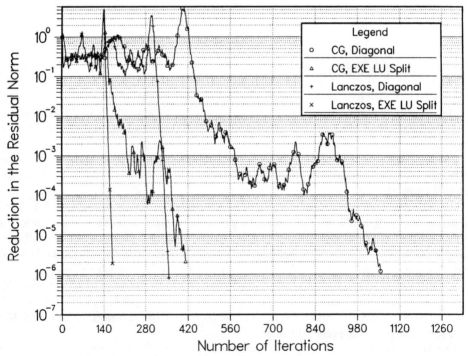

Figure 2.6 8 × 32 beam problem with $T = 1.0$ element aspect ratio = 4.

2.7 NUMERICAL EXAMPLES

Table 2.4 Result of tests using the beam in pure bending. + indicates that CPU time exceeded

			Lanczos with partial reorthogonalization				Conjugate gradients	
			E × E LU split		Diagonal		E × E LU split	Diagonal
Problem	Aspect ratio	κ	No. of iteration	Reorth. cost*	No. of iteration	Reorth. cost*	No. of iteration	No. of iteration
16 × 64 $n = 2142$	1	10^4	146	511	378	17 352	182	506
	2	2×10^5	199	1 208	542	106 880	344	1 027
	4	3×10^6	329	3 261	896	296 617	850	2 336
	8	5×10^7	586	47 667	1 132+	811 600+	2 714	6 000+
	40	3×10^{10}	886	217 883	1 648+	1 317 042+	6 000+	6 000+
4 × 16 $n = 151$	1	2×10^2	41	146	89	2 382	41	96
	2	3×10^3	56	196	111	5 625	71	236
	4	5×10^4	87	624	141	13 951	191	507
	8	7×10^5	114	2 312	151	17 482	605	1 532
	40	5×10^8	150	6 461	151	17 784	2 216	5 614

*Unit is one dot product and one SAXPY (vector plus a scalar times a vector).

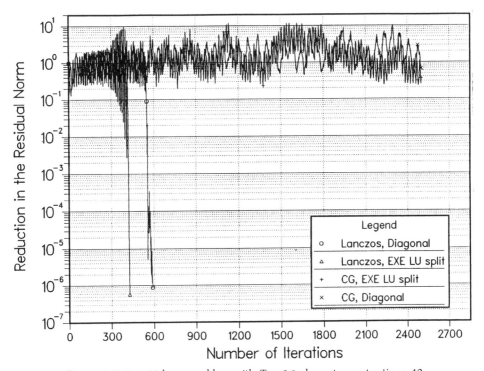

Figure 2.7 8 × 32 beam problem with $T = 0.1$ element aspect ratio = 40.

Table 2.5 Result of CG with S × S preconditioner for the beam in pure bending

Beam thickness	No. of substructures	Solution time (s)	Elimination time (s)	PCG time (s)	No. of iterations
1.0	4	91	82	9	58
	8	53	37	16	95
	16	40	14	26	148
	32	43	3	40	228
0.05	4	110	82	28	164
	8	80	37	43	253
	16	68	14	54	308
	32	76	3	73	410

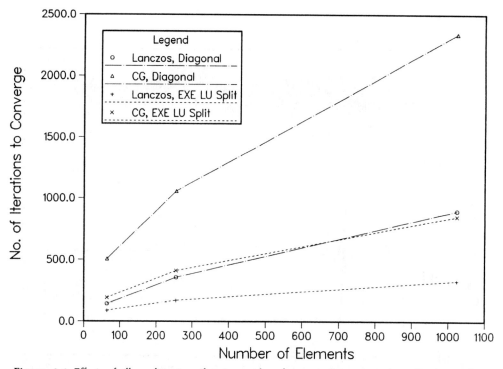

Figure 2.8 Effect of ill-conditioning due to mesh refinement: Lanczos and conjugate gradient methods for the beam problem.

2.7 NUMERICAL EXAMPLES   61

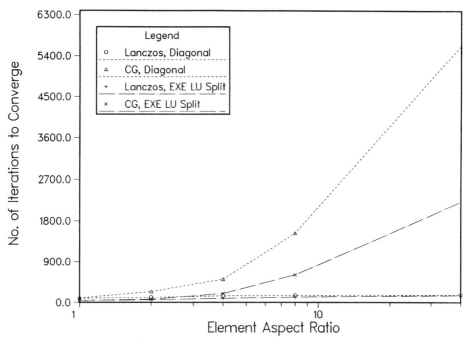

Figure 2.9 Effect of ill-conditioning due to element aspect ratio of Lanczos and conjugate gradient methods, 4 × 16 beam.

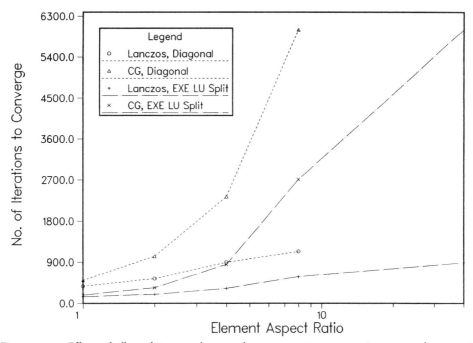

Figure 2.10 Effect of ill-conditioning due to element aspect ratio on Lanczos and conjugate gradient methods, 16 × 16 beam.

out using a direct method (profile solver). As the mesh is partitioned into more substructures, the number of unknowns in the interior decreases. As a result, the time for eliminating all the interior degrees of freedom becomes inversely proportional to the number of substructures. The substructures were created by partitioning the mesh through its thickness. Then the number of boundary nodes in each substructure remains constant. A direct consequence of this is that the total number of boundary nodes become directly proportional to the number of substructure.

2.8 CONCLUSIONS

The conjugate gradient method may be derived directly from the Lanczos algorithm by performing an implicit triangular factorization of the resulting reduced tridiagonal matrix. So long as this factorization is numerically stable the conjugate gradient will also be stable. For symmetric positive definite systems this factorization is stable. In the case of indefinite equations the success of the conjugate gradient method can not be guaranteed. However, the Lanczos will always converge for all symmetric systems.

Partial reorthogonalization improves the robustness of the Lanczos algorithm so that it always converges. However, for ill-conditioned systems the cost of partial reorthogonalization can be substantial. For well-conditioned problems where there is no tendency towards loss of orthogonality among the Lanczos vectors, partial reorthogonalization marginally increases the cost of the Lanczos algorithm. Therefore, partial reorthogonalization should always be used in conjuction with good preconditioners. It will pick up the slack for preconditioners.

The $S \times S$ preconditioner is introduced as an extension of the $E \times E$ preconditioner. It is a hybrid method combining direct elimination of degrees of freedom interior to substructures with iterative solution for the unknowns of the boundary nodes of substructure. The $S \times S$ preconditioners are always more effective than the $E \times E$ techniques.

ACKNOWLEDGEMENTS

The encouragement of Dr T. Shugar throughout this study is greatly appreciated. The work presented in this chapter was supported in part by the Naval Civil Engineering Laboratory under contract N47408-91-C-1216.

BIBLIOGRAPHY

[1] Lanczos, C. (1950) An iteration method for the solution of the eigenvalue problems of linear differential and integral operators, *J. Res. Nat. Bur. Standards*, **45**, 255–282.
[2] Hestenes, M.R. and Stiefel, E. (1952) Method of conjugate gradient for solving linear systems, *J. Res. Nat. Bur. Standards*, **45**, 409–436.
[3] Lanczos, C. (1952) Solution of systems of linear equations by minimized iteration, *J. Res. Nat. Bur. Standards*, **49**, 33–53.

[4] Lively, R.K. (1960) The analysis of large structural systems, *Computer J.*, **3**, 34–39.
[5] Zienkiewicz, O.C. (1977) *The Finite Element Method*, (3rd edn), Mc-Graw Hill, London.
[6] George, A. and Liu, J.W. (1981) *Computer Solution of Large Sparse Positive Definite Systems*, Prentice Hall, Englewood Cliffs, New Jersey.
[7] Reid, J.K. (1971) On the method of conjugate gradients for the solution of large sparse systems of linear equations, in J.K. Reid (ed.), *Large Sparse Sets of Linear Equations*, Academic Press, New York, pp. 231–254.
[8] Fox, R.L. and Stanton, E.L. (1968) Developments in structured analysis by direct energy minimization, *AIAA J.*, **6**, 1036–1042.
[9] Fried, I. (1969) More on gradient iterative methods in finite-element analysis, *AIAA J.*, **7**, 565–567.
[10] Hughes, T.J.R., Levit, I. and Winget, J. (1983) Implicit, unconditionally stable algorithms for heat conduction analysis, *ASCE J. Engg Mech. Div.*, **109**, 576–585.
[11] Ortiz, M., Pinsky, P.M. and Taylor, R.L. (1983) Unconditionally stable element-by-element algorithm for dynamic problems, *Comput. Meth. Appl. Mech. Engg*, **36**, 223–239.
[12] Hughes, T.J.R., Levit, I. and Winget, J. (1983) An element-by-element solution algorithms for problems of structural and solid mechanics, *Comput. Meth. Appl. Mech. Engg*, **36**, 241–254.
[13] Nour-Omid, B. and Parlett, B.N. (1985) Element preconditioning using splitting techniques, *SIAM J. Sci. Stat. Comput.*, **6**, 761–770.
[14] Winget, J.M. and Hughes, T.J.R. (1985) Solution algorithms for nonlinear transient heat conduction analysis employing element-by-element iterative strategies, *Comput. Meth. Appl. Mech. Engg*, **52**, 711–815.
[15] Kershaw, D.S. (1978) The incomplete Choleski-conjugate gradient method for solution of system of linear equations, *J. Comput. Phys.*, **26**, 43–65.
[16] Simon, H.D. (1984) The Lanczos algorithm with partial reorthogonalization, *Math. Comput.*, **42**, 115–142.
[17] Parlett, B.N. (1980) A new look at the Lanczos algorithm for solving symmetric systems of linear equations, *Linear Algebraic Applications*, **29**, 323–346.
[18] Golub, G.H. and Van Loan, C.F. (1983) *Matrix Computations*, The Johns Hopkins University Press, Baltimore.
[19] Parlett, B.N. (1980) *The Symmetric Eigenvalue Problem*, Prentice-Hall, Englewood Cliffs, New Jersey.
[20] Nour-Omid, B. (1984) A preconditioned conjugate gradient method for solution of finite element equations, *Innovative Methods for Nonlinear Problems*, ASME, New Orleans, December.
[21] Wachspress, E.L. (1963) Extended application of alternating direction implicit iteration model problem theory, *J. Soc. Industr. Appl. Math.*, **11**, 994–1016.
[22] Gourlay, A.R. (1977) Splitting methods for time dependent partial differential equations, in D. Jacobs (ed.), *The State of the Art in Numerical Analysis*, Academic Press, New York.
[23] Li, M.R., Nour-Omid, B. and Parlett, B.N. (1982) A fast solver free of fill-in for finite element problems, *SIAM J. Num. Anal.*, **19**(6), 1233–1242.
[24] Ortiz, M. and Nour-Omid, B. (1986) Unconditionally stable concurrent procedures for transient finite element analysis, *Computer Methods in Applied Mechanics and Engineering*, **58**, 151–174.
[25] Ortiz, M., Nour-Omid B. and Sotolino, E.D. (1988) Accuracy of a class of concurrent algorithms for transient element analysis, *Int. J. Num. Meth. Engg*, **26**, 379–391.
[26] Nour-Omid, B. and Ortiz, M. (1989) A family of concurrent algorithms for transient finite element analysis, *Proc. ASCE Structures Congress on Parallel Processing and Computational Strategies in Structural Engineering*, May.

[27] Nour-Omid, B., Raefsky, A. and Lyzenga, G. (1987) Solving finite element equations on concurrent computers, in A. Noor (ed.), *Parallel Computations and their Impact on Mechanics*, ASME, Boston, December, pp. 13–18.
[28] Nour-Omid, B. and Park, K.C. (1987) Solving structural mechanics problems on the Caltech hypercube machine, *Comput. Meth. Appl. Mech. Engg*, **61**, 161–176.

B. Nour-Omid
Scopus Technology, Inc.
1900 Powell St., Suite 900
Emeryville,
CA 94608
USA

3
Adaptive Iterative Solvers in Finite Elements

J. Mandel

University of Colorado at Denver, CO, USA

3.1 INTRODUCTION

Solution of large systems of equations arising from finite elements on modern supercomputers creates a need for innovative solution methods. The approach advocated here consists of a flexible and self-adaptive combination of direct and iterative methods, controlled using an *a priori* estimator of the condition number. One such flexible method, conjugate gradients preconditioned by an incomplete version of symmetric elimination, is presented. Results of its application to complicated linear elasticity problems in three dimensions are reported. Some general aspects of iterative methods for large structural problems are also discussed.

To obtain an efficient method, a problem-specific strategy is used. First the system is transformed by the elimination of selected non-zeros. The transformed system is then preconditioned by a block diagonal matrix that includes a coarse discretization of the same problem. The selection of the block diagonal is adaptive, which makes the method flexible and efficient for problems with a complicated geometry.

The method is illustrated on systems arising from the *p*-version finite element method for three-dimensional elasticity and a test implementation is described. Computational results show that the iterative solver outperforms significantly state-of-the-art direct sparse solvers for real-world test problems with distorted elements both in terms of CPU time and storage. The advantage of the iterative solver increases with the size of the problem. The largest problem solved so far was a complete aircraft fuselage frame discretized by 2755 serendipity elements of order 6, resulting in 412 197 degrees of freedom. For recent results with problems over 1 200 000 degrees of freedom see [44].

	Direct	**Iterative**
Solution of algebraic equations	exact	approximate
Performance depends on	non-zero structure	numerical values
Solution cost	predictable	unknown in advance
Additional right-hand side	cheap	repeat all
Irregular geometries	no problem	slows down
Solution time for large problems	seriously degrades $\approx \text{NDOF}^2$ or worse	smart methods OK $\approx \text{NDOF}$
Storage required for large problems	fill-in requires $\approx \text{NDOF}^{1.5}$	only original data + small additional data structures
Current usage	commercial standard use	academic special projects

Figure 3.1 Comparison of direct and iterative methods.

The method is related to multigrid methods and also to certain domain decomposition methods [1], with each element treated as a subdomain. An early formulation of the method for the p-version finite element method was presented in the conference paper [2]; for more theory and some other related methods, see also [3, 4]. This chapter is partially based on the conference presentation [5].

The basic approach to the development of adaptive iterative solvers is outlined in Section 3.2. Section 3.3 presents a formulation of the method and some theoretical results. Computational results are given in Section 3.4, and Section 3.5 discusses further applications and extensions.

3.2 DEVELOPMENT OF ROBUST ITERATIVE SOLVERS

3.2.1 Direct versus iterative solvers

Iterative solvers have yet to gain commercial acceptance for many reasons. A comparison of the main properties of iterative and direct sparse solvers is given in Figure 3.1.

While the use of computer resources for the generation of stiffness matrices is proportional to the number of elements, the solution of the resulting large, sparse systems of equations grows relatively more expensive due to *fill-in* as the model becomes larger: the triangular factors of the stiffness matrix have many more non-zero elements than the stiffness matrix itself. Various sparse matrix techniques have been developed to make direct solvers based on variants of Gaussian elimination more efficient [6], even on advanced vector architectures [7]. Unfortunately, fill-in cannot be avoided but only minimized by an ordering of variables. For symmetric, positive definite systems, elimination is stable without pivoting and so the operations performed do not depend on the numerical values of the data of the problem; thus performance of direct methods is independent of the numerical data in the symmetric, positive definite case.

Iterative methods do not involve any fill-in, but they use the original data of the problem or data structures that can be stored in about the same amount of memory as the original data. Because of this advantage, iterative methods are gaining acceptance in the finite element community in spite of their limitations [8–10]. The performance of iterative methods, however, does depend on the numerical values of the data of the problem to be solved. An important goal is thus to develop *robust* iterative methods that perform well for a wide range of data and exploit the special properties of the data at hand. Iterative methods gain efficiency by the use of information about the specific problem to be solved. *Multigrid methods* (see, for example, [11]) are especially efficient when the system (3.1) is a discretization of an elliptic partial differential equation and several discretizations with larger step size (or coarser elements) of the same problem are available. A similar method for high order elements using a hierarchy of discretizations for different degrees p was also studied [12]. The main problem with application of multigrid ideas here is that the solver must be able to handle arbitrary distorted elements and irregular meshes, so that it is not clear how to construct coarse grid problems.

3.2.2 Preconditioned conjugate gradients

The solution of large, sparse systems of linear equations

$$Ax = b \qquad (3.1)$$

presents a bottleneck for increasing the precision of three-dimensional finite elements models. Non-linear problems are also commonly reduced to the solution of a series of linear problems (3.1) by techniques such as Newton's method and its variants, which necessitates a repeated solution of linear systems (3.1).

The basic iterative method considered here is the method of preconditioned conjugate gradients [13], described in Figure 3.2. Let A and C be $N \times N$ symmetric positive definite matrices. To solve (3.1), preconditioned conjugate gradients call in each step for the evaluation of the matrix–vector product Ax and the solution of an auxiliary system

$$Cx = r \qquad (3.2)$$

for a given right-hand side r. The matrix C is called a *preconditioner*, and its choice is problem dependent. Some common choices for C are the diagonal of A or an incomplete LU decomposition of A [14, 15].

Step 1. Given $\varepsilon > 0$, $x_0 = 0$, let $k = 0$ and
$$r_0 = b - Ax_0 = b$$
$$h_0 = C^{-1}r_0$$
$$p_0 = h_0$$

Step 2. Do
$$\alpha_k = \frac{r_k^T p_k}{(Ap_k)^T p_k}$$
$$x_{k+1} = x_k + \alpha_k p_k$$
$$r_{k+1} = r_k - \alpha_k A p_k$$
$$h_{k+1} = C^{-1} r_{k+1}$$
$$\beta_{k+1} = \frac{r_{k+1}^T h_{k+1}}{r_k^T h_k}$$
$$p_{k+1} = h_{k+1} + \beta_k p_k$$

Step 3. Let c_k be an estimate of the relative condition number of C and A. If
$$c_k \frac{r_k^T h_k}{r_0^T h_0} \leq \varepsilon^2,$$
exit; else $k = k + 1$ and go to Step 2.

Figure 3.2 The preconditioned conjugate gradient algorithm.

Define the norms
$$\|x\| = \sqrt{x^T x}, \qquad \|x\|_A = \sqrt{x^T A x}$$

and the *relative condition number* κ of C and A as

$$\kappa = \frac{\max\limits_{x \neq 0} \dfrac{x^T A x}{x^T C x}}{\min\limits_{x \neq 0} \dfrac{x^T A x}{x^T C x}}.$$

Note that κ equals to the ratio of the extreme eigenvalues of $C^{-1}A$.

The particular version of the algorithm in Figure 3.2 follows the routine SPCG in CGCODE [16]. In Step 3, c_k is an estimate of κ, obtained from the sequence p_k by exploiting the relationship between conjugate gradients and the Lanczos algorithm for the matrix $C^{-1}A$. Estimates of extreme eigenvalues of $C^{-1}A$ are obtained as extreme eigenvalues of tridiagonal matrices built using the coefficients α_k and β_k [13]. This estimate is recomputed every few steps and always before the exit is actually taken.

Let x^* be the solution of (3.1). Since

$$\frac{\|x_k - x^*\|_A}{\|x^*\|_A} \leq \sqrt{\kappa \frac{r_k^T h_k}{r_0^T h_0}} \qquad (3.3)$$

(see [17]), the algorithm thus attempts to satisfy $\|x_k - x^*\|_A / \|x^*\|_A \leq \varepsilon$, that is, to achieve the relative error ε in the energy norm.

In exact arithmetic, this algorithm terminates and gives an exact solution in at most N steps [18]. We are interested in it as an iterative method, performing much fewer than N steps. The behaviour of the error depends on the distribution of the eigenvalues of $C^{-1}A$; see, for example, [19]. An upper bound on the error can be derived in terms of κ, see [20, p. 187]:

$$\|x_k - x^*\|_A \leq 2\left(\frac{\sqrt{\kappa} - 1}{\sqrt{\kappa} + 1}\right)^k \|x^*\|_A.$$

For more details and derivation of the preconditioned conjugate gradients algorithm, see, for example [13, 18].

The key to a fast preconditioned conjugate gradients method is thus the design of efficient preconditioner, which is not too expensive to apply and gives a low value of the condition number κ. Of course, these goals contradict each other: the lowest possible $\kappa = 1$ requires that $C = A$. One thus needs to strike a balance between the two objectives.

3.2.3 Design of practical iterative solvers

Efficient iterative methods of solving systems arising by discretizations of partial differential equations make extensive use of specific properties of the problem. But it is usually necessary that the discretization is created in a way that is suitable for the application of a fast iterative solver. This is true for both multigrid and domain decomposition methods [21, 11]. The algebraic multigrid method [22] uses unstructured discretizations, but then efficiency suffers. The successful software package PLTMG [23] for solving boundary value problems incorporates a fast iterative method as well as its own discretization and mesh generator.

In current practice, there is a need for iterative solvers that could be used with existing software packages. This means that the iterative solver has no control of the discretization, and has to take the discrete system as is. Requirements for a practical iterative method are as follows:

- *Fast*: It should be very fast for model problems and discretizations with a regular structure (uniform finite element meshes). It should be faster and require less storage than existing direct methods for a wide range of practically useful problems.
- *Robust*: It must handle distorted geometries, general meshes, strong anisotropies, etc. It may slow down for hard problems, but it must not fail.

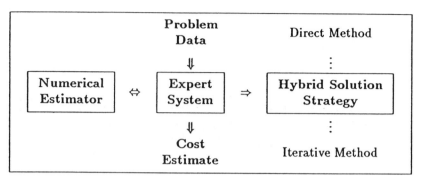

Figure 3.3 Scheme of an adaptive iterative solver.

- *Predictable*: It must give an *a priori* estimate of solution cost.
- *Fully automated*: It must select solution strategy using available data, transparently to the user.

To achieve these objectives, we propose using a scale of methods between purely iterative and strictly direct methods. The selection of a specific solution strategy is governed by a *heuristic (expert system)*, tuned for the specific class of problems at hand. The expert system makes its decisions based on an *a priori condition number estimator*. Such an estimator will be developed below. Figure 3.3 shows this approach graphically.

In the framework of the preconditioned conjugate gradient method, this means the use of a specific, problem dependent preconditioner C, which gives fast convergence of the iterations. The choice of C is *self-adaptive* to handle efficiently distorted finite element meshes, which occur often in practice.

3.3 ADAPTIVE SOLVER FOR THE *p*-VERSION FEM IN THREE DIMENSIONS

3.3.1 *The p-version finite element method*

We will consider the *p*-version in three dimensions as the conforming finite element method with hierarchical serendipity elements of high order p. Except for linear basis functions, the degrees of freedom are nodeless. The basis functions for the reference brick element $\hat{K} = [-1, +1]^3$ are of the form

$$(x \pm 1)(y \pm 1)(z \pm 1), \tag{3.4}$$

$$(x-1)(y-1)(1-z^2)L_n(z), \qquad n = 0,\ldots,p-2, \tag{3.5}$$

$$(x-1)(1-y^2)(1-z^2)L_m(y)L_n(z), \qquad m,n \geq 0, \qquad m+n \leq p-4, \tag{3.6}$$

$$(1-x^2)(1-y^2)(1-z^2)L_l(x)L_m(y)L_n(z), \qquad l, m, n \geq 0,$$

$$l + m + n \leq p - 6 \qquad (3.7)$$

associated with nodes, edges, faces, and the interior, respectively, with obvious permutations of the variables and changes of -1 to $+1$. In (3.5)–(3.7), L_n are polynomials of order n. Basis functions for other types of elements are defined similarly. For more details, see [24, 25]. Theoretical properties of the p-version FEM are discussed in [26, 27]. It is known that in two dimensions, exponential convergence of the discrete solutions can be achieved by increasing p and refining the elements at the same time, but little theory is known in three dimensions. The p-version codes we used to generate the problems and for direct solvers are MSC/PROBE version 4.1 (cf. [28] for basic information) and the STRIPE [29] version of Spring, 1991.

Our method is invariant to the particular choice of the polynomials L_n. It can be applied whenever the basis functions split into groups associated with the nodes, with each edge, each face, and with the interior as in (3.4)–(3.7). For example, functions associated with an edge are zero on all faces not adjacent to that edge.

3.3.2 Preconditioning as a transformation

Let M be a $N \times N$ non-singular matrix. Then the system (3.1) is equivalent to

$$\tilde{A}\tilde{x} = \tilde{b} \qquad (3.8)$$

where

$$\tilde{A} = MAM^T, \qquad \tilde{b} = Mb. \qquad (3.9)$$

After solving for $\tilde{x}$ from (8), the solution x can be obtained from

$$x = M^T \tilde{x}. \qquad (3.10)$$

The preconditioned conjugate gradients algorithm with A and $C = (MM^T)^{-1}$ is equivalent to the conjugate gradients without preconditioning (that is, $C = I$) applied with $\tilde{A}$ in the place of A (see, for example, [18]).

3.3.3 Description of the algorithm

Our method uses a transformation of the form (3.9) and $C \neq I$. The transformation matrix will be denoted $\bar{M}$ to avoid conflict with the notation in (3.9). Such a method is of course mathematically equivalent to one which uses only transformation or only preconditioning, but the use of both transformation and preconditioning results in extra flexibility in the formulation of the method and in its implementation.

Step 1. Transform the system $Ax = b$ to $\tilde{A}\tilde{x} = \tilde{b}$, with $\tilde{A} = \bar{M}A\bar{M}^T$, $\tilde{b} = \bar{M}b$. Keep the transformation matrix $\bar{M}$ (in a product form).

Step 2. Select the preconditioner C as a generalized block diagonal of the transformed matrix $\tilde{A}$.

Step 3. Assemble the matrix C and compute the Cholesky decomposition $C = LL^T$.

Step 4. Apply the preconditioned conjugate gradients method to the transformed system $\tilde{A}\tilde{x} = \tilde{b}$, using the known Cholesky decomposition of C from Step 3.

Step 5. Recover x by back transformation $x = \bar{M}^T \tilde{x}$.

Figure 3.4 The iterative method.

The scheme of the whole iterative method is outlined in Figure 3.4. We now give a more complete description of the steps in Figure 3.4.

Step 1. Transformation by incomplete elimination. This step consists of a sequence of the following elementary block transformations. Consider the matrix A written in a 3×3 symbolic block form

$$A = \begin{pmatrix} A_{11} & A_{12} & A_{13} \\ A_{21} & A_{22} & A_{23} \\ A_{31} & A_{32} & A_{33} \end{pmatrix} \tag{3.11}$$

and the transformation matrix X, defined by

$$X = \begin{pmatrix} I & 0 & 0 \\ -A_{21}A_{11}^{-1} & I & 0 \\ 0 & 0 & I \end{pmatrix}. \tag{3.12}$$

Using the symmetry of A, we have

$$XAX^T = \begin{pmatrix} A_{11} & 0 & A_{13} \\ 0 & A_{22} - A_{21}A_{11}^{-1}A_{12} & A_{23} - A_{21}A_{11}^{-1}A_{13} \\ A_{31} & A_{32} - A_{31}A_{11}^{-1}A_{12} & A_{33} \end{pmatrix}. \tag{3.13}$$

Thus, the first block of variables has been eliminated from the second block but not from the third.

The block form (3.11) is to be understood in a generalized sense; the blocks of variables can form an arbitrary disjoint decomposition of the set $\{1, \ldots, N\}$.

Note that one could eliminate the first block of variables from the third as well as the second block; a succession of N such steps, each with the first block consisting of just one variable eliminated from all the rest, is equivalent to Gaussian elimination, and reduces A to a diagonal matrix. In (3.13), we intentionally do not eliminate the first block completely from the rest of the system (that is, we do not force the blocks A_{31} and A_{13} to be zero), because we do not wish to change the block A_{33} and introduce new non-zero entries there.

This incomplete elimination was called *partial orthogonalization* of degrees of freedom in the first block to those in the second block in [2, 4], based on an interpretation of the matrix XAX^T as the stiffness matrix with respect to a new basis.

The algorithm proceeds by taking these incomplete elimination steps in the following succession:

1. All interior degrees of freedom corresponding to basis functions (3.7) with $m + n + l > 0$ are eliminated from the system.
2. For each face of the finite element structure, the face degrees of freedom (3.6) with $n + m > 0$ are eliminated from the degrees of freedom (3.5) with $n > 0$ for the adjacent edges.

This process does not introduce any fill-in, and it results in a transformed stiffness matrix with zero blocks corresponding to the face–edge interactions between basis functions of higher order than quadratic. There are, however, still non-zero blocks corresponding to the face–face and edge–edge interactions.

The transformation matrix $\bar{M}$ in Step 1 in Figure 3.4 is kept as the collection of the elementary transformation matrices X from (3.12), to be applied in the order in which they are generated.

Step 2. Selection of block diagonal preconditioner and adaptive strategies for distorted elements. Step 1 results in the system $\tilde{A}\tilde{x} = \tilde{b}$, which is equivalent to the original system, and has most face–edge interactions eliminated. The next step is to ignore as many other interactions as possible. In the simple case (no special treatment of distorted elements), the preconditioner C is selected to be the block diagonal of $\tilde{A}$ with the following blocks (some of which may be empty):

1. The first block consists of all degrees of freedom corresponding to linear and quadratic basis functions, that is, nodal basis functions (3.4)–(3.7) with $n = m = l = 0$.
2. For each edge, one block of the remaining degrees of freedom, corresponding to functions (3.5) with $n > 0$.
3. For each face, one block of the remaining degrees of freedom, corresponding to functions (3.6) with $n > 0$ or $m > 0$.
4. For each element, one block of the remaining degrees of freedom, corresponding to functions (3.7) with $n > 0$, $m > 0$, or $l > 0$.

The reason for including variables corresponding to the quadratic functions in the first block was explained in [2]. For three-dimensional problems, the convergence of the method suffers if the first block consists of linear degrees of freedom only. The first diagonal block of C is actually the stiffness matrix of the same problem using quadratic elements.

Step 1. Given element K and its transformed local stiffness matrix A_K, let V be the set of all edges and faces of K. Create a symmetric matrix $T = (t_{ij})$, $0 < t_{ij} \leq 1$, $t_{ii} = 1$, of size $|V| \times |V|$ of estimated strengths of couplings between all edges and faces. Set $t = 1$.

Step 2. Decrease t by a small amount and define G as the graph with an edge (i,j) if $t_{ij} \geq t$. Find all components of the graph G. If the number of components is too small, let $C_K = \tilde{A}_K$ and exit.

Step 3. Define C_K as the block diagonal park of $\tilde{A}_K$ with

1. one diagonal block for all linear and quadratic degrees of freedom;
2. one diagonal block for all degrees of freedom with $p > 2$ in each component of the graph G.

Step 4. Estimate the relative condition number of $\tilde{A}_K$ and C_K for the current decomposition of degrees of freedom of element K. If acceptable, exit, else go to Step 2.

Figure 3.5 Adaptive construction of the preconditioner for distorted elements.

The selection of C as above gives satisfactory convergence rates if the elements are not distorted. In essence, this selection of C ignores all face–face, face–edge, and edge–edge interactions not resolved in Step 1, except for the quadratic functions included in the first block. For distorted elements, ignoring all those interactions gives poor convergence rates. A heuristic procedure has been devised to enlarge adaptively the matrix C to include some of those interactions. For example, for an element with the aspect ratio $1:20:20$, the interaction between the long edges of a side with aspect ratio $1:20$ is quite strong and should be included in C. The transformation in Step 1 has the property that if A_K is the local stiffness matrix of element K understood to be embedded in a zero matrix of the size of the global stiffness matrix A, then the *transformed local stiffness matrix*

$$\tilde{A}_K = \bar{M} A_K \bar{M}^T \tag{3.14}$$

has non-zero entries only for the degrees of freedom of element K. The adaptive heuristic procedure selects a local preconditioner C_K based on $\tilde{A}_K$. C is then assembled from C_K just as the global stiffness matrix $\tilde{A}$ could be assembled form the local matrices $\tilde{A}_K$:

$$\tilde{A} = \sum_K \tilde{A}_K, \qquad C = \sum_K C_K. \tag{3.15}$$

The *adaptive procedure* is outlined in Figure 3.5. It uses an estimator of the relative condition number or A_K and C_K, based on the theoretical principles introduced in Section 3.3.4 below. Note that if selecting a smaller number of diagonal blocks does not yield an acceptably small condition estimate, the method will eventually give up and select $C_K = \tilde{A}_K$.

The matrix T of strengths of couplings is calculated from

$$t_{ij} = \rho(q_{ii}^{-1/2} q_{ij} q_{jj}^{-1/2}),$$

where ρ denotes spectral radius and q_{ij} is the 3 × 3 submatrix of the stiffness matrix for quadratic degrees of freedom and the interaction of face or edge i and face or edge j. Elements t_{ij} that correspond to interaction of edges on the opposite sides of a face or to interaction of opposite faces are slightly increased in order to give preference to exact resolution (i.e. inclusion in C_K) of those couplings in the adaptive process. Matrices C_K including such couplings were shown to give good condition numbers in numerical tests on a single element with high aspect ratios. The estimate of the relative condition number of C_K and $\tilde{A}_K$ was based on submatrices for quadratic degrees of freedom only.

The method guaranteed that the estimate of the condition number was never greater than 100.

Step 3. Assembly and Cholesky decomposition of C. The matrix constructed in Step 2 is well suited for solution by an envelope (variable band) method. If there are no distorted elements, C consists of a number of dense diagonal blocks for the faces and edges, and of one large, sparse block for the linear and quadratic degrees of freedom. An envelope method takes advantage of the block diagonal form of C automatically. If additional interactions are included in C, then C will not have a block diagonal structure, but, because distorted elements tend to occur in clusters, an ordering of variables can be found which results in a reasonably small envelope, and so the Cholesky decomposition of C can still be calculated efficiently.

Step 4. Preconditioned conjugate gradients. The method from Figure 3.2 is used with the transformed matrix $\tilde{A}$ playing the role of A.

Step 5. Back transformation of the solution. To apply (3.10), the matrices X^T from (3.13) are applied in the opposite order to that in which they were generated.

3.3.4 *Theoretical results and estimates of convergence*

Let inequality between symmetric matrices mean that their difference is positive semidefinite. The principal theoretical observation here is that if for all elements K,

$$m_1 C_K \leq \tilde{A}_K \leq m_2 C_K \tag{3.16}$$

then by summation over all elements and using (3.15), one has

$$m_1 C \leq \tilde{A} \leq m_2 C \tag{3.17}$$

which implies that the relative condition number of $\tilde{A}$ and C is at most m_2/m_1. The inequality (3.16) can be satisfied with $m_1 > 0$, $m_2 < \infty$ if and only if $\tilde{A}_K$ and C_K have the same null space [3]. This is fortunately the case here, because the null space of A_K and thus also of $\tilde{A}_K$ consists of vectors corresponding to small rigid body motions, which form a subspace of all linear functions. The ratio m_2/m_1 can be calculated numerically. If the null spaces of $\tilde{A}_K$ and C_K coincide and $\lambda_{1,K}$ and $\lambda_{2,K}$ are the minimal non-zero eigenvalue and the maximal eigenvalue of the generalized eigenvalue problem

$$\tilde{A}_K u_K = \lambda C_K u_K \qquad (3.18)$$

then (3.17) holds with

$$m_1 = \min_K \lambda_{1,K}, \qquad m_2 = \max_K \lambda_{2,K},$$

(see [3]). This simple observation is used in the present method as follows:

1. The specific version of the incomplete elimination algorithm and the construction of C_K described in Section 3.3.3 were selected based on the calculation of $\lambda_{2,K}/\lambda_{1,K}$ in many representative model situations [2]. Also, some weights t_{ij} in the adaptive algorithm were augmented so that the algorithm would produce desired, pre-tested block diagonal matrices C_K in model situations.
2. The adaptive selection of C_K in Figure 3.5 uses an estimate of $\lambda_{2,K}/\lambda_{1,K}$ from submatrices of $\tilde{A}_K$ and C_K. Calculating $\lambda_{1,K}$ and $\lambda_{2,K}$ exactly would be prohibitively expensive.

For a closely related method for two-dimensional problems, it can be shown that (see [30])

$$m_2/m_1 \leq \text{const.} \log^2 p.$$

For elements with high aspect ratios in two dimensions, the condition number $\lambda_{2,K}/\lambda_{1,K}$ grows like the square of the aspect ratio, and making C_K a suitable larger block diagonal part of $\tilde{A}_K$ was proved to inhibit this growth [31, 32].

3.4 COMPUTATIONAL RESULTS

3.4.1 Test implementation

The purpose of the test implementation was to assess the viability of the method and provide a tool for experimentation. The emphasis was on reliability of the results, flexibility, and programmer's convenience rather than efficiency, although gross inefficiencies were avoided wherever possible.

The code was written in FORTRAN 77 with the use of the C language macro preprocessor statements, and it has about 10 000 lines of source code (before macro expansions). The code also makes use of routines from the packages LINPACK, EISPACK, CGCODE, and SPARSPAK.

The program uses a simple stack memory management scheme so that it can handle problems of an arbitrary size. Element types are described using tables given

Table 3.1 Comparison of direct and iterative solver for MSC/PROBE

p	NDOF	Generating stiffness		Direct solver		Iterative solver		
		Disk	CPU	Disk	CPU	IT	Disk	CPU
5	9 687	18.0	1.7	38.9	0.9	25	20.7	0.17
6	15 090	36.9	4.7	77.4	2.4	29	42.6	0.37
7	22 386	70.3	11.2	144.5	5.7	38	83.5	0.98
8	31 758	125.8	22.0	253.5	12.3	45	154.0	3.51

Test problem CRANK1, automobile crankshaft, 107 elements (Figure 3.6). CPU in hours, disk space in MB, SUN Sparcstation 1. The disk space for the iterative solver includes the stiffness matrix data.

as input data. Arbitrary elements in any number of dimensions are allowed. The topology and connectivity data are handled in in-core data structures.

The global stiffness matrix is kept as a collection of local stiffness matrices and is never assembled. The transformations are reflected in the data of the local stiffness matrices, and only the pieces of the global stiffness matrix needed to calculate the transformations are assembled temporarily. The local stiffness matrices are kept in temporary files and/or memory buffers. A factored representation of the transformation matrix $\bar{M}$ is kept in several files.

The preconditioner C is kept and decomposed either in core or out of core. For the Cholesky decomposition of C, the variables are first reordered using SPARSPAK implementation of the RCM algorithm [33]. For the in-core version (used on SUN computers), a special version of the envelope Cholesky had to be developed, because the envelope algorithm from [33] resulted in excessive paging. The new routine splits the Cholesky decomposition into a number of stages. At each stage, the routine accesses only a 'window' in the envelope at random and the rest of the data is processed sequentially. The size of the window is given in a parameter, which should be smaller than the amount of RAM available to the program. A simple out-of-core assembler and envelope solver written by Dr Börje Andersson from the Aeronautical Research Institute of Sweden was used on CRAY-XMP. Since the assembled preconditioning matrix C is block diagonal with many full small diagonal blocks, such diagonal blocks were treated separately and vectorized across the blocks.

3.4.2 Comparison with direct solvers

All test problem in this section are linear elasticity in displacement formulation for isotropic material with Poisson ratio 0.3. The relative precision in the energy norm was monitored using (3.3), and it was better than 10^{-4} in all cases.

The results of computational tests with MSC/PROBE version 4.1 are in Tables 3.1 and 3.2. The direct solver was the multi-frontal solver distributed with MSC/PROBE. The data for the largest problem (CRANK2, 376 elements) and $p = 8$ could not be obtained because of disk space limitations. Also, the disk space required by the direct solver for this problem at $p = 6$ and $p = 7$ exceeded available space, while the iterative solver could be run without any problems. The finite element mesh for the problem CRANK1 is in Figure 3.6.

The amount of disk space and memory needed by the iterative solver is only

Table 3.2 Comparison of direct and iterative solver for MSC/PROBE

p	NDOF	Generating stiffness		Direct solver		Iterative solver		
		Disk	CPU	Disk	CPU	IT	Disk	CPU
3	10 992	12.5	0.61	51.37	1.6	14	14.3	0.54
4	19 671	30.8	2.0	124.3	5.0	19	36.1	1.72
5	32 478	67.8	5.6	309.1	18.5	25	77.0	2.09
6	50 541	137.2	15.7		(52)	27	155.9	3.41
7	74 988	259.3	55.5			32	301.6	7.36

Test problem CRANK2, automobile crankshaft, 376 elements. CPU in hours, disk space in MB, SUN 4/280. The disk space for the iterative solver includes the stiffness matrix data.

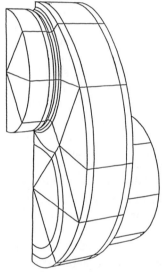

Figure 3.6 Crankshaft model with 107 elements. Test problem CRANK1 (Reproduced by permission of the MacNeal-Schwendler Corporation).

slightly more than for all local stiffness matrices. On the other hand, the disk space requirements of the direct solver grow explosively due to fill-in.

Table 3.3 contains a relative comparison of performance of direct and iterative solution for the problems CRANK1 and CRANK2 as well as two additional simple test problems BLOCK18W (3 × 3 × 2 cubes) and BLOCK64 (4 × 4 × 4 cubes). The middle columns of Table 3 give a breakdown of the CPU time for the main parts of the iterative solver. For problems with a large number of elements, the memory and time requirements of the iterative solver are dominated by the decomposition of the preconditioner C, which essentially involves a direct solution of the same problem on quadratic elements. The highest proportion of time spent in the decomposition was for the largest test problem (CRANK2). The proportion of the time spent in the decomposition decreases with larger p, because C then consists of a smaller number of non-zeros of $\tilde{A}$. Only the iterative part has to be repeated for the solution for an additional right-hand side.

3.4 COMPUTATIONAL RESULTS

Table 3.3 Relative performance of iterative solver on SUN4

Problem	p	NDOF	Transformation	Decomposition of C	Iterations	CPU time	Disk and memory
BLOCK18W	3	600	0.47	0.21	0.32	1.1	1.3
	4	1 065	0.35	0.22	0.43	1.8	1.4
	5	1 755	0.47	0.11	0.43	2.5	1.4
	6	2 724	0.57	0.05	0.38	2.3	1.4
	7	4 026	0.51	0.02	0.47	1.4	1.3
	8	5 679	0.54	0.01	0.45	1.1	1.2
BLOCK64	3	2 175	0.25	0.59	0.16	3.1	2.3
	4	3 795	0.20	0.55	0.26	4.8	2.3
	5	6 135	0.37	0.23	0.40	6.0	2.5
	6	9 387	0.47	0.12	0.42	7.1	2.7
	7	13 743	0.57	0.06	0.38	7.5	2.8
	8	19 395	0.66	0.03	0.31	7.1	2.6
CRANK1	5	9 687	0.44	0.21	0.35	5.3	1.9
	6	15 090	0.55	0.11	0.34	6.5	1.8
	7	22 386	0.64	0.05	0.31	5.8	1.7
	8	31 758	0.59	0.02	0.38	3.5	1.6
CRANK2	3	10 992	0.07	0.82	0.11	2.9	3.6
	4	19 671	0.04	0.82	0.14	2.9	3.4
	5	32 478	0.12	0.64	0.24	8.9	4.0
	6	50 541	0.27	0.39	0.34		
	7	74 988	0.38	0.26	0.37		

Proportions of CPU are the proportions of the CPU time the iterative method spends in the main parts of the algorithm. The transformation time includes the adaptive procedure (Figure 3.5). The back transformation (Step 4 in Figure 3.4) is included in the time of the iterations.
Improvement over direct solver gives the ratio of the CPU time and disk space/memory requirements of the direct solver compared to those of the iterative solver.

In the tests reported in Table 3.3, the iterative solver was faster in all cases. The biggest speed-up observed was 8.9 for the CRANK2 test problem and $p = 5$. The practical usefulness of the iterative solver depends on its cost relative to the rest of the computation, which was dominated by the generation of the stiffness matrix. For the larger problems, the cost of the direct solver exceeds many times that of generating the stiffness matrix, while the cost of the iterative solver remains relatively small.

The results of tests with STRIPE or CRAY-XMP are reported in Tables 3.4 and 3.5. The direct solver was out-of-core envelope, with a domain decomposition ordering. Since these test problems involved almost all elements heavily distorted and with high aspect ratios, the adaptive procedure has chosen many more couplings to be included in the matrix C that for the previous problems. The number of elements was also larger, resulting in a greater bandwidth of C. The extra disk space needed for the iterative solver was dominated by the storage of C. The number of iterations was also higher.

Still, the iterative solver was significantly faster than the direct solver. However,

Table 3.4 Comparison of direct and iterative solvers for STRIPE

p	NDOF	Generating stiffness		Direct solver		Iterative solver		
		Disk	CPU	Disk	CPU	IT	Disk	CPU
3	10 320	10	10	70	64	33	66	64
4	17 826	23	20	160	171	37	102	113
5	28 575	51	35	336	431	50	136	178
6	43 347	103	57	656	974	49	218	259
7	62 922	195	91	1 302	2 018	54	354	417
8	88 080	346	143	2 096	3 964	53	575	692

Test problem FRAME1, part of frame fuselage, 260 elements (Figure 3.7). CPU in seconds and disk space in MB. CRAY-XMP, 1 processor. The disk space for both the direct and iterative solver is the total space needed to run the solver, including the stiffness matrix data.

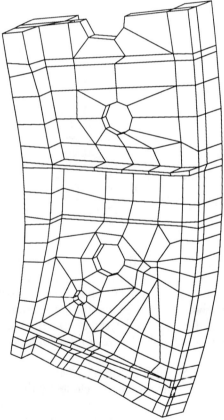

Figure 3.7 Part of frame fuselage, 260 elements. Test problem FRAME1 (Reproduced by permission of the Aeronautical Research Institute of Sweden).

since the iteration matrices were kept in disk files and because the CPU on CRAY-XMP is much faster than I/O, the CPU utilization was observed to be only at most

Table 3.5 Comparison of direct and iterative solvers for STRIPE

p	NDOF	Generating stiffness		Direct solver		Iterative solver		
		Disk	CPU	Disk	CPU	IT	Disk	CPU
4	165 534	0.25	0.06	3.05	1.2	76	1.04	1.0
5	268 758	0.55	0.10	(5.85)	(3.0)	117	1.44	1.6
6	412 197	1.1	0.17	(11.1)	(6.8)	124	2.24	2.0

Test problem FRAME2, complete frame fuselage, 2755 elements (Figure 3.8). CPU in hours and disk space in GB. CRAY-XMP, 1 processor. Numbers in parentheses are estimates. The disk space for both the direct and iterative solver is the total space needed to run the solver, including the stiffness matrix data.

5%. The code was heavily I/O bound in the transformation phase as well as in the iteration phase. The main advantage, however, was that using the iterative solver, it was possible to solve problem that before could not be solved at all with the available disk space.

3.5 DISCUSSION AND DIRECTIONS FOR FUTURE RESEARCH

3.5.1 Must iterative methods be I/O bound?

If the iteration matrix is stored on external memory, an iterative method becomes very I/O intensive. In computation of a sparse matrix–vector product

$$y_i = \sum_{j: a_{ij} \neq 0} a_{ij} x_j,$$

each a_{ij} is used only once, or at most twice if symmetry of the matrix is exploited. Even with perfect overlap of I/O and CPU, the wall clock time will be a large multiple of the CPU time. Especially on vector supercomputers, it takes much longer to get a word from a disk than to perform several floating operations with it. There are three possible remedies for this situation:

1. store the whole iteration matrix in RAM;
2. make one iteration of the method reduce the error so significantly that only a few iterations are needed;
3. calculate all matrix–vector products without calculating (and storing) the matrices.

For very large problems, it is not usually possible to keep the stiffness matrix in RAM. We will always want to solve larger problems than the size of our computers, so the concept of 'very large' is a moving target. On existing parallel machines, however, I/O in each iteration is simply out of the question, so variants 1 or 3 are the only options.

It is not clear at present how to spend a large amount of computational work

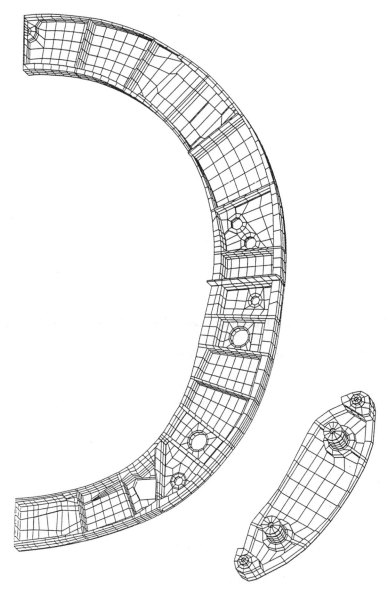

Figure 3.8 Complete frame fuselage, 2755 elements. Test problem FRAME2. The two parts are attached (Reproduced by permission of the Aeronautical Research Institute of Sweden).

locally at a part of the problem to achieve very large error reduction needed for alternative 2.

Alternative 3 is currently being investigated in the framework of the present method. A fast matrix–vector multiplication for the p-version FEM is available [34], and it can be combined with the use of special basis functions that have an effect similar to the use of the transformation in the present method [4]. A different fast multiplication was used in multigrid for spectral methods [35].

3.5.2 Variable degree elements

Adaptive refinement by varying the degree of elements is often used in practice (see, e.g., [25]). The formulation of the method described in this chapter can accommodate such elements without any change, and we expect that the preconditioner would perform equally well. However, this has not been implemented yet because of the complicated data structures and the substantial programming effort required.

3.5.3 Considerations for parallel architectures

The method described here is immediately parallelizable on the element level except for the decomposition of C.

The transformation requires the communication of parts of the matrices A_K only between adjacent elements. (This results in complicated programming, so other possibilities are being investigated; see Section 3.5.8.) The adaptive procedure is completely parallel. Preconditioned conjugate gradients require evaluation of inner products and of the matrix–vector products $\tilde{A}\tilde{x}$ and $C^{-1}r$. For the matrix–vector product $\tilde{A}\tilde{x}$, one only needs to communicate the values of $\tilde{A}_K\tilde{x}_K$ between adjacent elements. The back transformation can use transformation data stored locally.

The repeated evaluations of $C^{-1}r$ in preconditioned conjugate gradients and the preceding decomposition of C are a bottleneck to parallelism. Assembly and decomposition of C, of course, can be also parallelized, especially when the adaptive procedure enlarging C is not needed. But the adaptive procedure in general destroys the local character of C, so for some problems it will be more efficient not to do any adaptive changes of C. In a parallel computing environment, it may be more efficient to use several inner iterations of an easily parallelizable method to evaluate an approximation to $C^{-1}r$ in each step of preconditioned conjugate gradients rather than decompose C. For experience with parallel implementations of a related method in two dimensions, see [30, 36].

3.5.4 Elements with very high aspect ratios, shells and plates

The present method fails for thin elements with aspect ratios bigger than about 100. The heuristic does recognize that there is a problem and switches automatically to a direct method. The trouble is that the transformation phase is tuned for solid elements; a different type of transformation is needed for shells and plates.

3.5.5 Nearly incompressible materials

The p-version finite element method is known to avoid locking for nearly incompressible materials [37] and good engineering solutions can be obtained by post-processing [38]. The iterative method deteriorates for nearly incompressible materials and requires more iterations and/or a bigger matrix C, which is more expensive to decompose. One possible way to handle this problem is to use for preconditioning elements with artificial pressure or equivalent non-conforming elements.

Highly anisotropic materials pose a similar problem, which could be possibly handled by adding more additional variables.

In any case, the adaptive procedure (Figure 3.5) should be further developed to recognize such instances reliably and gradually switch to a direct solver, setting eventually $C = A$. The transformation should of course be avoided in such extreme cases. The performance of such a method would then depend on the difficulty of the problem; the method would be fast for nearly isotropic and well-compressible materials, while it would automatically reduce to a direct solver for difficult problems.

3.5.6 Computation of eigenvalues

The generalized eigenvalue problem

$$Au = \lambda Bu, \qquad (3.19)$$

with B the mass matrix, is commonly solved using the Lanczos method, which requires solving repeatedly linear systems with the matrix $A - \mu B$ for given shifts μ (see [39]). These systems need to be solved exactly, which would make iterative methods prohibitively expensive. However, combining inverse iterations with preconditioned conjugate gradient iterations may be a way to adapt an iterative method for the solution of linear systems to solve (3.19) (see [40, 41]).

3.5.7 Combination with incomplete LU decomposition (ILU)

The transformation by incomplete elimination is based on a similar principle to the well-known preconditioner by ILU. Incorporating the principles used here into the ILU framework could result in a superior method. The LU decomposition of C can be incorporated into the incomplete elimination phase as allowing some fill-in. The whole preprocessing phase could then be implemented as a modification of an existing direct solver by disallowing most of the fill-in arising in a direct solution of the whole system.

3.5.8 Element parallel transformation

The transformation by incomplete elimination operates on the *global* stiffness matrix, which requires significant communication between elements and results in rather complicated programming. A version of the method is being considered in which the transformation in a local stiffness matrix depends on the data of that matrix only.

3.5.9 Multilevel process

The method requires in each step the solution of an auxiliary system, which is the same problem with quadratic elements. The time for Cholesky decomposition of this subsystem dominates the computation for a large number of elements and limits the size of the problem if the decomposition is done in core. A fast iterative method for the solution of this auxiliary problem should be considered for very large problems, involving even smaller auxiliary problems. Another possibility may be to use a version of ILU, possibly combined with the incomplete elimination into one incomplete factorization process (cf. Section 3.5.7).

3.5.10 Application to other problems

The approach presented here can be applied to a large class of symmetric, positive definite linear problems arising from the finite element method. The basic requirements for its applicability and efficiency are:

1. The elements have a large number of degrees of freedom, which are naturally associated with the interior, faces, edges, and nodes of the elements.
2. A small subset of basis functions can be found that spans the null space of the local stiffness matrix. The system of equations derived from this subset of the basis functions is then solved in each iteration.

The p-version finite element method satisfies these requirements. For other finite element methods, condition 1 can be satisfied by forming *macroelements* by aggregating a number of existing elements, and condition 2 by a transformation of the degress of freedom of the macroelement.

Other currently considered areas of application include the Maxwell equations of electromagnetics discretized by finite elements in MSC/EMAS.

Complex problems with Hermitean positive definite matrices do not present any significant problems, and the existing theory applies. The principles and algorithms presented in this paper can be also applied to nonsymmetric and indefinite problems, but then the theory does not apply any more and the efficiency of the preconditioning in specific cases is yet to be investigated. Special versions of the conjugate gradients algorithm can be used for such problems [17].

3.5.11 Other adaptive solvers

Adaptive solvers based on other algorithms are being investigated. The main requirements are that the preconditioner is sufficiently flexible and controllable, and that its behaviour can be realistically and cheaply estimated for a given problem. Further development based on the Neumann–Neumann domain decomposition preconditioner [42, 43] could result in an efficient method that would be much easier to implement than the preconditioning method described here.

3.6 CONCLUSION

The iterative method presented here uses special properties of the system of linear equations arising from finite elements in three dimensions to achieve fast convergence and high efficiency. The method outperforms state of the art direct sparse solvers for large problems with complicated geometries and distorted elements.

Most computations parallelize naturally on the element level. Possible extensions of the method include the solution of eigenvalue problems, linear systems arising from singular perturbation problems such as elasticity for nearly incompressible or highly anisotropic materials, and general finite element systems for structural mechanics.

ACKNOWLEDGEMENTS

The work of the author reported here was partially supported by the National Science Foundation under grant DMS-8704169. The test problems reported in Tables 1, 2 and 3 as well as the use of MSC/PROBE were provided by the MacNeal-Schwendler Corporation. The test problems and results reported in Tables 4 and 5 and Cray-XMP time were provided by the Aeronautical Research Institute of Sweden. Special thanks are due to Dr. Börje Andersson of the Aeronautical Research Institute of Sweden for running the tests on Cray-XMP and for his help with interfacing the iterative solver to STRIPE.

BIBLIOGRAPHY

[1] Bramble, J.H., Pasciak, J.E. and Schatz, A.H. (1986) The construction of preconditioners for elliptic problems by substructuring, I. *Math. Comp.*, **47**(175), 103–134.
[2] Mandel, J. (1990) Hierarchical preconditioning and partial orthogonalization for the *p*-version finite element method, in T.F. Chan, R. Glowinski, J. Periaux and O.B. Widlund (eds.) *Third Int. Symp. Domain Decomposition Methods for Partial Differential Equations*, pp. 141–156, Philadelphia, SIAM.
[3] Mandel, J. (1989) Iterative solvers by substructuring for the *p*-version finite element method, *Comput. Meth. Appl. Mech. Engg.*, **80**, 117–128, 1990; Int. Conf. Spectral and High Order Methods for Partial Differential Equations, Como, Italy, June.
[4] Mandel, J. (1990) Two-level domain decomposition preconditioning for the *p*-version finite element method in three dimensions, *Int. J. Num. Meth. Engg.*, **29**(5), 1095–1108.
[5] Mandel, J. (1991) Fast iterative solver for finite elements using incomplete elimination, 1991 *MSC World Users Conference*, Los Angeles, CA, March, and *Fifth Copper Mountain Conference on Multigrid Methods*, April, 1991.
[6] Duff, I.S., Erisman, A.M. and Reid, J.K. (1986) *Direct Methods for Sparse Matrices*, Clarendon Press, Oxford.
[7] Ashcraft, C.C., Grimes, R.G., Lewis, J.G., Peyton, B.W. and Simon, H.D. (1987) Progress in sparse matrix methods for large linear systems on vector supercomputers, *Int. J. Supercomputer Applications*, **1**(4), 10–30.
[8] Crisfield, M.A. (1986) *Finite Elements and Solution Procedures for Structural Analysis*, Vol. I. *Linear Analysis*. Pineridge Press, Swansea.

[9] Farhat, C. (1989) A multigrid-like semi-iterative algorithm for the massively parallel solution of large scale finite element systems, in J. Mandel, S.F. McCormick, J.E. Dendy, Jr., C. Farhat, G. Lonsdale, S.V. Parter, J.W. Ruge and K. Stüben (eds.), *Proc. 4th Copper Mountain Conference on Multigrid Methods*, pp. 171–180, Philadelphia, SIAM.

[10] Lewis, J.G. and Pierce, D.J. (1989) Recent research in iterative methods at Boeing, *Comput. Phys. Commun.*, **53**, 213–221, 1989. Practical Iterative Methods for Large Scale Computations, Proceedings of the Minnesota Supercomputer Institute Workshop, 23–25 October 1988.

[11] McCormick, S.F. (ed.) (1987) *Multigrid Methods*, Vol. 3 of *Frontiers in Applied Mathematics*, SIAM, Philadelphia.

[12] Foresti, S., Brussino, G., Hassanzadeh, S. and Sonnad, V. (1989) Multilevel solution method for the *p*-version of finite elements, *Comput. Phys. Commun.*, **53**, 349–355.

[13] Concus, P., Golub, G.H. and O'Leary, D.P. (1976) A generalized conjugate gradient method for the numerical solution of elliptic partial differential equations, in J.R. Bunch and D.J. Rose (eds.), *Sparse Matrix Computations*, pp. 309–322, Academic Press.

[14] Manteuffel, T.A. (1980) An incomplete factorization technique for positive definite linear systems, *Math. Comput.*, **34**, 473–497.

[15] Meijerink, J.A. and Van der Vorst, H.A. (1977) An iterative solution method for linear equations systems of which the coefficient matrix is a symmetric M-matrix, *Math. Comput.*, **31**, 148–162.

[16] Ashby, S., Manteuffel, T.A. and Joubert, W. (1988) *CGCODE. Collection of conjugate gradients FORTRAN routines*, Los Alamos National Laboratory.

[17] Ashby, S., Manteuffel, T.A. and Saylor, P.E. (1990) A taxonomy for conjugate gradient methods, *SIAM J. Num. Anal.*, **27**, 1542–1568.

[18] Golub, G.H. and Van Loan, C.F. (1989) *Matrix Computations* (2nd edn), Johns Hopkins University Press.

[19] van der Sluis, A. and Van der Vorst, H.A. (1986) The rate of convergence of conjugate gradients, *Num. Math.*, **48**, 543–560.

[20] Luenberger, D.G. (1973) *Introduction to Linear and Nonlinear Programming*, Addison-Wesley, New York.

[21] Dryja, M. and Widlund, O.B. (1990) Towards a unified theory of domain decomposition algorithms for elliptic problems, in T. Chan, R. Glowinski, J. Périaux, and O. Widlund, (eds.), *3rd Int. Symp. Domain Decomposition Methods for Partial Differential Equations*, Houston, Texas, March 20–22, 1989. SIAM, Philadelphia, Pennsylvania.

[22] Ruge, J.W. and Stüben, K. (1987) Algebraic multigrid, in S.F. McCormick (ed.), *Multigrid Methods*, Vol. 5 of *Frontiers in Applied Mathematics*, Chapter 4, pp. 73–130. SIAM, Philadelphia, Pennsylvania.

[23] Bank, R.E. (1990) *PLTMG: A Software Package for Solving Elliptic Partial Differential Equations, User's Guide 6.0*, SIAM, Philadelphia, Pennsylvania.

[24] Szabó, B.A. (1990) The *p*- and *h-p*-version of the finite element method in solid mechanics. *Comput. Meth. Appl. Mech. Engg.*, **80**, 185–196; Int. Conf. Spectral and High Order Methods for partial Differential Equations, Como, Italy, June 1989.

[25] Szabó, B.A. and Babuskâ, I.M. (1991). *Finite Element Analysis*, John Wiley and Sons, New York.

[26] Babuška, I. and Suri, M. (1990) The *p*- and *h-p*-versions of the finite element method, An overview. *Comput. Meth. Appl. Mech. Engg.*, **80**, 5–26. Int. Conf. Spectral and High Order Methods for partial Differential Equations, Como, Italy, June 1989.

[27] Babuška, I., Szabó, B.A. and Katz, I.N. (1981) The *p*-version of the finite element method, *SIAM J. Num. Anal.*, **18**, 515–545.

[28] Szabó, B.A. (1985) *PROBE Theoretical Manual*, Noetic Technologies, St Louis, Montana.

[29] Andersson, B. and Falk, U. (1987) *Self-adaptive Analysis of Three-dimensional Structures*

Using the p-Version of the Finite Element Method, Technical Report FFA TN 1987–31, The Aeronautical Research Institute of Sweden, Bromma, Sweden.

[30] Babuška, I., Craig, A.W., Mandel, J. and Pitkäranta, J. (1991) Efficient preconditioning for the p-version finite element method in two dimensions, *SIAM J. Num. Anal.*, **28**, 624–662.

[31] Lett, G.S. (1990) *Iterative Solver for the p-Version Finite Element Method with Thin Rectangular Elements*, PhD thesis, Department of Mathematics, University of Colorado at Denver.

[32] Mandel, J. and Lett, G.S. (1991) Domain decomposition preconditioning for p-version finite elements with high aspect ratios, *Appl. Num. Math.*, **8**, 441–425.

[33] George, J.A. and Liu, J.W. (1981) *Computer Solution of Large Sparse Positive Definite Systems*, Prentice Hall, Englewood Cliffs, New Jersey.

[34] Andersson, B. (1991) In preparation.

[35] Rónquist, E.M. and Patera, A.T. (1987) Spectral element multigrid. I. Formulation and numerical results, *J. Sci. Comput.*, **2**, 389–406.

[36] Babuška, I. and Elman, H.C. (1989) Some aspects of parallel implementation of the finite element method on message passing architecture, *J. Comput. Appl. Math.*, **27**, 157–187.

[37] Vogelius, M. (1983) An analysis of the p-version of the finite element method for nearly incompressible materials, *Num. Math.*, **41**, 39–53.

[38] Szabó, B.A., Babuška, I.M. and Chayapathy, B.K. (1989) Stress computations for nearly incompressible materials by the p-version of the finite element method, *Int. J. Num. Meth. Engg*, **28**, 2175–2190.

[39] Cullum, J. and Willoughby, R.A. (1985) *Lanczos Algorithms for Large Symmetric Eigenvalue Computations*, Vol. 1. Theory, Birkhaüser, Boston.

[40] Szyld, D.B. (1983) *A Two-level Iterative Method for Large Sparse Generalized Eigenvalue Calculations*, PhD thesis, New York University.

[41] Szyld, D.B. (1988) Criteria for combining inverse and Rayleigh quotient iteration, *SIAM J. Num. Anal.*, **25**, 1369–1375.

[42] Bourgat, J.-F., Glowinski, R., Le Tallec, P. and Vidrascu, M. (1989) Variational formulation and algorithm for trace operator in domain decomposition calculations, in T. Chan, R. Glowinski, J. Périaux and O. Widlund (eds.), *Domain Decomposition Methods*, SIAM, Philadelphia, Pennsylvania.

[43] De Roeck, Y.-H. and Le Tallec, P. (1991) Analysis and test of a local domain decomposition preconditioner, in R. Glowinski, Y. Kuznetsov, G. Meurant, J. Périaux and O. Widlund (eds.), *4th Int. Symp. Domain Decomposition Methods for Partial Differential Equations*, SIAM, Philadelphia, Pennsylvania.

[44] Mandel, J. (to appear) Iterative solver for p-version finite element method in three dimensions, *Comp. Meth. Appl. Mech. Engg*.

J. Mandel
Computational Mathematics Group
University of Colorado at Denver
Denver
CO 80217-3364
USA

4
Solution of Large Boundary Element Equations

J. H. Kane and K. Guru Prasad

Clarkson University, NY, USA

SUMMARY

Multi-zone boundary element analysis (BEA) is known to produce overall system matrices with a sparse blocked character. Recent advancements in the direct and iterative solution of these special sets of algebraic equations are discussed. Static condensation is shown to dramatically reduce the computational burden associated with matrix fill-in in both the direct matrix factorization step and the subsequent forward reduction and backward substitution process for multiply connected BEA models. Preconditioned block interative unsymmetric equation solution schemes are shown to generally out-perform direct approaches for single- and multi-zone models. Colour (grey scale) plots of matrix populations associated with multi-zone BEA models are depicted to aid in explanation of the impact of preconditioning on the convergence of the iterative methods. Numerical examples are given with timing, storage, and accuracy statistics presented. A relatively extensive literature survey is given to help provide a general characterization of the state of the art in equation solving for BEA.

4.1 ESSENTIALS OF MULTI-ZONE BOUNDARY ELEMENT ANALYSIS

4.1.1 *Multi-zone boundary element analysis*

A brief account of multi-zone BEA follows to establish the notation and terminology used throughout. Multi-zone BEA is accomplished by first breaking up an entire

---

Solving Large-scale Problems in Mechanics, Edited by M. Papadrakakis
© 1993 John Wiley & Sons Ltd

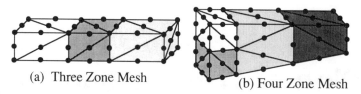

(a) Three Zone Mesh (b) Four Zone Mesh

Figure 4.1 Multi-zone boundary element models.

boundary element model into zones as illustrated, for example, by the three zone model shown in Figure 4.1(a). The surface of each zone totally encloses a sub-volume of the entire model and may generally contain boundary elements and nodes that belong exclusively in a single zone, and also boundary elements and nodes that are on an interface with other zones. The governing boundary integral relationship [1–3] can then be written for each zone. In elastostatics, for example, Somigliana's identity is the appropriate relationship. Substituting for the actual surface response, an approximate surface response interpolated from the node point values of traction and displacement using interpolation functions [1–3], one obtains the discretized boundary integral equation. By evaluating this expression at a set of locations of the load point of the fundamental solutions occuring in the boundary integral equation corresponding to the node point locations for the zone in question, one can generate a matrix system of equations for each zone:

$$[F^i]\{u^i\} = [G^i]\{t^i\} \tag{4.1}$$

In this equation, $[F^i]$ and $[G^i]$ are square and rectangular coefficient matrices, respectively, and $\{u^i\}$ and $\{t^i\}$ are column vectors of node point displacement and traction components, respectively, of conformable size. The rectangular nature of $[G^i]$ is due to the fact that jumps in specified tractions can exist when continuous boundary elements are employed. These additional tractions are collected and put at the end of $\{t^i\}$. The i superscript denotes that the superscripted quantities are those associated with the ith zone in a multi-zone BEA model.

Matrix relations written for individual zones can be put together for use in an overall analysis by considering conditions of displacement continuity and equilibrium of traction components at all zone interfaces. The continuity condition requires that nodal displacements, calculated for zone i at an interface between zone i and zone j, must equal the nodal displacement components calculated in zone j at that same interface. A similar relationship exists due to equilibrium considerations for components of the traction vector at interface nodes between two zones, except that a negative sign must be present to account for the opposite directions of the outward surface normals in the two zones in question:

$$\{u^i_{ij}\} = \{u^j_{ij}\} \tag{4.2}$$

$$\{t^i_{ij}\} = -\{t^j_{ij}\}. \tag{4.3}$$

The double subscript notation is used to convey that the vector in question is a

4.1 ESSENTIALS OF MULTI-ZONE BOUNDARY ELEMENT ANALYSIS

collection of components entirely on the interface between zone i and zone j. Expanding the size of the zone matrix equations to the size of the overall problem, bringing the unknown tractions at zone interfaces to the left-hand side of the equation, and using the continuity and equilibrium relations, one can form the boundary element system equations for the overall multi-zone BEA problem. For example, the equations for the three-zone problem shown in Figure 4.1(a) are given below:

$$\begin{bmatrix} [F^1_{11}] & [F^1_{12}] & [0] & -[G^1_{12}] & [0] & [0] & [0] & [0] & [0] \\ [0] & [F^2_{12}] & [0] & [G^2_{22}] & [F^2_{22}] & [F^2_{23}] & [0] & -[G^2_{23}] & [F^1_{12}] \\ [0] & [0] & [0] & [0] & [0] & [F^2_{23}] & [0] & [G^2_{23}] & [F^3_{33}] \end{bmatrix} \begin{Bmatrix} \{u^1_{11}\} \\ \{u^1_{12}\} \\ \{u^1_{13}\} \\ \{t^1_{21}\} \\ \{u^2_{22}\} \\ \{u^2_{23}\} \\ \{t^1_{31}\} \\ \{t^2_{32}\} \\ \{u^3_{33}\} \end{Bmatrix}$$

$$= \begin{bmatrix} [G^1_{11}] & [0] & [0] & [0] & [0] & [0] & [0] & [0] & [0] \\ [0] & [0] & [0] & [0] & [G^2_{22}] & [0] & [0] & [0] & [0] \\ [0] & [0] & [0] & [0] & [0] & [0] & [0] & [0] & [G^3_{33}] \end{bmatrix} \begin{Bmatrix} \{t^1_{11}\} \\ \{0\} \\ \{0\} \\ \{0\} \\ \{t^2_{22}\} \\ \{0\} \\ \{0\} \\ \{0\} \\ \{t^3_{33}\} \end{Bmatrix} \quad (4.4)$$

It should be noted that this model has no interface between zone 1 and zone 3. In this instance, the final multi-zone BEA system of equations can be produced by simply removing the blocks associated with this 1–3 interface shown in equation (4.4). This example points out the basic features of multi-zone boundary element matrix equations. The matrix equation shown below is actually a *hypermatrix* with matrices as its entries. Generally these matrix entries are called blocks or partitions. Likewise, the overall vectors shown have vectors for their entries and again these entries are referred to as blocks or partitions. The blocked sparsity characteristic of the matrices that result from the multi-zone BEA approach is clearly evident from the zero blocks present in equation (4.4).

Some important subtleties associated with multi-zone BEA deserve mention. The order of occurrence of the partitions shown in equation (4.4) affects the amount of computer storage and the number of computations required to solve the overall matrix equations by direct methods. This is due to the fact that during the block triangular factorization step, zero blocks above the main diagonal block but located below a populated block in a block column experience block fill-in. A similar block fill-in occurs in block rows to the left of the main diagonal. The order of the partitions presented in the above example has been purposely selected to cluster non-zero blocks as near to the main diagonal of the matrix as possible. For arbitrarily connected zones in multi-zone continuous BEA, nodes may very often be present in the model that are connected to more than two zones. Figure 4.1(b), for example, illustrates a situation where a multi-zone model has some nodes that are simultaneously connected to elements in three different zones. Each distinct combination of nodes sharing particular zones present in a multi-zone analysis determines a partition in the overall blocked matrix equations. Thus, it is generally required to employ an algorithm where it is possible to have nodes connected to as many zones as are present in the overall model, and yet still be able to automatically determine an optimal partition ordering scheme. At zone interfaces where sharp corners exist, jumps in the components of the traction vector exist, even though the components of the stress tensor must remain continuous. A general formulation for handling this situation in a consistent fashion is presented in Reference [30], along with an automatic block ordering scheme.

4.1.2 Boundary element substructures

The concept of condensation of degrees of freedom in BEA can be introduced by considering the matrix equations for a single zone. In the equations that follow, the superscript notation indicating the number of the particular zone being considered will be implied but explicitly omitted. Reordering the degrees of freedom and partitioning equation (4.1) into blocks that correspond to master degrees of freedom and also into blocks that correspond to degrees of freedom that *could* be condensed, one obtains

$$\begin{bmatrix} [F_{MM}] & [F_{MC}] \\ [F_{CM}] & [F_{CC}] \end{bmatrix} \begin{Bmatrix} \{u_M\} \\ \{u_C\} \end{Bmatrix} = \begin{bmatrix} [G_{MM}] & [G_{MC}] \\ [G_{CM}] & [G_{CC}] \end{bmatrix} \begin{Bmatrix} \{t_M\} \\ \{t_C\} \end{Bmatrix} + \begin{Bmatrix} \{f_M\} \\ \{f_C\} \end{Bmatrix}. \quad (4.5a, 4.5b)$$

In this equation, the additional right hand side vector is included to consistently account for any body force type of loading that might be present in the analysis, such as gravity, centrifugal, or thermal loading. Solving the matrix equation (4.5b) for $\{u_C\}$ gives

$$\{u_C\} = [F_{CC}]^{-1}([G_{CM}]\{t_M\} + [G_{CC}]\{t_C\} - [F_{CM}]\{u_M\} + \{f_C\}). \quad (4.6)$$

Substituting equation (4.6) into the matrix equation (4.5a) and collecting terms yields

$$[M_1]\{u_M\} = [M_2]\{t_M\} + [M_3]\{t_C\} + [M_4]\{f_C\} + \{f_M\}) \quad (4.7)$$

where

$$[M_1] = [F_{MM}] - [F_{MC}][F_{CC}]^{-1}[F_{CM}] \quad (4.8)$$

$$[M_2] = [G_{MM}] - [F_{MC}][F_{CC}]^{-1}[G_{CM}] \quad (4.9)$$

4.1 ESSENTIALS OF MULTI-ZONE BOUNDARY ELEMENT ANALYSIS

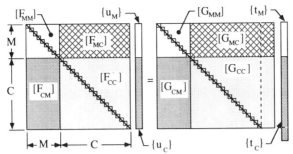

Assembled Zone Matrix Showing Master & Condensed Partitions

Typical Matrix Condensation
with Corresponding Matrix Sizes

Figure 4.2 Details of assembly, reordering, partitioning, and condensation of zone matrices.

$$[M_3] = [G_{MC}] - [F_{MC}][F_{CC}]^{-1}[G_{CC}] \tag{4.10}$$

$$[M_4] = -[F_{MC}][F_{CC}]^{-1}. \tag{4.11}$$

Equation (4.7) is called a condensed zone matrix equation, while equation (4.6) is referred to as a zone matrix expansion equation. This procedure is general, in that no restriction has been placed on which degrees of freedom are chosen as master or condensed degrees of freedom. Also, the condensation procedure embodied in the above equations is exact, in that no terms have been neglected, nor has any approximation been made. The zone matrix assembly procedure must obviously be able to determine the destinations of all element contributions in the partitions shown above.

Figure 4.2 illustrates the relative sizes of the matrices and vectors present in the zone substructuring process. This figure, along with the substructuring equations, can be used to help describe the individual zone, first-level assembly algorithm used in this overall analysis procedure, and also the subsequent, optional condensation step. The first-level assembly algorithm requires that accounting be performed to indicate the appropriate partition destinations for all boundary element contributions to the individual zone system matrices. These contributions can then be assembled with due regard for the boundary conditions present in the model, and by requiring that the individual zone's row sum be used to determine the 3 × 3 diagonal block entries in the [F] matrix shown in Figure 4.2. These diagonal entries correspond to the integration of the singular traction kernal, shape function, and Jacobian products that occur when the load point of the fundamental solutions used in BEA coincides with a node point of the element currently being integrated. Due to the possible use of specified displacement boundary conditions, corresponding column exchanges could necessitate the assembly of this row sum information into [G], instead of [F].

At first inspection, this substructuring process seems to require the *inversion* of the matrix block $[F_{CC}]$. A closer examination of the formulation, however, reveals that this is not the case. Whenever $[F_{CC}]^{-1}$ appears in these equations, it always premultiplies either a column vector or a rectangular matrix. As shown, below, *this use of matrix inverse notation is purely symbolic, and in the computer implementation of this substructuring approach, no matrix inversion is ever actually performed.* Instead, the triangular factorization of $[F_{CC}]$ is performed *once*, and subsequently these factors are used to solve matrix equations by forward reduction and backward substitution of the right hand side vector or group of vectors shown below. This is possible because the result of the product of $[F_{CC}]^{-1}$ and a vector or matrix can be premultiplied by $[F_{CC}]$ to yield a system of linear equations:

$$\{d\} = [F_{CC}]^{-1}\{v\} \Rightarrow [F_{CC}]\{d\} = \{v\} \tag{4.12}$$

and

$$\{D\} = [F_{CC}]^{-1}\{V\} \Rightarrow [F_{CC}]\{D\} = \{V\} \tag{4.13}$$

where

$$[D] = [\{d_1\},\{d_2\},\ldots,\{d_N\}]; \text{ and } [V] = [\{v_1\},\{v_2\},\ldots,\{v_N\}]. \tag{4.14}$$

In equation (4.13) the rectangular matrix $[D]$ is a collection of column vectors $\{d_i\}$ that are formed from forward reduction and backward substitution of the column vectors $\{v_i\}$ collected in the rectangular matrix $[V]$. As shown in Reference [30], it is possible to save significant computer storage space and data movement in the implementation of the above formulation by avoiding the storage of the rectangular matrices $[G_{MC}]$ and $[G_{CC}]$. If the symbolism is generalized to imply that $\{u_C\}$ contains the unknown node point values of the surface response (displacement or traction components) that may be condensed, and that $\{t_C\}$ contains the specified node point values of the surface boundary conditions (again these could be either displacement or traction components), then both $[G_{MC}]$ and $[G_{CC}]$ always premultiply column vectors of known quantities. This fact can be exploited to always form the column vector resulting from these multiplications and never actually assemble $[G_{MC}]$ and $[G_{CC}]$. In order to accomplish this, however, a procedure has to be developed to account for how these matrices *would be assembled* and perform the multiplications of the appropriate terms without actually assembling them. This strategy is reflected in the altered set of equations shown below. This strategy must also allow for the incorporation of the assembly of the row sum contribution from $[F]$ into the $[G_{CC}]$ matrix for all degrees of freedom with specified displacement boundary conditions.

$$[M_1]\{u_M\} = [M_2]\{t_M\} + \{v_C\} + [M_4]\{f_C\} + \{f_M\} \tag{4.15}$$

$$\{v_C\} = [G_{MC}]\{t_C\} - [F_{MC}][F_{CC}]^{-1}[G_{CC}]\{t_C\} \tag{4.16}$$

and

Given the zone matrix partitions $[F_{MM}]$, $[F_{CM}]$, $[F_{MC}]$, $[F_{CC}]$, $[G_{MM}]$, $[G_{CM}]$
and the vectors $\{v_1\} = [G_{CC}]\{t_C\}$, $\{v_2\} = [G_{MC}]\{t_C\}$, $\{f_M\}$, $\{f_C\}$
Form the triangular factorization of $[F_{CC}] = [L_{CC}][U_{CC}]$
Solve $[F_{CC}][D] = [F_{CM}]$ for $[D]$; and **Form** $[M_1] = [F_{MM}] - [F_{MC}][D]$
Solve $[F_{CC}][D] = [G_{CM}]$ for $[D]$; and **Form** $[M_2] = [G_{MM}] - [F_{MC}][D]$
Solve $[F_{CC}]\{d\} = \{v_1\}$ for $\{d\}$; and **Form** $\{v_C\} = \{v_2\} - [F_{MC}]\{d\}$
Solve $[F_{CC}]\{d\} = \{f_C\}$ for $\{d\}$; and **Form** $\{v_3\} = -[F_{CM}]\{d\}$
Form $\{v_C\} = \{v_C\} + \{v_3\} + \{f_M\}$

(a) Zone Condensation Operation

Given the zone matrix partitions $[F_{CM}]$, $[G_{CM}]$
and the triangular factorization of $[F_{CC}] = [L_{CC}][U_{CC}]$
and the vectors $\{v_1\}$, $\{t_M\}$, $\{u_M\}$, $\{f_C\}$
Form $\{v\} = [G_{CM}]\{t_M\} + \{v_1\} - [F_{CM}]\{u_M\} + \{f_C\}$
Solve $[F_{CC}]\{u_C\} = \{v\}$ for $\{u_C\}$

(b) Zone Expansion Operation

Figure 4.3 Zone condensation and expansion algorithm.

$$\{u_C\} = [F_{CC}]^{-1}([G_{CM}]\{t_M\} + \{v_1\} - [F_{CM}]\{u_M\} + \{f_C\}) \quad (4.17)$$

$$\{v_1\} = [G_{CC}]\{t_C\}. \quad (4.18)$$

Figure 4.3(a,b) illustrates the steps involved in the computer implementation of the zone condensation and expansion formulation. From this figure, it becomes clear that the fundamentally important operation involved is the triangular factorization of $[F_{CC}]$. This is followed by a series of forward and backward substitution operations, using this factorization, of the columns of the matrix blocks shown to form an intermediate matrix $[D]$. For each $[D]$, the matrix multiply and subtract operations indicated must also be performed.

4.1.3 Multi-zone analysis with substructures

Multi-zone BEA has been shown to produce overall matrix equations with a sparse blocked structure. Blocks can be optimally arranged according to the ways that the degrees of freedom share combinations of zones. An exact condensation scheme has also been shown that can eliminate degrees of freedom from an individual zone matrix, and an expansion formula presented to obtain the response at condensed degrees of freedom subsequent to the solution of the condensed system of equations. A natural way to combine substructuring with multi-zone BEA capability is to allow for the *possible* condensation of degrees of freedom that appear *exclusively* in any particular zone. In this case, the partitions to be eliminated by the condensation process coincide exactly with certain partitions already present in the multi-zone BEA procedure. This approach is also natural from a modelling perspective, since entire zones can be easily and arbitrarily identified for either condensation or no condensation, and also for subsequent expansion if they are to be condensed.

96 SOLUTION OF LARGE BOUNDARY ELEMENT EQUATIONS

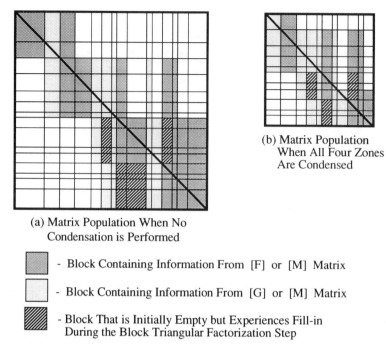

(a) Matrix Population When No Condensation is Performed

(b) Matrix Population When All Four Zones Are Condensed

- Block Containing Information From [F] or [M] Matrix

- Block Containing Information From [G] or [M] Matrix

- Block That is Initially Empty but Experiences Fill-in During the Block Triangular Factorization Step

Figure 4.4 Left-hand-side boundary element system matrix populations for the four-zone mesh shown in Figure 4.1.

As an example of this procedure, consider the four-zone boundary element model shown in Figure 4.1(b). A capability for automatic graphic depiction of sparse blocked matrix populations associated with any multi-zone model was used to produce the illustration of the left-hand side matrix associated with this four-zone model when no condensation is performed (Figure 4.4(a)). The left-hand side matrix associated with the same BEA model when all four zones are condensed is shown in Figure 4.4(b). These illustrations of matrix populations were generated in the identical scale. In the overall matrix with no condensation (Figure 4.4(a)), the block entries correspond to blocks present in the individual zone [F] and [G] matrices. In the overall matrix formed from the four condensed zone contributions (Figure 4.4(b)), the block entries correspond to the zone [M] matrices shown in equation (4.15). Note that many other combinations of zone condensation or no condensation are possible.

When zones can be arbitrarily selected for condensation, a second-level assembly procedure for the formation of the overall sparse blocked system of equations must be able to assemble condensed or uncondensed zone contributions to the overall matrix. The algorithm for this second assembly step is very similar to the assembly procedures described elsewhere except for the additional complication of dealing with both condensed and uncondensed zone contributions. This procedure takes each column of the zone matrix partitions shown in Figure 4.3, determines its block destination and column destination within the block of the overall system matrix, and proceeds to assemble this column. The overall system matrix is stored in a one-dimensional array with appropriate accounting arrays used to indicate the location

```
Start
  Start block triangular factorization procedure
  i = 0
Diag_loop
  i = i + 1
    if (I > N - 1) goto last_block
    Factor $A_{II} = L_{II}U_{II}$ using Gauss elimination with partial
      pivoting j = 0
Col_loop
  j = j + 1
    if (J > N) goto diag_loop
    if ($A_{JI}$ = 0) goto col_loop
    Solve $A_{II}D = A_{JI}$ for D by forward reduction & backward
      substitution of the columns of D using $L_{II}U_{II}K = I$
Row_loop
  K = K + 1
    if (K > N) goto col_loop
    if ($A_{KI}$ = 0) goto row_loop
    Form $A_{KJ} = A_{KJ} - A_{KI}D$
    goto row_loop
Last_block
  factor $A_{II} = L_{II}U_{II}$
End_proc
  Return
```

Figure 4.5 Sparse blocked matrix triangular factorization algorithm.

and size of individual blocks, along with indicators regarding whether each block is full, empty, or to be filled-in in the subsequent block triagular factorization step. Other information in these arrays includes block numbers associated with each zone, blocks associated with each node, order of each degree of freedom in a particular block, and whether a block is an [F], [G], or [M] type of block. The result of this second-level assembly process is an overall hypermatrix system of equations shown below:

$$[A]\{x\} = \{b\}. \tag{4.19}$$

4.2 EQUATION SOLVING IN BEA

4.2.1 *Block triangular factorization*

From Figure 4.4, it is easily seen that banded unsymmetric solvers would be totally ineffective at exploiting the blocked sparsity present in multi-zone BEA. Such solvers require two times the original bandwidth to perform their function. A blocked triangular factorization procedure is, however, quite effective in accomplishing this task. Blocked procedures can be understood as operating on the matrix entries of hypermatrices in a fashion completely analogous to standard triangular factorization procedures, with the exception that each operation is now a matrix operation equivalent to the scalar operation performed on the matrix with scalar entries. A typical blocked matrix triangular factorization procedure is shown in Figure 4.5. The procedure starts with the triangular factorization of the first diagonal block. This is performed using a Gauss elimination algorithm with partial pivoting. The triangular factors of this diagonal block are stored in the same location that the original diagonal block was located. This factorization is then used to alter the second block column. This is accomplished by forward reduction and backward substitution of the columns of $[A_{12}]$ to form $[D_{12}]$ (in the figure this matrix is symbolized by $[D]$). This matrix is then used to alter all the blocks below $[A_{12}]$ in block column 2 by the matrix multiplication and subtraction step shown in Figure 4.5. All blocks below $[A_{13}]$ in column 3 are then processed in a similar fashion, and then the fourth column and so forth until the entire matrix has been altered. This entire process is then repeated using the submatrix consisting all blocks except those in block row and column 1. The second major phase in the algorithm causes the alteration of the submatrix consisting of all blocks except those in block rows and columns 1 and 2. At every stage of this process, checks are made concerning the sparsity of the matrix. Any block operation that can be avoided due to the block sparsity present in the matrix, is avoided. Except for the need to store the largest $[D_{ij}]$ block that ever occurs in the overall procedure, the entire operation can be done *in place*, with the updated matrix blocks replacing the original blocks in computer memory.

4.2.2 Direct sparse blocked equation solving in BEA

Multi-zone BEA techniques giving rise to overall system matrices that are blocked, sparse, and unsymmetric are introduced in texts by Banerjee and Butterfield [1], Brebbia and Walker [2], and Brebbia et al. [3]. Papers by Lachat and Watson [4, 5], Lachat [6], Crotty [7], and Das [8], concerned with the assembly and solution of these types of matrix equations, have shown that multi-zone BEA significantly extends the range of model shapes that can be successfully treated, while simultaneously producing substantive computational economy in numerical integration, equation solving, and response recovery phases of the overall analysis process. All these presentations describe a direct matrix block triangular factorization process that exploits the substantial block sparsity present in multi-zone analyses. Bialecki and Nahlik [9] and Bialecki [10] describe an unsymmetric sparse blocked frontal equation-solving algorithm. Tomlin [11] and Butterfield and Tomlin [12] present similar treatments for nonhomogenous continuum problems modelled by multiple piecewise homogeneous BEA zones. Chang [13] employed multiple piecewise homogeneous zones and coupled these zones to finite elements to solve seepage problems.

Static condensation of degrees of freedom in BEA has also had exposure, with Beer [14], Mustoe [15], and Davies [16] describing procedures coupling boundary elements and finite elements in the same analysis and condensing the BEA matrices to just degrees of freedom on the boundary element — finite element interface. This strategy is also discussed by Jin et al. [17] and Margenov et al. [18] for solving elastic contact problems. Kane and co-workers [19–25] have shown that multi-zone sensitivity analysis significantly impacts the ability to exploit the additional matrix sparsity present in this step. Kane et al. [26] demonstrate how multi-zone methods facilitate the effect utilization of re-analysis techniques in shape optimization. Kline and co-workers [27, 28] developed a multi-zone analysis approach that always condenses the degrees of freedom exclusively in a single zone. The study by Bettess [29] on mixed direct — interative methods for equation solving in BEA is also noteworthy, although this treatment has not been extended to multi-zone analysis. Kane et al. [30, 31] describe a sparse blocked assembly and equation solving procedure that incorporates boundary element substructuring in a completely arbitrary fashion, allowing for both condensed and uncondensed zones to consistently coexist in the same multi-zone problem. In [32] Kane and Saigal and in [33] Kane and Wang extend to use of the condensation concept to the design sensitivity analysis process. Recently, Drake and Gray [34] presented details on a direct solution approach for BEA on an Intel vector hypercube system, using vector and parallel factorization and triangular equation solving techniques detailed in Eisenstat et al. [35], Geist and Romine [36], and Heath and Romine [37].

4.2.3 Iterative solution techniques in BEA

The effectiveness of iterative equation solving techniques within the overall continuum BEA context is described below. Iterative equation solution and preconditioning techniques are presented for single- and multi-zone BEA problems. Algorithms based on the conjugate gradient (CG) and generalized minimum residual (GMRES) technique are examined and their performances reported. Diagonal and block diagonal preconditioning approaches are described that accelerate the convergence of these methods. Convergence criteria are established that provide surface response and response gradient solutions as accurate as those obtained by direct methods. Through a sequence of example problems, the behaviour and effectiveness of these iterative approaches is demonstrated. The effect of the obviation of the block fill-in required in direct solution procedures is quantified.

In many disciplines, numerous technical articles have been published extolling the virtues of preconditioned iterative equation solving techniques. Typically, the matrices involved in the problems examined exhibit significant sparsity. It has been argued that these preconditioned iterative equation solution methods offer compelling promise vis-a-vis direct methods, for this class of problems, especially in regard to the following:

1. The number of floating-point arithmetic and I/O operations, and computer memory required may be less.
2. Accuracy may be more controllable.
3. These methods may prove more conductive to effective implementation on emerging computer systems with vector/parallel processing facilities.
4. They may lend themselves to more synergistic incorporation in the solution of evolving non-linear problems.

Preconditioned iterative methods have been described in the literature. In stark contrast to the typical applications receiving attention, BEA is generally perceived as a technique producing dense matrices, although multi-zone BEA with its blocked sparsity characteristics is known to be an exception. For the dense matrices associated with single-zone BEA, and the unique blocked sparsity patterns developed in multi-zone BEA, preconditioning strategies are presented, thus providing a quantitative demonstration of the capability of preconditioning in iterative equation-solution techniques in BEA.

4.2.4 *Treatments of general preconditioned solvers*

Textbook expositions of preconditioned conjugate gradient (CG) techniques include Ortega [38] and Golub and Van Loan [39]. Saad and Schultz [40] published the original paper describing the GMRES algorithm. Samuelsson et al. [41], and Angeleri et al. [42] report on the behaviour of preconditioned CG methods in finite element applications in structural (symmetric matrices) and structural and fluid flow (unsymmetric matrices) respectively. They noted that their convergence criterion might be too loose for some applications requiring accurate response prediction. Both papers conclude that the condition number (or, more generally, the spectral clustering) is crucial in determining the convergence of the method, with the latter publication citing poor performance for problems with unsymmetric matrices. Brussino and Sonnad [43–45] compared direct and several preconditioned iterative methods for very large sparse unsymmetric matrix equations, concluding that conjugate gradient squared (CGS) and GMRES were the best performers among the iterative methods studied, and suggested that block preconditioning techniques might produce further performance improvements. Poole and Overman [46] presented the results of extensive testing of the performance on finite element problems of direct and iterative equation solving techniques implemented on the NAS CRAY-2 vector supercomputer system and concluded that the iterative techniques have not yet been developed sufficiently to effectively utilize the processing capabilities of such a system. Application of preconditioned CG methods to unsymmetric systems using the normal equations (CGN) have been described by Zdatev and Nielsen [52] and Kincaid et al. [53].

4.2.5 *Unsymmetric conjugate gradient technique*

In [38, 39] the conjugate gradient technique for the iterative solution of square sets of algebraic equations is detailed. The convergence properties of this technique are well documented for symmetric positive definite matrices. The use of this method for unsymmetric matrices, however, requires special treatment. In the present work, a conjugate gradient iterative strategy was employed that operated on the normal equations produced *symbolically* by premultiplying equation (4.19) by the transpose of the left-hand side coefficient matrix:

$$[A]^T ([A] \{x\} = \{b\}) \Rightarrow [\bar{A}] \{x\} = \{\bar{b}\} \tag{4.20}$$

where

$$[\bar{A}] = [A]^T [A] \tag{4.21}$$

$$\{\bar{b}\} = [A]^T[A]. \tag{4.22}$$

It is important to note that *this step is purely symbolic* and the matrix product $[A]^T[A]$ is *never actually formed*. For a dense square matrix of size N, this operation would require N^3 operations. For sparse blocked multi-zone BEA matrices, this operation would also be prohibitively burdensome and generally result in a fully populated matrix. Instead of forming $[A]^T[A]$, the standard symmetric CG algorithm is modified to replace the first equation shown below with the second:

$$\{v\} = [\bar{A}]\{d\} \tag{4.23}$$

$$\{v_1\} = [A]\{d\}; \{v\} = [A]^T\{v_1\}. \tag{4.24}$$

The price of non-symmetry relative to the standard CG method is the requirement to perform the two matrix–vector multiplications indicated in equation (4.24) instead of the one multiplication shown in equation (4.23). The strategy of solving the normal equations also involves the use of an effective operator that has a condition number equal to the square of the condition number of $[A]$. This effect, however, is mitigated by the fact that our choice of preconditioner for this conjugate gradient normal equation (CGN) approach is also related to $[A]^T[A]$. Furthermore, as pointed out in Nachtigal *et al.* [47], CGN often performs well in practice in spite of its pessimistic theory.

4.2.6 *Preconditioner for the CGN technique*

The literature on preconditioning for conjugate gradient methods has been mainly concerned with its use on sparse matrices generally exhibiting significant sparsity within the generalized band (profile). BEA matrices (except those associated with multi-zone models of multiply connected domains) do not have the sparsity patterns normally encountered in accounts of existing preconditioning approaches. In particular, the use of standard incomplete factorization techniques in single-zone BEA is not directly applicable. The approach taken here was to seek a preconditioning matrix $[C]^2$ related to an additive decomposition (splitting) of $[A]^T[A]$ as

$$\left.\begin{array}{l}[C]^2 \approx [\bar{A}] \\ [C]^2 = [\bar{A}] + [R]\end{array}\right\} \tag{4.25}$$

where $[R]$ can be called a remainder matrix, representing the difference between $[A]^T[A]$ and $[C]$. The three main desirable features required of this decomposition are:

1. $[C]^2$ is easy to form,
2. It is easy to factor,
3. It is a good approximation to $[A]^T[A]$,

in the sense that it can be used to improve the conditioning of the preconditioned system of equations, as illustrated below, for left preconditioning:

$$[C]^{-2}[\bar{A}]\{x\} = [C]^{-2}\{\bar{b}\} \quad (4.26)$$

where

$$[C]^{-2}[\bar{A}] \approx [I]. \quad (4.27)$$

Various forms of $[C]^2$ were investigated by Kane et al. [54] including diagonal, block diagonal, and banded decompositions of $[A]^T[A]$. The most successful preconditioner found for use with fully populated equations resulting from single-zone BEA models was the diagonal preconditioning matrix:

$$[C_d]^2 = \text{diagonal of } [A]^T[A] \quad (4.28)$$

$$(c_{ii})^2 = \sum_{j=1}^{N} a_{ji}a_{ji}; \ (c_{ij})^2 = 0, \quad \text{for } i \neq j \quad (4.29)$$

$$(c_{ii})^{-2} = \frac{1}{(c_{ii})^2}. \quad (4.30)$$

The formation of this preconditioner requires N^2 operations, while its inverse requires only N further operations. The use of a banded generalization of this diagonal additive decomposition was also studied in [54]. In this approach the M nearest off-diagonals to the main diagonal of $[A]^T[A]$ are used to form the preconditioner. The performance of the banded preconditioner was found to be uncompetitive with the diagonal preconditioner. The steps in the preconditioned CGN algorithm are depicted in Figure 4.6. This figure can be used for both the diagonal and the banded preconditioning schemes by appropriate specialization of the factorization process.

For problems with sparse blocked matrices associated with multi-zone BEA models, Figure 4.6 can still be used to symbolize the operations involved in the CGN process. In this case, the matrix operations are sparse blocked matrix operations. For this problem, another preconditioner was found to be effective. This preconditioner was formed from the diagonal blocks of $[A]^T[A]$:

$$[C_{bd}]^2 = \text{block diagonal of } ([A]^T[A]) \quad (4.31)$$

where

$$[C_{ii}]^2 = \sum_{j=1}^{NB} [A_{ji}]^T[A_{ji}]; \ [C_{ij}]^2 = 0, \quad \text{for } i \neq j \quad (4.32)$$

where NB stands for the number of diagonal blocks present in the sparse blocked matrix. This method requires the storage of the block entries of $[C]^2$. Since $[A]^T[A]$

Determine an initial guess for $\{x\}$
   Form $[C]^2$
   Factor $[C]^2 = [U]^T[U]$
   Solve $[C]^2\{x\} = [A]^T\{b\}$ using $[U]^T[U]$ for $\{x\}$
Preliminaries to iteration loop
   Form $\{R_1\} = [A]\{x\} - \{b\}$ and $\{R\} = [A]^T\{R_1\}$
   Solve $[C]^2\{z\} = \{R\}$ using $[U]^T[U]$ for $\{z\}$
   Set $\{d\} = \{z\}$ and $k = 0$
   Form $\zeta = \{z\}^T\{R\}$
Begin loop
   Form $\{v_1\} = [A]\{d\}$ and $\{v\} = [A]^T\{v_1\}$
   Form $\delta = \{d\}^T\{v\}$ and $\alpha = -\zeta/\delta$
   Form $\{x\} = \{x\} + \alpha\{d\}$
   Form $\{R\} = \{R\} + \alpha\{v\}$
   Check convergence $\zeta < \varepsilon$ then **goto** quit
   Solve $[C]^2\{z\} = \{R\}$ using $[U]^T[U]$ for $\{z\}$
   Form $\zeta_1 = \{z\}^T\{R\}$ and $\beta = \zeta_1/\zeta$
   Form $\{d\} = \{z\} + \beta\{d\}$
   Set $\zeta = \zeta_1$ and $k = k + 1$
   Goto Begin loop

Figure 4.6 Details of the unsymmetric preconditioned conjugate algorithm.

is *symmetric*, only the upper triangles of these diagonal blocks need to be formed and the factorization process used can be a symmetric factorization. Also, the block factorization of $[C_{bd}]^2$ can be done in place, using the same space originally occupied by $[C_{bd}]^2$. Due to the symmetry of $[A]^T[A]$, it was convenient to make all of the preconditioning schemes used in the CGN approach belong to the class of left-right preconditioning procedures. Of all the preconditioning strategies described in [54] in conjunction with CGN, block diagonal preconditioning generally performed best.

4.2.7 Generalized minimum residual algorithm

The GMRES algorithm has been detailed in [40]. This algorithm works directly with the original set of unsymmetric matrix equations and therefore requires only one matrix–vector multiplication in each iteration. This method has been successfully adopted in a wide range of fluid mechanics applications, for example see Dervieux et al. [48], Navarra [49], Shakib et al. [50], and Wigton et al. [51]. Like any iterative method, the GMRES algorithm begins with an initial guess for the solution vector. A new solution vector is calculated once every k iterations, k being a preset parameter. In each set of k iterations, vectors $v_1, v_2, \ldots, v_k$ which form an orthonormal basis of the Krylov subspace K_k of $[A]$, based on the initial residual, are successively generated using a modified Gram–Schmidt procedure. The Krylov subspace is defined as

$$K_k = \text{span}\left(\{v_1\}, [A]\{v_1\}, [A]^2\{v_1\}, [A]^3\{v_1\}, \ldots, [A]^{k-1}\{v_1\}\right) \qquad (4.33)$$

where

$$\{v_1\} = \{r_0\}/\|r_0\|. \tag{4.34}$$

In this expression $\{r_0\}$ is the residual vector at the outset of the current set of k iterations. If $\{x_0\}$ is the initial guess vector, then a new vector $\{x\}$ is obtained in the form $\{x_0\} + \{z\}$, such that $\{x\}$ minimizes the residual norm over $\{z\}$ in K_k. It is shown in [40] that this amounts to finding the vector $\{y\}$ which minimizes the following quantity:

$$n = \|\beta\{e_1\} - [H_k]\{y\}\| \tag{4.35}$$

where β is the magnitude of the initial residual, $\{e_1\}$ is the first column of the identity matrix, and $[H_k]$ is an upper Hessenberg matrix with elements

$$[H_{ij}] = ([A]\{v_j\})^T\{v_i\}. \tag{4.36}$$

The new solution vector is then given by the expression $\{x\} = \{x_0\} + [V]\{y\}$, where $[V]$ is the matrix whose columns are the orthonormalized basis vectors v_1, $v_2, \ldots, v_k$. The minimization of n is done by accumulating a sequence of plane rotations to $[H_k]$ at every iteration. At the end of k iterations, $[H_k]$ has been transformed into an upper diagonal matrix $[R_k]$, yielding the system

$$[Q_k]\beta\{e_1\} - [R_k]\{y\} = \{0\}$$

or

$$\{g_k\} - [R_k]\{y\} = \{0\}. \tag{4.37}$$

This triangular system is solved by back substitution, where $[Q_k]$ is the matrix of plane (Jacobi) rotations applied to $[H_k]$. Thus, a modified Gram–Schmidt procedure is used to generate the orthonormal basis, while Jacobi rotations are utilized to form a $[Q][R]$ factorization of the Hessenberg matrix. This approach is a generalization of the MINRES [40] algorithm, to unsymmetric systems of equations.

The above procedure is often termed the *restarted* version of GMRES and referred to as GMRES(k), because a new Krylov subspace is built in every k iterations. The entire process is restarted with the last iterate as the initial guess for the next sequence of k iterations. Note that during one cycle through a set of k iterations, $[H]$ is being updated, $[Q]$ (the matrix of Jacobi rotations) is being calculated, and applied to $[H]$ as well as $\beta\{e_1\}$ to get $[R_k]$ and $\{g_k\}$, respectively. It can also be shown [40] that the last component of the vector $\{g_k\}$ is the norm of the residual at the end of that particular iteration. Hence, the residual can be monitored without computing the updated solution vector itself. Therefore, even though the new solution vector is calculated only once after k iterations, the size of the residual is being obtained and checked at every iteration. If this residual falls below a certain tolerance, there is no need to continue until k iterations have been performed, but instead, the iteration process is terminated. The new solution vector calculated at this stage is the converged solution.

1. **Form** Diagonal/blocked diagonal preconditioner $[C]$ = diag $[A]$
   i.e. $C_{ii} = A_{ii}$; $C_{ij} = 0$ if i not equal j (diagonal case)
   or $[C]$ = diagonal blocks of $[A]$ (blocked diagonal case)
   i.e. $[C]_{ii} = [A]_{ii}$; $[C]_{ij} = 0$ if i not equal j.
2. **Set** number of iterations before restart, i.e. set value of k.
3. **Compute** initial guess vector $\{x\} = [C]^{-1}\{b\}$. (symbolic, use factorization)
4. **Compute** initial residual $\{r_0\} = \{b\} - [A]\{x\}$ and $\|r_0\|$;
   and **Set** restartcode = 0.
5. **Check** for exact initial solution. If $\|r_0\|$ < eps quit.
6. **Set** $g(1) = \|r_0\|$ and $v_1 = r_0/\|r_0\|$
7. **Begin** Main loop.
   Do for $j = 1, 2 \ldots, k$, or **until** convergence
   Form $\{v_{j+1}\} = [A][C]^{-1}\{v_j\}$ (symbolic for blocked case, use factorization)
   Compute $h_{ij} = \{v_{j+1}\}^T\{v_i\}$ for $i = 1, 2, \ldots, j$
   Compute $h_{j+1,j} = \|v_{j+1}\|$; and $v_{j+1} = v_{j+1} - \sum_{i=1}^{j} h_{ij}\{v_i\}$
   Apply previous rotations $L = 1, 2, \ldots, j-1$
     to jth column of H matrix, where $H_{ij} = h_{ij}$
   Calculate rotations $\cos \theta(j) = h_{jj}/(h_{jj}^2 + h_{j+1,j}^2)^{1/2}$
     $\sin \theta(j) = -h_{j+1,j}/(h_{jj}^2 + h_{j+1,j}^2)^{1/2}$;
   Apply on $[H]$ as well as $\{g\}$ where $g(j + 1) = 0$
     (Note this affects h_{jj} and makes $h_{j+1,j} = 0$;
     also $g(j)$ and $g(j + 1)$ are changed.)
   Check for convergence. If $\|g(j + 1)\|$ < eps set $k = j$ and **goto** step 8.
   Check for end of loop. If $j < k$ increment j and **goto** step 7.
   Set restartcode = 1.
8. **Solve** equation $[H]\{y\} = \{g\}$ for $\{y\}$ by back substitution.
9. **Form** new solution vector $\{x\} = \{x\} + [C]^{-1}[V]\{y\}$ (symbolic for blocked case) where $[V]$ is matrix of vectors $v_1, v_2, \ldots, v_k$.
10. **Check** whether restart required; If restartcode = 0, **print** $\{x\}$ and **Stop**.
    Goto step 4.

Figure 4.7 Details of preconditioned GMRES algorithm.

If k is set to a relatively large number, so that convergence occurs before the iteration count reaches the value of k, even once, the resulting method is called *non-restarted* GMRES. The objective of keeping the value of k small enough so that there are restarts present stems partly from the desire to reduce the memory required for the storage of the orthonormal basis vectors and associated matrices, and partly from the desire to minimize the quadratic growth of the orthogonalization work in the iteration index. It was found that *the restarted version of GMRES was significantly inferior to non-restarted* GMRES, for the BEA matrices encountered in our experiments. The total number of iterations required, and hence the computing time, dramatically increased as k was reduced. If k was set sufficiently large, in order not to require a restart, the execution time was better than for any number of restarts.

4.2.8 Preconditioner for the GMRES technique

The GMRES algorithm used in this study is shown in detail in Figure 4.7. The procedure is fundamentally the same for both diagonal and blocked diagonal preconditioning, used with matrices resulting from single-zone and multi-zone BEA models respectively. In the case of diagonal preconditioning, the matrix multiplications involving $[C]^{-1}$ are trivial. In the case of block diagonal preconditioning, $[C]^{-1}$ is never explicitly computed. Instead, the triangular factors of the respective diagonal blocks of $[C]$ are computed once, and used in a block forward reduction and back substitution process to obtain any desired product vector $\{z\}$, shown below:

$$[C]^{-1}\{w\} = \{z\} \Rightarrow [C]\{z\} = \{w\}. \tag{4.38}$$

Because GMRES works with unsymmetric matrices, a number of different preconditioning strategies can be used. In the study reported in [54], left, right, and left–right preconditioning schemes were employed, with the right preconditioning scheme found to be the most effective. The right preconditioning approach also provides the most reliable convergence criteria for $\{x\}$, monitoring the preconditioned residual only. This preconditioning approach attempts to improve the effective condition number of the set of equations by operating on the modified system of equations as

$$[A]\{x\} = \{b\} \Rightarrow [A][C]^{-1}[C]\{x\} = \{b\} \Rightarrow [\hat{A}]\{\hat{x}\} = \{b\} \tag{4.39}$$

where

$$\left.\begin{aligned}[\hat{A}] &= [A][C]^{-1} \\ \{\hat{x}\} &= [C]\{x\}.\end{aligned}\right\} \tag{4.40}$$

As discussed for the CGN algorithm, preconditioning matrices are sought that approximate the original system matrix, and are easy to form and factor. In this case, diagonal preconditioning was investigated for the fully populated matrices associated with single-zone BEA models. Both diagonal and block diagonal preconditioning schemes were investigated for the sparse blocked matrices associated with multi-zone BEA models. For diagonal preconditioning, the entries in $[C]$ were taken to be the diagonal entries of $[A]$. For block diagonal preconditioning, the diagonal blocks of $[C]$ were chosen to be identical to the diagonal blocks of $[A]$. Note that block diagonal preconditioning requires that the individual blocks of $[C]$ be factored using an unsymmetic Gauss elimination procedure. Partial pivoting was employed in this procedure.

4.2.9 Convergence criterion

In solid mechanics, we seek *first* solutions for the primary unknowns appearing in the discretized boundary integral equations. These displacement and traction components are certainly of interest. In addition, however, we *subsequently* require that this response prediction be accurate enough to be able to compute the components of the stress tensor that are not surface tractions. These stress components are computed from the tractions, strains, and the stress–stain law. The strains utilized in this process are computed using displacement gradient information, which, in turn, involve differences of displacements. *Thus, accurate stress prediction requires that the response predicted in the BEA process be of extremely high quality.* All these considerations contribute to the observation that iterative equation solving techniques must have convergence criteria that ensure solutions with the above mentioned characteristics. In the problems examined in this study, it was always found that a relative error norm of 10^{-6} could be used for this purpose. In the CGN method, the relative error norm was taken as

$$\varepsilon = \|r_k\| / \|x_k\|. \tag{4.41}$$

In the GMRES approach x_k is replaced by x_0 because the solution vector is not explicitly computed in each step. *With this convergence criterion, the iterative techniques produced solutions that were virtually indistinguishable from those provided using direct methods.* When ε was set to larger values, however, stress predictions were computed that began to deviate, in some cases, from those found using solutions obtained with direct solvers. It should be noted that this requirement for highly accurate (converged) solutions is often absent in other published accounts of the convergence behaviour of these iterative methods, since the unknown vector, undifferentiated, is often itself the only quantity examined in these works.

4.3 NUMERICAL RESULTS

A series of test cases were executed to demonstrate the behaviour with respect to accuracy and efficiency of both the direct and preconditioned iterative equation-solving techniques discussed above and to characterize their implementation. In the test cases presented, either analytic solutions are used to verify the accuracy of the computer programs with iterative solvers, or separate analyses of models using direct solvers are employed. Timing results are presented to compare computational efficiencies and the computer memory required to store left-hand-side matrices given. Generally, the storage of this left-hand-side matrix accounts for 90% of the storage required in the overall BEA process. In the smaller example problems presented, the direct equation-solving step consumes less than 25% of the total CPU time required in the overall analysis process. In the larger problems presented, the direct equation solution process required approximately 50% of the total CPU time. All computations were carried out in double precision on the same computer with a Unix operating system and all programs were compiled using the same Fortran 77 compiler and compiler options. The test cases employed Young's modulus $E = 30 \times 10^6$ psi and Poisson's ratio $v = 0.3$.

4.3.1 *Performance of direct multi-zone procedures*

In order to illustrate the advantages associated with multi-zone analysis relative to single-zone boundary element analysis, an example geometry that presents definite pathological problems for single-zone analysis has been selected for study and modelled as both a single-zone and as a multi-zone model. A twenty-to-one aspect ratio (ratio of length to lateral dimension) beam problem was run as a single-zone

108 SOLUTION OF LARGE BOUNDARY ELEMENT EQUATIONS

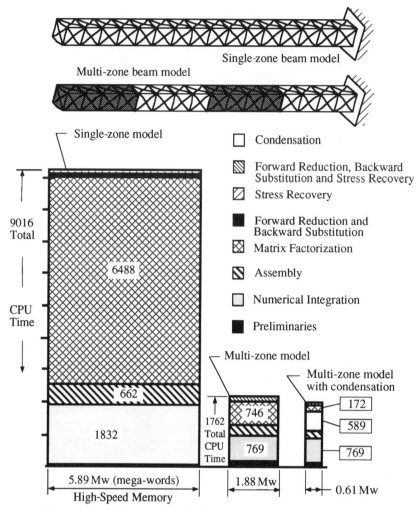

Figure 4.8 Performancer study of single and multi-zone BEA models.

BEA model and the resulting timings and storage information displayed in Figure 4.8. The same physical problem was modelled using a four-zone boundary element mesh also shown in Figure 4.8. Note that the four-zone model has alternating zones shown shaded. Corresponding computer storage space and CPU time required for the multi-zone analysis is also shown in this figure. In this figure, the CPU time required for the multi-zone analysis is also shown in this figure. In this figure, the CPU times for certain major operations are given inside the rectangular boxes shown. Comparison of the resources consumed during the equation solving step for both the single zone and the multi-zone analyses shows the dramatic improvements that can be obtained by employing the multi-zone approach in this class of problems. The single-zone analysis used about five times the CPU time and three times as much main memory as the multi-zone analysis. The third case shown in this figure depicts the results of a multi-zone analysis where the individual zone boundary element matrices are condensed. This third example ran in about the same time as the second, but the analysis was performed using considerably less computer memory due to the condensation strategy adopted.

The time spent in performing the matrix factorization step is shown to undergo the most significant improvement. In addition, numerical integration times tend also to be less for multi-zone models because only load points corresponding to nodes in the particular zone being integrated are used in the sequence of integrations used to form zone coefficient matrices. The accuracy of the boundary element method when applied to model such slender objects can also be dramatically improved in some instances when multi-zone techniques are utilized. In the beam problem described above, the accuracy of the predicted end deflection produced by the single-zone model was comparable to the corresponding multi-zone mesh. In a subsequent study, the same BEA models were 'stretched' to a forty to one aspect ratio shape. In these analyses, the multi-zone model remained stable, while the single-zone model yielded erroneous results. When one contrasts the resources corresponding to the first example (single-zone model) with the CPU time and memory resources utilized in the third example (multi-zone model utilizing condensation), the substantial impact of the techniques begins to become apparent. This example represents an extreme case of a long slender model. Counter examples where multi-zoning does not yield such dramatic improvements are presented subsequently.

4.3.2 *Rectangular bar*

The 12 to 1 aspect ratio rectangular bar shown in Figure 4.9, was run as both a single-zone and two-zone BEA model. This model was first built-in at the left end and subjected to a uniaxial tension at its right. The one-zone model contained 52 three-noded, isoparametric, continuous elements, 104 nodes, and 208 degrees of freedom. The left-hand side matrix associated with this model required 43 264 double precision words of computer memory. The corresponding two-zone model contained 56 elements, 107 nodes, and 224 degrees of freedom. The sparse blocked left-hand side matrix associated with this model required only 27 328 double-precision words of memory. Table 4.1 contains CPU timing and storage statistics, along with other pertinent statistics associated with these problems. Insights about the behaviour of the equation-solving techniques can be gained by consideration of these tabulated results.

1. The multi-zone model used considerably less storage than its single-zone counterpart, a characteristic that was repeated in several (but not all) of the examples studied. A small number of interfacial d.o.f. generally accompanies these memory savings.
2. In the single-zone analysis, GMRES with diagonal preconditioning was the fastest. GMRES with no preconditioning also outperformed the direct method. CGN methods were all inferior to the direct method for the one-zone model.
3. In the two-zone analysis, GMRES with diagonal preconditioning requiring 23 iterations was again the fastest, with GMRES with block diagonal preconditioning

110 SOLUTION OF LARGE BOUNDARY ELEMENT EQUATIONS

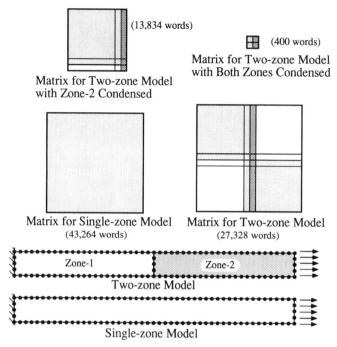

Figure 4.9 Rectangular bar BEA models and associated matrix populations.

Table 4.1 Statistics for 12 to 1 aspect ratio rectangular bar example

Solver	Preconditioner	Iterations	Form [C]	Factor [C]	Iterate	Total	Ratio*
\multicolumn{8}{c}{Single-zone model}							
				CPU seconds			
CGN	None	54	NA	NA	382.0	382.0	3.74
CGN	Diagonal	54	0.5	0.5	395.4	396.4	3.74
GMRES	None	30	NA	NA	119.0	119.0	2.08
GMRES	Diagonal	13	1.0	3.4	48.9	53.3	0.90
Direct	NA	1	0.0	272.1	0.0	272.1	NA
\multicolumn{8}{c}{Two-zone model}							
CGN	None	95	NA	NA	415.4	415.4	6.34
CGN	Diagonal	86	0.7	0.8	373.1	374.6	5.74
CGN	Block diagonal	14	90.7	30.6	89.7	211.0	0.93
GMRES	None	47	NA	NA	130.0	130.0	3.14
GMRES	Diagonal	23	0.5	1.2	62.4	64.1	1.53
GMRES	Block diagonal	16	1.1	56.0	42.8	99.9	1.06
Direct	NA	1	0.0	106.7	0.0	106.7	NA

*Ratio is equal to the number of iterations divided by the square root of the number of degrees of freedom.

using only 16 iterations also slightly outperforming the direct method. Thus, even though block diagonal preconditioning caused convergence in fewer iterations, the significant time spent forming (90 s), factoring (30 s), and using the blocked preconditioner was more than the time spent in the extra 7 iterations taken in the diagonally preconditioned approach. Note again that the CGN methods were all inferior to the direct method.

4. The direct equation solver for the two-zone model used less than half of the time required for the single-zone model. The two-zone model also spent significantly less time in the numerical integration phase of the analysis compared to the one-zone model.

5. Block-preconditioned CGN converged in 14 iterations and yet 16 iterations of blocked preconditioned GMRES were performed in less than half the time. This is because CGN requires two matrix–vector multiplications in each iteration and the formation time for the preconditioner in CGN is significant. Note further that the symmetric block factorization time in the block-preconditioned CGN method is roughly half of the time spent in the unsymmetric block factorization required in the block-preconditioned GMRES approach.

6. Multi-zone modelling improved the performance of the CGN approach, and did not improve the performance of the GMRES approach.

7. For two-dimensional sparse elliptic problems, effective preconditioned iterative solvers have been shown to converge in a number of iterations of the same order as the square root of the number of unknowns in the problem. The ratio of the number of iterations required over the square root of the degrees of freedom in the example problem was therefore tabulated, to quantify the comparison of the performance of these new boundary-based methods with analogous domain-based methods. It can be noted that the better preconditioned iterative approaches shown in Table 4.1 have ratios near unity.

8. In all cases, the converged solutions were as accurate as those derived using direct equation solution techniques.

In addition to the uniaxial tension loading case, the rectangular bar model was subjected to a transverse shear end load and the battery of computer runs described above were repeated. The performances of all the iterative solvers were marginally degraded (approximately 10%) in these analyses. This points out the fact that the characteristics of the right-hand side vector are also important in determining the convergence rates of these iterative solvers.

4.3.3 *Rectangular bar with substructuring*

A rectangular bar discussed above was also analysed using the zone condensation option discussed in previous sections. The two-zone model shown in Figure 4.9 was run with zone-2 condensed, and also with both zone-1 and zone-2 condensed. The overall left-hand-side matrix populations associated with these models are also shown in Figure 4.9. Also shown next to each matrix is the number of double-precision words required to store the matrix in computer memory. Note that all matrix populations shown in this figure have been drawn using a common scale. The results of these runs are collected in Table 4.2. The major conclusion drawn from this example is that BEA substructuring can be done in conjunction with preconditioned iterative equation solving. The timings shown in Table 4.2 indicate that this mixed direct/iterative approach to the equation-solving process merely transfers the use of computer resources. The CPU time spent in the zone condensation is saved in the iterative equation solution process operating on reduced-size matrices. *The number of iterations required in the iterative equation solving process was not significantly affected by condensation.*

4.3.4 *Tension strip with elliptical hole*

In order to critically examine the accuracy of the iterative solution techniques on a more realistic problem, a tension strip with a three-to-one elliptical hole shown in Figure 4.10 was studied. This problem was run as a quarter symmetry, single-zone problem, and also as a two- and three-zone problem. The two-zone model consisted of zone 1, shown in Figure 4.10, and a second zone comprising the regions occupied by zone 2 and zone 3 in this figure. Table 4.3 contains the statistics associated with these models. Examination of this table allows one to make the following observations:

1. This problem is not particularly conducive for multi-zone techniques. The memory to hold the left-hand-side matrix and the CPU time required to perform the direct matrix factorizations marginally increases as more zones are used to the model. This is due to the fact that the inter-zone boundaries add a significant number of degrees of freedom to the problem.
2. For the single-zone model, diagonal preconditioned GMRES out-performed the direct method, while CGN was uncompetitive.
3. For the two-zone model, block diagonal preconditioned GMRES was marginally faster than the direct method, and CGN improved but remained significantly slower than the direct method.
4. The three-zone model behaved quite similarly to the two-zone model except that the direct method became marginally faster than block diagonal preconditioned GMRES.

The values of the stresses predicted by the direct and iterative methods are displayed in Table 4.3, along with those obtained from the analytical solution. These s_{22} stress components for the sample points shown in Figure 4.10 are the components of the stress tensor that are not surface tractions. Figure 4.11 shows the local coordinate system associated with these stress components. As discussed in the section on the convergence criterion, this normal stress component is a derived stress that is generally the least accurately predicted in BEA. Table 4.3 shows that the convergence criterion used in the iterative solution techniques is stringent enough to produce computed stress response predictions as accurate as those found using direct methods.

4.3 NUMERICAL RESULTS

Table 4.2 Statistics for rectangular bar example with zone condensation

Solver	Precond.	Iterations	Condensed zone	CPU seconds Condense	Form [C]	Factor [C]	Iterate	Total
CGN	None	80	2	51.3	NA	NA	175.8	227.1
CGN	Diagonal	79	2	51.3	0.4	0.4	175.5	227.5
CGN	Block diagonal	12	2	51.3	44.2	15.0	41.8	152.3
GMRES	None	43	2	51.3	NA	NA	79.5	130.8
GMRES	Diagonal	22	2	51.3	0.2	0.4	30.1	82.0
Direct	NA	1	2	51.3	NA	57.3*	1.5†	110.1
CGN	Diagonal	11	1, 2	93.2	0.3	0.3	0.6	94.4
CGN	Block diagonal	9	1, 2	93.2	0.2	0.1	0.9	94.2
GMRES	Diagonal	10	1, 2	93.2	0.1	0.2	0.6	94.1
Direct	NA	1	1, 2	93.2	NA	0.6*	0.1†	93.9

*This entry is for factorization of assembled condensed overall left-hand-side matrix.
†Iteration time is for one forward reduction and back substitution.

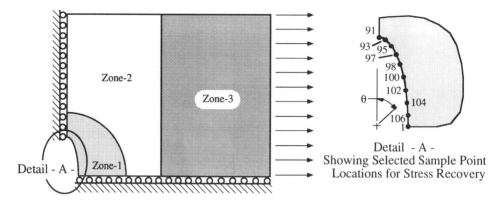

Figure 4.10 Tension strip with elliptical hole BEA model.

4.3.5 Multiply connected domain model with fill-in

Figure 4.12 shows a seven-zone BEA model used to study the behaviour of solution techniques on matrices that exhibit fill-in during direct matrix factorization. The BEA model shown has 122 elements, 217 nodes, and 488 degrees of freedom. Due to the manner in which the zones are connected, a significant number of matrix blocks will experience block fill-in in a direct Gauss elimination procedure. Figure 4.12 also shows the population of the left-hand side matrix for this problem. The blocks shown with the solid shading are the ones that experience this fill-in. The storage space required by the matrix without fill-in is 42 336 double-precision words. The amount of memory required for this matrix allowing for fill-in is 74 512 (1.76 times as much). *The significant savings in the memory required using interative methods for such BEA models is thus demonstrated by this problem.* This example problem is the first problem discussed to exhibit block fill-in. The matrix encountered here therefore has characteristics that allow for the use of a preconditioner based on an incomplete $[L][U]$ decomposition (ILUD). This type of preconditioning scheme takes $[C]$ to be the incomplete factorization of $[A]$. This factorization is formed by ignoring the fill-in that occurs in the complete factorization process for all zero blocks within the profile of the original matrix. Table 4.4 contains the statistics associated with a number of test runs made using this model. As can be seen in Figure 4.12, this incomplete factorization can be accomplished in significantly less time (92.1 s versus 223.7 s) than the complete factorization process. From these results we observe that the GMRES algorithm with ILUD preconditioning, requiring only 4 iterations, was the fastest method by far, followed by the direct method. The non-restarted GMRES algorithm with block diagonal preconditioning, requiring 77 iterations begins to require a substantial amount of computer resources in the last few iterations. With the introduction of an improved preconditioning scheme, and an effective restarted GMRES algorithm, it seems very plausible that the computer resources used by this blocked diagonal preconditioned iterative equation-solving method will eventually also be less than that required by the direct method.

A number of experiments can be made using this 7-zone model by condensing one or more zones. The overall time for the equation solution portion of the analysis

4.3 NUMERICAL RESULTS **115**

Table 4.3 Stresses computed in tension strip with elliptical hole

Point	Angle	Single-zone Direct	Single-zone CGS	Single-zone GMRES	Two-zone Direct	Two-zone CGS	Two-zone GMRES	Three-zone Direct	Three-zone CGS	Three-zone GMRES
91	0.0	7.3480	7.3556	7.3514	7.3514	7.3515	7.3515	7.3514	7.3519	7.3514
92	1.92	7.1085	7.1109	7.1073	7.1112	7.1113	7.1113	7.1112	7.1115	7.1112
93	3.89	6.4132	6.4073	6.4094	6.4153	6.4153	6.4154	6.4153	6.4151	6.4153
94	5.98	5.3220	5.3219	5.3224	5.3237	5.3237	5.2327	5.3237	5.3237	5.3237
95	8.28	4.2295	4.2303	4.2293	4.2309	4.2309	4.2309	4.2309	4.2308	4.2309
96	10.89	3.2056	3.2050	3.2057	3.2066	3.2066	3.2066	3.2066	3.2066	3.2066
97	14.04	2.3532	2.3530	2.3534	2.3540	2.3540	2.3540	2.3539	2.3539	2.3539
98	18.09	1.6061	1.6062	1.6062	1.6066	1.6067	1.6066	1.6066	1.6066	1.6066
99	23.96	1.0404	1.0394	1.0405	1.0408	1.0408	1.0408	1.0408	1.0407	1.0408
100	33.78	0.53181	0.53188	0.53185	0.53200	0.52020	0.53199	0.53198	0.53194	0.53198
101	36.19	0.17216	0.17293	0.17224	0.17224	0.17222	0.17225	0.17223	0.17214	0.17222
102	39.06	−0.17880	−0.17891	−0.17878	−0.17887	−0.17884	−0.17888	−0.17888	−0.17892	−0.17888
103	42.57	−0.40301	−0.40365	−0.40287	−0.40312	−0.40315	−0.40314	−0.40313	−0.40320	−0.40314
104	47.02	−0.60855	−0.60862	−0.60237	−0.60880	−0.60879	−0.60883	−0.60881	−0.60884	−0.60881
105	58.65	−0.67249	−0.67243	−0.67251	−0.67281	−0.67284	−0.67284	−0.67281	−0.67286	−0.67282
106	66.86	−1.0653	−1.0657	−1.0658	−1.0665	−1.0665	−1.0665	−1.0665	−1.0666	−1.0664
1	90.00	−1.0021	−1.0021	−1.0035	−1.0031	−1.0031	−1.0031	−1.0031	−1.0031	−1.0031

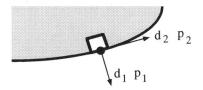

Figure 4.11 Local coordinate system used in surface stress recovery.

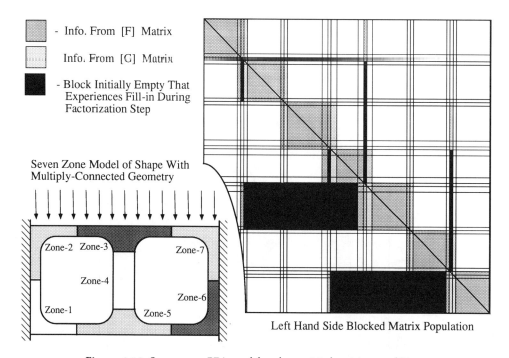

Figure 4.12 Seven-zone BEA model and associated matrix population.

Table 4.4 Statistics for seven-zone problem

Solver	Preconditioner	Iterations	CPU seconds			Total	Ratio
			Form [C]	Factor [C]	Iterate		
CGN	Block diagonal	147	68.2	18.9	1277.7	1364.8	6.65
GMRES	Block diagonal	77	2.0	30.3	531.2	563.5	3.48
GMRES	ILUD	4	—	92.1	57.1	149.2	0.18
Direct	NA	1	NA	217.2*	8.8†	226.0	NA

*This entry is for factorization of assembled condensed overall left-hand-side matrix.
†Iteration time is for one forward reduction and back substitution.

was 226.0 s with no zone condensation, whereas the same operation required only 148.1 s when all zones were condensed and expanded. The number of master degrees of freedom in this problem was 112, which is much less than the original 488 degrees of freedom when no condensation was done. It can be seen that the overall matrix

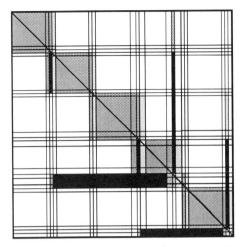

Figure 4.13 Matrix with zone-5 and zone-7 condensed.

factorization time was reduced from 214.5 to 9.7 s, but the time required for zone condensation is only 127.8 s. The time spent solving by forward reduction and back substitution, and then expansion was usually of the order of that taken by the forward reduction and back substitution during a regular run with no condensation. Other timings obtained by condensing different combinations of zones show that condensation always helps, and that the most significant gains occur when the zones responsible for fill-in are condensed. This is the case, for example, when only zones 5 and 7 are condensed. In this case, the final solution vector is of considerable size compared to original solution vector size. The number of condensed degrees of freedom is a small number in this case and the time for condensation is thus relatively small. The important thing to note here is that the factorization of the final matrix corresponding to the master degrees of freedom, is reduced by a factor much higher than the ratio of corresponding vector sizes. The matrix populations associated with the model (Figure 4.13) clearly indicate why this happens. For problems with such significant block fill-in, we thus see that zone condensation makes the direct blocked factorization procedure extremely competitive with preconditioned GMRES in terms of both storage and CPU time.

4.3.6 Three-dimensional example

The three-dimensional (3D) quarter symmetry model of a flat plate with a central circular hole subject to uniaxial tension shown in Figure 4.14 was used to help gauge the performance of the multi-zone solvers on larger-scale problems. This two-zone model had 282 6-noded triangular boundary elements and 541 nodes. The BEA left-hand-side matrix required 1 591 632 double-precision words of computer memory. Zone condensation was not employed in this example. The direct blocked factorization and subsequent forward reduction and backward substitution for this problem took 1442.7 s. Remarkably, the diagonal preconditioned GMRES approach did the equation

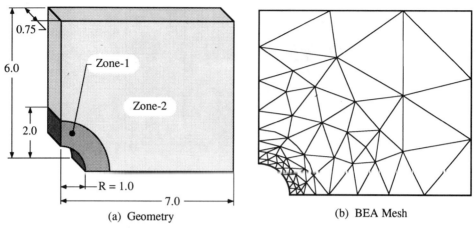

Figure 4.14 Tension strip with elliptical hole.

solving task in only 566.2 s, of which 14.7 s was spent in the formation of the preconditioner. 59 iterations were needed for this problem to converge. The blocked diagonal preconditioned GMRES method solved the BEA equations in 1164.6 s, with 903.2 s consumed in the formation and subsequent factorization of the preconditioner. This approach required 28 iterations. The superior performance of the diagonal preconditioner over the blocked diagonal preconditioner in this example is due to the fact that the diagonal blocks represent a substantial portion of the overall matrix (similar to the matrix shown in Figure 4.9 for the case of the two-zone model with no condensation). The convergence tolerance used in this example was 10^{-7}, and again the surface displacement, traction and derived stress component predictions obtained from the iterative technique were identical to those computed using the direct solver. It is interesting to note that zone condensation would not contribute to the computational efficiency of the direct blocked factorization procedure for this model because there is no fill-in. On emerging vector and parallel computer systems, however, the reduction in data movement (communication) associated with the zone condensation approach has proved to be quite useful [55].

4.3.7 Convergence rates and mesh quality

In the course of the numerical experiments involved with this study, it was observed that convergence rates of the iterative techniques were closely related to the element proportions used in the BEA mesh. To examine this behaviour, three BEA models of a rectangular bar were employed. These models were made arbitrarily finer than required simply to provide systems of equations of reasonable size. The first model was composed of elements that are exactly the same size. In the second and third models, the elements are not the same size. The second model has equally spaced elements along its vertical sides and these elements were 1.25 times the size of the horizontal elements. The last model has unequally sized elements along its vertical sides with one element 1.50 times as long as the horizontal elements. The number of iterations required for convergence of the diagonally preconditioned CGN solver for these models was 37, 31, and 63 respectively. The number of iterations taken by the diagonally preconditioned GMRES approach was 9, 9, and 19 respectively. This experiment was repeated with similar four-zone models with results that indicated an identical trend.

4.3.8 *Visualization of entry magnitudes in matrix populations*

Figure 4.15 contains a grey scale image of the entries in the BEA left-hand-side matrix associated with the single- and two-zone rectangular bar example problem. Each entry in the matrix is plotted with a shade that corresponds to its absolute value. Thus the darker spots in these matrices are locations of the large values. The diagonal clustering of these large values is seen to be quite prominent in both the single- and two-zone cases. There are, however, some interesting patterns of somewhat large entries in off-diagonal positions as well. The grey scale plot, however, does not exactly convey the relative sizes of these off diagonal entries. In actuality, these grey dots represent entries with a magnitude of about 20% of the corresponding diagonal entries. The consideration of these populations can lead to an appreciation of why the diagonal preconditioners perform so well in the example problems. Further study of these populations may lead to the development of even more effective preconditioning methodologies that are designed to exploit the special characteristics of these BEA matrices.

4.4 CONCLUSIONS

A series of equation-solving techniques has been presented for the unsymmetric matrix equations occurring in BEA. Generally, the GMRES approach outperformed the CGN and direct methods. For multi-zone problems, block diagonal preconditioning generally led to convergence in the least iterations, although sometimes diagonal preconditioning resulted in the consumption of less CPU time. The three-dimensional problem illustrated the fact that the iterative solvers can exhibit quite significant performance improvements relative to the direct approach. For problems with block fill-in, the ILUD preconditioned GMRES algorithm was shown to be especially effective. Preconditioners used in these studies were related to additive decompositions (splittings) of the effective left-hand-side matrix. These preconditioners allowed for the convergence of the iterative techniques in the order of $\sqrt{N}$ iterations. For both single-zone and multi-zone problems, the preconditioned iterative approaches were generally faster than direct methods. For multi-zone problems where significant fill-in occurs in the direct factorization step, it has been demonstrated that the iterative methods have the potential to obviate the requirement to store these extra blocks. For such problems, it has also been shown that boundary element zone substructuring can be done to further reduce storage requirements while still performing the overall equations solution step via an iterative procedure. Zone condensation was also shown to represent an extremely competitive feature of the direct blocked factorization approach to BEA equation solving for problems exhibiting significant fill-in. For

120 SOLUTION OF LARGE BOUNDARY ELEMENT EQUATIONS

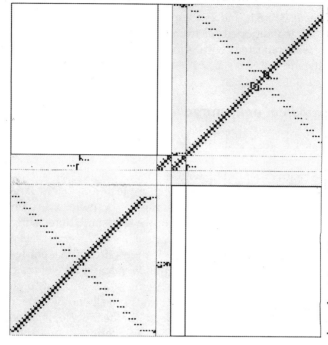

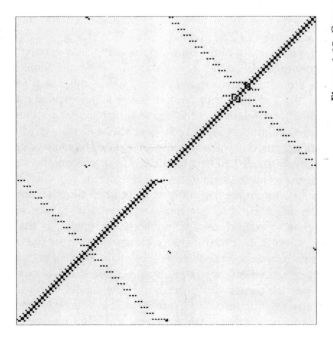

Figure 4.15 Grey scale plots of values of matrix entries.

extremely large problems, and for evolving non-linear problems with localized non-linear effects, the simultaneous utilization of condensation and iterative equation solution will certainly become important.

ACKNOWLEDGEMENTS

Portions of the work discussed herein have been supported by grants from the US National Science Foundation (DDM-8996171) and NASA Lewis Research Center (NAG 3-1089) to Clarkson University. Any opinions, findings, and conclusions or recommendations expressed in this publication are those of the authors and do not reflect the views of these other organizations.

BIBLIOGRAPHY

[1] Banerjee, P.K. and Butterfield, R. (1981) *Boundary Element Methods in Engineering Science*, McGraw-Hill, London.
[2] Brebbia, C.A. and Walker, S. (1980) *Boundary Element Technics in Engineering*, Newnes Butterworths, London and Boston.
[3] Brebbia, C.A., Telles, J.C.F. and Wrobel, L.C. (1984) *Boundary Element Techniques*, Springer Verlag, Berlin and New York.
[4] Lachat, J.C. and Watson, J.O. (1975) A second generation boundary integral program for three dimensional elastic analysis, chapter in *Boundary Integral Equation Method: Computational Applications in Applied Mechanics* (eds.) T.A. Cruse and F.J. Rizzo, Appl. Mech. Div., ASME, Vol. 11, New York, 1975.
[5] Lachat, J.C. and Watson, J.O. (1900) Progress in the use of boundary integral equations, illustrated by examples, *Comput. Meth. Appl. Mech. Engg*, **10**, 273–289.
[6] Lachat, J.C. (1979) *Further Developments of the Boundary Integral Technique for Elastostatics*, Ph.D. Thesis, Southampton University.
[7] Crotty, J.M. (1982) A block equation solver for large unsymmetric matrices arising in the boundary element method, *Int. J. Num. Meth. Engg*, **18**, 997–1017.
[8] Das, P.C. (1978) A disc based block elimination technique used for the solution of non-symmetrical fully populated matrix systems encountered in the boundary element method, *Proc. Int. Symp. Recent Developments in Boundary Element Methods*, Southampton University, pp. 391–404.
[9] Bialecki, R. and Nahlik, R. (1987) Linear equations solver for large block matrices arising in boundary element methods, *Boundary Elements IX, Vol. 1*, Computational Mechanics Publications, Springer-Verlag, New York.
[10] Bialecki, R. (1987) Nonlinear equations solver for large equation sets arising when using BEM in homogenous regions of nonlinear material, *Boundary Elements IX*, Vol. 1, Computational Mechanics Publications, Springer-Verlag, New York.
[11] Tomlin, G.R. (1972) *Numerical Analysis of Continuum Problems in Zoned Anisotropic Media*, Ph.D. Thesis, Southampton University.
[12] Butterfield, R. and Tomlin, G.R. (1971) Integral techniques for solving zoned anisotropic continuum problems, *Proc. Int. Conf. Variational Methods in Engineering*, Southampton University, pp. 9/31–9/51.
[13] Chang, O.V. (1979) *Boundary elements applied to seepage problems in zoned anisotropic soils*, MSc. Thesis, Southampton University.

[14] Beer, G. (1986) Implementation of Combined Boundary Element-Finite Element Analysis with Applications in Geomechanics, Chapter 7 in P.K. Banerjee and J.O. Watson (eds.), *Developments in Boundary Element Methods*, Elsevier, London and New York, pp. 191–225.

[15] Mustoe, G.G.W. (1982) *A Combination of the Finite Element Method and Boundary Solution Procedures for Continuum Problems*, Ph.D. Thesis, University of Wales, University College, Swansea.

[16] Davies, T.G. (1979) *Linear and Nonlinear Analysis of Pile Groups*, Ph.D. Thesis, University of Wales, University College, Cardiff.

[17] Jin, H., Runesson, K. and Samuelsson, A. (1987) Application of the boundary element method to contact problems in elasticity with a nonclassical friction law, *Boundary Elements IX*, Computational Mechanics Publications, Springer-Verlag, Southampton and Boston, Vol. 2, pp. 397–415.

[18] Margenov, S., Georgiev, K., Hadjikov, L. and Novakova, M. (1987) An effective approach for boundary element method application to friction contact problems, *Boundary Elements IX*, Vol. 1, Computational Mechanics Publications, Springer-Verlag, Southampton and Boston, pp. 439–445.

[19] Kane, J.H. (1986) Shape optimization utilizing a boundary element formulation, *BETECH 86, Proceedings*, Computational Mechanics Publications, Springer-Verlag, Southampton and Boston, pp. 781–803.

[20] Kane, J.H. and Saigal, S. (1988) Design sensitivity analysis of solids using BEM, *J. Engg Mech.*, ASCE, **114**(10), 1703–1722.

[21] Saigal, S. and Kane, J.H. (1990) A boundary element shape optimization system for aircraft components, *AIAA J.*, **28**(7), 1203–1204.

[22] Saigal, S., Aithal, R. and Kane, J.H. (1989) Conforming boundary elements in plane elasticity for shape design sensitivity, *Int. J. Num. Meth. Engg*, **28**, 2795–2811.

[23] Saigal, S., Kane, J.H. and Aithal, R. (1989) Semi-analytical structural sensitivity formulation using boundary elements, *AIAA J.*, **27**(11), 1615–1621.

[24] Saigal, S., Borggaard, J.T. and Kane, J.H. (1989) Boundary element implicit differentiation equations for design sensitivities of axisymmetric structures, *Int. J. Solids and Structures*, **25**(5), 527–538.

[25] Kane, J.H. and Saigal, S. (1988) Design sensitivity analysis of boundary element substructures, *2nd NASA/Air Force Symp. Recent Experiences in Multi Disciplinary Analysis and Optimization*, September, 1988, Proceedings published as NASA CP-3031, vol. 2, pp. 777–799.

[26] Kane, J.H., Kumar, B.L.K. and Gallagher, R.H. (1990) Boundary element iterative reanalysis for continuum structures, *J. Engg Mech.*, ASCE, **116**, 2293–2309.

[27] Hodous, M.F., Bozek, D.G., Ciarelli, D.M., Ciarelli, K.J., Katnik, R.B. and Kline, K.A. (1983) Vector processing applied to boundary element algorithms on the CDC Cyber 205, *Bulletin de la Direction des Etudes et des Recherches*, Serie C, No. 1, 87–94.

[28] Kline, K.A., Tsao, N.K. and Friedlander, C.B. (1985) Parallel processing and the solution of boundary element equations, in T.A. Cruse, A.B. Pifko, and H. Armen (eds.), *Advanced Topics in Boundary Element Analysis*, ASME AMD Publication.

[29] Bettess, J.A. (1987) Solution techniques for boundary integral matrices, in R.W. Lewis, E. Hinton, P. Bettess, and B.A. Schrefler, (eds.) *Numerical Methods in Transient and Coupled Problems*, Wiley-Interscience, pp. 123–147.

[30] Kane, J.H., Kuman, B.L.K. and Saigal, S. (1990) An arbitrary condensing, noncondensing solution strategy for large scale, multi-zone boundary element analysis, *Computer Methods in Applied Mechanics and Engineering*, **29**(2), 219.

[31] Kane, J.H., Wang, H. and Kumar, B.L.K. (1990) Nonlinear thermal analysis with boundary element zone condensation, *Computational Mechanics*, **7**, 107–122.

[32] Kane, J.H. and Saigal, S. (1990) An arbitrary multi-zone condensation technique for boundary element design sensitivity analysis, *AIAA J.*, **28**(7), 1277–1284.
[33] Kane, J.H. and Wang, H. (1991) Boundary element shape sensitivity analysis formulations for thermal problems with nonlinear boundary conditions, *AIAA J.*, **29**, 1978–1989.
[34] Drake, J.B. and Gray, L.J. (1988) *Parallel Implementation of the boundary Element Method on the iPSC2 Hypercube for Electroplating Applications*, Oak Ridge National Laboratory Report ORNL-6515, Oak Ridge, Tennessee, December.
[35] Eisenstat, S.C., Heath, M.T., Henkel, C.S. and Romine, C.H. (1988) Modified cyclic algorithms for solving triangular systems on distributed-memory multiprocessors, *SIAM J. Sci. Statist. Comput.*, **9**(3), 589–600.
[36] Geist, G.A. and Romine, C.H. (1988) LU factorization algorithms on distributed-memory multiprocessor architectures, *SIAM J. Sci. Statist. Comput.*, **9**(4), 639–649.
[37] Heath, M.T. and Romine, C.H. (1988) Parallel solution of triangular systems on distributed-memory multiprocessors, *SIAM J. Sci. Statist. Comput.*, **9**(3), 558–588.
[38] Ortega, J.M. (1988) *Introduction to Parallel and Vector Solution of Linear Systems*, Plenum Press, New York.
[39] Golub, G.H. and Van Loan, C.F. (1983) *Matrix Computations*, Johns Hopkins University Press, Baltimore MD.
[40] Saad, Y. and Schultz, M.H. (1986) GMRES: a generalized minimal residual algorithm for solving nonsymmetric linear systems, *SIAM J. Sci. Statist. Comput.*, **7**(3), 856–869.
[41] Samuelsson, A., Wiberg, N.E. and Bernsang, L. (1986) Study of the efficiency of iterative methods in linear problems in structural mechanics, *Int. J. Num. Meth. Engg*, **22**, 209–218.
[42] Angeleri, F., Sonnad, V. and Bathe, K.J. (1989) Studies of finite element procedures η an evaluation of preconditioned iterative solvers, *Computers and Structures*, **32**(3/4), 671–677.
[43] Brussino, G. and Sonnad, V. (1989) A comparison of direct and preconditioned iterative techniques for sparse, unsymmetric systems of linear equations, *Int. J. Num. Meth. Engg*, **28**, 801–815.
[44] Brussino, G. and Sonnad, V. (1986) *A Comparison of Preconditioned Iterative Techniques for Sparse, Indefinite, Unsymmetric Systems of Linear Equations*, IBM Document Number KGN-83, Kingston, New York.
[45] Brussino, G. and Sonnad, V. (1987) *A Comparison of Preconditioned Iterative Techniques for Sparse, Unsymmetric Systems of Linear Equations: Part II*, IBM Document Number KGN-102, Kingston, New York.
[46] Poole, E.L. and Overman, A.L. (1988) *The Solution of Linear Systems of Equations with a Structural Analysis Code on the NAS CRAY-2*, NASA Report CR4158, Washington, DC.
[47] Nachtigal, N.M., Reddy, S.C. and Trefethen, L.N. (1992) How fast are nonsymmetric matrix iterations? *Proc. Copper Mountain Conference on Iterative Methods, SIAM J. Matrix Anal. Appl.*, **13**, 778–795.
[48] Dervieux, A., Fezoui, L., Steve, H., Periaux, J. and Stoufflet, B. (1989) Low-storage implicit upwind-FEM schemes for the Euler equation, *Proc. 11th International Conference on Numerical Methods in Fluid Dynamics*, Springer-Verlag, Berlin, pp. 215–219.
[49] Navarra, A. an application of GMRES to indefinite problems in meteorology, *Applied Numerical Mathematics*, to appear.
[50] Shakib, F., Hughes, T.J.R. and Johan, Z. (1989) Element-by-element Algorithms for Nonsymmetric Matrix Problems Arising in Fluids, in J.H. Kane, A.D. Carlson and D.L. Cox (eds.), *Solution of Superlarge Problems in Computational Mechanics*, Plenum, New York, pp. 1–33.
[51] Wigton, L.B., Yu, N.J. and Young, D.P. (1985) GMRES acceleration of computational fluid dynamics codes, *Proc. AIAA 7th Computational Fluid Dynamics Conference*, AIAA

Washington DC, Paper no. 85, p. 1494.
[52] Zdatev, Z. and Nielsen, H.B. Solving large and sparse linear least squares problems by conjugate gradient algorithms, *Computer Mathematics and Applications*, **15**(3), 185–202.
[53] Kincaid, D.R., Oppe, T.C. and Joubert, W. (1989) An introduction to the NSPCG Software Package, *Int. J. Num. Meth. Engg*, **27**, 589–608.
[54] Kane, J.H., Keyes, D.E. and Guru Prasad, K. (1991) Iterative equation solution techniques in boundary element analysis, *Int. J. Num. Meth. Engg*, **31**, 1511–1536.
[55] Kane, J.H., Keshava Kumar, B.L., Wilson, R.B. and Srinivasan, A.V. Data Parallel (SIMD) Boundary Element Stress Analysis, *Int. J. Num. Meth. Engg*, Submitted.

J.H. Kane and K. Guru Prasad
Mechanical and Aeronautical Engineering Department
Clarkson University
Potsdam
NY 13699
USA

5
Solution of Large Eigenvalue Problems

G. Gambolati

Università degli Studi di Padova, Italy

5.1 INTRODUCTION

Iterative methods to process large sparse matrices are currently subject to a critical review and a continuous development consistent with the advancing computer technology and the new generations of processors which allow for the treatment of numerical models of increasing size.

It is known that several problems in engineering, physics, chemistry and environmental sciences can be effectively addressed with the aid of matrix techniques. In this context a very powerful and widespread tool is represented by eigenanalysis, where a generalized eigenproblem:

$$\mathbf{A}\mathbf{v} = \lambda \mathbf{B}\mathbf{v} \tag{5.1}$$

needs to be solved. Frequently the matrix pencil **A**, **B** is symmetric and positive definite with a high degree of sparsity.

To take a few significant examples, we note that determining the natural modes **v** and the corresponding frequencies λ satisfying equation (5.1) is a very common task in plasma physics (Gruber [1]; Chance *et al.* [2]; Grimm *et al.* [3]), vibrational analysis of mechanical structures (Craig and Bampton [4]; Anderson *et al.* [5]; Bathe and Wilson [6]), buckling analysis of elastic compounds (Clough and Penzien [7]), hydrodynamics (Gambolati *et al.* [8]), and lightwave technology (Yamamoto *et al.* [9]; Silvester and Ferrari [10]; Zoboli and Bassi [11]).

Solving Large-scale Problems in Mechanics, Edited by M. Papadrakakis
© 1993 John Wiley & Sons Ltd

The number of degrees of freedom for typical eigenproblems in the aforementioned applications is between 1000 and 10 000, but there is a need to overcome this limit in some cases. Very often only several of the leftmost eigenpairs are needed, and the fulfilment of this task for medium- to large-size matrices may require much CPU time even on modern powerful mainframes. Hence the search for new and cost-effective numerical procedures to solve equation (5.1) is still in progress.

While there are numerous good quality software packages for solving (5.1) for full matrices (IMSL, NAG, Smith et al. [12]; Garbow et al. [13]), relying on the triangular factorization of either $\mathbf{A}$ or $\mathbf{B}$ and the subsequent application of the QL algorithm (Parlett [14]), efficient software for sparse matrices is not as readily available in the literature or on the market.

One approach that is widely used by structural engineers for extracting some eigenpairs in a preset interval is subspace iteration (SI, Parlett [14]), which was originally programmed in the routine SSPACE (Bathe and Wilson [15]) and is implemented in several SAP codes. Recent ADINA packages based on an accelerated SI version (Bathe and Ramaswamy [16]) are available.

The SI technique factorizes matrix $\mathbf{E}$:

$$\mathbf{E} = \mathbf{A} - \sigma \mathbf{B} = \mathbf{L}_\sigma \mathbf{D} \mathbf{L}_\sigma^T \tag{5.2}$$

where $\mathbf{D}$ is diagonal and σ is a shift factor selected to locate the partial interval over which the eigenvalues and eigenvectors of (5.1) are sought. SI looks for the characteristic values μ of matrix $(\mathbf{L}_\sigma^{-1})^T \mathbf{D}^{-1} \mathbf{L}_\sigma^{-1} \mathbf{B}$ which are related to the λs of problem (5.1) by the equation

$$\mu = \frac{1}{\lambda - \sigma}. \tag{5.3}$$

Another method, currently attracting much attention, is the Lanczos [17] method which was revisited some years ago in a sparse matrix context by Paige [18] and Cullum and Willoughby [19, 20]. A block generalization was developed by Cullum and Donath [21] and Golub and Underwood [22] while Ericsson and Ruhe [23] proposed the spectral transformation technique which involves the triangular factorization of several matrices $\mathbf{E}$ for various shift parameters σ.

A third family of iterative methods is based on optimization by gradient and conjugate gradient schemes. The quantity to be optimized may be either the trace (Sameh and Wisniewski [24]) or a penalty function (Beliveau et al. [25]) or, more commonly, the Rayleigh quotient:

$$R(\mathbf{x}) = \frac{\mathbf{x}^T \mathbf{A} \mathbf{x}}{\mathbf{x}^T \mathbf{B} \mathbf{x}}. \tag{5.4}$$

The latter approach requires both $\mathbf{A}$ and $\mathbf{B}$ to be symmetric and positive definite. The procedures which rely on conjugate gradients are particularly attractive since they are simple to implement and well suited to vectorization.

The minimization of $R(\mathbf{x})$ by a gradient (steepest descent) algorithm was first presented by Hestenes and Karush [26, 27] (see also Hestenes [28]). Unfortunately the steepest descent converges very slowly unless the problem is very small. A more reliable gradient scheme, referred to as 'coordinate relaxation', was subsequently devised by Faddeev and Faddeeva [29] and analysed by other authors as well (Nesbet [30]; Bender and Shavitt [31]; Falk [32]; Shavitt et al. [33]). Nisbet [34], Ruhe [35], Schwarz [36] and Papadrakakis [37] suggested an improvement of this method by defining an acceleration parameter ω, in complete analogy to the SOR scheme for solving iteratively linear difference equations arising from the integration of elliptic boundary value problems. It should be remarked, however, that there is no general theory for the *a priori* assessment of the best over-relaxation factor ω_{opt}. Therefore ω_{opt} needs to be determined empirically for any given pencil **A**, **B**, and this greatly reduces the interest in coordinate relaxation techniques for engineering applications.

An important modification of the steepest descent method is the conjugate gradient (CG) technique originally developed by Hestenes and Stiefel [38] for the solution of linear symmetric positive definite systems. In line with this approach are the contributions by Bradbury and Fletcher [39], Fried [40], Gerardin [41] and Ruhe [42]. Papadrakakis [43] has combined the CG algorithm with symmetric coordinate over-relaxation, but a limitation of his approach is again the need for improving the rate of convergence through an optimal acceleration factor which is problem dependent and generally unknown *a priori*.

Iterative schemes based on convergent splittings were introduced by Ruhe [44] and later reanalyzed by Evans and Shanehchi [45] who proposed a better splitting choice. It is to be noted, however, that in this approach as well, convergence needs to be accelerated by a relaxation parameter which is difficult, if not impossible, to evaluate for matrices with an arbitrary sparsity structure.

All of these well-established gradient techniques may fail to work in practice when the size of **A** and **B** becomes large or even moderately large. Some years ago Sameh and Wisniewski [24] stated that 'no efficient method for simultaneously obtaining several eigenvalues and eigenvectors is available'.

Iterative procedures for the simultaneous computation of the leftmost eigenpairs relying on the multiple Rayleigh quotient optimization were developed by Longsine and McCormick [46] and Schwarz [47]. Unfortunately they do not converge for real unstructured matrices of large dimension (Sartoretto *et al.* [48]).

In recent years a modification of the CG method has proved very effective in ameliorating the performance of this scheme as applied to the solution of large sparse sets of symmetric positive definite equations (Concus *et al.* [49]; Axelsson [50]; Meijerink and van der Vorst [51]; Kershaw [52]; Gambolati [53, 54]; Manteuffel [55]; Gambolati and Perdon [56]). With reference to the product between the error and the residual, the number M of iterations required to achieve a relative accuracy δ is expressed by (Axelsson [50]):

$$M = \text{int}\left(\frac{1}{2}\sqrt{\xi}\ln\frac{2}{\delta} + 1\right)$$

where ξ is the spectral condition number of the CG iteration matrix. The previous equation holds for ξ values not too close to 1 and may be safely used for ξ larger than 3 or 4.

For any given system ξ is very much dependent on the preconditioning matrix. The preconditioned CG method has proved a very cost-effective algorithm for solving large systems if a good preconditioner can be obtained (relatively) inexpensively.

The idea of preconditioning has recently been extended to the CG optimization of the Rayleigh quotient (Papadrakakis and Yakoumidakis [57], Gambolati et al. [58]; Sartoretto et al. [48]). The results have been very encouraging, particularly for arbitrarily structured matrices, which convergence being achieved after only a few iterations compared to the eigenproblem size.

Since the Lanczos approach has been demonstrated to be superior to the SI method [59], the latter is not considered in the analysis that follows.

In this paper we initially review some classical engineering applications which result naturally in an eigenproblem.

The accelerated conjugate gradient method for the evaluation of the leftmost part of the eigenspectrum of equation (5.1) is described first, along with a discussion of effective preconditioners. The Lanczos method is then presented in a sparse finite element context, and it is shown that CG schemes play an important role within the Lanczos approach as well. Finally both methods are applied to partially solve five realistic finite element eigenproblems with an arbitrary sparsity structure and dimension up to almost 5000. The eigenproblems arise from the finite element integration of diffusion type PDEs. The performance of the two methods is discussed and compared for the problem of evaluating the 40 leftmost eigenpairs.

5.2 EIGENPROBLEMS IN ENGINEERING

We restrict our analysis to the symmetric eigenproblems where matrices **A** and **B** arise from the finite element (or finite difference) integration of self-adjoint initial boundary value problems. Some classical examples leading to equation (5.1) in engineering applications are described in the sequel.

A large interconnected mechanical system, for which **K** and **M** represent the stiffness and mass matrices, respectively, is governed by the dynamic equation:

$$\mathbf{M}\ddot{\mathbf{x}} + \mathbf{K}\mathbf{x} = \mathbf{0} \tag{5.5}$$

where **x** is the vector of displacements from static equilibrium and $\ddot{\mathbf{x}}$ denotes the second time derivative (acceleration). Assuming all **x** components oscillate at the same frequency, we write:

$$\mathbf{x} = \boldsymbol{\phi} \sin \omega t. \tag{5.6}$$

Replacing (5.6) in (5.5) yields

$$(\mathbf{K} - \omega^2 \mathbf{M})\boldsymbol{\phi} = \mathbf{0} \tag{5.7}$$

which is a generalized eigenproblem of type (5.1) with $\lambda = \omega^2$.

5.2 EIGENPROBLEMS IN ENGINEERING

If the system response is influenced by damping and is driven by an external load **p**(*t*), equation (5.5) is substituted by the more general dynamic equation:

$$\mathbf{M\ddot{x}} + \mathbf{C\dot{x}} + \mathbf{Kx} = \mathbf{p}(t) \qquad (5.8)$$

where **C** is the damping matrix. In the classical mode superposition method for solving equation (5.8), a coordinate transformation matrix $\mathbf{\Phi}_m$ is required which contains the *m* leftmost eigenvectors of the undamped eigenproblem (5.7):

$$\mathbf{K}\mathbf{\Phi}_m = \mathbf{M}\mathbf{\Phi}_m\mathbf{\Lambda}_m \qquad (5.9)$$

where $\mathbf{\Lambda}_m$ is a diagonal matrix containing the *m* smallest frequencies (squared) of the structural system (i.e. the *m* smallest eigenvalues of equation (5.7)).

A similar result is obtained for systems which are continuous in space, such as elastic bodies, after being discretized into finite elements. In this case the pencil **K** and **M** may be very large and sparsely structured.

Another eigenproblem occurs in the buckling analysis of elastic compounds. We write

$$(\mathbf{K} - \lambda \mathbf{K}_G)\mathbf{\phi} = \mathbf{0} \qquad (5.10)$$

where $\mathbf{K}_G$ is the so-called geometric stiffness matrix. The vectors $\mathbf{\phi}$ describe the unstable configuration of the structure and the scalars λ provide useful information on the critical loads which lead to structural collapse.

In the analysis of optical waveguides, the complex wave equation, also referred to as curl-curl equation, needs to be solved:

$$\bar{\nabla} \times (\bar{\bar{\mathbf{k}}}^{-1} \bar{\nabla} \times \bar{\mathbf{h}}) - \omega^2 \varepsilon_0 \mu_0 \bar{\mathbf{h}} = \mathbf{0} \qquad (5.11)$$

where $\bar{\mathbf{h}}$ is the complex magnetic vector field intensity, $\bar{\bar{\mathbf{k}}}$ is the relative electric permittivity tensor, ε_0 is the vacuum permittivity and μ_0 is the vacuum magnetic permeability. In equation (5.11) × and $\bar{\nabla}$ denote cross product and gradient vector, respectively. After integrating equation (5.11) over an optical waveguide by a finite element method, we are left with a generalized eigenvalue problem of type (5.1) where **A** and **B** are real symmetric matrices. Moreover **B** is positive definite, and so is **A** if the magnetic signals propagate within lossless media. The eigensolution allows for the determination of the natural frequencies of the waveguide along with the fundamental magnetic modes which are of paramount importance in the proper design of efficient optical communication devices.

Very similar examples, from a mathematical viewpoint, are represented by a vibrating string, a membrane, and an oscillatory water basin. The corresponding two-dimensional differential equation may be written as

$$\frac{\partial^2 h}{\partial x^2} + \frac{\partial^2 h}{\partial y^2} + K^2 h = 0. \qquad (5.12)$$

Here h is the deflection from a rest position and K is inversely proportional to the wavelength. Equation (5.12) is often solved by finite elements under various boundary conditions, with either h or its normal derivative set equal to zero.

Again we obtain a matrix equation of type (5.1) where λ is equal to K^2 and $\mathbf{A}$ and $\mathbf{B}$ are the stiffness and capacity matrices, respectively, both of which are symmetric positive definite.

Equations (5.11) and (5.12) could be solved through a variational formulation (Berk [60]):

$$\omega^2(\bar{\mathbf{h}}) = \frac{\int_R (\bar{\nabla} \times \bar{\mathbf{h}})^+ \bar{\bar{\mathbf{k}}}^{-1}(\bar{\nabla} \times \bar{\mathbf{h}}) \, dR}{\int_R \bar{\mathbf{h}}^+ \varepsilon_0 \mu_0 \bar{\mathbf{h}} \, dR} = \text{stationary} \qquad (5.13a)$$

$$K^2(h) = \frac{\int_R \frac{1}{2}\left[\left(\frac{\partial h}{\partial x}\right)^2 + \left(\frac{\partial h}{\partial y}\right)^2\right] dR}{\int_R \frac{1}{2} h^2(x, y) \, dR} = \text{stationary} \qquad (5.13b)$$

subject to the same boundary conditions as the original equations. It is well known that (5.13a) and (5.13b) are stationary for any function $\bar{\mathbf{h}}$ or h coinciding with an eigenfunction of equation (5.11) or (5.12), with $\omega^2(\bar{\mathbf{h}})$ or $K^2(h)$ the associated eigenvalue.

Similarly the eigenpairs of eigenproblem (5.1) may be found by looking for the stationary points of the Rayleigh quotient (5.4), which represents the numerical (finite element) approximation of equations (5.13). The eigensolutions λ and $\mathbf{v}$ are the approximate eigenvalues and eigenfunctions of equations (5.11) and (5.12).

The semi-discrete system of first-order ordinary differential equations governing linear heat diffusion and groundwater flow,

$$\mathbf{P}\dot{\mathbf{u}} + \mathbf{H}\mathbf{u} = \mathbf{f} \qquad (5.14)$$

is obtained through a finite element integration in space of the corresponding initial boundary value problems. Here $\mathbf{P}$ and $\mathbf{H}$ are the $N \times N$ capacity and conductivity matrices, respectively, and $\mathbf{f}$ is the heat or flow supply vector. The variable $\mathbf{u}$ is a vector of nodal temperatures or hydraulic potentials. The Rayleigh–Ritz method for solving equation (5.14) leads to a generalized eigenproblem of type (5.1) where $\mathbf{A} = \mathbf{H}$ and $\mathbf{B} = \mathbf{P}$ (Biot [61]).

In the frequency analysis of equations (5.11) and (5.12) and in the reduction methods for solving equation (5.14) we usually are interested in the p smallest eigenvalues and eigenvectors of the associated generalized eigenproblem. The highest eigenpairs are not physically meaningful since they correspond to the shortest wavelengths. Therefore we require the p leftmost eigenpairs of equation (5.1), or, alternatively, the p leftmost stationary points of ratio (5.4).

In the following we refer to matrices **A** and **B** as the stiffness and capacity matrices, respectively, regardless of the particular applications under consideration. Our analysis will focus on large sparse finite element eigenproblems of arbitrary structure. Although not strictly required by the Lanczos method, the pencil **A**, **B** is also assumed to be positive definite.

5.3 OPTIMIZATION OF RAYLEIGH QUOTIENTS BY ACCELERATED CONJUGATE GRADIENTS

The method for the partial eigensolution of equation (5.1) presented in this section is an improvement of two optimization algorithms recently developed by Gambolati et al. [58] and Sartoretto et al. [48].

The former (Gambolati et al. [58]) requires the assessment of several shift parameters, which are related to the distribution of the leftmost characteristic values and are therefore problem dependent. The latter (Sartoretto et al. [48]) performs a simultaneous optimization of (5.4) using a set of vectors which are constantly kept **B**-orthogonal, and involves the periodic execution of a Ritz projection step to ensure convergence. Both approaches achieve the desired eigensolution by minimizing $R(\mathbf{x})$ via the accelerated conjugate gradients.

The scheme described here is based on a successive sequential optimization of Rayleigh quotients by conjugate gradients preconditioned with a fixed preconditioner.

There are two essential ideas underlying this method:

1. acceleration of traditional CG by a good and inexpensive preconditioner;
2. successive minimization of $R(\mathbf{x})$ in subspaces of decreasing size $N - j$ **B**-orthogonal to the j engenvectors already found.

Assume that the j smallest eigenvectors of (5.1) are known. They form the $N \times j$ matrix V_j. Since the eigenvectors are **B**-orthogonal and **B**-normalized, $\mathbf{V}_j$ satisfies the equation

$$\mathbf{V}_j^T \mathbf{B} \mathbf{V}_j = \mathbf{I}_j$$

where $\mathbf{I}_j$ is the $j \times j$ identity matrix.

In the preconditioned CG method we use the matrix transformation

$$\mathbf{y} = \mathbf{X}\mathbf{x} \qquad (5.15)$$

where **X** is a symmetric matrix. Substituting (5.15) in (5.4) yields a transformed Rayleigh quotient $R_1(\mathbf{y})$:

$$R_1(\mathbf{y}) = \frac{\mathbf{y}^T \mathbf{X}^{-1} \mathbf{A} \mathbf{X}^{-1} \mathbf{y}}{\mathbf{y}^T \mathbf{X}^{-1} \mathbf{B} \mathbf{X}^{-1} \mathbf{y}} = \frac{\mathbf{y}^T \mathbf{G} \mathbf{y}}{\mathbf{y}^T \mathbf{F} \mathbf{y}} \qquad (5.16)$$

where $\mathbf{G} = \mathbf{X}^{-1} \mathbf{A} \mathbf{X}^{-1}$ and $\mathbf{F} = \mathbf{X}^{-1} \mathbf{B} \mathbf{X}^{-1}$.

In the preconditioned CG method $R_1(\mathbf{y})$ is optimized by classic conjugate gradients and $\mathbf{x}$ is finally obtained from $\mathbf{y}$ using (5.15). The $(j+1)$st leftmost eigenpair λ_{N-j}, $\mathbf{v}_{N-j}$ is obtained through the following steps (where the equations are expressed in terms of the original variable $\mathbf{x}$. See Gambolati et al. [62]).

1. Start with an initial vector $\mathbf{x}_0$ such that $\mathbf{V}_j^T \mathbf{B} \mathbf{x}_0 = 0$, i.e. $\mathbf{x}_0$ is taken to be **B**-orthogonal to the subset $\mathbf{V}_j$ of the j eigenvectors previously evaluated. Vector $\mathbf{x}_0$ is obtained from an arbitrary vector $\mathbf{x}$ made **B**-orthogonal to $\mathbf{V}_j$ by a Gram–Schmidt **B**-orthogonalization process:

$$\mathbf{x}_0 = \mathbf{x} - (\mathbf{x}^T \mathbf{B} \mathbf{v}_N) \mathbf{v}_N - \ldots - (\mathbf{x}^T \mathbf{B} \mathbf{v}_{N-j+1}) \mathbf{v}_{N-j+1}. \qquad (5.17)$$

Set $k = 0$ (iteration index) and $\mathbf{p}_{-1}$ as an arbitrary vector.

2. If $k = 0$ then set $\beta_k = 0$, otherwise compute β_k by (Gambolati et al. [58])

$$\beta_k = -\frac{\mathbf{p}_{k-1}^T \mathbf{A} \mathbf{K}^{-1} \mathbf{g}_k}{\mathbf{p}_{k-1}^T \mathbf{A} \mathbf{p}_{k-1}} \qquad (5.18)$$

where

$$\mathbf{g}_k = \frac{2}{\mathbf{x}_k^T \mathbf{B} \mathbf{x}_k} [\mathbf{A} \mathbf{x}_k - R(\mathbf{x}_k) \mathbf{B} \mathbf{x}_k] \qquad (5.19)$$

is the gradient of $R(\mathbf{x})$ computed at the current iterate $\mathbf{x}_k$ and $\mathbf{K}^{-1} = \mathbf{X}^{-1}\mathbf{X}^{-1}$ is the (symmetric positive definite) preconditioning matrix.

3. Calculate

$$\tilde{\mathbf{p}}_k = \mathbf{K}^{-1} \mathbf{g}_k + \beta_k \mathbf{p}_{k-1}. \qquad (5.20)$$

4. Evaluate vector $\mathbf{p}_k$ by **B**-orthogonalizing $\tilde{\mathbf{p}}_k$ against $\mathbf{V}_j$ by a Gram–Schmidt **B**-orthogonalization process (equation (5.17)).

5. Compute the coefficient α_k by minimizing the Rayleigh quotient (Perdon and Gambolati [63]):

$$R(\mathbf{x}_k + \alpha_k \mathbf{p}_k)$$

which yields:

$$\alpha_k = \frac{nd - mb + \sqrt{\Delta}}{2(bc - ad)} \qquad (5.21)$$

with

$$\Delta = (nd - mb)^2 - 4(bc - ad)(ma - nc)$$

$$a = \mathbf{p}_k^T \mathbf{A} \mathbf{x}_k \qquad b = \mathbf{p}_k^T \mathbf{A} \mathbf{p}_k \qquad c = \mathbf{p}_k^T \mathbf{B} \mathbf{x}_k$$
$$d = \mathbf{p}_k^T \mathbf{B} \mathbf{p}_k \qquad m = \mathbf{x}_k^T \mathbf{B} \mathbf{x}_k \qquad n = \mathbf{x}_k^T \mathbf{A} \mathbf{x}_k. \qquad (5.22)$$

6. Set

$$\tilde{\mathbf{x}}_{k+1} = \mathbf{x}_k + \alpha_k \mathbf{p}_k \tag{5.23}$$

and form the new approximation vector $\mathbf{x}_{k+1}$ by **B**-normalizing $\tilde{\mathbf{x}}_{k+1}$:

$$\mathbf{x}_{k+1} = \tilde{\mathbf{x}}_{k+1}/\sqrt{(\tilde{\mathbf{x}}_{k+1}^T \mathbf{B} \tilde{\mathbf{x}}_{k+1})}. \tag{5.24}$$

Increment the iteration counter and go back to step 2. The iteration is completed whenever k is larger than the allowed number of iterations or when

$$e_r = \frac{|R(\mathbf{x}_{k+1}) - R(\mathbf{x}_k)|}{R(\mathbf{x}_k)} \tag{5.25}$$

is smaller than a preset tolerance TOLL.

7. If $e_r <$ TOLL, $R(\mathbf{x}_{k+1})$ and $\mathbf{x}_{k+1}$ are the $(j+1)$st smallest eigenvalue and corresponding eigenvector of (5.1).

Note that the last eigenpair λ_N, $\mathbf{v}_N$ can also be computed by this general procedure by setting $\mathbf{x}_0 = \mathbf{x}$ in step 1 and $\mathbf{p}_k = \tilde{\mathbf{p}}_k$ in step 4. Since the eigenvalues and eigenvectors already determined are used (in steps 1 and 4) to deflate the original eigenproblem and since the convergence rate of the characteristic value approximation is twice that of the eigenvector approximation (Fried [40] and Ruhe [42]), each iteration of the accelerated CG cycle should be terminated when e_r achieves a very small value.

An exit criterion based on the residual error might also be appropriate. In such a case the relative residual:

$$r_r = \frac{|\mathbf{A}\mathbf{x}_k - R(\mathbf{x}_k)\mathbf{B}\mathbf{x}_k|}{|\mathbf{A}\mathbf{x}_k|} \tag{5.26}$$

where $|\cdot|$ stands for the Euclidean norm, can be used in place of (5.25) to ensure eigenvector accuracy.

Let us now discuss the convergence properties of the CG method. We consider the evaluation of the leftmost eigenpair λ_N, $\mathbf{v}_N$. Denote by $\mathbf{H}(\mathbf{x})$ the Hessian of $R(\mathbf{x})$:

$$\mathbf{H}(\mathbf{x}) = (h_{ij}(\mathbf{x})) = \left(\frac{\partial^2 R(\mathbf{x})}{\partial x_i \partial x_j}\right) \quad i, j = 1, \ldots, N. \tag{5.27}$$

At $\mathbf{x} = \mathbf{v}_N$, $R(\mathbf{v}_N) = $ min and $\mathbf{g}(\mathbf{v}_N) = 0$. Hence in a suitable neighbourhood of $\mathbf{v}_N$, $R(\mathbf{x})$ is approximated by

$$R(\mathbf{x}) \simeq \lambda_N + \frac{1}{2}\left(\frac{\partial}{\partial x_1}h_1 + \frac{\partial}{\partial x_2}h_2 + \cdots + \frac{\partial}{\partial x_N}h_N\right)^2 R(\mathbf{v}_N) \qquad (5.28)$$

$$= \lambda_N + \frac{1}{2}(\mathbf{x} - \mathbf{v}_N)^T \mathbf{H}(\mathbf{v}_N)(\mathbf{x} - \mathbf{v}_N)$$

where $h_i = x_i - v_{N,i}$, $i = 1, \ldots, N$, are incremental scalar quantities. $\mathbf{H}(\mathbf{v}_N)$ is positive semi-definite and therefore the surface

$$\varphi(\mathbf{x}) = (\mathbf{x} - \mathbf{v}_N)^T \mathbf{H}(\mathbf{v}_N)(\mathbf{x} - \mathbf{v}_N) = \text{const} \qquad (5.29)$$

geometrically represents a family of hyper-ellipsoids in the $(N-1)$-dimensional subspace orthogonal to $\mathbf{v}_N$. In fact, as will be shown below, $\mathbf{h}(\mathbf{v}_N)$ has vector $\mathbf{v}_N$ as the eigenvector associated with the zero eigenvalue. Asymptotic convergence of the classic CG method is controlled by the roundness of hyper-ellipsoid (5.29) which is measured on first approximation by the ratio between the extreme hyper-ellipsoid axes a_2 and a_N. It may be shown (Gambolati [54]) that a_2 and a_N are proportional to the inverse square root of the largest and smallest (different from zero) eigenvalues of $\mathbf{H}(\mathbf{v}_N)$, i.e.

$$a_2 = \sqrt{\left(\frac{\text{const}}{\lambda_{N-1}[\mathbf{H}(\mathbf{v}_N)]}\right)} \qquad a_N = \sqrt{\left(\frac{\text{const}}{\lambda_1[\mathbf{H}(\mathbf{v}_N)]}\right)}.$$

Therefore the asymptotic convergence is controlled by the square root of the spectral condition number of $\mathbf{H}(\mathbf{v}_N)$:

$$\frac{a_2}{a_N} = \sqrt{\left(\frac{\lambda_1[\mathbf{H}(\mathbf{v}_N)]}{\lambda_{N-1}[\mathbf{H}(\mathbf{v}_N)]}\right)} = \sqrt{(\xi[\mathbf{H}(\mathbf{v}_N)])}. \qquad (5.30)$$

A similar result has been shown to hold in the CG solution of linear systems (where $\mathbf{H}$ is replaced by the coefficient matrix).

By taking the second partial derivatives of $R(\mathbf{x})$ and recalling expression (5.19) for the gradient $\mathbf{g}(\mathbf{x})$ of $R(\mathbf{x})$, we readily obtain

$$\mathbf{H}(\mathbf{x}) = \frac{2}{\mathbf{x}^T \mathbf{B} \mathbf{x}}[\mathbf{A} - R(\mathbf{x})\mathbf{B} - \mathbf{B}\mathbf{x}[\mathbf{g}(\mathbf{x})]^T - \mathbf{g}(\mathbf{x})\mathbf{x}^T \mathbf{B}] \qquad (5.31)$$

$$\mathbf{H}(\mathbf{v}_N) = 2(\mathbf{A} - \lambda_N \mathbf{B}). \qquad (5.32)$$

It may be easily verified that $\mathbf{A} - \lambda_N \mathbf{B}$ is positive semi-definite with eigenvector $\mathbf{v}_N$ corresponding to the zero eigenvalue.

When the preconditioned CG method is used, optimization is performed on $R_1(\mathbf{y})$ whose Hessian is similar to (5.32):

$$\mathbf{H}(\mathbf{y}_N) = 2(\mathbf{G} - \lambda_N \mathbf{F}) \tag{5.33}$$

where $\mathbf{y}_N = \mathbf{X}\mathbf{v}_N$. Asymptotic convergence is now controlled by the square root of the spectral condition number of $\mathbf{G} - \lambda_N \mathbf{F}$. By a similarity transformation we obtain for the eigenvalues μ of $\mathbf{G} - \lambda_N \mathbf{F}$:

$$\mu(\mathbf{G} - \lambda_N \mathbf{F}) = \mu(\mathbf{X}^{-1}\mathbf{A}\mathbf{X}^{-1} - \lambda_N \mathbf{X}^{-1}\mathbf{B}\mathbf{X}^{-1})$$
$$= \mu[\mathbf{K}^{-1}(\mathbf{A} - \lambda_N \mathbf{B})]. \tag{5.34}$$

Hence convergence for preconditioned CG may be analyzed by $\xi(\mathbf{H}_1)$ where $\mathbf{H}_1$ reads:

$$\mathbf{H}_1 = \mathbf{K}^{-1}(\mathbf{A} - \lambda_N \mathbf{B}). \tag{5.35}$$

For $\mathbf{K}^{-1} = \mathbf{I}$, $\mathbf{H}_1$ obviously becomes $\frac{1}{2}\mathbf{H}(\mathbf{v}_N)$. Note that $\mathbf{H}_1$ is a singular matrix and therefore $\xi(\mathbf{H}_1)$ is to be formed again by the use of its leftmost (different from zero) eigenvalue.

5.4 PRECONDITIONING

The purpose of preconditioning the classic GC method is to round the hyper-ellipsoid associated with the Hessian of $R(\mathbf{x})$, thus leading to a reduction of the Hessian condition number.

Let us start by showing that an excellent preconditioner would be $\mathbf{K}^{-1} = \mathbf{A}^{-1}$. In fact in this case $\mathbf{H}_1$ becomes

$$\mathbf{H}_1 = \mathbf{I} - \lambda_N \mathbf{A}^{-1}\mathbf{B} \tag{5.36}$$

which possesses the same eigenvectors as eigenproblem (5.1) and eigenvalues $\mu_i = 1 - \lambda_N/\lambda_i$, $i = 1, 2, \ldots$. Hence we have

$$\mu_1 = 1 - \lambda_N/\lambda_1$$
$$\mu_{N-1} = 1 - \lambda_N/\lambda_{N-1}$$
$$\xi(\mathbf{H}_1) = \frac{\lambda_1 - \lambda_N}{\lambda_{N-1} - \lambda_N} \frac{\lambda_{N-1}}{\lambda_1}. \tag{5.37}$$

Assume $\mathbf{B} = \mathbf{I}$ and denote $\mathbf{H}(\mathbf{v}_N)$ simply by $\mathbf{H}$. Then

$$\xi(\mathbf{H}) = \frac{\lambda_1 - \lambda_N}{\lambda_{N-1} - \lambda_N}$$

$$\xi(\mathbf{H}_1)/\xi(\mathbf{H}) = \lambda_{N-1}/\lambda_1 \ll 1$$

since for large finite element matrices $\mathbf{A}$ arising from engineering practice the characteristic values are scattered in a wide spectral interval with $\lambda_{N-1} \ll \lambda_1$. If $\mathbf{B} \neq \mathbf{I}$ this simple result no longer holds, but experience indicates that in any case $\xi(\mathbf{H}_1)$ is much smaller than $\xi(\mathbf{H})$ and is not much larger than 1 which is the theoretically best value (see Table 5.1 in Section 5.6).

In practical applications for moderate to large values of N, computation of $\mathbf{A}^{-1}$ is too expensive and we may use instead an approximation to $\mathbf{A}^{-1}$, as is effectively provided by one of the several incomplete Cholesky decompositions of $\mathbf{A}$ (Evans [64]; Tuff and Jennings [65]; Axelsson [66]; Jennings and Malik [67]; Gustafsson [68, 69]; Kershaw [52]; Manteuffel [55]; Eisenstat [70]; Appleyard and Cheshire [71]; Axelsson and Gustaffsson [72]; Jacobs [73]; Jennings [74]; Ajiz and Jennings [75]; Nour-Omid [76]; Jackson and Robinson [77]; Tismenetsky and Efrat [78]; Wong et al. [79]; Zyvoloski [80]).

Although sophisticated polynomial (Johnson et al. [81]; Saad [82]) and multistep (Adams [83]) preconditioners have been developed, mainly for the efficient implementation of parallel algorithms, the diagonal scaling and the pointwise incomplete Cholesky factorization of Kershaw [52], referred to as ICCG(0) by Meijerink and van der Vorst [51], represent two of the most cost-effective choices in an FE context, as was also shown by Gambolati et al. [84] and Pini and Gambolati [85].

The preconditioner implemented and experimented with in the present contribution is ICCG(0), which turned out to be a very reliable and efficient preconditioner for the partial FE eigensolution on scalar machines (Gambolati et al. [84]).

The preconditioning matrix $\mathbf{K}^{-1}$ is expressed as:

$$\mathbf{K}^{-1} = (\mathbf{L}\mathbf{L}^{\mathrm{T}})^{-1} \tag{5.38}$$

$\mathbf{L}$ being the lower triangular factor obtained from the incomplete factorization of $\mathbf{A}$. The sparsity pattern of $\mathbf{L}$ is chosen to match with that of the lower part of $\mathbf{A}$. Extensive experience (Gambolati and Perdon [56]) indicates that $\xi\,[\mathbf{A}(\mathbf{L}\mathbf{L}^{\mathrm{T}})^{-1}]$ may be several orders of magnitude smaller than $\xi(\mathbf{A})$, thus providing empirical evidence that (5.38) is to be regarded as a good (and inexpensive) approximation to $\mathbf{A}^{-1}$.

Our disadvantage of the incomplete factorization as it is used in the present analysis is that its success is not theoretically guaranteed for FE matrices with an arbitrary sparsity structure (negative roots may result). However, over several years of intensive use of CG methods, this inconvenience has occurred only once, in connection with a severely ill-conditioned stiffness matrix.

5.5 LANCZOS METHOD

The Lanczos method can be thought of as a recursive scheme to construct a $\mathbf{B}$-orthogonal vector basis $\mathbf{q}_1, \ldots, \mathbf{q}_N$, known as the Lanczos vector basis, for the Krylov space:

$$\sigma_N(\mathbf{q}_1) = \mathrm{span}\,[\mathbf{q}_1, \mathbf{A}^{-1}\mathbf{B}\mathbf{q}_1, (\mathbf{A}^{-1}\mathbf{B})^2\mathbf{q}_1, \ldots, (\mathbf{A}^{-1}\mathbf{B})^{N-1}\mathbf{q}_1]. \tag{5.39}$$

In 1950 Lanczos [17] proposed the use of these vectors in a Rayleigh–Ritz procedure to solve the generalized symmetric eigenproblem (5.1).

The process starts from any arbitrary vector $\mathbf{q}_1 = \mathbf{r}_0$ and makes use of the recursive relationship:

$$\gamma_{j+1}\mathbf{q}_{j+1} = \mathbf{r}_j = \mathbf{A}^{-1}\mathbf{B}\mathbf{q}_j - \alpha_j\mathbf{q}_j - \beta_j\mathbf{q}_{j-1}, \qquad 1 \leq j \leq N-1 \qquad (5.40)$$

with $\beta_1 = 0$.

If we multiply both sides of equation (5.40) by $\mathbf{q}_j^T\mathbf{B}$ and $\mathbf{q}_{j-1}^T\mathbf{B}$ and require $\mathbf{q}_{j+1}$ to be **B**-orthogonal to both $\mathbf{q}_j$ and $\mathbf{q}_{j-1}$, we obtain the following expression for α_j and β_j:

$$\alpha_j = \mathbf{q}_j^T\mathbf{B}\mathbf{A}^{-1}\mathbf{B}\mathbf{q}_j / \mathbf{q}_j^T\mathbf{B}\mathbf{q}_j \qquad (5.41)$$

$$\beta_j = \mathbf{q}_{j-1}^T\mathbf{B}\mathbf{A}^{-1}\mathbf{B}\mathbf{q}_j / \mathbf{q}_{j-1}^T\mathbf{B}\mathbf{q}_{j-1}. \qquad (5.42)$$

The parameter γ_{j+1} is defined arbitrarily and may be used to increase stability, for instance by **B**-normalizing the vector $\mathbf{q}_{j+1}$, which is accomplished by setting

$$\gamma_{j+1} = \sqrt{(\mathbf{r}_j^T\mathbf{B}\mathbf{r}_j)}. \qquad (5.43)$$

It is easily shown that the new Lanczos vector $\mathbf{q}_{j+1}$ supplied by (5.40) with α_j and β_j given by (5.41) and (5.42) is **B**-orthogonal to $\mathbf{q}_{j-2}, \ldots, \mathbf{q}_1$ as well. Because of the **B**-orthogonality of the vectors $\mathbf{q}_j$, the Lanczos process terminates for some $j = m \leq N$.

With the parameters γ_{j+1}, α_j and β_j, we form the tridiagonal matrix **T**:

$$\mathbf{T}_m = \begin{pmatrix} \alpha_1 & \beta_2 & & & & \\ \gamma_2 & \alpha_2 & \beta_3 & & & \\ & \gamma_3 & \alpha_3 & \beta_4 & & \\ & & \ddots & \ddots & \ddots & \\ & & & & & \beta_N \\ & & & & \gamma_N & \alpha_N \end{pmatrix}. \qquad (5.44)$$

It may be recognized that equations (5.40) can be written in matrix form as

$$\mathbf{A}^{-1}\mathbf{B}\mathbf{Q} = \mathbf{Q}\mathbf{T} \qquad (5.45)$$

where **Q** is the Lanczos vector matrix $\mathbf{Q} = (\mathbf{q}_1, \mathbf{q}_2, \ldots, \mathbf{q}_N)$. Premultiplication of both sides of (5.45) by $\mathbf{Q}^T\mathbf{B}$ yields

$$\mathbf{Q}^T\mathbf{B}\mathbf{A}^{-1}\mathbf{B}\mathbf{Q} = \mathbf{Q}^T\mathbf{B}\mathbf{Q}\mathbf{T} = \mathbf{T} \qquad (5.46)$$

since $\mathbf{Q}^T\mathbf{B}\mathbf{Q} = \mathbf{I}$. Hence **T** is obtained from $\mathbf{A}^{-1}\mathbf{B}$ through a similarity transformation and has the same eigenvalues as $\mathbf{A}^{-1}\mathbf{B}$. These are the inverse eigenvalues of equation (5.1). Concerning the eigenfactors, if **y** is an eigenvector of **T** corresponding to the eigenvalue μ, we have

$$\mathbf{Ty} = \mathbf{Q}^T\mathbf{BA}^{-1}\mathbf{BQy} = \mu\mathbf{y} \tag{5.47}$$

and premultiplication of both sides by $\mathbf{Q}$ gives

$$\mathbf{A}^{-1}\mathbf{B}(\mathbf{Qy}) = \mu(\mathbf{Qy}) \tag{5.48}$$

i.e. eigenvector $\mathbf{v}$ of (5.1) is linked to eigenvector $\mathbf{y}$ of $\mathbf{T}$ by the equation

$$\mathbf{v} = \mathbf{Qy}. \tag{5.49}$$

It has been recognized by several authors (see for instance Bathe [86] and Hughes [87]) that the extreme (largest) eigenvalues of the upper leading minor $\mathbf{T}_m$ of $\mathbf{T}$ represent very accurate approximations of the inverse of the smallest eigenvalues of $\mathbf{Av} = \lambda\mathbf{Bv}$ even for values of $m \ll N$.

Thus in practice the Lanczos process may be terminated in much less than N steps. According to equation (5.46) the reduced matrix $\mathbf{T}_m$ is obtained by projecting $\mathbf{BA}^{-1}\mathbf{B}$ on the space spanned by the first m Lanczos vectors $\mathbf{Q}_m$:

$$\mathbf{T}_m = \mathbf{Q}_m^T\mathbf{BA}^{-1}\mathbf{BQ}_m. \tag{5.50}$$

The eigenpairs $\mu^{(m)}$ and $\mathbf{y}^{(m)}$ of $\mathbf{T}_m$ are computed by available standard routines and those of (5.1) are evaluated by the equations

$$\lambda_{N-i+1} = \frac{1}{\mu_i^{(m)}} \qquad i = 1, 2, \ldots. \tag{5.51}$$

$$\mathbf{v}_{N-i+1} = \mathbf{Q}_m \mathbf{y}_i^{(m)}$$

In the Paige-style Lanczos variant [18] the parameter γ_{j+1} is defined by equation (5.43). This implies that the Lanczos vector $\mathbf{q}_{j+1}$ is $\mathbf{B}$-normalized, i.e. $\mathbf{q}_{j+1}^T\mathbf{Bq}_{j+1} = 1$. and β_j simplifies to

$$\beta_j = \mathbf{q}_{j-1}^T\mathbf{BA}^{-1}\mathbf{Bq}_j = \mathbf{q}_j^T\mathbf{BA}^{-1}\mathbf{Bq}_{j-1} \tag{5.52}$$
$$= \mathbf{q}_j^T\mathbf{B}(\gamma_j\mathbf{q}_j + \alpha_{j-1}\mathbf{q}_{j-1} + \beta_{j-1}\mathbf{q}_{j-2}) = \gamma_j.$$

Thus matrix $\mathbf{T}$ becomes now symmetric and the upper leading minor $\mathbf{T}_m$ which interests us here reads

$$\mathbf{T}_m = \begin{pmatrix} \alpha_1 & \beta_2 & & & & \\ \beta_2 & \alpha_2 & \beta_3 & & & \\ & \beta_3 & \alpha_3 & \beta_4 & & \\ & & \ddots & \ddots & & \\ & & & & & \beta_m \\ & & & & \beta_m & \alpha_m \end{pmatrix}. \tag{5.53}$$

The Paige-style Lanczos process may be written as

$$\beta_{j+1}\mathbf{q}_{j+1} = \mathbf{r}_j = \mathbf{A}^{-1}\mathbf{B}\mathbf{q}_j - \alpha_j\mathbf{q}_j - \beta_j\mathbf{q}_{j-1} \qquad j = 1, 2, \ldots \tag{5.54}$$

with

$$\beta_1 = 0, \quad \mathbf{q}_1 = \mathbf{r}_0/\sqrt{(\mathbf{r}_0^T\mathbf{B}\mathbf{r}_0)} \quad \text{and} \quad \alpha_j = \mathbf{q}_j^T\mathbf{B}\mathbf{A}^{-1}\mathbf{B}\mathbf{q}_j \tag{5.55}$$

$$\beta_{j+1} = \mathbf{q}_{j+1}^T\mathbf{B}\mathbf{r}_j = \sqrt{(\mathbf{r}_j^T\mathbf{B}\mathbf{r}_j)}. \tag{5.56}$$

If the starting vector $\mathbf{r}_0$ is **B**-orthogonal to s eigenvectors of equation (5.1), in exact arithmetic the sequence (5.54) leads to $\mathbf{q}_{j+1} = 0$ with $j = N - s$. The same holds true if the matrix pencil **A, B** has s multiple eigenvalues.

When we find $\mathbf{q}_{j+1} = 0$, the Lanczos process breaks down and consequently the order of **T** can be smaller than N. As a major consequence **T** (and hence $\mathbf{T}_m$) does not allow for detection of multiple eigenvalues of the eigenproblem (5.1). In practical calculations undetectable multiplicity occurs at machine accuracy. Experience indicates that, using double precision arithmetic, no difficulties arise in distinguishing between eigenvalues which are multiple up to eight significant decimal digits.

In actual computations the main difficulty with the Lanczos scheme is a progressive loss of **B**-orthogonality of the currently generated vector with respect to earlier vectors due to round-off errors and cancellation. A more accurate matrix $\mathbf{T}_m$ might be produced by **B**-reorthogonalizing $\mathbf{q}_{j+1}$ against $\mathbf{q}_j$ (and $\mathbf{q}_{j-1}$) in each step. It is suggested by van Kats and van der Vorst [88] that in most cases it is sufficient to **B**-reorothogonalize only $\mathbf{q}_2$ against $\mathbf{q}_1$. However a more systematic approach to monitor the loss of **B**-orthogonality has been suggested by Simon [89] by computing at each iteration of the Lanczos process the quantity $\eta_{j+1,i} = \mathbf{q}_{j+1}^T\mathbf{B}\mathbf{q}_i$. Setting

$$\beta_{i+1}\mathbf{q}_{i+1} = \mathbf{A}^{-1}\mathbf{B}\mathbf{q}_i - \alpha_i\mathbf{q}_i - \beta_i\mathbf{q}_{i-1} \tag{5.57}$$

left-multiplying equations (5.54) and (5.57) by $\mathbf{q}_i^T\mathbf{B}$, and $\mathbf{q}_j^T\mathbf{B}$, respectively, and then subtracting, yields

$$\beta_{j+1}\eta_{j+1,i} = \beta_{i+1}\eta_{j,i+1} + (\alpha_i - \alpha_j)\eta_{j,i} + \beta_i\eta_{j,i-1} - \beta_j\eta_{j-1,i}$$

$$j = 2, 3, \ldots; \quad i = j - 1, \ldots, 2, 1; \quad \eta_{j,j} = 1; \quad \eta_{j,j-1} = \varepsilon \tag{5.58}$$

where ε denotes the unit round-off error of the computer in use and $\eta_{j,0} = 0$. Note that $\eta_{j,i} = \eta_{i,j}$.

The recursion formula (5.58) can be viewed as a scheme to construct, row by row, the $(j+1) \times (j+1)$ symmetric matrix:

$$\begin{pmatrix} 1 & & & & \\ \varepsilon & 1 & & & \\ \eta_{3,1} & \varepsilon & 1 & & \\ \vdots & \vdots & & \ddots & \\ \eta_{j+1,1} & \eta_{j+1,2} & \cdots & \varepsilon & 1 \end{pmatrix}. \tag{5.59}$$

If $\eta_{j+1,i}$ is found to be larger than the threshold value $\sqrt{\varepsilon}$, loss of **B**-orthogonality has occurred and the newly generated Lanczos vector $\mathbf{q}_{j+1}$ is **B**-orthogonalized against the preceding vectors $\mathbf{q}_i, \mathbf{q}_{i-1}, \ldots, \mathbf{q}_1$ using a Gram–Schmidt type procedure. The partial **B**-reorthogonalization just described is used by Nour-Omid and Clough [90], Hughes [87], Coutinho *et al.* [91] and Dunbar and Woodbury [92]. Other techniques meant to preserve **B**-orthogonality have been discussed by Paige [93], Parlett and Scott [94] and Cullum and Willoughby [95].

Borrowing a result from the selective orthogonalization of Parlett and Scott [94], Simon [89] suggests that whenever $\mathbf{q}_{j+1}$ is **B**-reorthogonalized against all previous vectors $\mathbf{q}_i$, then also $\mathbf{q}_{j+2}$ should be **B**-reorthogonalized since 'one reorthogonalization by itself does not help very much to reduce the size of $\mathbf{q}_{j+2}^T \mathbf{B} \mathbf{q}_i$' (Simon [89], p. 126). In addition Simon [89] gives some indication of how to reduce the number of previous vectors against which $\mathbf{q}_{j+1}$ and $\mathbf{q}_{j+2}$ should be **B**-reorthogonalized. This would require the introduction of a new parameter η controlling the size of the neighbourhood of $\eta_{j+1,i}$ which has been found larger than η, and would make the management of partial **B**-reorthogonalization quite laborious.

In view of the fact that partial (in place of full) **B**-reorthogonalization saves only a very modest fraction of the overall processing time required to generate the Lanczos sequence (5.54), where most of the cost is associated with performing the product between $\mathbf{A}^{-1}$ and $\mathbf{B}\mathbf{q}_j$, Simon's [89] suggestion of locating an appropriate interval of $\eta_{j+1,i}$ is not worth the additional programming effort. Therefore in the examples that follow, when conditions needed for **B**-reorthogonalization are satisfied, $\mathbf{q}_{j+1}$ as well as $\mathbf{q}_{j+2}$ are fully **B**-reorthogonalized against all the previous vectors.

Finally, note that whenever **B**-reorthogonalization occurs relationship (5.58) breaks down and both $\eta_{j+1,i}$ and $\eta_{j+2,i}$ are to be set equal to ε (actually equal to the products $\mathbf{q}_{j+1}^T \mathbf{B} \mathbf{q}_i$, and $\mathbf{q}_{j+2}^T \mathbf{B} \mathbf{q}_i$, but this is expensive and is not done in practice).

We conclude this section by noting that block versions of the Lanczos algorithm have also been developed (Cullum and Donath [21] and Golub and Underwood [22]). These block generalizations lead to a block tridiagonal matrix and are more powerful than the pointwise Lanczos process, but the mathematics involved is more complicated and the algorithms are computationally more expensive. According to Nour-Omid *et al.* [59] small blocks (of size 1, 2 or 3) are the most convenient (see Grimes *et al.* [96] for successful block implementations).

5.6 NUMERICAL RESULTS

Preconditioned CG and Lanczos algorithms as described in the previous sections are used to partially solve five generalized eigenproblems of increasing size N up to almost 5000. All examples arise from the FE integration of the two- or three-dimensional diffusion equation, and the resulting matrices have an arbitrary sparsity structure. The matrix dimensions are $N = 222, 441, 812, 1952$ and 4560 with degrees of sparsity 96.9, 98.5, 99.2, 99.6 and 99.7%, respectively.

The distribution of the 40 leftmost eigenvalues is shown in Figure 5.1. Note that for the $N = 222$ and $N = 1952$ cases the characteristic values occur in well-separated small clusters, while in the other sample cases the eigenspectrum is more uniform with its density increasing rightward. The lengths of the partial eigenintervals are quite different for the five examples: $\lambda_{N-40}/\lambda_N = 240.9$ ($N = 222$), 205.8 ($N = 441$), 20636.4 ($N = 812$), 40.1 ($N = 1952$) and 14.8 ($N = 4560$). The much larger (partial) condition number occurring for $N = 812$ is a probable indication that this problem is ill-conditioned, as will be discussed later.

We start by analysing the convergence behaviour of the CG scheme. As was shown in Section 5.3, the asymptotic convergence toward λ_N and $\mathbf{v}_N$ is controlled by $\xi(\mathbf{H}_1)$ with the Hessian $\mathbf{H}_1$ given by equation (5.35).

The condition numbers of $\mathbf{H}_1$ are supplied in Table 5.1 for three choices of preconditioner: (a) $\mathbf{K}^{-1} = \mathbf{I}$ (no preconditioning, i.e. $\mathbf{H}_1 = 1/2\mathbf{H}$); $\mathbf{K}^{-1} = (\mathbf{L}\mathbf{L}^T)^{-1}$ (pointwise incomplete Cholesky preconditioner), and (c) $\mathbf{K}^{-1} = \mathbf{A}^{-1}$ (best possible preconditioner, but expensive). Table 5.1 also gives the ξ values of matrices $\mathbf{A}(\mathbf{L}\mathbf{L}^T)^{-1}$ and $\mathbf{B}^{-1}\mathbf{A}$.

Note in Table 5.1 that for $\mathbf{K}^{-1} = \mathbf{A}^{-1} \xi(\mathbf{H}_1)$ is very close to 1 and thus a very high rate of convergence should be expected in this case. Also note that the condition number of each eigenproblem is quite large (up to 7.1×10^9 for $N = 1952$) and $\xi(\mathbf{H}_1)$ with $\mathbf{K}^{-1} = (\mathbf{L}\mathbf{L}^T)^{-1}$ is smaller than $\xi[\mathbf{A}(\mathbf{L}\mathbf{L}^T)^{-1}]$, i.e. smaller than the condition number of the CG iteration matrix.

Figure 5.2 shows the convergence profiles in the calculation of the minimal eigenpair with $\mathbf{K}^{-1} = (\mathbf{L}\mathbf{L}^T)^{-1}$ and a starting vector $\mathbf{x}$ set equal to the one adopted by Sartoretto et al.[81], i.e. $\mathbf{x}_j = 0$ for $2 \leq j \leq 40$, and $\mathbf{x}_j = 1$ otherwise.

Several features of Figure 5.2 are worth stressing. First, convergence is not monotonic. Second, convergence is faster than in the preconditioned CG execution of $\mathbf{A}^{-1}\mathbf{B}\mathbf{q}_j$ (as is accomplished in the present Lanczos algorithm, see below). Third, the behaviour of the profiles indicates a good correlation with $\xi(\mathbf{H}_1)$ (Table 5.1). Fourth, a value of r_r on the order of 10^{-4}, as is finally achieved in Figure 5.2, represents a quite accurate solution for both the eigenvalue and the eigenvector. Experience indicates that a larger r_r (between 10^{-3} and 5×10^{-3}) may also provide satisfactory results up to the 40th leftmost eigenpair for all of the problems.

As the deflation proceeds the Hessian becomes different from that of equation (5.35) and thus the rate of convergence towards the higher eigenpairs may deviate significantly from the profiles of Figure 5.2. This may be seen in Table 5.2 (left columns) giving the number k of iterations required to meet the exist test $r_r = 6 \times 10^{-3}$ in the computation of each eigenpair.

Initial guess vectors are again the same as those used by Sartoretto et al. [48]. Table 5.2 reveals that k does not increase monotonically with the eigenvalue level j and exhibits quite irregular behaviour. Note that for some eigenpairs convergence slows down significantly, thus leading to a degradation of the overall performance of the deflation CG algorithm. Occasional low convergence may be accounted for by the non-monotonic behaviour of r_r. It may happen that r_r (as well as e_r) originally

142 SOLUTION OF LARGE EIGENVALUE PROBLEMS

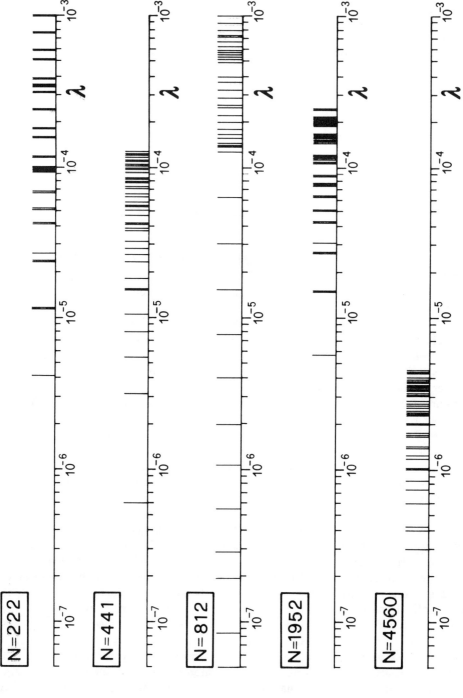

Figure 5.1 Distribution of the 40 leftmost eigenvalues for the various sample problems.

5.6 NUMERICAL RESULTS

Table 5.1 Spectral condition number of the Hessian $\mathbf{H}_1$ for various preconditioning matrices $\mathbf{K}^{-1}$. The condition number of $\mathbf{A}(\mathbf{LL}^T)^{-1}$ and that of the original eigenproblem are also given

		$\xi(\mathbf{H}_1)$			
N	$\mathbf{K}^{-1} = \mathbf{I}$ (i.e. $\mathbf{H}_1 = \frac{1}{2}\mathbf{H}$)	$\mathbf{K}^{-1} = (\mathbf{LL}^T)^{-1}$	$\mathbf{K}^{-1} = \mathbf{A}^{-1}$	$\xi[\mathbf{A}(\mathbf{LL}^T)^{-1}]$	$\xi(\mathbf{B}^{-1}\mathbf{A})$
222	2.2×10^2	1.7×10^1	1.6	7.8×10^1	1.5×10^9
441	3.8×10^2	2.9×10^1	1.3	1.2×10^2	4.6×10^3
812	4.4×10^6	7.6×10^2	2.3	6.9×10^3	1.4×10^8
1952	2.1×10^3	2.7×10^2	1.6	1.3×10^3	7.1×10^9
4560	8.3×10^5	5.6×10^1	3.8	8.8×10^1	3.3×10^6

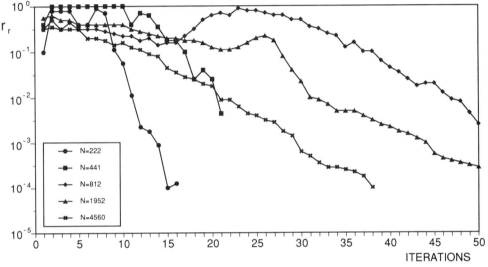

Figure 5.2 Convergence profiles in the computation of the smallest eigenpair for the various sample problems.

decrease to a minimum value, indicating that a stationary point of $R(\mathbf{x})$ is achieved. If the iteration is continued, r_r and e_r initially rise and then decrease again to finally meet the exit test at the smallest optimal point of $R(\mathbf{x})$. This mechanism may require a large number of iterations if a very small TOLL is prescribed (Table 5.2, right-hand columns).

By contrast if TOLL is set up at an appropriate value the CG iteration may be completed at a characteristic value which is not the optimal one, yet is very close to it. In this way the deflation procedure and the CG scheme converge to some eigenpair which is not minimal at current level j but is minimal at level $j+1$ or $j+2$ and a local reordering may be in order at the end of the calculation. This behaviour was indeed observed for $r_r = 6 \times 10^{-3}$ in the sample problems with $N = 222$ and 1952 during eight and five deflation steps, respectively, and led to a great saving of computer time. Table 5.2 (right-hand columns) emphasizes that a

Table 5.2 Number of iterations k required by the deflation-preconditioned CG approach to meet the acceptability criterion ($r_r = 6 \times 10^{-3}$ or $e_r = 10^{-8}$) in the evaluation of the 40 leftmost eigenpairs for the various sample problems. Values of k greater than 100 are printed in boldface

Level j	N=222		441		812		1952		4560	
1	12	16	21	21	49	55	39	61	24	39
2	11	19	34	44	23	32	39	55	24	50
3	12	17	16	21	25	32	45	59	17	31
4	28	40	25	33	17	26	49	98	15	34
5	8	55	21	27	16	26	72	**117**	17	40
6	7	16	13	**208**	15	24	28	42	23	37
7	11	22	33	35	15	25	22	49	39	**142**
8	22	34	25	33	14	24	27	64	20	37
9	13	22	29	44	13	23	34	52	24	52
10	18	28	35	55	12	20	42	66	39	51
11	13	22	51	74	11	24	**118**	**274**	20	73
12	20	30	9	31	9	23	30	59	19	35
13	63	**208**	24	34	23	60	90	**140**	**116**	**283**
14	79	**339**	35	66	15	**272**	26	**200**	16	60
15	71	**133**	99	**232**	**122**	**100**	19	39	23	40
16	75	**171**	23	39	42	99	16	44	18	67
17	13	15	31	50	83	55	25	51	25	42
18	10	21	32	62	64	44	31	71	83	**167**
19	25	31	28	**439**	47	37	69	**107**	51	**105**
20	12	25	66	61	40	35	59	91	40	**104**
21	30	47	31	53	33	33	65	**110**	50	86
22	14	21	33	96	99	**299**	21	50	43	63
23	22	27	34	52	24	36	35	60	31	65
24	16	25	39	77	16	31	42	89	30	48
25	23	31	97	**113**	14	31	59	**113**	44	67
26	70	**126**	87	**263**	29	58	**136**	**264**	**125**	**176**
27	12	**159**	12	43	72	28	23	**306**	45	92
28	68	**329**	**108**	**213**	**136**	**166**	66	41	23	71
29	10	14	24	42	**108**	36	13	38	45	**116**
30	9	26	61	94	23	**181**	21	43	71	**169**
31	10	19	**105**	**310**	**111**	**136**	67	**143**	39	**214**
32	16	34	17	37	36	**103**	85	**187**	87	**193**
33	30	33	25	**500**	47	59	31	69	25	50
34	28	49	37	58	70	65	81	**242**	83	**237**
35	20	20	41	78	44	42	**151**	**346**	39	77
36	17	27	59	**141**	**115**	**190**	**112**	**196**	45	94
37	13	25	32	**375**	86	57	21	**205**	27	53
38	20	31	34	69	20	38	31	50	49	**500**
39	67	**129**	36	88	34	49	58	**102**	89	**108**
40	50	**162**	26	73	**111**	**184**	95	**157**	25	**341**

very small tolerance ($e_r = 10^{-8}$) yields the correct eigenpair at each level, at the expense, however, of an increased number of iterations.

The Hessian condition numbers (Table 5.1) seem to suggest that convergence

Table 5.3 Number of iterations k required to meet the exit test ($e_r = 10^{-8}$) in the evaluation of the 20 leftmost eigenpairs for the sample problem with $N = 222$, using the two different preconditioners

Level j	$\mathbf{K}^{-1} = (\mathbf{LL}^T)^{-1}$	$\mathbf{K}^{-1} = \mathbf{A}^{-1}$
1	16	4
2	19	4
3	17	6
4	40	13
5	55	11
6	16	4
7	22	17
8	34	21
9	22	12
10	28	30
11	22	42
12	30	19
13	208	43
14	339	228
15	133	23
16	171	15
17	15	158
18	21	25
19	31	44
20	25	29

Table 5.4 Number of CG iterations required to meet the exit test ($e_r = 10^{-8}$) in the evaluation of the three leftmost eigenpairs for various sample problems, using the two different preconditioners

	\multicolumn{8}{c}{N}							
	441		812		1952		4560	
Level j	$(\mathbf{LL}^T)^{-1}$	$\mathbf{A}^{-1}$	$(\mathbf{LL}^T)^{-1}$	$\mathbf{A}^{-1}$	$(\mathbf{LL}^T)^{-1}$	$\mathbf{A}^{-1}$	$(\mathbf{LL}^T)^{-1}$	$\mathbf{A}^{-1}$
1	21	5	55	9	61	6	39	14
2	44	14	32	8	55	13	50	47
3	21	7	32	13	59	8	31	15

might be greatly improved by the use of $\mathbf{A}^{-1}$ as the CG preconditioner. In fact the CG iterations are reduced, as may be seen from Tables 5.3 and 5.4. On balance there is no computational advantage, however, since the increase in the rate of convergence is totally offset by the greater expenditure needed to compute $\mathbf{K}^{-1}\mathbf{g}_k = \mathbf{A}^{-1}\mathbf{g}_k$ in equation (5.20). This product is performed by the CG scheme preconditioned with $(\mathbf{LL}^T)^{-1}$ (Gambolati and Perdon [56]) and requires a number of iterations which is not counterbalanced by the fewer CG iterations of the eigenvalue calculation.

As far as the storage requirement is concerned, apart from the core memory needed for $\mathbf{A}$, $\mathbf{B}$, $\mathbf{L}$, stored in compressed mode, and for working vectors $\tilde{\mathbf{p}}_k, \mathbf{x}_k, \mathbf{g}_k$, the deflation CG procedure basically requires extra storage for p eigenvectors.

Now let us consider the Lanczos method. We recall that in the Lanczos recursive

Table 5.5 Average number of CG iterations required to determine the product $\mathbf{A}^{-1}\mathbf{g}_k$ of equation (5.18) or the product $\mathbf{A}^{-1}\mathbf{Bq}_j$ of equation (5.54) at machine accuracy

N	Iterations
222	25
441	41
812	53
1952	90
4560	75

equation the product between $\mathbf{A}^{-1}$ and the vector $\mathbf{Bq}_j$ is obtained as the solution to the linear system $\mathbf{Ax} = \mathbf{Bq}_j$ via the preconditioned CG scheme (Gambolati and Perdon [56]) with quite fast convergence (Table 5.5). Hence in the present analysis the Lanczos approach actually incorporates the CG algorithm, which may allow for the cost-effective treatment of very large matrices with an arbitrary sparsity structure (e.g. the problem with $N = 4560$).

When analysing the Lanczos technique a few considerations must be kept in mind. First, the Lanczos approach is not iterative in the classical sense. Second, convergence to the leftmost eigenpairs in the Lanczos subspace $\mathbf{Q}_m$ occurs in a much more irregular manner than it does with the deflation CG method. Some eigenvector can quickly be found in the subspace $\mathbf{Q}_m$ while others may have an incorrect position within the partial eigenspectrum. Third, the only representative accuracy measure is the relative residual r_r.

With the Lanczos method there is no way to control the magnitude of r_r against m as could be done in a classical iterative scheme. Hence, depending on machine accuracy, as m increases, r_r for some eigenpair may become much smaller than is actually required. In light of the above, and to make a meaningful comparison with the deflation CG approach, any eigenpair satisfying the acceptability criterion (i.e. a prescribed r_r value) should be accepted, regardless of the fact that some Lanczos results may be much more accurate.

Figure 5.3 shows the convergence of the Lanczos method vs m for the problems with $N = 222$ (Figure 5.3(a)), 1952 (Figure 5.3(b)), and 4560 (Figure 3(c)). In all of the examples $\mathbf{q}_1$ is set equal to $\mathbf{A}^{-1}[1, \ldots, 1]^T$ and is $\mathbf{B}$-normalized. Figures 5.3(a,b) reveal that r_r may display non-monotonic convergence and, more importantly, that the number m_1 of eigenvectors which satisfy the tolerance for a given size m of the Lanczos subspace are not necessarily the m_1 smallest eigenvectors of (5.1). As may be seen from Figure 5.3(a), v_{N-5} and v_{N-16} are determined prior to v_{N-1}, v_{N-2}, v_{N-3}, and v_{N-4}, while v_{N-3} is determined prior to v_{N-2}. Note that in Table 5.6 the three largest eigenvalues of $\mathbf{T}_6$ for the problem with $N = 222$ give an excellent approximation of λ_N, λ_{N-5} and λ_{N-16}. From the foregoing, it turns out that we may have some difficulty recognizing whether the eigenpairs which satisfy the acceptability criterion are actually the leftmost ones. Also note in Figures 5.3(a) and 5.3(b) that an eigenpair determined with good accuracy is not necessarily stable, and the same eigenpair may be affected by a much larger error in a higher subspace $\mathbf{Q}_m$ (in Figure 5.3(a) $r_r(v_{N-1})$ is 3×10^{-12} at $m = 15$ and 2×10^{-3} at $m = 18$). A major practical

5.6 NUMERICAL RESULTS

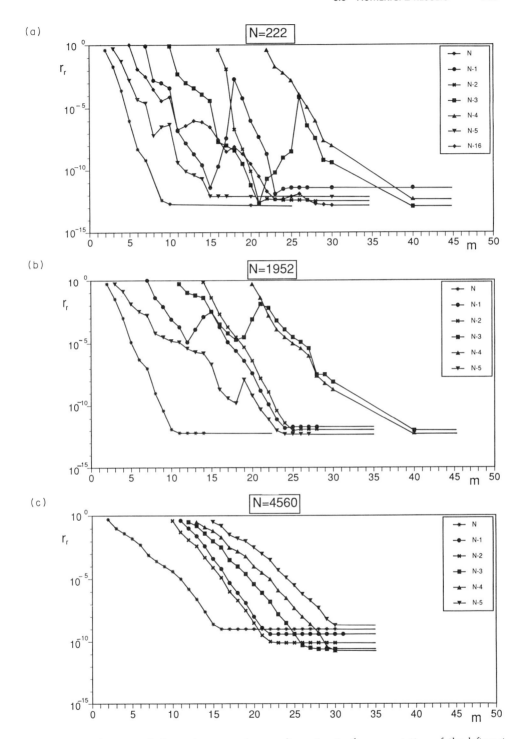

Figure 5.3 Relative residuals vs. Lanczos subspace dimension in the computation of the leftmost eigenpairs by the Lanczos methods for three sample problems.

Table 5.6 Eigenvalue estimate and relative eigenvector error r_r by the Lanczos method with $m = 6$ for the sample problem with $N = 222$

j	$1/\mu^{(6)}_{7-j}$	r_r
1	$0.219\ 10^{-2}$	$0.102\ 10^{1}$
2	$0.596\ 10^{-3}$	$0.974\ 10^{0}$
3	$0.317\ 10^{-3}$	$0.620\ 10^{0}$
4	$0.978\ 10^{-4}$	$0.157\ 10^{-1}$
5	$0.270\ 10^{-4}$	$0.691\ 10^{-5}$
6	$0.415\ 10^{-5}$	$0.611\ 10^{-8}$

Table 5.7 Number m_1 of eigenpairs for which $r_r < 0.6 \times 10^{-2}$ vs. size m of the Lanczos space V_m; m_2 indicates the subset of the smallest (ordered) eigenpairs satisfying the test

	\multicolumn{10}{c}{N}
	222		441		812		1952		4560	
m	m_1	m_2	m_1	m_2	m_1	m_2	m_1	m_2	m_1	m_2
2	0	0	0	0	0	0	0	0	0	0
3	0	0	0	0	0	0	0	0	0	0
4	1	1	0	0	0	0	1	1	0	0
5	2	1	1	1	0	0	1	1	0	0
6	2	1	1	1	1	1	2	1	1	1
7	3	1	1	1	2	2	2	1	1	1
8	5	2	2	2	2	2	2	1	1	1
9	5	2	2	2	4	4	2	1	1	1
10	5	2	3	3	5	5	4	2	1	1
20	11	4	8	8	12	12	8	4	6	6
30	15	7	13	13	13	13	10	7	11	11
40	22	12	17	17	22	21	19	18	15	15
50	34	19	23	23	27	27	24	18	18	18
60	45	28	29	29	36	36	34	26	25	25
70	58	41	34	33	42	42	39	29	27	27
80	68	57	42	41	42	42	41	31	36	36

consequence of this is that there may be the need for repeating the Lanczos eigenvalue calculation for increasingly higher values of m for as long as r_r and the leftmost part of the estimated eigenspectrum have stabilized with no new additions of intermediate eigenpairs. Figure 5.3 shows the high accuracy attained by the Lanczos method, which is actually comparable to that of a direct technique, and may be higher than is required in practice.

The number m_1 of eigenpairs satisfying the tolerance $r_r < 6 \times 10^{-3}$ is shown in Table 5.7 for several m values up to $m = 80$ and for the various problems. The subset m_2 of eigenpairs which are correctly ordered (i.e. the m_2 eigenpairs which occupy without gaps the leftmost part of the eigenspectrum) is also given in Table 5.7. It is worth noting that m_2 is significantly less than m_1 for the $N = 222$ and

5.6 NUMERICAL RESULTS

Table 5.8 Values of ε for each sample problem as provided by the product $\mathbf{q}_1\mathbf{Bq}_2$ with $\mathbf{q}_2$ B-reorthogonalized with respect to $\mathbf{q}_1$ in double precision arithmetics on the IBM 9370/30 computer

N	222	441	812	1952	4560
	0.21×10^{-16}	0.30×10^{-15}	0.12×10^{-14}	0.94×10^{-15}	0.14×10^{-22}

Table 5.9 Loss of B-orthogonality in the Lanczos procedure without B-reorthogonalization in double precision arithmetics on the IBM 9370/30 computer (problem with $N = 222$)

j	$\mathbf{q}_1^T\mathbf{Bq}_j$
2	0.15×10^{-14}
3	0.11×10^{-13}
4	0.58×10^{-12}
5	0.88×10^{-10}
6	0.40×10^{-7}
7	0.10×10^{-5}
8	0.11×10^{-3}
9	0.47×10^{-1}
10	0.62×10^{0}

$N = 1952$ sample cases. Inspection of Table 5.7 indicates that, as a rule of thumb, 80 Lanczos vectors are required to correctly assess 40 eigenpairs, i.e. $2m_2$ Lanczos vectors should suffice to calculate m_2 eigenpairs. We may need more, however, if m_2 is less than 20 (Table 5.7).

Partial B-reorthogonalization of the current Lanczos vector against the previous vectors is effective for all problems, except for the example with $N = 812$, and leads to a saving of up to 15% of CPU time compared to the full B-reorthogonalization case. B-reorthogonalization of two adjacent vectors occurs every three to six Lanczos iterations, depending on the problem, with an ε value set equal to

$$\varepsilon = |\mathbf{q}_1^T\mathbf{Bq}_2| \tag{5.60}$$

where, following a suggestion by van Kats and van der Vorst [88], $\mathbf{q}_2$ is always B-reorthogonalized against $\mathbf{q}_1$. Table 5.8 provides the values of ε obtained from (5.60) in double precision arithmetics on an IMB 9370/30 computer. The loss of B-orthogonality in the Lanczos procedure occurs quickly after few initial steps and the eigenpair accuracy without B-reorthogonalization deteriorates irreparably as m exceeds 10 (Table 5.9).

Concerning the storage requirement of the pointwise Lanczos method, we note that this needs more core memory than the deflation CG procedure, primarily to store approximately $2p$ Lanczos vectors (against p eigenvectors of the CG scheme), in addition to the memory needed for matrix (5.59) and for the eigenvectors of $\mathbf{T}_m (m \sim 2p)$.

For the sample problem with $N = 812$, however, the partial B-reorthogonalization

Table 5.10 CPU times (s) on the IBM 9370/30 computer required to evaluate the 40 leftmost eigenpairs at the accuracy level $r_r = 6 \times 10^{-3}$ by the accelerated conjugate gradients (a) and the Lanczos method with partial **B**-reorthogonalization (b) for the various sample problems

Number of eigenpairs	N									
	222		441		812		1952		4560	
	(a)	(b)	(a)	(b)	(a)	(b)	(a)	(b)	(a)	(b)
1	1.2	5	3.9	19	16.6	57	34	158	79	584
5	8.4	24	27	50	56	90	254	762	422	1976
10	18	45	67	86	91	158	519	1036	967	2998
20	73	60	172	168	407	342	1071	1741	2855	5093
40	171	100	489	334	1537	657	2953	3732	7831	10968

fails to work for m beyond 40 (i.e. it is effective only up to $m = 40$ with $m_2 = 21$ (Table 5.7)). For $m = 80$ the eigenpairs are all miscalculated and full **B**-reorthogonalization proves necessary. A similar difficulty is experienced with the deflation CG approach as well, where we obtain good estimates up to the 20th eigenpair while the subsequent ones are increasingly incorrect. To maintain the desired accuracy up to $j = 40$, **B**-reorthogonalization of $\tilde{\mathbf{x}}_{k+1}$, equation (5.23), against the previous eigenvectors is required, in addition to the **B**-reorthogonalization of $\tilde{\mathbf{p}}_k$, equation (5.20). It is likely that the anomalous behaviour of the two algorithms in this example is accounted for by the relative ill-conditioning of matrix **A** (in fact we have $\lambda_{812}(\mathbf{A}) = 6 \times 10^{-10}$ while $10^{-5} < \lambda_N(\mathbf{A}) < 10^{-2}$ for the other problems).

Finally, Table 5.10 compares the CPU times required by the CG optimization and Lanczos methods for the computation of the 40 leftmost eigenpairs. The codes were run on an IBM 9370/30 computer using double precision and an acceptability criterion $r_r = 6 \times 10^{-3}$. The eigenproblems involving $\mathbf{T}_m$ were solved by routine IMTQL2 of the EISPACK package (Argonne National Laboratory).

Careful inspection of Table 5.10 reveals that the deflation CG approach is much less expensive (up to a factor 5) in the assessment of the 20 smallest characteristic values, irrespective of the problem size, and is still more convenient in the calculation of 40 eigenpairs for the examples with $N = 1952$ and 4560.

On balance it appears that the Rayleigh quotient optimization procedure is more advantgeous than the Lanczos technique in the evaluation of several of the smallest eigenpairs to a fairly high accuracy. Moreover, deflation and conjugate gradients represent a more straightforward method than the Lanczos scheme. The latter, in fact, may require the repeated eigensolutions over several Lanczos subspaces of increasing size to ensure that convergence to the complete partial eigenspectrum has occurred.

5.7 CONCLUSION

The numerical analysis performed on five finite element eigenproblems of varying size N shows that the deflation-CG optimization technique is less expensive than the CG-Lanczos method in the assessment of the $p(p \leqslant 20)$ leftmost eigenpairs to fairly high accuracy (relative residual tolerance on the order of 5×10^{-3}), irrespective of N. For a large number of mode shapes ($p = 40$) the Lanczos approach was found to converge faster, except for the examples with $N = 1952$ and $N = 4560$, for which the break-even point of the two methods is likely to occur at $p > 40$. If a more severe tolerance is prescribed, however, the Lanczos scheme may be superior in terms of CPU times even for small p.

The Lanczos approach with partial **B**-orthogonalization requires a subspace of m Lanczos vectors which is roughly twice the number of desired eigenpairs (for $m \geqslant 40$). Hence storage requirements generally favour the Rayleigh quotient optimization approach, by a factor up to 2 or more for very sparse matrix pencils and relatively large p.

If some of the eigenvalues are very closely clustered or are equal in the first few significant decimal digits (as was the case with the $N = 222$ and $N = 1952$ examples), then the eigenpairs obtained from the CG-Lanczos scheme, and satisfying the acceptability criterion, are not necessarily the leftmost ones. In this case the Lanczos eigensolution is to be repeated over several Lanczos vector subspaces of increasing size m, until the procedure converges to the complete (without gaps) leftmost part of the eigenspectrum.

This may involve extra computational effort and cost for the CG-Lanczos algorithm, depending on the actual structure of the spectrum. This fact emphasizes the reliability, robustness, and ease of computation of the deflation–optimization method compared to the Lanczos procedure.

ACKNOWLEDGEMENTS

The author is very grateful to Manolis Papadrakakis, Claudio Paniconi, Giorgio Pini and Flavio Sartoretto for their critical revision of the manuscript and many useful suggestions.

The present work has partly been supported with funds from the Progetto Finalizzato 'Sistemi Informatici e Calcolo Parallelo' Sottoprogetto 'Calcolo Scientifico per Grandi Sistemi' of the Italian CNR.

BIBLIOGRAPHY

[1] Gruber, R. (1978) Finite hybrid elements to compute the ideal magnetohydrodynamic spectrum of an axisymmetric plasma, *J. Comput. Phys.*, **26**, 379.
[2] Chance, M.S., Greene, J.M., Grimm, R.C. et al. (1978) Comparative numerical studies of ideal magnetohydrodynamics instabilities, *J. Comput. Phys.*, **28**, 1.
[3] Grimm, R.C., Greene, J.M. and Johnson, J.L. (1978) Computation of the magnetohydrodynamic spectrum in axisymmetric toroidal confinement systems, in *Methods of Computational Physics*, 16, Academic Press, New York.
[4] Craig, M.M. and Bampton, M.C.C. (1968) Coupling of substructures for dynamic analysis, *AIAA J.*, **6**, 1313–1319.

[5] Anderson, R.G., Irons, B.M. and Zienkiewicz, O.C. (1968) Vibration and stability of plates using finite elements, *Int. J. Solids and Structures*, **4**, 1031.
[6] Bathe, K.J. and Wilson, E.L. (1972) Large eigenvalue problems in dynamic analysis, *ASCE J. Engg. Mech. Div.*, **98**, 1471.
[7] Clough, R.W. and Penzien, J. (1975) *Dynamics of Structures*, McGraw-Hill, New York.
[8] Gambolati, G., Pini, G. and Sartoretto, F. (1986) Solution to large symmetric eigenproblems by an accelerated conjugate gradient method, *Envir. Soft.*, **1**, 31–39.
[9] Yamamoto, S., Koyamada, Y. and Makimoto, T. (1972) Normal mode analysis of anisotropic and gyrotropic thin film waveguides for integrated optics, *J. App. Phys.*, **43**(12), 5090–5097.
[10] Silvester, P.P. and Ferrari, R.L. (1990) *Finite Elements for Electrical Engineers*, Cambridge University Press.
[11] Zoboli, M. and Bassi, P. (1992) The finite element method for anisotropic optical waveguides, in C.G. Someda and G. Stegeman (eds.), *Anisotropic and Nonlinear Optical Waveguides*, Elsevier, Amsterdam, 77–116.
[12] Smith, B.T., Boyle, J.M., Dongarra, J.J. and Moler, C.B. (1976) *Matrix Eigensystem Routines-EISPACK Guide*, Lectures Notes in Computer Science 6 (2nd edn.), Springer-Verlag, New York.
[13] Garbow, B.S., Boyle, J.M., Dongarra, J.J. and Moler, C.B. (1977) *Matrix Eigensystem Routines-EISPACK Guide Extension*, Lecture Notes in Computer Sciences 51, Springer-Verlag, New York.
[14] Parlett, B.N. (1980) *The Symmetric Eigenvalue Problem*, Prentice-Hall, Englewood Cliffs, New Jersey.
[15] Bathe, K.J. and Wilson, E.L. (1976) *Numerical Methods in Finite Element Analysis*, Prentice-Hall, Englewood Cliffs, New Jersey.
[16] Bathe, K.J. and Ramaswamy, S. (1980) An accelerated subspace iteration method, *J. Comp. Methods Appl. Mech. Engg*, **23**, 313–331.
[17] Lanczos, C. (1950) An iteration method for the solution of the eigenvalue problem of linear differential and integral operators, *J. Res. Nat. Bur. Standards*, **45**, 255–282.
[18] Paige, C.C. (1972) Computational variants of the Lanczos method for the eigenproblem, *J. Inst. Math. Applics.*, **10**, 373–381.
[19] Cullum, J. and Willoughby, R.A. (1978) Lanczos and the computation in specified intervals of the spectrum of large, sparse, real symmetric matrices, *Symp. Sparse Matrix Computations*, Knoxville, Tennessee.
[20] Cullum, J. and Willoughby, R.A. (1979) Fast modal analysis of large, sparse but unstructured symmetric matrices, *Proc. 17th IEEE Conf. Decision and Control*, San Diego, California.
[21] Cullum, J. and Donath, W.E. (1974) A block Lanczos algorithm for computing the q algebraically largest eigenvalues and a corresponding eigenspace of large, sparse, real symmetric matrices, *Proc. IEEE Conf. Decision and Control*, Phoenix, Arizona.
[22] Golub, G.H. and Underwood, R.R. (1977) The block Lanczos method for computing eigenvalues, in J.R. Rice (ed.), *Mathematical Software* III, Academic Press, New York, pp. 361–377.
[23] Ericsson, T. and Ruhe, A. (1980) The spectral transformation Lanczos method for the numerical solution of large sparse generalized symmetric eigenvalue problems, *Math. Comput.*, **35**, 1251–1268.
[24] Sameh, A.H. and Wisniewski, J.A. (1982) A trace minimization algorithm for the generalized eigenvalue problem, *SIAM J. Num. Anal.*, **19**, 1243–1259.
[25] Beliveau, J-G., Lemieux, P. and Soucy, Y. (1985) Partial solution of large symmetric generalized eigenvalue problems by nonlinear optimization of a modified Rayleigh quotient, *Comp. & Struct.*, **21**, 807–813.

[26] Hestenes, M.R. and Karush, W. (1951) A method of gradients for the calculation of the characteristic roots and vectors of a real symmetric matrix, *J. Res. Nat. Bur. Stand.*, **47**, 45–61.
[27] Hestenes, M.R. and Karush, W. (1951) Solution of $\mathbf{Ax} = \lambda\mathbf{Bx}$, *J. Res. Nat. Bur. Stand.*, **47**, 471–478.
[28] Hestenes, M.R. (1953) Determination of eigenvalues and eigenvectors of matrices, *Appl. Math. Series, Nat. Bur. Stand.*, **29**, 89–94.
[29] Faddeev, D.K. and Faddeeva, V.N. (1963) *Computational Methods of Linear Algebra*, Freeman, San Francisco.
[30] Nesbet, R.K. (1965) Algorithm for diagonalization of large matrices, *J. Chem. Physics*, **43**, 311–312.
[31] Bender, C.F. and Shavitt, I. (1970) An iterative procedure for the calculation of the lowest real eigenvalue and eigenvector of a nonsymmetric matrix, *J. Comp. Phys.*, **6**, 146–149.
[32] Falk, S. (1973) Berechnung von eigenwerten und eigenvektoren normaler matrizenpaare durch Ritz-iterationen, *Z. Angew. Math. Mech.*, **53**, 73–91.
[33] Shavitt, I., Bender, C.F., Pipano, A. and Hosteny, R.P. (1973) The iterative calculation of several of the lowest or highest eigenvalues and corresponding eigenvectors of very large symmetric matrices, *J. Comput. Phys.*, **11**, 90–108.
[34] Nisbet, R.M. (1972) Acceleration of the convergence in Nesbet's algorithm for eigenvalues and eigenvectors of large matrices, *J. Comput. Phys.*, **10**, 614–619.
[35] Ruhe, A. (1974) SOR methods for the eigenvalue problem with large sparse matrices, *Math. Comp.*, **28**, 695–710.
[36] Schwarz, H.R. (1974) The eigenvalue problem $(\mathbf{A} - \lambda\mathbf{B})\mathbf{x} = 0$ for symmetric matrices of high order, *Comp. Methods App. Mech. Engg.*, **3**, 11–28.
[37] Papadrakakis, M. (1981) On the estimation of the optimum accelerator of SCOR for eigenvalue problem, *Comp. & Struct.*, **14**, 157–161.
[38] Hestenes, M.R. and Stiefel, E. (1952) Methods of conjugate gradients for solving linear systems, *J. Res. Nat. Bur. Stand.*, **49**, 409–436.
[39] Bradbury, W.W. and Fletcher, R. (1966) New iterative methods for solution of the eigenproblem, *Num. Math.*, **9**, 259–267.
[40] Fried, I. (1972) Optimal gradient minimization schema for finite element eigenproblems, *J. Sound Vibr.*, **20**, 333–342.
[41] Gerardin, M. (1971) The computational efficiency of a new minimization algorithm for eigenvalue analysis, *J. Sound Vibr.*, **19**, 319–331.
[42] Ruhe, A. (1977) Computation of eigenvalues and eigenvectors, in V.A. Barker (ed.), *Sparse Matrix Techniques*, Springer-Verlag, p. 130–184.
[43] Papadrakakis, M. (1984) Solution of the partial eigenproblem by iterative methods, *Int. J. Num. Methods Engg.*, **20**, 2283–2301.
[44] Ruhe, A. (1975) Iterative eigenvalue algorithms based on convergent splittings, *J. Comp. Phys.*, **19**, 110–120.
[45] Evans, D.J. and Shanehchi, J. (1982) Preconditioned iterative methods for the large sparse symmetric eigenvalue problem, *Comp. Meth. Appl. Mech. Engg*, **31**, 251–264.
[46] Longsine, D.E. and McCormick, S.F. (1980) Simultaneous Rayleigh quotient minimization methods for $\mathbf{Ax} = \lambda\mathbf{Bx}$, *Lin. Alg. Appl.*, **34**, 195–234.
[47] Schwarz, H.R. (1982) Simultaneous Rayleigh-quotient iteration methods for large sparse generalized eigenvalue problems, in J. Hinze (ed.), *Numerical Integration of Differential Equations and Large Linear Systems*, Lecture Notes in Mathematics, 968, Springer-Verlag, Berlin-N.Y., pp. 384–398.
[48] Sartoretto, F., Pini, G. and Gambolati, G. (1989) Accelerated simultaneous iterations for large finite element eigenproblems, *J. Comp. Phys.*, **81**, 53–69.

[49] Concus, P., Golub, G.H. and O'Leary, D.P. (1976) A generalized conjugate gradient method for the numerical solution of elliptic partial differential equations, in J.R. Bunch and D.J. Rose (eds.), *Sparse Matrix Computations*, Academic Press, New York, pp. 309–332.
[50] Axelsson, O. (1976) A class of iterative methods for finite element equations, *Comput. Meth. Appl. Mech. Engg*, **9**, 123–137.
[51] Meijerink, J.A. and van der Vorst, H.A. (1977) An iterative solution method for linear systems of which the coefficient matrix is a symmetric M-matrix, *Math. Comput.*, **31**, 148–162.
[52] Kershaw, D.S. (1978) The incomplete Cholesky-conjugate gradient method for the iterative solution of systems of linear equations, *J. Comput. Phys.*, **26**, 43–65.
[53] Gambolati, G. (1980) Fast solution to finite element flow equations by Newton iteration and modified conjugate gradient method, *Int. J. Num. Meth. Engg*, **15**, 661–675.
[54] Gambolati, G. (1980) Perspective on a modified conjugate gradient method for the solution of linear sets of subsurface equations, *Proc. IV Int. Conf. Finite Elements in Water Resources*, Mississippi University, S.Y. Wang et al. (eds.) CML Publications, Southampton, 1980, pp. 2.15–2.30.
[55] Manteuffel, T.A. (1980) An incomplete factorization technique for positive definite linear systems, *Math. Comput.*, **24**, 473–480.
[56] Gambolati, G. and Perdon A.M. (1984) The conjugate gradients in subsurface flow and land subsidence modelling, in J. Bear and M.Y. Corapcioglu (eds.), *Fundamentals of Transport Phenomena in Porous Media*, pp. 953–983, Martinus Nijoff, The Hague.
[57] Papadrakakis, M. and Yakoumidakis, M. (1987) On the preconditioned conjugate gradient method for solving $(\mathbf{A} - \lambda\mathbf{B})\mathbf{x} = 0$, *Int. J. Num. Meth. Engg*, **24**, 1355–1366.
[58] Gambolati, G., Pini, G. and Sartoretto, F. (1988) An improved iterative optimization technique for the leftmost eigenpairs of large symmetric matrices, *J. Comput. Phys.*, **74**, 41–60.
[59] Nour-Omid, B., Parlett, B.N. and Taylor, R.L. (1983) Lanczos versus subspace iteration for solution of eigenvalue problems, *Int. J. Num. Meth. Engg*, **19**, 859–871.
[60] Berk, A.D. (1956) Variational principles for electromagnetic resonators and waveguides, *IRE Tras. Antennas Prop.*, **AP-4**, 104–111.
[61] Biot, M.A. (1957) New methods in heat-flow analysis with application to flight structures, *J. Aero. Sci.*, **24**, 857–865.
[62] Gambolati, G., Sartoretto, F. and Florian, P. (1992) An orthogonal accelerated deflation technique for large symmetric eigenproblems, *Comput. Meth. App. Mech. Engg*, **94**, 13–23.
[63] Perdon, A.M. and Gambolati, G. (1986) Extreme eigenvalues of large sparse matrices by Rayleigh quotient and modified conjugate gradients, *Comput. Meth. Appl. Mech. Engg*, **56**, 251–264.
[64] Evans, D.J. (1967) The use of preconditioning in iterative methods for solving linear equations with symmetric positive definite matrices, *J. Inst. Math. Applics.*, **4**, 295–314.
[65] Tuff, A.D. and Jennings, A. (1973) An iterative method for large systems of linear structural equations, *Int. J. Num. Meth. Engg*, **7**, 175–183.
[66] Axelsson, O. (1977) Solution of linear systems of equations: iterative methods, in V.A. Barker (ed.), *Sparse Matrix Techniques*, Springer, Berlin, pp. 1–51.
[67] Jennings, A. and Malik, G.M. (1977) Partial elimination, *J. Inst. Math. Applics.*, **20**, 307–316.
[68] Gustafsson, I. (1978) A class of first order factorizations, *BIT*, **18**, 142–156.
[69] Gustafsson, I. (1980) On modified incomplete factorization methods, in A. Dold and B. Eckman (eds.), Lecture Notes in Mathematics, 96, Springer-Verlag, New York, pp. 334–351.

[70] Eistenstat, S.C. (1981) Efficient implementation of a class of preconditioned conjugate gradient methods, *SIAM J. Sci. Stat. Comput.*, **2**, 1–4.
[71] Appleyard, J.R. and Cheshire, I.M. (1983) Nested factorization, *7th SPE Symp. Reservoir Simulation*, San Francisco, paper 12264.
[72] Axelsson, O. and Gustafsson, I. (1983) Preconditioning and two-level multigrid methods of arbitrary degree of approximation, *Math. Comput.*, **40**, 219–242.
[73] Jacobs, D.A.H. (1983) Preconditioned conjugated gradient algorithms for solving finite difference systems, in D.J. Evans (ed.), *Preconditioning Techniques in Numerical Solution of Partial Differential Equations*, Gordon & Breach, New York, pp. 509–536.
[74] Jennings, A. (1983) Development of an ICCG algorithm for large sparse systems, Preconditioning Techniques in D.J. Evans (ed.), *Numerical Solution of Partial Differential Equations*, Gordon & Breach, New York, pp. 426–438.
[75] Ajiz, M.A. and Jennings, A. (1984) A robust incomplete Cholesky-conjugate gradient algorithm, *Int. J. Num. Meth. Engg*, **20**, 949–966.
[76] Nour-Omid, B. (1984) A preconditioned conjugate gradient method for solution of finite element equations, *Proc. Int. Conf. Innovative Methods for Nonlinear Problems*, W.K. Lin, T. Belytschko and K.C. Park (eds.), Pineridge Press, Swansea, UK, pp. 17–40.
[77] Jackson, C.P. and Robinson, P.C. (1985) A numerical study of various algorithms related to the preconditioned conjugate gradient method, *Int. J. Num. Meth. Engg*, **21**, 1315–1338.
[78] Tismenetsky, M. and Efrat, I. (1986) An efficient preconditioning algorithm and its analysis, *Lin. Alg. Appl.*, **80**, 252–256.
[79] Wong, Y.S., Zang, T.A. and Hussaini, M.Y. (1986) Preconditioned conjugate gradient residual methods for the solution of spectral equations, *Comp. & Fluids*, **14**(2), 85–95.
[80] Zyvoloski, G. (1986) Incomplete factorization for finite element methods, *Int. J. Num. Meth. Engg*, **23**, 1101–1109.
[81] Johnson, O.G., Micchelli, C.A. and Paul, G. (1983) Polynomial preconditioners for conjugate gradient calculations, *SIAM J. Num. Anal.*, **20**, 362–376.
[82] Saad, Y. (1985) Practical use of polynomial preconditioning for the conjugate gradient methods, *SIAM J. Sci. Stat. Comput.*, **6**, 865–881.
[83] Adams, L. (1985) m-step preconditioned conjugate gradient methods, *SIAM J. Sci. Stat. Comput.*, **6**, 452–463.
[84] Gambolati, G., Pini, G. and Zilli, G. (1988) Numerical comparisons of preconditionings for large sparse finite element problems, *Num. Meth. Partial Differ. Eqs*, **4**, 139–157.
[85] Pini, G. and Gambolati, G. (1990) Is a simple diagonal scaling the best preconditioner for conjugate gradients on supercomputers? *Adv. Water Resour.*, **13**, 147–153.
[86] Bathe, K.J. (1982) *Finite Element Procedures in Engineering Analysis*, Prentice-Hall, Englewood Cliffs, New Jersey.
[87] Hughes, T.J.R. (1987) *The Finite Element Method*, Prentice-Hall, Englewood Cliffs, New Jersey.
[88] van Kats, J.M. and van der Vorst, H.A. (1976) Numerical results of the Paige-style Lanczos method for the computation of extreme eigenvalues of large sparse matrices, TR-3, *A.C.C.U.-Reeks*, **18**, Utrecht, The Netherlands.
[89] Simon, H.D. (1984) The Lanczos algorithm with partial reorthogonalization, *Math. Comput.*, **42**, 115–142.
[90] Nour-Omid, B. and Clough, R.W. (1984) Dynamic analysis of structures using Lanczos coordinates, *Earthquake Engg Struct. Dyn.*, **12**, 565–577.
[91] Coutinho, A.L.G., Landau, L., Lima, E.C.P. and Ebecken, N.F.F. (1987) The application of Lanczos mode superposition method in dynamic analysis of offshore structures, *Comput. Struct.*, **25**, 615–625.
[92] Dunbar, W.S. and Woodbury, A.D. (1989) Application of the Lanczos algorithm to the solution of the groundwater flow equation, *Water Resour. Res.*, **25**, 551–558.

[93] Paige, C.C. (1976) Practical use of the symmetric Lanczos process with re-orthogonalization, *BIT*, **10**, 183–195.
[94] Parlett, B.N. and Scott, D.S. (1979) The Lanczos algorithm with selective re-orthogonalization, *Math. Comput.*, **33**, 217–238.
[95] Cullum, J.K. and Willoughby, R.A. (1985) *Lanczos Algorithms for Large Symmetric Eigenvalue Computations*, Vol. 1, *Theory*, Birkhauser, Boston.
[96] Grimes, R.G., Lewis, J.G. and Simon, H.D. (1986) The implementation of a block shifted and inverted Lanczos algorithm for eigenvalue problems in structural engineering, *Appl. Math. Tech. Rep.*, ETA-TR-39, Boeing Comp. Services.

G. Gambolati
Dipartimento di Metodi e Modelli Matematici per le Scienze Applicate
Università degli Studi di Padova
35 131 Padova
ITALY

6
Lanczos Eigensolution Method for High-performance Computers

S. W. Bostic

NASA, Hampton, VA, USA

6.1 INTRODUCTION

One of the most computationally intensive tasks in large scale mechanics problems is the solution of the eigenproblem. Eigenproblems occur in virtually all scientific and engineering disciplines. This chapter will discuss a particular method, the Lanczos method, for the solution of this problem. A brief discussion of the theory of the method will be followed by the computational analysis of the method and the implementation on parallel-vector computers. Two structural analysis applications will be presented: the buckling of a composite panel, and the free vibration analysis of an entire aircraft.

Several efficient eigenvalue solvers are widely used in the structural analysis community, examples of which include: the QR and QL methods [1–3], the inverse power method [1–3], subspace or simultaneous iteration [1–3], determinant search [4], and the sectioning method [5]. Each of these methods has advantages for certain classes of problems and limitations for others. Many of the most popular methods, such as the QR and QL methods, solve the complete system of equations rather than a reduced set. For very large problems, these methods prove to be inefficient. In contrast, recent studies indicate a growing acceptance of the Lanczos method as a basic eigenvalue analysis procedure for large-scale problems. Compared to subspace iteration, one of the most widely used algorithms, the Lanczos method is as accurate

---

Solving Large-scale Problems in Mechanics, Edited by M. Papadrakakis
© 1993 John Wiley & Sons Ltd

and more efficient and has the advantage that information previously computed is preserved throughout the computation [6–11]. The Lanczos method shares the rapid convergence property of the inverse power and subspace iteration methods but is more efficient when only a few eigenvalues of a large order system are required. The single vector Lanczos procedure is the focus of this chapter, although block Lanczos algorithms are being examined by many researchers. Block Lanczos algorithms compute several vectors simultaneously and are effective for cases where multiple roots are expected [12–15].

The implementation of the Lanczos method and some techniques that optimize the solution process by exploiting the vector and parallel capabilities on today's high-performance computers are discussed in this chapter.

6.2 APPLICATION TO STRUCTURAL PROBLEMS

The large-scale mechanics problems to be addressed in this chapter are the free vibration problem and the buckling problem. In the example problems, the finite element method is used to discretize the structure; that is, the structure is approximated by many 'finite' elements joined together at 'nodes'. The fact that the elements can be connected in a variety of ways means that they can represent exceedingly complex shapes. The finite character of the structural connectivity makes the analysis by algebraic (or matrix) equations possible. All material properties of the original system are retained in the individual elements. The element properties, represented by the stiffness and mass matrices, are then assembled into global matrices. Matrix equations then express the behaviour of the entire sturcture. For detailed discussion of the finite element method as applied to structural dynamics see References [16–18].

The finite element method was originally developed for the aerospace industry to provide a solution for extremely complex configurations. The simultaneous availability of high-speed digital computers permitted the application of the method to a large range of engineering problems, and by the early 1970s it was the method of choice for the numerical solution of continuum problems. Today there exist many large finite element codes capable of solving large-scale problems on individual workstations as well as large mainframe computers. Vibration and buckling problems are representative of the types of problems that require efficient algorithms as well as fast computation rates for timely solutions.

To determine the dynamic structural response, a free vibration analysis is carried out to find the natural frequencies and mode shapes. The natural frequencies are those at which a system oscillates in free vibration or without external forces. In free vibration the motion of the structure is maintained by gravitational or elastic restoring forces. The natural frequencies of a system are related to its eigenvalues and must be known to prevent resonance which occurs when the natural frequencies coincide with the frequencies in the applied dynamic loads. They are also used in aeroelastic analysis and flexible deformation control. In the buckling problem, the buckling load is related to the eigenvalue. In these problems, the response of a system is represented by a set of eigenvectors and eigenvalues.

The Lanczos method solves the standard eigenvalue problem, $Av = \lambda v$, by a recursion formula. The application of this recursion results in a set of vectors, the

Lanczos vectors, and elements of a tridiagonal matrix, T. The original large eigenvalue problem is transformed into a small tridiagonal problem which can easily be solved to obtain a small subset of eigenvalues and eigenvectors. These solutions can then be used to find the eigenpairs of interest of the original problem. The vibration and buckling structural analyses discussed in this chapter result in generalized eigenvalue problems and must be transformed to the standard eigenvalue problem for the Lanczos method. This transformation from the generalized eigenvalue problem to the standard eigenvalue problem necessary for the application of the Lanczos method is a computationally expensive operation. Because the eigenvalues of interest for this class of problems are either the smallest eigenvalues, or those eigenvalues in a given range of the spectrum, the problem undergoes an inversion transformation as well. In order to make the algorithm even more general, a shift parameter is introduced into the original problem to allow solution for the eigenvalues closest to the value of the shift.

6.2.1 *Vibration analysis*

For the vibration problem, the transformation process from the generalized eigenvalue problem to the standard eigenvalue problem to be solved by the Lanczos method is as follows.

The generalized eigenvalue problem for free vibration is

$$Kx = \omega^2 Mx \tag{6.1}$$

where K is a symmetric positive semi-definite stiffness matrix and M is a symmetric positive definite matrix and represents either a banded consistent mass matrix or a diagonal mass matrix, where the masses of the elements are lumped at the nodes. The vectors x_i represent the eigenvectors, or vibration mode shapes and ωs are the eigenvalues or the vibration frequencies. The solution of equation (6.1) by the Lanczos method would yield the largest eigenvalues. For the vibration and buckling problems implemented here, the smallest eigenvalues are the ones of interest. Therefore, a shift, σ, close to the eigenvalues of interest, is introduced and then the problem is inverted. The computations necessary to convert the original problem to an equivalent shifted inverse form and then transform the generalized eigenvalue problem into the standard Lanczos form, $Av = \lambda v$, for the vibration problem follow.

Introducing a shift, σ, and inverting by letting

$$\omega^2 = 1/\lambda + \sigma \tag{6.2}$$

then substituting (6.2) in equation (6.1), gives

$$Kx = (1/\lambda + \sigma) Mx. \tag{6.3}$$

Multiplying each side by λ yields

$$K\lambda x = Mx + \sigma \lambda Mx \tag{6.4}$$

and rearranging terms gives

$$\lambda[K - \sigma M]x = Mx. \tag{6.5}$$

Finally, multiplying both sides by $[K - \sigma M]^{-1}$ produces

$$\lambda x = [K - \sigma M]^{-1} Mx. \tag{6.6}$$

Letting

$$A = [K - \sigma M]^{-1} M \tag{6.7}$$

substituting (6.7) in equation (6.6) and rearranging term yields

$$Ax = \lambda x, \text{ or the standard eigenvalue equation.} \tag{6.8}$$

The implementation of the Lanczos algorithm requires the computation of the vector quantity Av for a given v. It is important to avoid the expensive computation of finding the inverse of the matrix in equation (6.7), which would result in a full matrix, losing the advantage of the banded, sparse, symmetric matrix. The following procedure is thus implemented.

Let

$$\bar{K} = [K - \sigma M] \tag{6.9}$$

1. Factor $\bar{K}$

$$\bar{K} = LDL^T \tag{6.10}$$

where L is a lower triangular matrix with unit diagonal, D is a diagonal matrix and L^T, or the transpose of L, due to symmetry, is the upper triangular matrix.

2. Then rearrange terms and introduce y

$$Av = (LDL^T)^{-1} Mv \tag{6.11}$$

$$\text{or } Av = (L^T)^{-1}(LD)^{-1} Mv \tag{6.12}$$

$$L^T Av = (LD)^{-1} Mv = y \tag{6.13}$$

3. Solve for y

$$LDy = Mv. \tag{6.14}$$

4. Then solve for Av

$$L^T(Av) = y. \tag{6.15}$$

6.2.2 Buckling analysis

Similarly, the transformation for the buckling problem is carried out as follows. The generalized buckling problem is,

$$Kx = -\delta K_g x \tag{6.16}$$

where K is the linear stiffness matrix, K_g is the geometric stiffness matrix, x is a buckling mode shape and δ is the buckling load. Because the geometric, or differential, stiffness matrix K_g may be an indefinite matrix, a different shifting and inverting strategy is required.

In this case let

$$\delta = \sigma\lambda/(1-\lambda) \tag{6.17}$$

then substituting (6.17) into equation (6.16) gives

$$Kx = -(\sigma\lambda/(1-\lambda))K_g x. \tag{6.18}$$

Multiplying each side by $(1-\lambda)$ yields

$$Kx - K\lambda x = -\sigma\lambda K_g x \tag{6.19}$$

then

$$Kx = [K - \sigma K_g]\lambda x. \tag{6.20}$$

Multiplying each side by $(K - \sigma K_g)^{-1}$ gives

$$\lambda x = [K - \sigma K_g]^{-1} Kx. \tag{6.21}$$

Therefore, for the buckling problem

$$A = [K - \sigma K_g]^{-1} K \tag{6.22}$$

or, in standard form,

$$Ax = \lambda x. \tag{6.23}$$

Each multiplication by the mass matrix, M, in the vibration case is replaced by a multiplication by the stiffness matrix in the buckling case. To form the matrix A, the geometric stiffness matrix is multiplied by the shift parameter in the buckling case in place of the mass matrix as in the vibration case. The eigenvalue problems are real symmetric problems and the matrices, which result from the finite element method, are symmetric, sparse and banded. Symmetry of the matrices, where the upper triangle of the matrix is identical to the lower triangle matrix, reduces the storage and computation requirements. Sparsity refers to the number of non-zeros in the matrix. A banded matrix is one where the non-zeros are clustered close to the diagonal. For the vibration problem, the mass matrix (M) can be either a diagonal mass matrix where the mass is taken to be at the nodes, or the consistent mass matrix containing the distributed mass associated with the elements.

6.3 LANCZOS METHOD

The Lanczos method was first introduced in 1950 by Cornelius Lanczos [19]. When the method was applied to real problems, using finite arithmetic, the method did not behave in accordance with the theoretical properties, numerical instabilities arose and the method was not widely accepted. In recent years, due to the research of many analysts [3, 12–15, 20–25], these instabilities have been understood and eliminated. As a result, new and innovative approaches have been developed to implement the method. The basic procedure uses a recursion to produce a set of vectors, referred to as the Lanczos vectors, and scalars that form a tridiagonal matrix. This tridiagonal matrix can then be easily solved for its eigenvalues, which are used to compute a few of the eigenvalues of the original problem.

The basic Lanczos algorithm solves the standard eigenvalue problem:

$$Ax = \lambda x \tag{6.24}$$

using the basic Lanczos recursion described below, which results in a reduced eigenvalue problem:

$$Tq = \lambda q \tag{6.25}$$

where T is a tridiagonal matrix consisting of αs on the diagonal and βs on the off diagonals:

$$T = \begin{bmatrix} \alpha_1 & \beta_2 & & & \\ \beta_2 & \alpha_2 & \beta_3 & & \\ & \beta_3 & \alpha_3 & \beta_4 & \\ & & \beta_4 & \alpha_4 & \beta_5 \\ & & & \cdots & \end{bmatrix}.$$

The steps in this transformation process to tridiagonal form are:

1. Initialization

    (a) Choose a starting vector v_1, where v_1 is normalized., $|v_1| = 1$.
    (b) Set $\beta_1 = 0$ and $v_0 = 0$.

2. Iteration

    for $i = 1, 2, 3, \ldots, m$ as follows:

$$w = Av_i - \beta_i v_{i-1} \qquad (6.26)$$

Then,

$$\alpha_i = v_i^T w \qquad (6.27)$$

$$c = w - \alpha_i H v_i \qquad (6.28)$$

$$\beta_{i+1} = [c^T H c]^{1/2} \qquad (6.29)$$

$$v_{i+1} = c/\beta_{i+1} \qquad (6.30)$$

where for the vibration case H is M and for the buckling case H is K, w and c are temporary vectors, the α vector is the diagonal term of the resulting tridiagonal matrix T and the β vector is the off-diagonal term. The vectors $v_1, v_2, \ldots, v_m$ are the set of Lanczos vectors.

The order N of A may be 10 000 or more while order m is typically equal to twice the number of eigenvalues and eigenvectors desired, usually less than 50.

For each eigenvalue, λ, of T_m, a corresponding eigenvector, q, is computed such that,

$$T_m q = \lambda q. \qquad (6.31)$$

The frequencies of the vibration problem are found by

$$\omega^2 = \sigma + 1/\lambda \qquad (6.32)$$

and the eigenvalues of the buckling problem are found by

$$\delta = \sigma(\lambda/(1 - \lambda)). \qquad (6.33)$$

The m eigenvectors X_m of equation (6.1) are then found by

$$X_m = V_m Q_m. \qquad (6.34)$$

The eigenvalues of the tridiagonal matrix can be easily obtained using readily available library routines such as the QL algorithm, a fast, efficient method for the solution of tridiagonal matrices.

6.3.1 Reorthogonalization of the Lanczos vectors

When the Lanczos method was first put into practice, it was found that due to finite arithmetic calculations, the vectors tend to lose their orthogonality. Extra eigenvalues, labelled 'spurious', may appear, as well as redundant values of the 'good' eigenvalues. One of the on-going topics of research concerning the Lanczos method involves finding robust ways to overcome this deficiency. One approach is to reorthogonalize at each step, thus eliminating the effects of the previous vectors on the succeeding ones.

This has been considered 'expensive' and shortcuts, such as selective reorthogonalization or partial reorthogonalization, have been proposed by Parlett and others [15, 21–25].

Another approach is proposed by Cullum and Willoughby [13] which involves no reorthogonalization but uses an indentification test to select those approximations which are to be accepted. By comparing the eigenvalues found using the complete tridiagonal matrix to the eigenvalues found using the submatrix obtained by deleting the first row and column of the tridiagonal matrix, a decision can be made as to which eigenvalues are approximate enough to be considered accurate. In the examples that follow, the effect of reorthogonalization will be shown. An example of the unacceptable eigenvalues that appear when no reorthogonalization is performed will be presented and the cost of reorthogonalization will be tabulated. No attempt will be made to compare or promote the various reorthogonalization techniques.

6.4 COMPUTATIONAL ANALYSIS

The preceding sections outlined the computational steps to be carried out when implementing the Lanczos method. An efficient algorithm must take into consideration the architecture of the computer on which it will be implemented. The next sections will discuss the characteristics of high-performance computers and some of the techniques and tools available to improve the efficiency of the Lanczos method.

6.4.1 High-performance computers

The computational power of today's high-performance computers now makes it possible to solve large, complex problems which were prohibitively expensive to solve in the past. This computational power in turn requires new computational algorithms that address the present problems now of interest as well as take advantage of the capabilities of the latest generation of computers. The vector capabilities of these computers offer speed-ups in computation of several magnitudes over sequential computers. When this vector capability is coupled with the capability to perform computations in parallel which is now available on many different types of architectures, the potential for solving larger problems substantially increases. This increase in computational power yields a more accurate and efficient solution to the eigenproblem.

The computation rate on high performance computers is commonly measured in millions of floating point operations per second or Megaflops (Mflops). The computation rate of the most powerful supercomputers has surpassed a billion floating point operations, or Gigaflops, and rates will soon be measured in Teraflops, or a trillion floating point operations per second.

The example problems cited in this chapter were executed on the Convex C220, the CRAY-2, and the CRAY Y-MP multicomputers. The Convex C220 at the NASA

Langley Research Center consists of two central processing units, each of which can compute at a rate of from 20 to 40 Mflops for a computationally intensive calculation. The CRAY-2 at NASA Langley Research Center has four central processing units (CPUs), while the CRAY Y-MP at NASA Ames Research Center has eight. Optimized code running on one CPU of the CRAY computers typically generates results in the 100–200 Mflops range. Each CPU in the Convex and CRAY has multiple vector functional units which access very large main memories though eight high-speed vector registers. These functional units can operate simultaneously and the maximum performance rate is obtained when both the addition and multiplication functional units are operating simultaneously.

6.4.2 Vectorization optimization

The vectorization of code must be fully optimized before considering any parallel processing on parallel–vector computers. For maximum performance, software must be tuned to best exploit the hardware capabilities. The high performance of vector computers is due to 'vector units', designed to perform such computations as adds and multiplies simultaneously. Arithmetic operations are 'pipelined' into these vector units. Pipelined arithmetic units allow for operations to be overlapped as in an assembly line. Several specialized subsections work together to execute an operation. When the first section completes its processing on a set of operands, the results are passed to the next section, and a new set of operands enters the pipe. To carry out such operations there must be no data dependency. In other words, DO loops must be avoided where a result depends on completion of a previous iteration of the loop, such as in the recursion: $A(I) = A(I - 1) + B(I)$. By efficient vectorization, speed-ups of 10–20 can be achieved.

On vector computers, three factors that influence the vector computational rate are: the number of memory accesses per computation, the length of the vectors, and the vector stride, which is the spacing in memory between elements of the vectors involved. Long vectors reduce the ratio of the overhead and initial memory access time to the amount of computation. Vectors of stride one, or vectors whose elements are contiguous in memory, are the fastest to access. The number of memory accesses can be reduced using different techniques, the most rewarding being loop unrolling.

Loop unrolling

A useful technique to enhance vector performance is loop unrolling. An example of loop unrolling to level 6 is shown in Figure 6.1. The example is a matrix–vector multiply, $C = A*B$, where A is an $n \times n$ matrix and B and C are $n \times 1$ vectors, the iterations of the inner loop are decreased by a factor of 6 by the explicit inclusion of the next five iterations. In the loop below, the vector C is accessed once for the 6 multiply and add operations. The vector multiply, vector add, and vector access from memory are, for the most part, carried out concurrently.

```
DO 10 j = 1, n, 6
DO 10 i = 1, n
  C(i) = C(i) + A(i, j)* B(j)
              + A(i, j + 1)* B(j + 1)
              + A(i, j + 2)* B(j + 2)
              + A(i, j + 3)* B(j + 3)
              + A(i, j + 4)* B(j + 4)
              + A(i, j + 5)* B(j + 5)
10 CONTINUE
```

Figure 6.1 Loop unrolling to level 6.

Compiler directives

High-performance computers have sophisticated compilers which can detect vectorizable and sometimes parallelizable code. There are situations, however, when the compiler cannot optimize code because of unknown conditions, such as data dependencies. Compiler directives can be used by the analyst in these situations. When the analyst knows that the variable in question will never have a value that would create a dependency conflict, vectorization of a loop can be forced. The use of this directive requires an intimate knowledge of the algorithm and problem in order to maintain the integrity of the data, but can result in significant reductions in computation time.

Local memory

The latest generation of computers often have local memory, or caches, which can be used to store data which is accessed repeatedly in a computation sequence. The number of memory accesses can be decreased dramatically when this option is available. An example of the savings on the CRAY-2 using local memory storage for data is shown in the examples to follow. This option is utilized in the factorization of the matrix to store the multipliers used to update the columns.

6.4.3 Parallel processing

There are many types of parallel architectures now available. An early classification of computer architectures consisted of four types of architectures: single-instruction/single data, a scalar computer consisting of one processor working on one stream of data, single-instruction/multiple data, which defines vector machines where a single stream of instructions operate on separate elements of an array simultaneously, multiple-instruction/single data which implies that several instructions are operating on a data stream simultaneously and multiple-instruction/multiple data computers, where several processors concurrently act on multiple data streams [26]. The computers used in this study belong to the class labelled multiple-instruction multiple data (MIMD) computers with shared memory, although some of the principles will apply when programming for other architectures. The implementations to be presented were designed for computers with a few, powerful processors, as opposed to the class of computers referred to as massively parallel processors which have thousands of processors, each capable of performing a small amount of computation.

The use of multiple processors to execute portions of a program simultaneously offers significant speed-ups in computation. However, these speed-ups can be difficult to achieve in practice. All programs have a portion of work that must be executed sequentially, or duplicated in other processors, however, and it is rare when large portions of the work can be equally divided among the number of processors available. There is also overhead associated with parallel processing, particularly synchronization, or the process of coordinating the tasks within the parallel regions. Parallel tasks must execute independently, in any order and without regard to the number of processors available at running time. The main purpose of executing computations in parallel is to decrease wall clock time for a particular solution. The actual computation time, which is the sum of time expended by all processors, will increase. One must determine the potential wall clock time to be saved versus the programming effort involved and the increase in total CPU time to evaluate the benefits of parallel processing.

Software written for parallel processing requires more analysis, with particular attention given to data dependencies, the scope of the data, critical regions where communication must be synchronized and load balancing, which implies an equal distribution of work to each available processor. Local, or private data, such as loop indices, are only accessible to the defined task. Shared, or common data is accessible to all tasks. Shared data must be protected and the proper communication and synchronization provided.

In some cases load balancing can be determined before run time, in which case it is referred to as static load balancing. If the work must be distributed during execution, dynamic load balancing is required. Parallel codes are more difficult to test and debug than sequential or vectorized codes.

Granularity

The level of parallel processing depends on the granularity of the computation. A high level of granularity refers to executing large sections of code, such as complete subroutines, concurrently. The initial parallel processing software mechanism on the CRAY, referred to as macrotasking, had a high overhead and required a high level of granularity to be efficient. If the computation can be divided into large independent tasks that are equal and can be performed simultaneously, macrotasking can be invoked with a minimum amount of overhead.

The computations necessary for eigenvalue analysis typically do not have tasks that can be carried out concurrently at the subroutine level. For the small granularity inherent in these algorithms, microtasking is used to process tasks within a subroutine. For instance, in a loop that will be generated many times, the number of times through the loop can be divided up among the available processors.

Autotasking is a feature now available on CRAY systems and other computers with sophisticated compilers that detects parallel regions in a pre-processing phase. The autotasking capability detects regions that can be microtasked and automatically generates code to assign tasks to all processors that are available. The programming effort in this case is minimal, as is the computational overhead.

Dedicated versus batch mode

The decrease in wall clock time depends on many factors, one of which is the mode in which the program is executed. The high-performance computers used in research labs and in production evironments are typically heavily utilized. Parallel programs running in a batch mode are competing with all the other programs in the system for hardware resources. In a batch mode programs use those processors available at the time and there is no guarantee that more than one processor will be assigned to a given execution. For particular high priority, long-running jobs, where it is important to get the answers as fast as possible, such as weather prediction, jobs may run in a dedicated mode where all processors are assigned to the one job. In a dedicated mode all processors will be available to the program and the work will be divided as directed. In both cases a decrease in wall-clock time should result for the execution of a program, but when run in batch mode, the decrease could be significantly less. In the examples following, statistics will be presented showing the difference between running in the two modes.

6.5 IMPLEMENTATION OF THE LANCZOS METHOD

The first step in the algorithm development process is to identify the time-consuming calculations. Software tools are available on today's high-performance computers to analyse the computations in a program. The Lanczos algorithm was analysed using the flowtrace capability on the CRAY-2. This utility computes the percentage of time spent in each section of the code. For the structural analysis problems presented in this study, the three dominant computational steps are: factorization of the matrix $\bar{K}$, as in equation (6.10), the forward/backward solution steps in equations 6.14 and 6.15, and the matrix–vector multiplications in equations 6.28 and 6.29. For typical structural analysis problems, the factorization and forward/backward solution steps combined in the Lanczos method take over 50 % of the total computation time and the matrix-vector multiplications take another 20–25 % of the computation time [6].

The following sections will address some of the issues involved in exploiting the architectural capabilities of high-performance computers in order to decrease the computation time of the Lanczos eigensolver. The first section will describe the direct Cholesky solver for variable-band matrices that is used in the implementation of the Lanczos method presented earlier in this chapter.

6.5.1 Variable band Cholesky solver

An important area of research on parallel–vector computers has been the solution of linear systems of equations. In many algorithms, the equation solution, including the factorization of the matrix and the forward/backward solves, is the dominant factor in terms of the amount of computation. This is particularly true in the Lanczos algorithm, as the matrix to be factored can be extremely large and the forward/backward solution is repeated many times. Comparisons have been made between direct and indirect solvers and various implementations have been tested. Memory requirements, the number of floating point arithmetic operations required and the speed at which the operations can be performed influence the choice of which solver to use.

The decomposition of a symmetric, positive-definite matrix into lower and upper triangular matrices which can then be used in the forward and backward solution steps, is attributed to Cholesky [2]. The solver used in the example problems that follow is a variation of the Cholesky solver, described as a variable band Cholesky solver. The decomposition, or factorization, used in the Lanczos eigensolver is an LDL^T factorization, which results in a lower triangular matrix, L, a diagonal matrix, D, and an upper triangular matrix L^T which is the transpose of the lower triangular matrix. This solver has been shown to be efficient and accurate, and outperformed iterative solvers, as well as sparse solvers, on a vector computer for representative structural analysis problems [27, 28]. As previously stated, the matrices that results from the finite element method in structural analysis are often large, sparse and banded. The amount of computation involved in the factorization of the matrix $\bar{K}$ and the equation solution steps depend on the size of the problem and the bandwidth of the matrices. In the variable band storage scheme used in the described implementation, the number of degrees of freedom of the finite element model determine the number of rows in the stiffness and mass matrices. The length of each row (or column, as the matrices are symmetric) or bandwidth is determined by the connectivity of the elements. The number of rows in the matrices or the number of degrees of freedom for a complex aircraft or space station model can be several hundred thousand. For these large problems the issue of data storage and access is most important in determining the efficiency of the implementation. The Cholesky factorization and equation solver to be described uses column-oriented variable-band storage.

Storage schemes

The most efficient type of data storage is determined by the computation algorithm to be implemented. For sparse, banded matrices a choice must be made between storing the banded matrix which contains zero elements but results in long, efficient vector operations and storing only the non-zero elements, referred to as sparse storage, which conserves storage and reduces the amount of computation but often seriously decreases the computation rate. Poole and Overman compare banded equation solvers with sparse equation solvers in Reference [27]. Results vary, depending on both the problem to be solved and the hardware on which the program is executing. For the typical structural analysis problems on a CRAY-2 the variable band Cholesky equation solver was the fastest in terms of the megaflop rate and computation time. The high computation rate for this solver more than made up for the increase in memory due to storing the zero elements within the band and the resulting increase in the number of arithmetic floating point operations.

One other advantage of the variable band solver is the type of computation dictated by the algorithm. The two vector computations encountered most in the factorization and forward/backward solve are the inner product $(x^T x)$ and the saxpy, or $\Sigma(ax_i + y_i)$, where x and y are vectors and a is a scalar. On vector machines the saxpy is by far the more efficient operation. With proper use of loop unrolling, the saxpy operation allows overlapping of memory accesses with simultaneous use of both the add and multiply functional units. The variable band storage scheme's efficiency is in part due to the fact that it enables the use of the saxpy operations.

Reordering of nodes

When using a banded solver, the amount of computation involved in the factorization and forward/backward solves is directly proportional to the semi-bandwidth. It is, therefore, important to decrease the semi-bandwidth, or the distance from the diagonal to the last non-zero, as much as possible. The non-zero quantities in the stiffness matrix represent the connectivity of the elements in the finite element model. Often, the numbering of the nodes is done by a computer program, or an analyst to whom the structure of the resulting matrix is not of concern. In some cases, rows or columns of the matrix may be exceptionally long and have large semi-bandwidths, but contain mostly zeros. For maximum efficiency the nodes of the finite element model often need to be renumbered to reduce the semi-bandwidth of the matrices. The particular method used to renumber the nodes for the applications discussed in this chapter was a reverse Cuthill–McKee [29] profile minimizer. Such algorithms can significantly reduce the semi-bandwidth of the matrices and for the example problems, a significant amount of storage and computation is saved by using this reordering scheme.

Column storage versus skyline storage

For the variable band Cholesky algorithm, the lower triangular part of the symmetric matrix is stored by columns, beginning with the main diagonal down to the last non-zero entry in each column, including zeros. This data storage scheme is in contrast to the skyline or profile schemes which store the upper triangular part of the matrix by columns beginning with the main diagonal and storing all coefficients up to the first non-zero in each column. The advantage of the skyline storage scheme is that it requires less storage. One advantage of using the variable band storage scheme is the type of floating point operations associated with the method, particularly the saxpy operation. The vector lengths are also longer, which helps to offset the fact that more total computations are required. A schematic of the storage

```
                1
                2  5
                3  6  8
                4  7  9
                      10 14
                      11 15
                      12 16
```

Figure 6.2 Variable band storage of lower triangle by columns.

scheme is shown in Figure 6.2. The numbers in Figure 6.2 indicate the order in which the elements of the matrix are stored.

Reference [27] describes the variable band Cholesky solver method in detail. This method is able to exploit key architectural features of vector computers and runs well in excess of 100 Mflops on the CRAY-2 and CRAY Y-MP computers. The storage scheme allows the factorization routine to be carried out with stride one vectors, or those with elements stored in contiguous locations. To increase the speed of the factorization, an immediate update strategy is used where each column is used to update the other columns as soon as it is computed. The forward solution uses a column sweep approach, thus accessing the data in the most efficient way. The variable band storage format results in using the efficient saxpy operation in the factorization, allowing addition and multiplication to be performed simultaneously.

The lower triangular matrix L and the diagonal matrix D are stored in the location previously occupied by $\bar{K}$, as this original matrix is not needed again. A byproduct of the factorization $\bar{K}$ is that the number of eigenvalues less than the given shift (σ) can be found. The number of negative entries in the diagonal matrix D produces this information.

6.5.2 Matrix-vector multiplication

Another time-consuming operation in the solution process was the matrix–vector operation which is carried out three times for each iteration. For this calculation, it proved more efficient to eliminate the zeros in the variable band matrix and to store only the non-zero coefficients of the lower triangular part of the matrix by columns in a single dimensioned array.

Two integer pointer arrays are used to store the beginning location of each column and the length of each column. The matrix–vector multiplication takes advantage of the fast saxpy operation, explained previously. This storage scheme can effectively shorten the vector lengths, so a trade-off exists between storing only the non-zeros and the variable band storage. Statistics comparing the sparse matrix–vector multiplier versus a banded matrix–vector multiplier will be shown in the applications section.

6.6 APPLICATIONS

Predicting the structural response of the next generation of aerospace structures will place great demands on available computational power. The complexity of these structures dictates finite element models of small granularity which result in very large problems to be solved. The applications to be addressed here are representative, although on a smaller scale than what can be realistically expected.

The first example, a blade-stiffened panel with cutout, is a representative component of an aerospace vehicle. The second example is a preliminary model of a high-speed civil transport. The examples are used to best determine the most efficient exploitation of parallel-vector computers.

6.6.1 Vectorized Lanczos implementations

Since the benefits of vectorization, in terms of reducing computation time, are much greater than the benefits of multitasking on the parallel–vector computers used in this study, the Lanczos algorithm was first vectorized using the techniques described in the previous section. The effects on the computation time of these optimization techniques, including automatic vectorization, compiler directives and loop unrolling are shown with the buckling of the laminated blade-stiffened panel problem as an example.

Panel problem

The finite element model of a graphite–epoxy blade-stiffened compression panel with a discontinuous stiffener is depicted in Figure 6.3.

This graphite–epoxy panel represents a generic class of laminated composite structures whose properties must be understood before being incorporated into future aerospace vehicles. This problem was selected as an example because experimental results are available and the characteristics, such as a discontinuity, large displacements, and a brittle material system, are representative of practical composite structures [30]. The panel skin is a 25-ply laminate and each blade stiffener is a 24-ply laminate. The panel was loaded in axial compression. The loaded ends of the panel are clamped and the sides are free. The Lanczos method found the buckling load of this stiffened panel with cut-out, which agreed with those found by the subspace iteration method. The Lanczos method was an order of magnitude faster.

The first buckling mode is shown in Figure 6.4.

A discretization of the panel which resulted in 6426 degrees of freedom was used as a model for this study and the first five buckling modes were computed. The Lanczos eigensolver was first coded and run in a scalar mode, with no optimization, on the Convex 220. The total computation time for this problem was 181.6 s, with the factorization of the matrix taking 55% of the time and the forward/backward solutions taking 30% of the time. The automatic vectorization option of the compiler was then exercised and the total time was decreased to 64 s. The loops that the

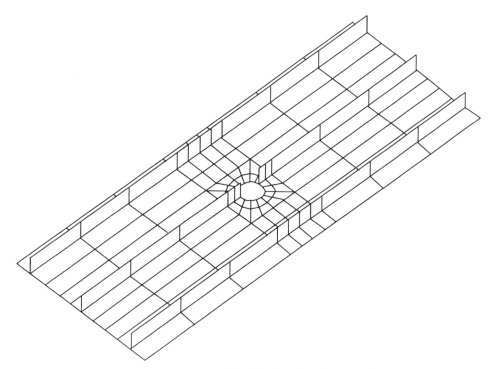

Figure 6.3 Finite element model of blade-stiffened panel with cut-out.

compiler could not vectorize without intervention of the analyst were studied for data dependencies. Compiler directives were inserted where applicable, reducing the computation time to 21 s. The main loop in the factorization routine was unrolled to an optimal level 6 decreasing the computation time to 14.1 seconds. The total savings in execution time for the panel buckling problem on the Convex obtained by automatic vectorization, compiler directives and loop unrolling are shown in Figure 6.5. An overall speed-up of almost 13 is achieved for the optimized vector code over the sequential implementation for this representative problem.

Another major time-consuming calculation for the solution to the panel problem was determined to be the matrix–vector multiplications. A comparison was made between using the variable band storage scheme for this operation and converting the matrix to a sparse storage, or eliminating the zeros within the columns, thereby reducing the number of floating-point operations but at the same time decreasing the vector length. It was found that the sparse storage scheme resulted in a significant decrease in computation time even though the megaflop rate was decreased. The comparison of the number of operations, the computation times and the megaflop rate (Mflops) between a variable band matrix multiply and a sparse matrix multiplication for the panel problem on the CRAY-2 are shown in Table 6.1.

As shown, the time to multiply the stiffness matrix in the sparse format by a vector was 0.013 s while the time to multiply the same matrix stored in variable band format was 0.042 s. The megaflop rate is over four times faster using the

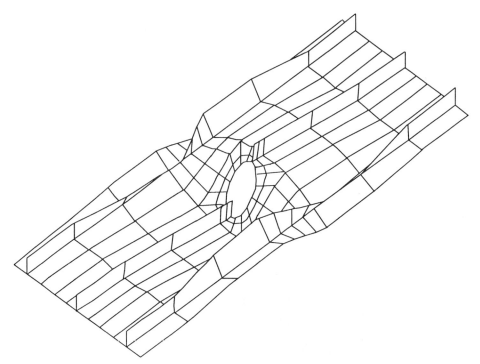

Figure 6.4 First buckling mode of blade-stiffened panel.

variable band algorithm, but the fewer floating-point operations of the sparse storage scheme results in reducing the overall execution time by 69%. This matrix–vector multiplication was performed over three times for each Lanczos step, thus even a small reduction in time results in a significant saving in overall computation time.

The use of local memory on the CRAY-2 can speed up calculations by decreasing the number of memory references. In the factorization step, those vectors that will be accessed consecutively many times are stored in the local memory. The comparison of the computation times for factoring the matrix for the panel problem on the CRAY-2 using local memory and not using local memory are shown in Table 6.2.

High speed civil transport on CRAY-2

Considerable research in the aerospace field is being directed toward the development of supersonic civil transport aircraft. A finite element model for the preliminary design studies of a high speed civil transport is shown in Figure 6.6. The symmetric half of the structure is composed of 2851 nodes, 5189 two-noded rod elements, 4329 four-noded quadrilateral elements and 114 three-noded triangular elements. This structure has 17 106 degrees of freedom. Eliminating the constrained degrees of freedom results in 16 146 active degree of freedom, resulting in stiffness and mass matrices of that size. After resequencing the node numbering for minimal bandwidth, the maximum semi-bandwidth of the problem was 594 and the average semi-

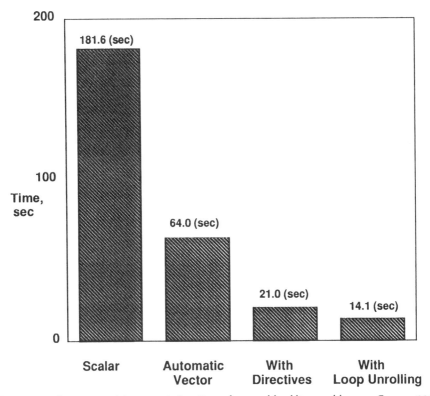

Figure 6.5 Improvement in computation times for panel-buckling problem on Convex 220.

Table 6.1 Variable band matrix–vector multiplication vs. sparse matrix–vector multiplication for panel problem with 6423 degrees of freedom

Type of Matrix	Number of operations	Time (s)	Mflops
Variable band	4 129 044	0.042	97.8
Sparse (non-zeros, only)	300 512	0.013	22.7

Table 6.2 Effects of using local memory in factorization step for panel problem with 6423 degrees of freedom on a CRAY-2

	Time (s)	Mflops
No local memory	1.57	127.0
Local memory	1.49	133.7

bandwith was 319.

Timing results for the high-speed civil transport problem when run on the CRAY-2 with the optimized variable band factor and solve routines and the sparse

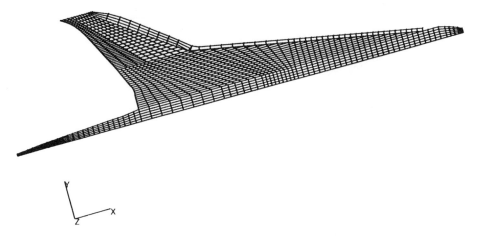

Figure 6.6 Finite element model of symmetric half of high-speed civil transport.

Table 6.3 Solution time for vectorized HSCT problem on CRAY-2

Computation step	Time (s)	Number of floating point operations	Mflops
Factorization	8.5	1 345 852 840	158
Forward/backward solve	11.7	1 260 167 646	107
Reorthogonalization	1.4	114 313 680	80
Matrix–vector multiply	15.3	379 409 549	25
Total CPU time	40.3		

matrix–vector routine as described previously and without any parallelization are shown in Table 6.3. The value of m, or the number of approximate eigenvalues to be calculated, was 60. This results in 30 'acceptable' eigenvalues and associated eigenvectors. This input value was held constant for all of the examples that follow. The size of the matrices resulted in long vector lengths, making the vector operations efficient and the megaflop rate large. As shown in the table, the megaflop rate for the factorization step was 158.

6.6.2 Parallel Lanczos implementation

The vectorized code was next parallelized using the CRAY autotasking capability with compiler directives inserted where data dependencies could not be resolved by the compiler. The HSCT problem was run on the CRAY-2 and the CRAY Y-MP and performance measurements were made.

Multitask performance measurement

There are many different types of measurement tools available on the CRAY systems. One of these, the job accounting report, lists multitasking time usage. Several timing

Table 6.4 Solution time for high-speed civil transport problem on the CRAY Y-MP

Concurrent CPUS	*Connect seconds	= CPU seconds
1	2.650	2.650
2	0.037	0.074
3	0.135	0.404
4	0.792	3.167
5	0.554	2.770
6	1.649	9.894
7	0.212	1.482
8	0.332	2.652
3.63	6.3587	23.0925

Table 6.5 Wall clock time for matrix factorization on the CRAY-2

Number of processors operating concurrently	Time (s)	Actual speed-up	Theoretical speed-up
1	7.9		
2	4.4	1.8	2
3	3.2	2.5	3
4	2.5	3.2	4

routines are available that report CPU time, wall clock time and the number of clock ticks used for each job, or section of a job. Even with these measurement tools, the performance of a multitasked program is sometimes difficult to measure and the timing results vary from run to run, particularly when run in a batch mode. An example of timing statistics obtained when running the high-speed transport problem on the CRAY Y-MP in a batch mode is shown in Table 6.4. In this case an average of 3.63 processors of the eight processors were used. The total CPU time was 23 s, the time on only one processor was 2.65 s and the time spent using more than one processor was 3.71 s. The CPU time is obtained by multiplying the number of processors used (column 1) by the amount of time spent using those processors concurrently (column 2).

The purpose of multitasking is to decrease the wall clock time for a particular computer run. The CPU time will increase due to overhead associated with the parallelization. In the Lanczos algorithm, the matrix factorization step was the calculation that benefited most from the parallelization. Computer runs were made on the CRAY-2 in a dedicated mode on two, three and four processors, respectively. Table 6.5 shows the wall clock time taken for the factorization for these cases. The actual speed-up of 3.2 on four processors represents a considerable decrease in wall clock time for this computational step. Although not shown in the table, a megaflop rate of 826 was obtained using four processors concurrently in the factorization step.

Table 6.6 Frequencies and effect of reorthogonalization for HSCT problem

Eigenvalues found with total reorthogonalization	Eigenvalues found with no reorthogonalization
(rad/s)	
0.02331	0.02331
0.36548	0.02331
0.60044	0.02331
0.70849	0.02331
0.74632	0.02331
0.90365	0.02331
0.91478	0.02331
1.0005	0.02331
1.1048	0.36548
1.1271	0.36548
1.3130	0.36548
1.3760	0.60044

	CONVEX		CRAY-2		CRAY Y-MP	
	Time (s)	Mflops	Time (s)	Mflops	Time (s)	Mflops
Reorthogonalization	8.0	50	1.4	80	0.5	227
Total solution	728.0		40.0		24.0	

6.6.3 Effects and timing of reorthogonalization

Without any reorthogonalization of the Lanczos vectors, repeated and/or spurious eigenvalues will appear. The HSCT problem is used to demonstrate the loss of orthogonality that occurs when implementing the Lanczos method. The first twelve natural frequencies of the HSCT are shown in the left-hand column of Table 6.6 with the vectors reorthogonalized at every step. The values in the right-hand column represent the eigenvalues found with no reorthogonalization. Using a value of m equal to 60, 30 eigenvalues were accepted as approximations to the eigenvalues of the system. When no reorthogonalization was performed, the recursion converged to those eigenvalues on the right. The first eigenvalue was repeated 8 times before the second eigenvalue was found, which was then repeated three times. The computation times and megaflop rates for the total solution and for total reorthogonalization on the CRAY-2 are shown in the table.

The results shown in Table 6.6 confirm the problem inherent in the Lanczos algorithm when no reorthogonalization is performed. The computation times to perform total reorthogonalization on three computers are presented. The computation is highly vectorizable and high megaflop rates, up to 227 on one processor on the CRAY Y-MP, were achieved. The time to reorthogonalize took from 1 to 3% of the computational time on the three computers. For the example problems, a minimum number of Lanczos steps are performed, as only a few of the smallest eigenvalues are of interest. For very large problems, when many eigenvalues are needed, the cost of total reorthogonalization may be prohibitive. Many sophisticated algorithms are being proposed to reduce this overhead and maintain semi-orthogonality among the Lanczos vectors. The loss of orthogonality can be tracked, and reorthogonalization only performed at certain intervals. The number of vectors used can also be decreased. An explanation of the loss of orthogonality and comparison of reorthogonalization algorithms can be found in Reference [25]. The size of the problem, the number of Lanczos steps to be taken, and storage requirements are variables that influence the computational cost of the algorithms.

6.7 SUMMARY

The Lanczos method is an efficient method for solving the eigenvalue problem and is adaptable to vectorization and parallelization. This method is being incorporated into large finite element codes to solve the vibration and buckling problems where only a few of the natural frequencies and mode shapes are needed. Many adaptations and enhancements to the method as originally proposed are being developed to increase the efficiency and reliability of the eigenvalue and eigenvector approximations. Examples include reorthogonalization algorithms to ensure numerical stability and block Lanczos methods to overcome the difficulties in determining multiple roots. Incorporating efficient equation solvers and optimizing the algorithm to take full advantage of the architectural characteristics of the hardware can significantly improve the computational rate.

The many vector operations inherent in the Lanczos method exploit the vector capabilities of the Convex and CRAY computers. The automatic vectorization of the Convex compiler resulted in a 65% decrease in computation time for the example panel buckling problem. Further reductions in computation time were achieved using compiler directives and loop unrolling. The computationally intensive step of factoring the large matrix benefited most from the parallelization on the CRAY computers. The speed-up in computation time for the factorization of the matrix in the transport example problem was 1.8 on two processors and 3.2 on four processors.

To solve very large problems on other than supercomputers, storage demands may become an important issue. Storage reduction schemes discussed include: reducing the bandwidth of the matrices by reordering the nodes, using a diagonal mass matrix instead of the full consistent mass matrix when appropriate and storing the matrices in a sparse format, thus eliminating all zero elements. Other approaches that are being investigated include 'out-of-core' solvers and domain decomposition schemes to partition the problem into sections. Such algorithms will be necessary for large applications on work stations and distributed-memory parallel processors.

The Lanczos vectors can be used as basis vectors in reduced basis methods for structural dynamics, flexible body vibration control and transient thermal problems. Their many uses make it important to have the most efficient and accurate algorithm possible on the latest generation of high-performance computers and on the next generation of parallel–vector computers. Ongoing research is aimed at improving the algorithm and employing the vectors in diverse types of problems. To analyse the large, complex aerospace structures now being designed will require powerful computers and efficient algorithms that can use the available computational power to the best advantage.

BIBLIOGRAPHY

[1] Wilkinson, J.H. (1965) *The Algebraic Eigenvalue Problem*, Oxford University Press.
[2] Jennings, A. (1977) *Matrix Computation for Engineers and Scientists*, J Wiley, Chichester.
[3] Parlett, B. (1980) *The Symmetric Eigenvalue Problem*, Prentice-Hall, Englewood Cliffs, New Jersey.
[4] Bathe, K.J. and Wilson, E.L. (1972) Large eigenvalue problems in dynamic analysis, *ASCE J. Eng. Mech. Div.*, **98**, 1471–1485.
[5] Jensen, P.S. (1972) The solution of large symmetric eigenproblems by sectioning, *SIAM J. Num. Anal.*, **9**, 534–545.
[6] Bostic, S.W. (1990) A vectorized Lanczos eigensolver for high-performance computers, *Proc. AIAA/ASME/ASCE/AHS 31st Structures, Structural Dynamics and Materials Conference*, AIAA Paper No. 90-1148, Long Beach, California, April 2–4, pp. 652–662.
[7] Bostic, S.W. and Fulton, R.E. (1987) A Lanczos eigenvalue method on a parallel computer, AIAA Paper 87-0725-CP, *Proc. AIAA/ASME/ASCE/AHS 28th Structures, Structural Dynamics and Materials Conference*, Monterey, California, April 6–8, pp. 123–135.
[8] Bostic, S.W. and Fulton, R.E. (1985) A concurrent processing implementation for structural vibration analysis, AIAA Paper 85-0783-CP, *Proc. AIAA/ASME/ASCE/AHS 26th Structures, Structural Dynamics and Materials Conference*, Orlando, Florida, April 15–17, pp. 566–572.
[9] Bostic, S.W. and Fulton, R.E. (1986) Implementation of the Lanczos method for structural vibration analysis, AIAA Paper 86-0930-CP, *Proc. AIAA/ASME/ASCE/AHS 27th Structures, Structural Dynamics and Materials Conference*, San Antonio, Texas, May 19–21, pp. 400–410.
[10] Jones, M.T. and Patrick, M.L. (1989) *The Use of Lanczos's Method to Solve the Large Generalized Symmetric Definite Eigenvalue Problem*, NSA CR-181914, ICASE Report 86-69, September.
[11] Storaasli, O., Bostic, S., Patrick, M., Mahajan, U. and Ma, S. (1990) Three parallel computation methods for structural vibration analysis, *J. Guidance, Control, and Dynamics*, **13**(3), 555–561.
[12] Gupta, V.K. and Newell, J.F. (1991) Band Lanczos vibration analysis of aerospace structures, *Proc. Symp. Parallel Methods on Large-Scale Structural Analysis and Physics Applications*, Pergamon Press, New York, NY.
[13] Cullum, J. and Willoughby, R.A. (1985) *Lanczos Algorithms for Large Symmetric Eigenvalue Computations*, vol. 1, Theory, Birkhauser, Boston.
[14] Komzsik, L. (ed.) (1990) *Handbook for Numerical Methods*. MSC/NASTRAN version 66, The Macneal-Schwendler Corporation, Los Angeles, California, April, pp. 4.4-1, 4.4-12.
[15] Parlett, B. (1986) *The State-of-the-Art in Extracting Eigenvalues and Eigenvectors in Structural Mechanics Problems*, Department of Mathematics, University of California, Berkeley.
[16] Huebner, K.H. and Thornton, E.A. (1982) *The Finite Element Method for Engineers*, J Wiley, New York.
[17] Craig, R.R. Jr. (1981) *Structural Dynamics, An Introduction to Computer Methods*, J Wiley, New York.
[18] Hibbeler, R.C. (1983) *Engineering Mechanics, Dynamics*, Macmillan, New York.
[19] Lanczos, C. (1950) An iteration method for the solution of the eigenvalue problem of linear differential and integral operators, *J. Res. Natl. Bureau of Standard*, **45**, 255–282.
[20] Ojalvo, I.U. and Newman, M. (1970) Vibration modes of large structures by an automatic matrix reduction method, *AIAA J.*, **8**, 1236–1239.
[21] Paige, C.C. (1972) *Accuracy and Effectiveness of the Lanczos Algorithm for the Symmetric $Ax = \lambda Bx$ Problem*, Rep. STAN-CS-72-270, Stanford University Press, Palo Alto, California.

[22] Scott, D.S. (1982) The advantages of inverted operators in Rayleigh–Ritz approximations, *SIAM J. Sci. Stat. Comput.*, **3**(1), 68–75.
[23] Simon, H.D. (1984) The Lanczos algorithm for solving symmetric linear systems, Center for Pure and Applied Mathematics, University of California at Berkeley, distributed by Defense Technical Information Center, Alexandria, Virginia, February.
[24] Nour-Omid, B. and Clough, R.W. (1984) Dynamic analysis of structures using Lanczos coordinates, *Earthquake Engineering and Structural Dynamics*, **12**, 565–577.
[25] Simon, H.D. (1984) The Lanczos algorithm with partial reorthogonalization, *Mathematics of Computation*, **42**, 115–142.
[26] Hockney, R.W. and Jesshope, C.R. (1981) *Parallel Computers*, Adam Hilger, Bristol, pp. 27–29.
[27] Poole, E.L. and Overman, A.L. (1991) Parallel variable-band Choleski solvers for computational structural analysis applications on vector multiprocessor supercomputers, *Proc. Symp. Parallel Methods on Large-Scale Structural Analysis and Physics Applications*, Pergamon Press, New York, N.Y.
[28] Poole, E.L. (1991) Comparing direct and iterative equation solvers in a large structural analysis software system, in *Computing Systems in Engineering*, Pergamon Press, Oxford, England, September.
[29] George, A. and Liu, J.W.-H. (1981) *Computer Solution of Large Sparse Positive Definite Systems*, Prentice-Hall, Englewood Cliffs, New Jersey.
[30] Knight, N.F. and Stroud, J.W. (1985) *Computational Structural Mechanics: A New Activity at the NASA Langley Research Center*, NASA TM 87612, September.

S.W. Bostic
National Aeronautics and Space Administration
Structural Mechanics Division
Computational Mechanics Branch
Langley Research Center
Hampton
VA 23681-0001
USA

7
Solving Large-scale Non-linear Problems in Solid and Structural Mechanics

M. Papadrakakis

National Technical University of Athens, Greece

7.1 INTRODUCTION

Significant advances have been made in the last decade in the development and application of non-linear solution procedures. This has been brought about primarily for two reasons: a better understanding of the physical properties of complex structures and the ever-increasing advances in computer technology with increased computational capacity and greater availability. Of course, this rapid development has been somewhat accelerated by the need to understand what happens to complex structures when subjected to intense loading conditions.

One of the central features in performing non-linear analysis is the proper selection of solution algorithms, which is still as much an art as a science. The nature of structural non-linearities is generally quite diverse, when both kinematic and material effects are included, which makes structural response quite unpredictable. In this context, the selection of a reliable as well as computationally efficient solution strategy is a very demanding task. Unlike linear problems, it is extremely difficult to develop a single methodology of general validity which can be used to handle the diversity of non-linear structural behaviour.

The classification of the non-linear solution procedures is not an easy task since all may be considered iterative or can be used in combination with an incremental procedure. A more distinct classification approach is based on the level of mathematical

---

Solving Large-scale Problems in Mechanics, Edited by M. Papadrakakis
© 1993 John Wiley & Sons Ltd

formulation. Using the principle of virtual work or setting to zero the first variation of the potential energy, if available, the condition of total equilibrium, which states that the external and internal forces must balance, may be directly expressed. The next level of equations, expressing the condition of incremental equilibrium, can be obtained using the incremental form of the virtual work equation, or the second variation of the total potential energy.

Usually the first class of methods do not employ a Hessian (stiffness) matrix and they are referred to as explicit methods, while the second class of methods which use both levels of formulation do involve the computation and handling of a stiffness matrix and they are referred to as implicit methods. Many solution procedures, however, using the implicit approach have evolved without requiring the formation of a stiffness matrix, while explicit methods may also be applied in connection with a stiffness matrix. The key difference between these two classes of methods is that in explicit methods there is no need to solve a linearized problem in each non-linear iteration as is required in implicit methods.

This chapter reviews solution procedures for non-linear finite element analysis. Solution methods that are restrictive in scope or with which the author has had no experience are not covered. More emphasis is given to Newton-like methods in connection with iterative solution methods for the linearized equations and the arc-length method for tracing equilibrium paths.

7.2 EXPLICIT METHODS

Discrete equilibrium equations arising from finite element non-linear formulations may be presented in the general compact form $g(u, p, \theta) = 0$, where u denotes the unknown vector of generalized displacements, p is an array of control parameters, θ is a functional of past history of the generalized deformation gradients, and g is the residual vector of out-of-balance generalized forces. If the system is conservative, g is the gradient of the total potential energy. In many applications the state and control parameters may be segregated and under fixed loading may be expressed as

$$g(u) = F(u) - P = 0 \tag{7.1}$$

where $F(u)$ is the internal force vector and P the applied external force vector.

A class of explicit methods based on equation (7.1) can be expressed by the three-term recursion formula introduced by Engeli et al. [1], for solving linear equations. The general form of the recursive expression is [2]

$$u_{i+1} = -\frac{g_i}{q_i} + \left(\frac{e_{i-1}}{q_i} + 1\right)u_i - \frac{e_{i-1}}{q_i}u_{i-1} \tag{7.2}$$

where $g_i = g(u_i)$, q_i and e_{i-1} are iteration coefficients characteristic of the method. It can be shown [3] that according to the expressions which define the coefficients q_i and e_{i-1}, the recursive formula of equation (7.2) can be transformed to the conjugate gradient iterative method, the second-order Richardson process, the Chebyshev semi-iterative method or the dynamic relaxation method. A characteristic property of explicit methods based on equations (7.1) and (7.2) is that they can be employed in the same form for the solution of a system of linear as well as non-linear equations, by using the same iterative process. For this reason they are also called pure iterative methods or vector iteration methods since they involve only vector operations.

7.2.1 Non-linear conjugate gradient (CG) method

The CG method can be described by the three-term recursion formula with the following correspondence between the iteration parameters of equation (7.2) and those of the CG algorithm (1.3) of Reference [4]:

$$q_i = \frac{1}{\alpha_i}, \quad e_{i-1} = \frac{\beta_{i-1}}{\alpha_{i-1}} \quad \text{for} \quad i \geq 1 \tag{7.3}$$

with $e_{i-1} = 0$ for $i = 0$.

The only significant modification of the non-linear CG is that the step length α_i must be computed by an iterative process, called line search, rather than in closed form as is obtained in linear applications of the method. This may have a tremendous influence on the performance of the method depending on the line search routine employed and the cost of residual calculations. Different algorithms based on non-linear versions of the CG method have been proposed in the field of unconstrained optimization, which maintain the convergence characteristics of the method with inaccurate line searches. In structural non-linear problems, however, the nature of non-linearities and the computational effort may bear no direct and simple relationship with that for mathematical test functions. As a result of this the conclusions regarding the effectiveness of these algorithms for solving non-linear problems in structural analysis are likely to be different. This was confirmed in a study performed in Reference [5] where it was observed that the basic non-linear CG of Fletcher and Reeves [6] combined with an effective line search routine is a good choice among the different versions of non-linear CG algorithms. When strong non-linearities are present, however, it was found that the expression for β

$$\beta_i = g_{i+1}^T (g_{i+1} - g_i) / g_i^T g_i \tag{7.4}$$

proposed by Polak and Ribière [7] is a better choice.

7.2.2 Dynamic relaxation (DR) method

Dynamic relaxation is a robust iterative method for solving highly non-linear systems. The algorithm solves the non-linear equations (7.1) by viewing them as the steady-state solution of the second-order pseudo-dynamic problem $M\ddot{u} + C\dot{u} + F(u) - P = 0$, where M and C are properly selected fictitious mass and damping matrices. When integrating the pseudo-dynamic problem with the central difference scheme, the method can be represented by equation (7.2) where the characteristic parameters are functions of the lower and higher pseudo-frequencies of the pseudo-dynamic problem and take the values

$$q_i = \frac{(\omega_{max} + \omega_{min})^2}{4}, \quad e_{i-1} = \frac{(\omega_{max} - \omega_{min})^2}{4} \quad \text{for} \quad i \geq 1 \tag{7.5}$$

with $q_i = (\omega_{max}^2 + \omega_{min}^2)/2$ and $e_{i-1} = 0$ for $i = 0$. Adaptive procedures for the selection of the DR parameters are proposed by Papadrakakis [8] and Underwood [9], which enhance the convergence properties of the method.

Explicit methods are very effective in highly non-linear problems in which the cost of the residual calculation is not expensive. They also possess the advantage of being more susceptible to parallelization and vectorization. Studies performed in References [2, 3, 10, 11] showed the superiority of explicit over implicit methods and among the former that of explicit DR over non-linear CG. Dynamic relaxation is especially attractive in problems with rough non-linearities characterized by discontinuous field equations, involving inequality constraints and path-dependent non-smooth local responses [12–14]. Implicit methods may behave poorly in such cases where the stiffness matrix may oscillate wildly as the solution changes by small amounts. DR on the other hand is less sensitive to these changes since it does not linearize the response within a load or time step and its convergence path may to a certain extent be controlled by the iteration parameters. Overall, DR is an easily programmable method and very robust non-linear solution procedure that is considerably less sensitive to this type of non-linearity than the stiffness-based implicit methods are.

Since the overhead associated with the residual calculation in explicit methods may be excessive in structures with complex elements and smooth non-linearities, because explicit methods require many more iterations to converge than implicit methods, it is advisable to use implicit methods for these types of problems. However, DR may be coupled with implicit methods in problems where smooth and rough non-linearities coexist in different parts of the structure.

7.3 IMPLICIT METHODS

Newton-like methods are the most popular class of implicit methods for the solution of equation (7.1). Following Fletcher [15], the terminology 'Newton–Raphson method' is more appropriate to general multidimensional non-linear equations, while the term 'Newton method' can be reserved for function minimization problems.

7.3.1 Conventional Newton–Raphson (CNR) method

Let the system of non-linear equations be given by equation (7.1), where u and g are both N-dimensional vectors, and u_i is an approximation to the root u^* of equation (7.1) such that

$$u^* = u_i + x_i. \tag{7.6}$$

Then, using Taylor's theorem, the conventional Newton–Raphson iteration is obtained by

$$g(u^*) = 0$$

or

$$g(u_i + x_i) = g_i + \left[\frac{\partial g}{\partial u}\right]_i x_i + \cdots \approx g_i + K_i x_i \qquad (7.7)$$

where K_i is the Jacobian of g or the tangent stiffness matrix evaluated at u_i. An approximate value for x_i can be obtained by solving the equation

$$K_i x_i = -g_i. \qquad (7.8)$$

In some cases equation (7.8) may only be used to give the direction of the step at each iteration, whereas the length of the step is obtained from a line search procedure. A step length α_i is chosen to improve the solution along the direction x_i and the next approximation to u^* is u_{i+1} such that

$$u_{i+1} = u_i + \alpha_i x_i. \qquad (7.9)$$

An energy-based formulation can also be adopted for describing these methods by taking the first and second variation of the total potential energy Φ of the system. Although such formulation is strictly applicable to conservative systems, it can be useful because it allows the adoption of various solution algorithms and techniques developed in the field of mathematical programming or unconstrained optimization. From this perspective the solution methods may be viewed as optimizing a twice continuous differentiable function $\Phi(u)$ with g_i and $[\partial g/\partial u]_i$ being the gradient and the Hessian of Φ, respectively.

CNR takes full advantage of first and second derivative information of the function to be optimized and possesses, in general, ideal characteristics of local convergence and stability. It is well known, however, that for large-scale problems the CNR may be too costly due to the formulation and factorization of the tangent stiffness matrix at each iteration. To reduce the cost in memory, operations or both, various modifications to the previously described generic algorithm have been proposed. The simplest variation of this kind is the modified Newton–Raphson method (MNR), in which the same matrix K_0 is used for all iterations at each loading step. Normally K_0 is the tangent stiffness matrix computed at the beginning of the step. The convergence of the method is linear rather than quadratic and may run into serious difficulties if the step brings large changes in the kinematics or material properties of the structure. Another variant of Newton's method called conjugate-Newton, was proposed by Irons and Elsawaf [16] after preconditioning the non-linear CG algorithm with the tangent stiffness matrix K_0. The resulting expression for x_i becomes

$$x_i = \delta_i + \beta_{i-1} x_{i-1} \qquad (7.10)$$

with

$$\beta_{i-1} = \frac{\delta_i^T(g_i - g_{i-1})}{\delta_{i-1}^T g_{i-1}} \quad \text{and} \quad \delta_i = -K_0^{-1} g_i. \tag{7.11}$$

7.3.2 Quasi-Newton (QN) methods

The idea behind QN, or variable metric methods, is to start with an initial approximation K_0 of $[\partial g/\partial u]_0$ or an approximation on K_0^{-1} of $[\partial g/\partial u]_0^{-1}$ and then to use a simple correction at every iteration step to update the matrix

$$K_i = K_{i-1} + V_i \tag{7.12}$$

in order to ensure the secant property

$$K_i s_i = y_i \tag{7.13}$$

(correspondingly, $K_i^{-1} = K_{i-1}^{-1} + U_k$ satisfying $K_i^{-1} y_i = s_i$), with

$$s_i = u_i - u_{i-1}, \quad y_i = g_i - g_{i-1}. \tag{7.14}$$

The attractive feature of this class of methods is that by using these updates the expensive recomputation and factorization of the stiffness matrix is avoided, while it accounts to some degree for changes due to non-linearities of the problem. Thus, QN methods attempt a compromise between CNR and MNR, by providing cost-effective updates to the stiffness matrix compared with CNR and accomplishing improved convergence properties compared with MNR.

Satisfaction of the quasi-Newton equation (7.13) can be achieved via many different formulations. Most common methods utilize rank-one or rank-two updates. The rank-one updates involve the generation of updates using one correction term. Typical updates performed on the inverse Hessian are the so-called Broyden [17] and Davidon [18] updates. The rank-two updates are generated using two correction terms and provide many potential forms. The most widely used rank-two updates, belong to the Broyden β-family [19], where the most important members of this family are the DFP (Davidon–Fletcher–Powell) and BFGS (Broyden–Fletcher–Goldfarb–Shanno) updates. As these updates accumulate, the BFGS method renders at its limit a stiffness matrix resembling tangential stiffness.

Much work has been devoted to the development of this type of methods in the fields of mathematical programming and unconstrained optimization. Variable metric methods were originally introduced by Davidon [18] and subsequently clarified by Fletcher and Powell [20]. Extensive treatment of the methods are provided by Wolfe [21], Fletcher [15], Gill *et al.* [22] and Dennis and Schnabel [23], whereas Dennis and Moré [24] present an extensive survey of the underlying theory. One of the earliest applications of quasi-Newton methods of finite element problems was that of Fox and Stanton [25]. It was, however, the paper by Matthies and Strang [26] which was influential in attracting attention to the application of quasi-Newton methods in non-linear finite element analysis. Some of the first implementations of quasi-Newton methods in structural mechanics were reported by Bathe and Cimento [27], Geradin *et al.* [28], Pica and Hinton [29], Crisfield [30, 31], followed by References [32–7]. An extensive study on the performance of quasi-Newton methods was presented by Schweizerhof [38]. The BFGS update is considered, in the mathematical programming literature, as the most effective of the variable metric updates [24,39] and has been adopted in a number of studies in structural mechanics [26, 27, 29, 34–7]. It was observed, however, in some structural analysis applications [28, 32], that rank-two schemes do not result in better convergence rates and rank-one updates are preferable due to their lower cost.

Early work on quasi-Newton methods did not influence the development of solution procedures in non-linear finite element applications because of the destruction of the sparsity pattern of the stiffness matrix away from equilibrium during the updating procedure. During the late 1970s and early 1980s much research effort was focused on the development of sparsity-preserving quasi-Newton updates. Efficient sparsity-preserving QN algorithms were proposed by Nocedal [40] and Buckley and LeNir [41] in the mathematical programming literature; and also by Matthies and Strang [26] in structural finite element applications. In these applications, the stiffness matrix, or its inverse, is not directly modified and banding properties are not destroyed.

Sparsity-preserving updates for Broyden and Davidon rank-one updates and for DFP and BFGS rank-two updates can be expressed in the following forms:

1. Broyden's inverse update:

$$K_i^{-1} = K_{i-1}^{-1} + \frac{(s_i - K_{i-1}^{-1} y_i) s_i^T K_{i-1}^{-1}}{s_i^T K_{i-1}^{-1} y_i} \tag{7.15a}$$

2. Davidon's inverse update:

$$K_i^{-1} = K_{i-1}^{-1} + \frac{(s_i - K_{i-1}^{-1} y_i)(s_i - K_{i-1}^{-1} y_i)^T}{(s_i - K_{i-1}^{-1} y_i)^T y_i} \tag{7.15b}$$

3. DFP inverse update:

$$K_i^{-1} = K_{i-1}^{-1} + \varrho_i s_i s_i^T - \frac{K_{i-1}^{-1} y_i y_i^T K_{i-1}^{-1}}{y_i^T K_{i-1}^{-1} y_i} \tag{7.15c}$$

4. BFGS inverse update:

$$K_i^{-1} = (I - \varrho_i s_i y_i^T) K_{i-1}^{-1} (I - \varrho_i y_i s_i^T) + \varrho_i s_i s_i^T \tag{7.15d}$$

with

$$\varrho_i = \frac{1}{s_i^T y_i} \tag{7.16}$$

or under the general expression

$$K_i^{-1} = U_{QN}(K_{i-1}^{-1}, s_i, y_i) = U_{QN}(K_0^{-1}, s_1, y_1, \ldots, s_i, y_i). \tag{7.17}$$

One can thus define K_i^{-1} implicitly instead of by means of K_0^{-1} and a set of updates, each update consisting of two vectors s_i, y_i. Then, the solution $x_i = K_i^{-1} g_i$ may be evaluated without having K_i^{-1} explicitly available.

In the following we restrict our implementation to the BFGS update. The solution vector x_i for the BFGS becomes

$$x_i = -\left\{ \left[\prod_{j=i}^{1}(I - \varrho_j s_j y_j^T) \right] K_0^{-1} \left[\prod_{j=1}^{i}(I - \varrho_j y_j s_j^T) \right] \right\} g_i \tag{7.18}$$

$$- \sum_{k=1}^{i-1} \left\{ \left[\prod_{j=i}^{k+1}(I - \varrho_j s_j y_j^T) \right] \varrho_k s_k s_k^T \left[\prod_{j=k+1}^{i}(I - \varrho_j y_j s_j^T) \right] \right\} g_i - \varrho_i s_i s_i^T g_i.$$

This technique of computing x_i leads to vectorized BFGS or QN updates which involve vector operations only, except one matrix–vector operation with the initial update K_0^{-1}, which constitutes the solution to the linearized problem inside each non-linear iteration. Recursive expressions for efficient computation of the solution x_i are presented in References [26, 28, 30, 38, 40-2]. Similar expressions can be established for the other types of QN updates.

The original form of equation (7.15(d)) for the BFGS update can be expressed more elegantly in the product form [43]

$$K_i^{-1} = (I + w_i v_i^T) K_{i-1}^{-1} (I + v_i w_i^T) = A_i^T K_{i-1}^{-1} A_i \tag{7.19}$$

and

$$v_i = -\left(\frac{s_i^T y_i}{s_i^T K_{i-1} s_i} \right)^{1/2} K_{i-1} s_i - y_i, \qquad w_i = \frac{s_i}{s_i^T y_i}. \tag{7.20}$$

The product form is applicable to positive-definite matrices and is suitable for monitoring the behaviour of K_i^{-1} by the easily computable condition number of the updating matrix A_i. This form has been used in References [26, 27, 29].

The iterative procedure of an implicit method for the solution of equation (7.1) can therefore be briefly described in the following algorithmic form:

Choose u_0 and K_0 in the domain of attraction of u^*
for $i = 0, 1, \ldots$
1. Compute $g_i = F(u_i) - P$

2. Update K_{i-1} or K_{i-1}^{-1} (for $i \geq 1$)
3. Solve $K_i x_i = -g_i$ for x_i (7.21)
4. Find appropriate α_i (line search) and set $u_{i+1} = u_i + \alpha_i x_i$.
5. If the required accuracy is attained: $\|g_i\|/\|P\| \leq \varepsilon$ then stop with u_{i+1} as the approximation to u^*, else increment i to $i+1$ and go to step 1.

7.3.3 Limited memory quasi-Newton (LMQN) implementations

Despite the efficient storage handling properties of the vectorized BFGS update, the accumulation of two vectors per iteration that have to be stored may substantially increase the storage requirements for the implementation of the method. To further reduce the storage requirements, limited memory quasi-Newton updates have been proposed which occasionally discard all vectors and replace them with new ones.

Four ways of employing limited memory updates are presented in this study. The first and simplest has been introduced by Matthies and Strang [26]. They use an LDL^T factorization of the tangent stiffness matrix K_0 and after every Mth iteration they discard all previous updates and start a new cycle with $K_i = K_0$. This approach will be denoted 'MSTR'.

The second LMQN implementation is a modification to the algorithm proposed by Buckley and LeNir [41] in which after every Mth iteration each new update replaces the previous one, i.e.

$$K_i^{-1} = U_{BFGS}(K_{M-1}^{-1}, s_i, y_i) \qquad (7.22)$$

until a restart is forced by Powell's criterion $|g_{i+1}^T K_0 g_i| > 0.2|g_i^T K_0 g_i|$ at iteration $i + 1 = r$. Upon restart, all previous updates are discarded and K_r^{-1} is defined as

$$K_r^{-1} = U_{BFGS}(K_0, s_r, y_r). \qquad (7.23)$$

The process continues as if K_r^{-1} were K_1^{-1}. This implementation is abbreviated to 'BLNR'. This special treatment of the updates makes the algorithm behave in two distinct ways: For M-1 iterations after restart, it is a common LMQN; but from that point on and until next restart, it behaves like an alternative form of the preconditioned CG with preconditioning matrix K_{r+M}^{-1} [39, 44].

The remaining two implementations have been suggested by Nocedal [40]. The first one is the usual LMQN except that after M iterations each new update replaces the earliest one. Thus, the typical update for $i > M$ becomes

$$K_i^{-1} = U_{BFGS}(K_0^{-1}, s_{i-M+1}, y_{i-M+1}, s_{i-M+2}, y_{i-M+2}, \ldots, s_i, y_i). \qquad (7.24)$$

This approach will be known as Special Quasi-Newton (SQN) after Nocedal. The other implementation is a preconditioned non-linear conjugate gradient algorithm according to Nazareth [39], with K_{i-1}^{-1} acting as a preconditioner:

$$x_i = -K_{i-1}^{-1} g_i + \beta_{i-1} x_{i-1} \tag{7.25}$$

and

$$\beta_{i-1} = y_i^T K_{i-1}^{-1} g_i / y_i^T x_{i-1} \tag{7.26}$$

where K_{i-1}^{-1} is produced in the same way as in SQN. It is worth noting that it is K_{i-1}^{-1} and not K_i^{-1} that it is used to generate x_i as in QN methods. This approach will be known henceforth as Special Conjugate Gradient (SGG). Similar LMQN implementations can be established for the other types of QN updates.

Perry [45] and Shanno [46] first considered LMQN updates (with $K_0 = I$) and subsequently Shanno called such methods 'memoryless QN methods'. Since these methods require non-negligible storage, Gill et al. [44] considered the name 'limited memory QN methods' as more appropriate. Research in the field of mathematical programming has established links between the BFGS method and CG procedures and has developed hybrid procedures [39, 41, 44, 47]. In fact the BFGS update may be interpreted as a CG algorithm for which the metric, instead of being fixed, is updated at each step and produces the same direction vector in linear problems under exact line searches [39].

In structural mechanics Crisfield, in a number of papers [30, 31, 48, 49], developed a range of 'faster MNR' or 'secant-Newton techniques' which are closely related to the LMQN procedure. The MSTR of Matthies and Strang, and SQN of Nocedal with one update vector ($M = 1$), are identical to the three-vector iterative update of Reference [30]. The SCG with $M = 1$ coincides with the conjugate Newton method of Irons and Elsawaf [16] (equations (7.10), (7.11)). Similar methods have been proposed by Papadrakakis and Gantes [50] based on CG or BFGS updates using different initial metrics for K_0 as preconditioners. Finally, the QN method with $M = 0$ becomes the MNR if K_0 is the tangent stiffness matrix.

If the number of equations is N and the number of updates stored is M, then the limited memory BFGS methods need to store $M(2N + 1)$ more numbers $(s_1, ..., s_M, y_1, ..., y_M, \varrho_1, ..., \varrho_M)$ and perform $2M(2N + 1)$ more operations per iteration than the MNR with the same K_0. The extra cost of storage and computation is proportional to M. The total memory available can be readily partitioned between the choice of K_0 and the number of update vectors.

Another approach to extend Newton-like methods to large-scale problems is through the use of finite differencing and sparse matrix techniques. Sparsity-preserving updates, in the symmetric case, were first suggested by Powell [51], Toint [52] and Shanno [46] and applied in structural analysis problems by Kamat et al. [53], together with techniques for replacing the Hessian by a sparse finite difference approximation [54, 55]. Finally, techniques for updating the factors of the initial Hessian rather than the Hessians themselves have been proposed [56, 57]. However, for reasons of efficiency and simplicity all these techniques have not attracted much attention since the LMQN implementations emerged.

Numerical experience has shown that in some iterations it is better to avoid the quasi-Newton update and replace the solution vector by the MNR or CNR directions. Undesirable quasi-Newton updates can be avoided by establishing admissible domains for the eigenvalues of A_i of equation (7.19) [37], or for some characteristic parameters of the methods [30]. Thus, different 'cut-out' criteria have been devised which are more or less problem dependent and are similar to that of the re-start in the CG method [58].

7.3.4 Truncated Newton-like methods

Within an implicit strategy the solution of non-linear equations is obtained through the generation of a series of linearized problems. It has been observed that even for non-linear three-dimensional structural mechanics problems the linear equation solution represents the major cost of the analysis. With moderately sized problems of several tens of thousands of degrees of freedom, traditional direct linear equation solutions use up to 70–80% of the total CPU time. It is therefore imperative that the cost of linear equation solutions be reduced in both storage and CPU time if large-scale non-linear implicit computations are to be economically feasible. The use of iterative schemes for the solution of the linearized system of equations at each non-linear step has two desirable advantages which result in cost-effective solutions: (i) it efficiently exploits the sparsity of the involved matrices, as described in Reference [4]; and (ii) it provides a means of controlling the accuracy of the linearized solution.

Since Newton-like equations are based on a Taylor series expansion near the solution u^*, there is no guarantee that the search direction they compute will be as important far away from u^*. At the beginning of the solution process a reasonable approximation to the Newton-like direction may be almost as effective as the accurate direction itself. It is only gradually, as the solution is approached, the Newton-like direction takes on more and more meaning. Solving exactly the linearized problem at each implicit iteration is, therefore, not justified when far away from the solution. Instead, an inaccurate solution to the linearized problem is computed using an iterative method in order to minimize the effort while distant from the non-linear solution, and then gradually increase it as the non-linear solution is approached. Notice that this solution strategy is doubly iterative: there is an outer or major 'non-linear' iteration to improve the solution u of the non-linear problem, and a truncated inner or minor 'linear' iteration to compute the search direction x.

The asymptotic convergence rate of the resulting algorithm is controlled by a user-specified parameter which specifies the accuracy of the linear (inner) solution. In this context $x_i^{(j)}$ is acceptable as the solution of the system $K_i x_i = -g_i$ of equation (7.8), after j iterations of the linear solver, if the truncated termination condition criterion $\|r^{(j)}\|/\|g_i\| < \eta_i$ is satisfied with $r^{(j)} = K_i x_i^{(j)} + g_i$. The parameter η_i is the forcing sequence that controls the required accuracy of non-linear (outer) step i. An expression for η_i is adopted in this study [59]:

$$\eta_i = \min\left\{\eta_0, \left(\frac{\|g_i\|}{\|g_0\|}\right)^t\right\} \quad \text{with} \quad 0 < t < 1 \quad (7.27)$$

which is motivated by similar ones proposed by Dembo and Steihaug [60] and by Nour-Omid et al. [61], and is scale invariant. The optimal choice of the parameters η_0 and t is problem-dependent and can be adjusted accordingly. This forcing sequence will result in an adaptive algorithm for solving the non-linear problem. When far away from the solution, $\|g_i\|$ is large and hence minimal work is required to obtain a direction which satisfies the truncated termination condition. As the non-linear solution is approached $\|g_i\|$ becomes small and the linear solution is obtained with increased accuracy.

The technique combining an iterative solver in the context of solving non-linear equations was first applied by Sherman [62] using the successive overrelaxation method as a linear solver. The truncated Newton method to unconstrained minimization problems has been considered in References [60, 63–6] using CG or Lanczos iterative solvers. In non-linear finite element problems preconditioned conjugate gradient (PCG) and Lanczos methods have been combined with MNR or QN non-linear iterations in a number of applications [61, 67–72], while in Reference [59] PCG has been combined with a number of LMQN updates. The use of PCG with truncation interpolates between the preconditioned non-linear CG and MNR or QN direction. A finite difference formula was also used to approximate the matrix–vector multiplication (Ku) required at each CG iteration in the truncated Newton-like methods. This operation is performed by differencing the gradient along the vector u [60, 63–5, 73].

7.4 TRACING EQUILIBRIUM PATHS

7.4.1 Incremental-Iterative formulation

In the previous section it was assumed that the analyst prescribes the various load levels for which equilibrium configurations are to be calculated. This can be difficult without an appropriate knowledge of the load-carrying capacity of the structure. There are also cases in structural mechanics were load–displacement curves exhibit unstable branches followed by stable equilibrium paths. The true response in Figure 7.1 is dynamic. Under load or displacement control the dynamic response would follow the dashed lines a or b of Figure 7.1, respectively. The coupling of non-linear static and dynamic solution procedures may be cumbersome and in most cases a static solution is used to follow the unstable paths. For some problems all that may appear to be required is the load level at the first limit point. However, without analysis techniques that allow the critical points to be passed even this information may be unavailable or unreliable. 'Collapse loads' are often associated with a failure to achieve convergence of the iterative solution procedure. But iterative failure may occur for a number of other reasons connected with the implementation of the method and not with the strength of the structure. When studying the sensitivity of a structure against imperfections, the effects of a concentrated load on a restricted part of the structure, or the behaviour of structures with softening materials, it is important to obtain information on the nature of load shedding after the occurrence of a local instability in order to assess the behaviour of the whole structure. It is therefore necessary that the whole load–displacement path be traced using a reliable automatic incremental procedure.

Since conventional load control is not generally sufficient to trace unstable paths, more general incremental control strategies are adopted with iterations performed

7.4 TRACING EQUILIBRIUM PATHS

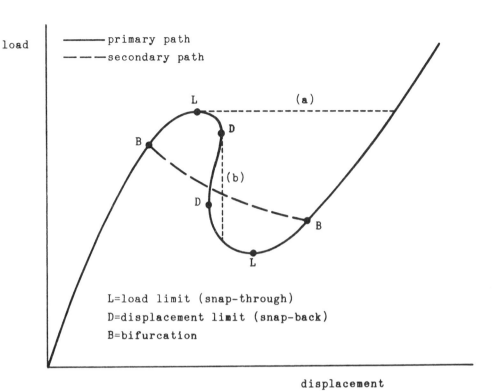

Figure 7.1 Load–displacement paths.

on the load as well as on the displacement space. Most path-following solution strategies are based on a combination of incremental steps followed by equilibrium iterations in the non-linear space. If we assume that the structure is subjected to a loading λP, where the vector P is a fixed external loading and the scalar λ is a load-level parameter, equation (7.28) defines a state of proportional loading in which the loading pattern remains fixed:

$$P^m_{i+1} = P^m_i + \delta\lambda_{i+1} P. \tag{7.28}$$

The superscript m denotes the load step number and subscript i the non-linear iteration. The solution procedures described in the previous section may constitute a part of an overall incremental-iterative solution strategy in which the fundamental equilibrium equation (7.8) is given by

$$\alpha_i K^m_i x_i = P^m_{i+1} - F^m_i \tag{7.29}$$

and the total displacements inside the increment are given by

$$\Delta u_{i+1} = \Delta u_i + \alpha_i x_i \tag{7.30}$$

196 SOLVING NON-LINEAR PROBLEMS IN SOLID AND STRUCTURAL MECHANICS

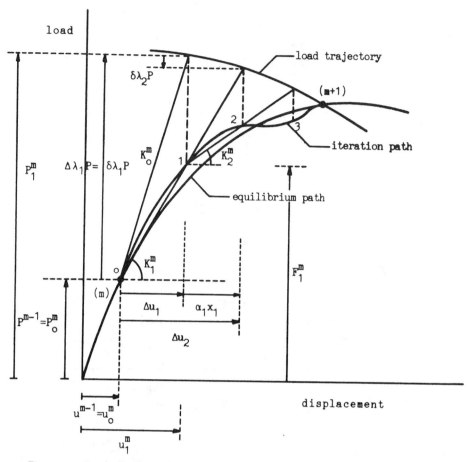

Figure 7.2 Load–displacement equilibrium and non-equilibrium paths. Nomenclature.

with initial conditions as shown in Figure 7.2. Non-linear iterations must now be performed in the load–displacement space, where the load factor is considered as a new variable in conjunction with the displacement vector. A constraint of the form

$$f(\delta\lambda, \Delta u) = 0 \tag{7.31}$$

must therefore be added to the equilibrium equations, where f represents the 'hypersurface' in the load–displacement space on which the iteration process is constrained. As a result of this additional equation, the load varies inside the increment and iterations are performed under variable loading (see Figure 7.2). The augmented Jacobian is neither symmetric nor banded, and instead of solving the augmented coefficient matrix in one step, a two-step procedure is usually adopted [74–6].

Denoting the gradient or the unbalanced force vector $g(u, \delta\lambda)$ as

$$g_i = F_i^m - P_i^m \tag{7.32}$$

the solution vector is obtained from

$$\alpha_i x_i = \delta\lambda_{i+1} x'_i + x''_i \tag{7.33}$$

with

$$x'_i = -K_i^{-1} P \quad \text{and} \quad x''_i = -K_i^{-1} g_i \tag{7.34a,b}$$

where the superscript m has been omitted for clarity. The solution vector consists now of two parts: x'_i is a multiple of a solution vector due to the externally applied load, whereas x''_i stems from the out-of-balance forces. Equation (7.33) has now to be combined with the constraint equation (7.31) and a line search procedure in order to give the incremental displacements Δu_{i+1}.

The incremental–iterative procedure of an implicit method under variable loading may be briefly described in the following algorithmic form:

- a. Incremental phase
  - a1. Compute $\delta\lambda_1, \Delta u_1 = (K_0^m)^{-1} \delta\lambda_1 P$ (predictor step)
- b. Non-linear iterative phase (corrector steps) ($i = 1, 2, \ldots$)
  - b1. Compute F_i^m, g_i
  - b2. Test for convergence: $\|g_i^m\| / \|P_i^m\| \leq \varepsilon$
  - b3. Define K_i (or K_i^{-1})
- c. Solution phase (7.35)
  - c1. Solve for x'_i and x''_i
  - b4. Compute $\delta\lambda_{i+1}$ from the constraint equation and set $\Delta\lambda_{i+1} = \Delta\lambda_i + \delta\lambda_{i+1}$
  - b5. Compute α_i from a line search routine
  - b6. Update displacements: $\Delta u_{i+1} = \Delta u_i + \alpha_i x_i$
    and $u_{i+1}^m = u_i^m + \Delta u_{i+1}$
  - a2. Update the external load: $P_0^{m+1} = P_0^m + \Delta\lambda P$

7.4.2 Constrained implicit methods

A variable loading implementation of CNR and QN methods requires the two solutions of equations (7.34) in each non-linear iteration. This is avoided in MNR methods where x'_i remains constant inside the increment and only x''_i needs to be evaluated in each iteration. With K_i or K_i^{-1} to be updated at each iteration CNR or QN methods require some additional computational work to cope with the different solution x'_i per iteration. Two formulations are presented regarding QN methods which avoid the two complete solutions of equations (7.34) by elaborating on the two expansions of $K_i^{-1} P$ and $K_i^{-1} g_i$ resulting from equation (7.18).

Following the nomenclature of Figure 7.2, with positive $\delta\lambda_i$, as shown in the figure, the secant property involves the vectors

$$s_i = \Delta u_i - \Delta u_{i-1} = \alpha_{i-1} x_{i-1} \tag{7.36}$$

$$y_i = g_i - g'_{i-1} \tag{7.37}$$

with

$$g'_{i-1} = g_{i-1} - \delta\lambda_i P. \tag{7.38}$$

Equations (7.34) in relation to equations (7.17) become

$$x'_i = U_{\text{BFGS}}(K_0^{-1}, s_1, y_1, \ldots, s_i, y_i)P \tag{7.39}$$

$$x''_i = -U_{\text{BFGS}}(K_0^{-1}, s_1, y_1, \ldots, s_i, y_i)g. \tag{7.40}$$

Analytic expressions for equations (7.39) and (7.40) may be written after introducing the auxiliary solution vectors

$$x = K_0^{-1}P, \qquad x^*_i = K_0^{-1}g_i \qquad \text{and} \qquad z_i = K_0^{-1}(g_i - g'_{i-1}) \tag{7.41}$$

$$\begin{aligned}
x''_i = &-\left\{\prod_{j=i}^{1}(I - \varrho_j s_j y_j^T)\right\}\left\{x^*_i - \varrho_i z_i s_i^T g_i\right. \\
&- \left[\sum_{k=1}^{i-1}\varrho_k z_k s_k^T\left[\prod_{j=k+1}^{i}(I - \varrho_j y_j s_j^T)\right]g_i\right] \\
&- \sum_{k=1}^{i-1}\left[\left[\prod_{j=i}^{k+1}(I - \varrho_j s_j y_j^T)\right]\varrho_k s_k y_k^T\left[\prod_{j=k+1}^{i}(I - \varrho_j y_j s_j^T)\right]\right]g_i - \varrho_i s_i s_i^T g_i
\end{aligned} \tag{7.42}$$

and

$$\begin{aligned}
x'_i = &-\left\{\prod_{j=i}^{1}(I - \varrho_j s_j y_j^T)\right\}\left\{x_i - \varrho_i z_i s_i^T P\right. \\
&- \left[\sum_{k=1}^{i-1}\varrho_k z_k s_k^T\left[\prod_{j=k+1}^{i}(I - \varrho_j y_j s_j^T)\right]P\right] \\
&+ \sum_{k=1}^{i-1}\left[\left[\prod_{j=i}^{k+1}(I - \varrho_j s_j y_j^T)\right]\varrho_k s_k y_k^T\left[\prod_{j=k+1}^{i}(I - \varrho_j y_j s_j^T)\right]\right]P + \varrho_i s_i s_i^T P.
\end{aligned} \tag{7.43}$$

Further elaboration on the equations (7.42) and (7.43) may produce a recursive formula for x'_i, which is computationally more efficient [38, 42]. Introducing the auxiliary vector w_i and the scalar parameter γ_i, with

$$w_i = K_{i-1}^{-1}(g_i - g'_{i-1}) = K_0^{-1}g_i - \sum_{j=1}^{i-1}(\varrho_j s_j w_j^T + \varrho_j w_j s_j^T)g_i$$

$$+ \left[\sum_{j=1}^{i-1}\gamma_j s_j s_j^T\right]g_i + \frac{s_i}{\alpha_{i-1}} \tag{7.44}$$

$$\gamma_i = \varrho_i[\varrho_i w_i^T(g_i - g'_{i-1}) + 1] \tag{7.45}$$

the two solution vectors become

$$x_i'' = -w_i + \frac{s_i}{\alpha_{i-1}} + \varrho_i s_i w_i^T g_i + \varrho_i w_i s_i^T g_i - \gamma_i s_i s_i^T g_i \tag{7.46}$$

$$x_i' = x_{i-1}' - \varrho_i s_i w_i^T P - \varrho_i w_i s_i^T P + \gamma_i s_i s_i^T P. \tag{7.47}$$

Both implementations require the storage of a third update vector, namely z_i or w_i, per iteration in addition to s_i, y_i.

The corresponding expression to equation (7.33) for the conjugate-Newton method becomes

$$\alpha_i x_i = \delta\lambda_{i+1} x_i' + x_i'' + \beta_{i-1} x_{i-1} \tag{7.48}$$

with

$$\beta_{i-1} = \frac{(x_i'' + \delta\lambda_{i+1} x_i')^T(g_i - g'_{i-1})}{(x_{i-1}')^T g'_{i-1}}. \tag{7.49}$$

The same expressions define the SCG of Nocedal [40] with y_i replacing g'_{i-1} in the denominator of β_{i-1}.

Recursive expressions similar to equations (7.46) and (7.47) can be easily established for the LMQN, based on Matthies and Strang [26] (MSTR) implementation, and after some minor modifications [42] for the BLNR of Buckley and LeNir [41]. The use of a recursive expression for the SQN and SCG LMQN updates is not possible because after M iterations each new update replaces the oldest one, and thus the value of w_i of equation (7.44) is completely changed. However, the computational overhead from the use of the product-form expressions (7.42) and (7.43) is expected to be negligible if the number of update vectors M is not large.

Quasi-Newton methods for tracing equilibrium paths have been used in a variety of structural analysis problems [32, 35, 38, 77, 78]. Applications of LMQN with variable loading for tracing unstable branches have been reported by Crisfield [30, 31, 79, 80] using different LMQN versions with one update and Runesson et al. [34] using MSTR with one or two updates. Limited memory conjugate and secant-

Newton methods have been tested in References [72, 81] in an explicit-type formulation without performing a complete factorization of the stiffness matrix.

7.4.3 Iteration control inside an increment

Iteration control inside an increment is achieved with the constraint equation (7.31) which permits the determination of the load parameter $\delta\lambda_{i+1}$ of equation (7.33). Various constraint equations have been proposed in the literature which perform iterations at constant displacement, arc-length, external work, at minimum unbalanced displacement or force norm, and at constant weighted response. The arc-length concept is considered a very successful constraint and was originally introduced by Riks [82] and Wempner [83]. The generalized constraint relationship has the form

$$\Delta u_{i+1}^T \Delta u_{i+1} + b\Delta\lambda_{i+1}^2 P^T P = \Delta l^2 \tag{7.50}$$

where Δl is the arc-length radius. The scalar b is a scaling parameter and Park [84] has proposed that it should be related to Bergan's [85] current stiffness parameter. In this way as the stiffness of the structure increases, so the contribution of the load term in the constraint equation tends towards standard load control. The case of $b = 1$ is the hyperspherical arc-length method of Crisfield [75] and Ramm [76], while numerical experience revealed that the loading terms in equation (7.50) had little effect, and they suggested the hypercylindrical constraint by setting b to zero. This constraint results in a very successful general-purpose iterative technique with a large domain of attraction at critical points. It is this version of arc-length constraint that has been used in this study for testing the performance of the non-linear algorithms.

Having obtained $\alpha_i x_i$ from equation (7.33), the new incremental displacements are $\Delta u_{i+1} = \Delta u_i + \delta\lambda_{i+1} x_i' + x_i''$, where $\delta\lambda_{i+1}$ is the only unknown. Substitution of the new incremental displacements into the constraint equation (7.50), with $b = 0$, leads to the scalar quadratic equation

$$a_1 \delta\lambda_{i+1}^2 + a_2 \delta\lambda_{i+1} + a_3 = 0 \tag{7.51}$$

with

$$a_1 = (x_i')^T x_i'$$

$$a_2 = 2(\Delta u_i + x_i'')^T x_i' \tag{7.52}$$

$$a_3 = 2\Delta u_i^T x_i'' + (x_i'')^T x_i''$$

and

$$\delta\lambda_1 = \pm \left(\frac{\Delta l^2}{(\Delta u_1)^T \Delta u_1} \right)^{1/2} \tag{7.53}$$

which can be solved for $\delta\lambda_{i+1}$ so that the new incremental displacements are defined.

The two roots of equation (7.51) correspond to intersections of the constraining hypercylinder with the equilibrium path. If the roots are real, then the appropriate root is chosen by ensuring that the scalar product of Δu_{i+1} and Δu_i is a positive quantity. When both choices of $\delta\lambda_{i+1}$ yield a positive value, the correct root is the one nearest the solution $\delta\lambda_{i+1} = -a_3/a_2$. This procedure has been found to be successful in preventing the solution path doubling back on itself [75]. Complex roots will occur if $a_2^2 - 4a_1 a_3 < 0$. This situation can arise if the initial load increment is too large in relation to the current degree of non-linearity of the structure along the equilibrium path or if multiple secondary paths are present. In the event of such a case, Crisfield [86] proposed to control the roots via the step length parameter α_i or to bisect Δl and restart non-linear iterations at the last established equilibrium configuration. If Δl becomes too small then another type of constraint equation must be used [87, 88].

Modifications to the arc-length method using linearized forms of the constraint equation have been considered in References [35, 76, 89], while more robust hyperelliptical constraints [84, 90–2] or the orthogonal trajectory approach [93] have been also proposed. In References [87, 94] iterations are performed at constant external work. Survey articles and computational tests are presented on the performance of different constraints [78, 95–8], while a detailed formulation of the hypercylindrical arc-length method is provided by Crisfield [99].

7.4.4 Incremental control along the equilibrium paths

In addition to the selection of the proper corrector iteration strategy and constraint equation, the user is faced with the task of selecting the proper size of the initial load increment and of detecting the type of critical points for proper advancement along the equilibrium path. The choice of the initial load increment is important for the efficiency of the numerical procedure and should be self-adapting to the local non-linearity of the equilibrium path. Large load increments may cause slow convergence or no convergence at all, while small increments may lead to unnecessary equilibrium configurations rather than those required for an adequate definition of the load–displacement response. A number of methods for automatic incrementation have been presented based on the current stiffness parameter [100], or parabolic approximation to the load-deflection response [101], and on load incrementation strategies based on the ratio I_d/I_{m-1} [75, 76, 87, 96, 102], where I_d is the desired number of iterations and I_{m-1} is the number of iterations required for convergence at step $m - 1$. The recurrent formulae for automatically selecting the step-size work, in arc-length procedures, are based upon the fact that each increment converges within a constant desired iteration number. A re-solution is necessary if convergence is not attained within the maximum specified number of iterations or if the solution appears to be diverging.

These techniques work fairly adequately for problems with smooth non-linearities, but exhibit certain deficiencies in handling problems with sudden or strong changes of the solution path [103]. Jeusette et al. [102] have added a weighting factor to the classical formulae that allows the influence of solution path curvature in the step size evaluation to be taken into account:

$$\Delta l^{(m+1)} = \Delta l^{(m)} (I_d/I_m)^{1/2} \frac{\|u^{m-1} - u_1^{m-1}\|}{\|u^m - u_1^m\|} \tag{7.54}$$

where u^{m-1}, u^m are the two previous solutions and u_1^m, u_1^{m-1} are the predictor values at the two preceding increments using a forward Euler tangential predictor. Relation (7.54) is initialized for the first two steps as follows: $\Delta l^{(1)} = \max\{A, 1/A\}$ with $A = \|u^{(1)}\|/\|u_1^{(1)}\|$ and $\Delta l^{(2)} = \Delta l^{(1)} (I_d/I_1)^{1/2}$. The quotient A is a measure of the system initial non-linearity and the initial displacement vector is obtained by a conventional load control method.

Having defined the arc-length radius for each increment, the corresponding initial load increment is obtained from equation (7.53) by properly choosing the sign of the square root. The selection of the correct sign that assures no backtrack in the solution path is based on a combination of the current stiffness parameter [85]

$$S_p = \frac{P^T \Delta u_1^m}{P^T \Delta u_1^{(1)}} \tag{7.55}$$

and the lower eigenvalues of the tangent stiffness matrix at the beginning of the increment. The current stiffness parameter changes sign immediately after a load or displacement limit point has been passed, while the determinant of the tangent stiffness matrix changes sign after a load limit or bifurcation point has been passed. It is assumed that K_T is accurately formulated in order to yield information on the state of the stability of the structure. Based on these observations the existence of only one negative eigenvalue of K_T when associated with a negative S_p is connected with a limit point and a change of sign is required: $\text{sign}(\Delta \lambda_1^{m+1}) = -\text{sign}(\Delta \lambda_1^m)$. However, a single negative eigenvalue with a positive S_p may relate to either a bifurcation or a snap-back point. If the load is increasing it is assumed that the former is occurring and no change of sign is occurring. If we have a change of sign in S_p followed by a change of sign in an eigenvalue, then the sign must also change: $\text{sign}(\Delta \lambda_1^{m+1}) = -\text{sign}(\Delta \lambda_1^m)$.

Bifurcation points are associated in an incremental procedure with the existence of one negative eigenvalue with positive S_p or with more than one negative eigenvalue of K_T. The determination of secondary branches associated with bifurcation points is achieved with special algorithms such as branch switching or related techniques. A relatively simple method which has been frequently used in non-linear structural analysis is performed by examining the eigenvalues and corresponding eigenvectors of the tangent stiffness matrix near the critical point. Then, the likely failure modes are identified with regard to the imposed load and are subsequently used for a perturbation of the solution at bifurcation points. This perturbation is performed by adding the scaled eigenvector to the deformed configuration as follows:

$$u_j = u_c + \xi_j \frac{\varphi_j}{\|\varphi_j\|} \tag{7.56}$$

where u_c is the displacement vector of the last converged solution, φ_j is the jth eigenvector of the corresponding negative eigenvalue and ξ_j is a scaling factor. Branch switching and critical point detecting techniques associated with an incremental solution are discussed in References [97, 98, 102–7] while direct computation of critical points are presented by Wriggers *et al.* [108] and Wriggers and Simo [109].

7.4.5 Tracing equilibrium paths with truncated Newton-like algorithms

In order that a solution algorithm may allow us to recognize the types of critical points and follow the corresponding paths, it is necessary to investigate the lower eigenvalue spectrum of the tangent stiffness matrix at the predictor step. Whenever a critical point is passed, a change of sign in the determinant or in an eigenvalue of K_0^m is observed. The calculation of det K_0^m is possible without additional effort if a complete factorization of the matrix is performed. It is well known that det $K_0^m = d_1 d_2 \ldots d_N = \prod_i d_i$, where d_i are the factorized diagonal elements of the LDL^T factorization. From the Sturm sequence property, the inertia of K_0^m (number of positive, zero, and negative eigenvalues), is reflected in the factorized diagonal elements d_i. In other words, the number of positive, zero and negative d_i corresponds to the same number of positive, zero and negative eigenvalues of K_0^m.

If the conjugate gradient method is employed for the solution of the linearized problem inside each non-linear iteration, then a separate eigenproblem routine has to be incorporated in order to obtain information about the minimum eigenvalues of the tangent matrix [102]. This can be done fairly efficiently by minimizing the finite element discretized form of the Rayleigh quotient via a PCG technique [110, 111]. Such a minimization algorithm can be easily inserted in a general non-linear equation solver keeping the automatic character of the solution procedure. If the Lanczos method, however, is used for the linearized equations, then the required eigenvalue information is automatically extracted. The main feature of the Lanczos method is that certain properties of K are gradually transferred to T_j and Q_j as iterations proceed (equation (1.5) of Reference [4]). Since the Lanczos algorithm may also be described in exact arithmetic as the process of factorizing K into the product of an orthonormal matrix multiplied by a tridiagonal multiplied by an orthonormal,

$$K = Q_N T_N Q_N^T, \qquad Q_N Q_N^T = I \qquad (7.57\text{a,b})$$

a combination of equations (7.57a,b) gives a similarity transformation which states that K and T_N have the same eigenvalues. Thus, by monitoring the sign of the diagonal factors d_j of T_j during the predictor step, as computed in the algorithm (1.11) of Reference [4], we can monitor automatically the change of sign of the det K and we therefore, can trace the appearance of critical points on the load–deflection path. Another advantage of the Lanczos method is that it can be applied without modification to stable and unstable branches since the method is numerically stable for indefinite problems as well. Different implementations of the Lanczos method for the solution of non-linear problems are reported by Papadrakakis and Kalathas [112].

The solution of indefinite problems by the CG method needs special care. The convergence of the method is not assured since division by a too small number may occur. The minimal residual method [113], which minimizes the norm of the residual vector over the Krylov subspace K_j, produces a monotone convergence and can be implemented with minor alterations to the regular CG algorithm. Different modifications to the original CG algorithm have been proposed by Luenberger [114, 115] and Paige and Saunders [113]. A comparison of a number of iterative methods for indefinite problems is presented by Saad [116], while a taxonomy of all types of PCG algorithms is given by Ashby et al. [117].

The preconditioning matrix also needs special care for path following truncated Newton-like procedures. The linear iterative solver may break down unless the preconditioning matrix is positive definite. Non-positive definite preconditioners, based on incomplete Cholesky factorization, may occur even if K is positive definite [4]. There are different approaches that can be employed to produce a positive definite preconditioner. A simple, but crude, intervention is to replace the pivots which become negative, or near zero, during incomplete factorization by a positive value and proceed with the incomplete factorization. A more reliable approach is to modify K to a positive definite matrix that is close to K. A very simple remedy is to add αI to the original matrix, where α is a scalar, such that $\tilde{K} = K + \alpha I$ is positive definite and then perform the incomplete factorization on $\tilde{K}$. Other more involved approaches have been used [64–6,118] based on ideas for solving indefinite problems by direct methods [73, 119–21].

The modified positive definite preconditioner, if it is not properly selected, may sometimes substantially impair the convergence rate of the method. For this reason we keep the unmodified preconditioner for the corrector steps and only in the event of a break-down of the algorithm do we switch to the modified positive definite one. For the predictor step, however, a modified preconditioner must be combined with the Lanczos method in order to be able to trace the inertia of K. This is explained by the fact that the diagonal factors of T_j computed by the Lanczos algorithm correspond to the transformed stiffness

$$\bar{K} = B^{-1}K. \tag{7.58}$$

According to the Sylvester theorem [122], the inertia of a matrix remains unchanged after a congruent transformation. Such a transformation in our case is possible only when B is positive-definite and can thus be expressed as $B = WW^T$. This observation can be proved by substituting B in equation (7.58) with WW^T and performing a similarity transformation, followed by a congruent transformation.

7.5 THE LINE SEARCH

7.5.1 *Fixed loading–stable equilibrium configurations*

Line search is a systematic procedure to find an optimum scalar step-length parameter α which scales the solution vector x. The magnitude of α is obtained in such a way as to cancel the projection of the gradient-residual vector in the direction x. In energy terms this coincides with making the total potential energy Φ at u_{i+1} stationary in the direction x_i, i.e. $(\partial \Phi(u_i + \alpha x_i)/\partial \alpha = \Phi'(\alpha))$

$$\Phi'(\alpha) = x_i^T g(u_i + \alpha x_i) = 0. \tag{7.59}$$

As explicit expressions for α are not available in non-linear problems, iterative schemes are employed using a one-dimensional optimization routine. The exact evaluation of the step-length is an expensive operation in finite element analysis since it may involve numerous calculations of the gradient-residual vector. For this reason equation (7.59) is only approximately satisfied.

The classical approach, which was introduced by Fletcher and Reeves [6], proceeds first with the bracketing of the solution along the search direction and then with the search for optimum α by an iterative scheme. A crucial factor in the performance of the line search routine is the evaluation of the initial step size by which the bracketing of the solution is attempted. A heuristic approach is proposed by Smith et al. [123] where the first step size at iteration i is given by

$$\alpha_{i,1} = c_i \|x_i\|_\infty \tag{7.60}$$

in which c_i is a parameter dependent on the history of all $i-1$ line searches previously performed. Starting with $c_0 = 1$, $c_i = (2)^{r-1} c_{i-1}$, where r is the number of increments required to bracket the root at iteration $i-1$. The purpose of this procedure is to bracket the root with a minimum effort subject to the constraint that the resulting interval be reasonably small. The initial step size evaluated from equation (7.60) is then checked to find whether $\Phi'(\alpha_{i,1}) \geq 0$. If this inequality is satisfied, then the minimum is bounded, otherwise $\Phi'(\alpha)$ is examined at the points $2\alpha_{i,1}, 4\alpha_{i,1}, \ldots, \alpha_1, \alpha_2$ where α_2 is the first of these values at which $\Phi'(\alpha)$ is non-negative. This technique was found to be effective in non-linear CG applications where a large number of iterations are performed [3, 5]. In PCG-like methods or QN methods, even with weak initial metric, this approach ceases to be effective since the best initial estimate is usually $\alpha_{i,1} = 1$ (see Reference [70]).

Once the interval is determined an iterative scheme is applied based on curve fitting (quadratic or cubic) or a linear interpolation (or extrapolation) [124] scheme. The *regula falsi* approximation is a linear method which has been found in a number of cases [3, 5, 10] to perform efficiently compared with curve-fitting techniques. This method requires only gradient evaluations and achieves a much better rate of convergence than the curve-fitting techniques, especially in these cases where the directional derivatives (7.59) cross zero with a steep slope.

The line search algorithm may be described as follows:

1. $\alpha_3 = \alpha_1 - \Phi'(\alpha_1) \left(\dfrac{\alpha_1 - \alpha_2}{\Phi'(\alpha_1) - \Phi'(\alpha_2)} \right)$

2. Test for convergence
3. If $\Phi'(\alpha_3)\Phi'(\alpha_2) < 0$, then $\alpha_1 = \alpha_3$, else $\alpha_2 = \alpha_3$ (7.61)

4. $\alpha_3 = \frac{1}{2}(\alpha_1 + \alpha_2)$, test for convergence
5. Repeat Step 3, go to Step 1.

Steps 3 and 5 insure that the interpolation points always bracket the root and the bisection Step 4 ensures against slow convergence for strongly non-linear problems.

A second approach replaces the one-variable minimization routine by a stability test which insures convergence of the method. If the stability criterion

$$|x_i^T g(u_i + \alpha_i x_i)| < \mu |x_i^T g_i| \tag{7.62}$$

is satisfied then α_i is accepted as the step-length parameter. Otherwise, depending on the sign of the directional derivative, either α_i is doubled or a *regula falsi* step is performed and a stability check is repeated with the new α_i. The value of μ is taken in the range (0–1). A slightly modified *regula-falsi* method, called the Illinois algorithm, has been used in connection with the BFGS method [26, 34, 125]. Crisfield [126] has also applied similar techniques with the secant-Newton method. In mathematical programming literature the stability criterion (7.62) is supplemented by a sufficient reduction in Φ [22, 124]. The development of these particular rules are due to the initial work by Goldstein [127] and Armijo [128].

Line searches are indispensable in CG-like methods and less vital in quasi-Newton methods. In the latter, line searches play a more important role when weak initial approximations to the Hessians are used (i.e. $K_0 = I$ or a preconditioner R). Since LMQN is in many ways intermediate between PCG and quasi-Newton methods an intermediate line search should be used in terms of the approximate solution of equation (7.59). When strong initial metrics are used then line searches may increase the robustness of the iterative technique. In other words on highly non-linear increments the iterative procedure will very often converge with line searches where it would diverge without them. In Newton's method the necessity for a line search would arise only when the Newton step lies outside the region within which $g(u_i + x_i)$ is well approximated by the linear model $Lm(x) = g_i + K_i x_i$.

7.5.2 Variable loading–unstable equilibrium configurations

The line search procedure is traditionally applied to problems involving stable equilibrium states or to optimization of a function in mathematical programming problems. In such circumstances the search direction is normally a descent direction with the directional derivative $\Phi'(\alpha = 0)$ being negative. However, when slack line searches are employed, truncated iterations are performed for the linearized solutions, or a non-tangent initial metric is used, then it is possible, even in stable branches, for the directional derivative to be found positive. In such cases stable equilibrium states may be defined with some negative step-lengths [69, 81]. There are also reports in the mathematical programming literature where a modified positive definite matrix is used whenever K_i is found to be indefinite [22, 41, 64, 65, 129].

An equilibrium state in the unstable path corresponds to a saddle point of the total potential energy. This means that uphill directions may be present as well and the line search procedure must be capable of deciding in each direction whether to maximize or minimize along the direction x. The second directional derivative $\Phi''(\alpha) = x_i^T K_i x_i$, at $\alpha = 0$, gives information about the nature of the energy surface. Specifically, if $\Phi''(\alpha) \geq 0$ then minimize otherwise maximize in the direction x_i. Under this condition the algorithm tries to find the nearest stationary point of the function Φ. When the tangent stiffness matrix K_j is not available, the second directional derivative is approximated with first directional derivatives and the decision condition checks whether $|\Phi'(\alpha = 0)| > |\Phi'(\alpha = \varepsilon)|$ with $\varepsilon > 0$. If this inequality holds then find $\alpha_i > 0$, else find $\alpha_i < 0$ with $\Phi'(\alpha) \approx 0$. This line search routine has been applied in optimization problems by Biegler-König [130] and in structural mechanics by Papadrakakis and Theoharis [81].

The first application of line searches in combination with the arc length method was presented by Crisfield [126]. Assuming that the load level parameter at $\alpha_{i,1} = 1$ is approximated by the final value of $\delta\lambda_{i+1}$, then

$$\alpha_{i,1}(=1)x_i = \delta\lambda_{i+1}x_i' + x_i''. \tag{7.63}$$

A combination of equations (7.63) with (7.30) and (7.50) gives the quadratic equation (7.51) for $\delta\lambda_{i+1}$ and the coefficients now take the form

$$a_1 = \alpha_{i,j}(x_i')^T x_i$$

$$a_2 = 2(\Delta u_i + \alpha_{i,j} x_i'')^T x_i' \tag{7.64}$$

$$a_3 = 2\Delta u_i^T x_i'' + \alpha_{i,j}(x_i'')^T x_i''.$$

In introducing line searches into the arc length method we need to proceed as follows: for the jth trial step length $\alpha_{i,j}$ at the ith iteration of the mth increment calculate $\delta\lambda_{i+1,j}$ from equations (7.51) and (7.64) and the new direction $x_{i,j}$ from equation (7.63). The displacements are adjusted, as in Step b6 of algorithm (7.35), and the internal forces $F_{i+1,j}$ are evaluated from the force–displacement relations. The new external loading is given by $P_{i+1,j} = P_i + \delta\lambda_{i+1,j}P$ and the gradient-residual vector $g_{i+1,j}$ is obtained from equation (7.32). Thus, the new directional derivative is defined by $\Phi'(\alpha_{i,j}) = x_{i,j}^T g_{i+1,j}$ and the stability criterion (7.62) is tested. In this implementation both the load level and the iterative direction vary inside the line search. Simplified line search schemes with arc length methods are presented in References [38, 95].

In the vicinity of the solution, the step length should have a value close to unity. When α is too small, the iteration is not effective because the solution vector retracts to the old value, while an excessively large value of α may cause numerical instability. In both cases, the energy function is not well approximated by a quadratic function involving the old stiffness matrix and thus, either a restart procedure is used or a new tangent stiffness matrix is evaluated.

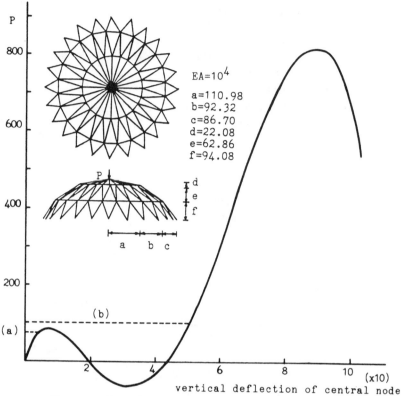

Figure 7.3 Example (i). Geometry and load–displacement path.

7.6 NUMERICAL EXAMPLES

In this section the behaviour and efficiency of different non-linear methods are examined for three benchmark examples: (i) the 168-member articulated dome [2] of Figure 7.3 with $N = 147$ d.o.f.; (ii) the 48-member frame dome [59] of Figure 7.4 with $N = 222$ d.o.f.; and (iii) the clamped quadratic shell of Figure 7.5 with 3×3 elements in each quarter and 129 d.o.f. In the first two examples elastic behaviour is assumed, while in the shell example both geometric and material non-linearities are taken into account. The formulation of the equilibrium equations for the frame is based on a beam–column approach [2, 131] and for the shell on a 9-node Heterosis degenerated isoparametric element [132] with 42 d.o.f. and 6 layers through its thickness. The loading is applied, either fixed or variable, with the arc length method in each increment. Non-linear iterations are terminated when

$$\|g'^m\|/\|P_i^m\| < 10^{-4}.$$

The required accuracy of the normalized residual vector for the Lanczos algorithm in the solution phase is specified to $\varepsilon = 10^{-4}$. When the truncated version is used then $\eta_0 = 0.1$ and $t = 0.2$ of the forcing sequence (7.27).

7.6 NUMERICAL EXAMPLES

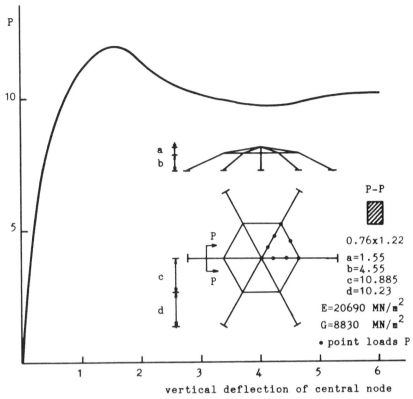

Figure 7.4 Example (ii). Geometry and load–displacement path.

Owing to the large number of parameters involved in each method, the following appellation is used to describe the methods:

$$AAA(M) - BB(\mu) - TR.$$

The symbol AAA(M) corresponds to the type of non-linear iterations used and becomes CNR, MNR, CN, if the conventional, modified Newton–Raphson and conjugate Newton methods are applied, or MSTR(M), BLNR(M), SQN(M), SCG(M), if the corresponding LMQN methods are used with M updates. The symbol BB(μ) corresponds to the line search used, and it takes the form CLS if the conventional line search is used with *regula falsi* iterations of algorithm (7.61) or ST(μ) if the stability criterion (7.62) is applied with tolerance μ. The symbol TR denotes the application of a truncated version of the method.

Figure 7.6 depicts the performance of different update formulae without truncation for the three examples considered in the following cases: fixed and variable loading, strict and loose line searches. The performance of the methods is demonstrated with respect to the normalized CPU time. Figure 7.7 shows the performance of SQN with different number of updates compared with the CNR and MNR for the truss example. Two levels of fixed loading showed in Figure 7.3 and variable loading with 30

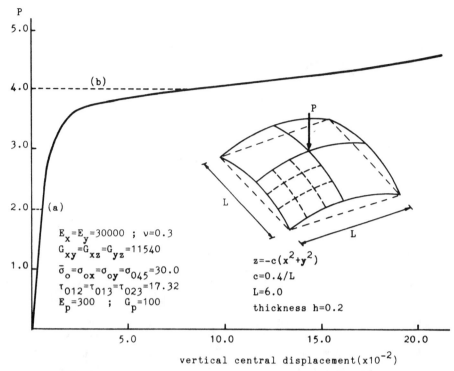

Figure 7.5 Example (iii). Geometry and load–displacement path.

increments are considered. Figure 7.8 shows the corresponding comparisons for the frame example, and Figure 7.9 for the shell example.

Figure 7.10 shows some tests of the truncated versions of the methods in relation to the truss example, while Figures 7.11 and 7.12 depicts the performance of the truncated versions for the frame and shell examples. In all cases the SSOR preconditioner [4] is used for the preconditioned Lanczos method. Finally, Figure 7.13 shows a comparison of SQN(M) with two different implementations in relation to the predictor and corrector steps. The first is the usual implementation, with a truncated solution at each linearized step, and the tangent stiffness matrix as K_0 in equation (7.17). The second implementation avoids the linearized solution by approximating K_0 with a preconditioning matrix R. In the latter application the SSOR preconditioning matrix is also used.

7.7 CONCLUDING REMARKS

It is apparent that comparing a group of non-linear algorithms is not an easy task. The variability of the methods, combined with the fact that their performance is dependent on the type of problem and the non-linearities involved, make it difficult to draw universal conclusions about the behaviour of non-linear methods. However, certain conclusions, or trends, can be drawn on the basis of experience and the numerical results presented in this chapter.

7.7 CONCLUDING REMARKS

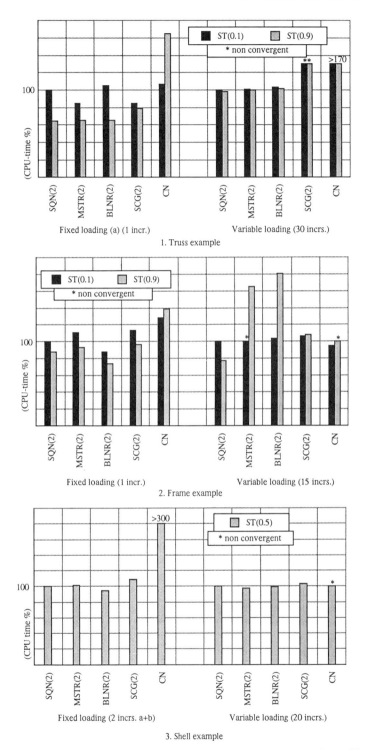

Figure 7.6 Relative performance of different LMQN methods for fixed and variable loadings.

212 SOLVING NON-LINEAR PROBLEMS IN SOLID AND STRUCTURAL MECHANICS

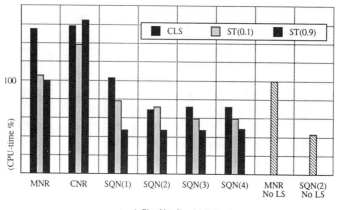
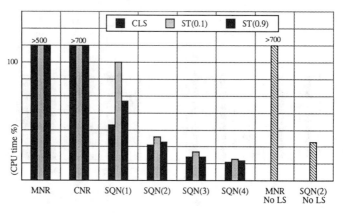
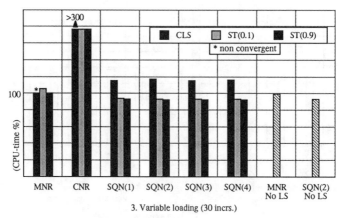

Figure 7.7 Truss example. Relative performance of SQN(M) versus CNR and MNR for fixed and variable loadings.

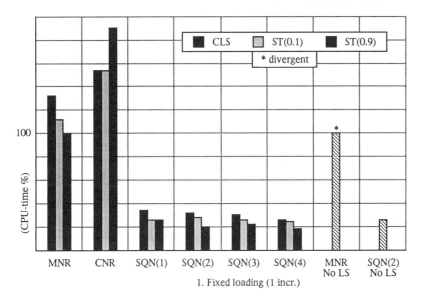

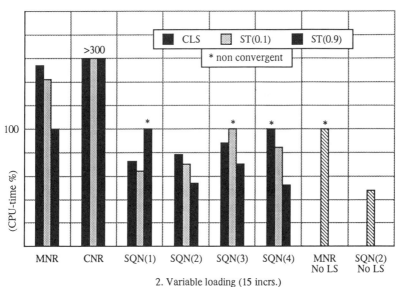

Figure 7.8 Frame example. Relative performance of SQN(M) versus CNR and MNR for fixed and variable loadings.

Since the solution of linear equations represents a major cost of implicit non-linear analysis, the use of iterative solvers, as opposed to direct solvers for the solution of the linearized equations, allows the efficient solution, in both computing time and storage, of large-scale non-linear problems. Iterative solvers can also be combined with the concept of truncation leading to a further improvement in the efficiency of the methods. The use of truncation may increase the total number of non-linear iterations, but the decrease of effort of the linear solver has an overall beneficial

SOLVING NON-LINEAR PROBLEMS IN SOLID AND STRUCTURAL MECHANICS

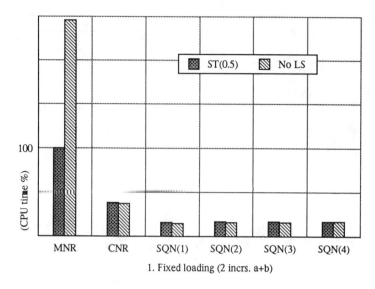

1. Fixed loading (2 incrs. a+b)

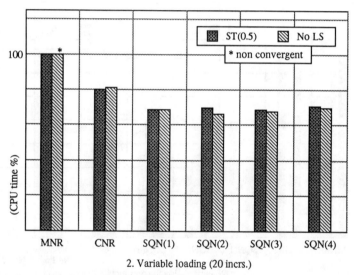

2. Variable loading (20 incrs.)

Figure 7.9 Shell example. Relative performance of SQN(M) versus CNR and MNR for fixed and variable loadings.

effect on the performance of the non-linear solution scheme. For problems with an expensive gradient evaluation and strong non-linearities, it is advisable to use a more strict forcing sequence parameter in order to reduce the number of additional gradient evaluations.

It seems that there is no clear winner among the limited memory quasi-Newton methods SQN, MSTR and BLNR. However, SQN has an edge over the other two, because it exhibits greater robustness and better performance under slack line searches. SCG and conjugate-Newton methods are more sensitive to the accuracy

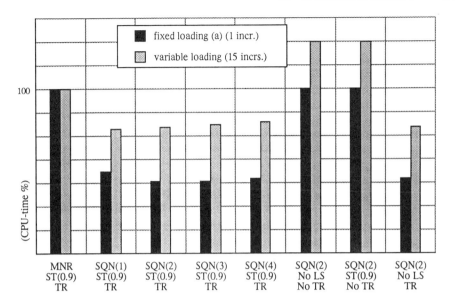

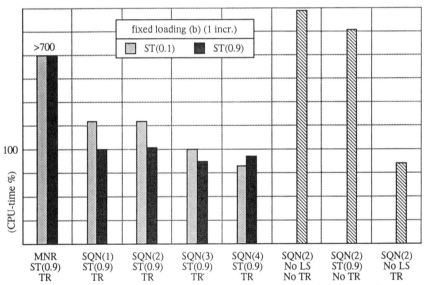

Figure 7.10 Truss example. Relative performance of SQN(M) and MNR with and without truncation.

of line searches and should be avoided. The optimum number of update vectors depends on the degree of current non-linearity of the problem and is in the range of 1 to 4. A tangential initial metric for the predictor step, combined with a truncated iterative solution, is a much better choice than the implementation of a preconditioning matrix to approximate the initial metric and avoiding the solution of a linearized problem in each iteration.

The influence of line searches on the performance of the methods is dependent

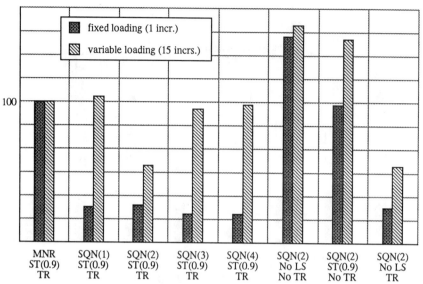

Figure 7.11 Frame example. Relative performance of SQN(*M*) and MNR with and without truncation.

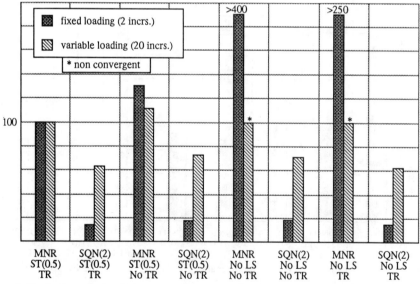

Figure 7.12 Shell example. Relative performance of SQN(*M*) and MNR with and without truncation.

on the choice of the predictor and the iteration matrix of the corrector steps, and on the type and degree of non-linearity of the structure. Explicit methods with preconditioning, implicit methods of the conjugate type (SCG, CN) or of the secant type (quasi-Newton, BLNR, SQN, MSTR) with weak initial metrics in the predictor step, are more sensitive to the accuracy of step-length evaluation. As a result, these

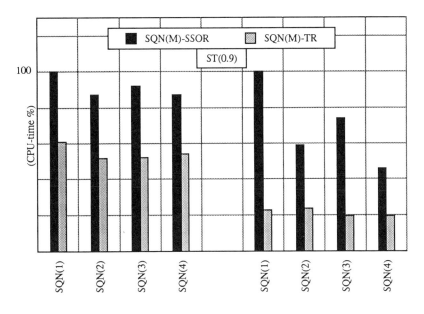

Figure 7.13 Relative performance of SQN(M) with the SSOR initial metric versus the tangential initial metric followed by truncated iterations.

methods are generally not the correct choice for problems with expensive evaluation of the gradients.

Secant-type implicit methods, on the other hand, with tangential predictors can be combined with inexpensive slack line searches and increase the efficiency of the methods, particularly for highly non-linear problems. The more significant gain from line searches is that they increase the robustness of the iterative procedure. It seems, however, that the choice of the iteration matrix is clearly more important than an accurate line search as a method of achieving overall computational efficiency.

There is substantial superiority of LMQN over CNR and MNR in problems with fixed loading and a high degree of non-linearity. Although a drastic improvement is observed on MNR with the arc-length constraint, and on CRN in plasticity problems, the overall computational efficiency of LMQN is maintained. There are cases, however, that MNR or CNR may be more effective than LMQN in some portions of the load–displacement path. It was also observed that some of the methods may run into difficulties while trying to locate an equilibrium configuration, because the iteration path, followed by this particular method, is disturbed by the existence of multiple neighbouring equilibrium states. For these reasons, the computational strategy must be equipped with restart procedures and with an adaptive scheme in the implementation of the non-linear methods. Thus, according to the degree of non-linearities and the peculiarities of the specific problem, the accuracy of the line searches may be adjusted, the number of updates may vary and a recomputation of the tangent stiffness matrix may be necessary inside each increment.

ACKNOWLEDGEMENTS

The author is very grateful to S. Bitzarakis and G. Pantazopoulos for their cooperation in implementing the algorithms and obtaining the numerical results presented in this chapter.

BIBLIOGRAPHY

[1] Engeli, M., Ginsburg, T., Rutishauser, H. and Stiefel, E. (1959) *Refined Iterative Methods for Computation of the Solution and the Eigenvalue of Self-Adjoint Boundary Value Problems*, Birkhauser Verlag, Basel/Stuttgart.

[2] Papadrakakis, M. (1981) Post-buckling analysis of spatial structures by vector iteration methods, *Comput. Struct.*, **14**(5–6), 393–402.

[3] Papadrakakis, M. (1982) A family of methods with three-term recursion formulae, *Int. J. Num. Meth. Engg*, **18**, 1785–1799.

[4] Papadrakakis, M., Solving large-scale linear problems in solid and structural mechanics. This volume.

[5] Papadrakakis, M. and Ghionis, P. (1986) Conjugate gradient algorithms in nonlinear structural analysis problems, *Comput. Meth. Appl. Mech. Engg*, **59**, 11–27.

[6] Fletcher, R. and Reeves, C.M. (1969) Function minimization by conjugate gradients, *Comput. J.*, **6**, 149–154.

[7] Polak, E. and Ribière, G. (1969) Note sur la convergence de méthods de directions conjugées, *Comput. J.*, **3**, 35–43.

[8] Papadrakakis, M. (1981) A method for the automatic evaluation of the dynamic relaxation parameters, *Comp. Meth. Appl. Mech. Engg*, **25**, 35–48.

[9] Underwood, P.G. (1983) Dynamic relaxation techniques: A review, in T. Belytscko and T.J.R. Hughes (eds.), *Computational Methods for Transient Analysis*, North-Holland, Amsterdam, pp. 245–267.

[10] Papadrakakis, M. (1979) Methods with three-term recursion formulae for the analysis of cable structures, *IASS World Congress on Shell and Spatial Structures*, Madrid.

[11] Shugar, T.A. (1991) Automated dynamic relaxation solution method for compliant structures, in I.D. Parsons and B. Nour-Omid (eds.), *Iterative Equation Solvers for Structural Mechanics Problems*, Winter Annual Meeting, ASME, CED-Vol. 4, Atlanta, GA., pp. 1–12.

[12] Felippa, C. (1984) Dynamic relaxation under general increment control, in W.K. Liu, T. Belytscko and K.C. Park (eds.), *Innovative Methods for Nonlinear Problems*, Pineridge Press, Swansea, UK, pp. 103–133.

[13] Farhat, C. and Grivelli, L. (1989) A general approach to nonlinear FE computations on shared-memory multiprocessors, *Comput. Meth. Appl. Mech. Engg*, **72**, 153–171.

[14] Farhat, C. (1989) Parallel computational strategies for large space and aerospace flexible structures: Algorithms, implementations and performance, in P. Melli and C.A. Brebbia (eds.), *Supercomputing in Engineering Structures*, Springer-Verlag, pp. 109–132.

[15] Fletcher, R. (1980) *Practical Methods of Optimization-Vol. 1: Unconstrained Optimization*, Wiley, New York.

[16] Irons, B. and Elsawaf, A. (1977) The conjugate-Newton algorithm for solving finite element equations, in K.J. Bathe, T.J. Oden and W. Wunderlich (eds.), *Proc. U.S.-German Symp. on Formulations and Computational Algorithms in Finite Element Analysis*, MIT Press, pp. 656–672.

[17] Broyden, C.G. (1965) A class of methods for solving nonlinear simultaneous equations,

Math. Comput., **19**, 577–593.
[18] Davidon, W.C. (1959) *Variable Metric Method for Minimization*, Report ANL-5990 Rev, Argonne Nat. Lab., Argonne, Illinois.
[19] Broyden, C.G. (1967) Quasi-Newton methods and their application to function minimization, *Math. Comput.*, **21**, 368–381.
[20] Fletcher, R. and Powell, M.J.D. (1963) A rapidly convergent descent method for minimization, *Comput. J.*, **6**, 163–168.
[21] Wolfe, M.A. (1975) *Numerical Methods for Unconstrained Optimization, An Introduction*, Van Nostrand/Rheinhold, London.
[22] Gill, P.E., Murray, W. and Wright, M.H. (1981) *Practical Optimization*, Academic Press, London.
[23] Dennis, J.E. and Schnabel, R. (1983) *Numerical Methods for Unconstrained Optimization and Nonlinear Equations*, Prentice-Hall, Englewood Cliffs, New Jersey.
[24] Dennis, J.E. and Moré, J.J. (1977) Quasi-Newton methods, motivation and theory, *SIAM Review*, **19**, 46–89.
[25] Fox, L. and Stanton, E. (1968) Developments in structural analysis by direct energy minimization, *AIAA J.*, **6**, 1036–1042.
[26] Matthies, H. and Strang, G. (1979) The solution of nonlinear finite element equations, *Int. J. Num. Meth. Engg*, **14**, 1613–1626.
[27] Bathe, K.J. and Cimento, A.P. (1980) Some practical procedures for the solution of nonlinear finite element equations, *Comput. Meth. Appl. Mech. Engg*, **22**, 59–85.
[28] Geradin, M., Idelsohn, S. and Hogge, M. (1981) Computational strategies for the solution of large nonlinear problems via quasi-Newton methods, *Comput. Struct.*, **13**, 73–81.
[29] Pica, A. and Hinton, E. (1980) The quasi-Newton BFGS method in the large deflection analysis of plates, in C. Taylor, E. Hinton and D.J.R. Owen (eds.), *Numerical Methods for Non-linear Problems*, Pineridge Press, Swansea, UK, pp. 355–365.
[30] Crisfield, M.A. (1980) Incremental/iterative solution procedures for nonlinear structural analysis, in C. Taylor, E. Hinton and D.J.R. Owen (eds.), *Numerical Methods for Non-linear Problems*, Pineridge Press, Swansea, UK, pp. 261–290.
[31] Crisfield, M.A. (1984) Accelerating and damping the modified Newton–Raphson method, *Comput. Struct.*, **18**, 395–407.
[32] Ramm, E. and Matzenmiller, A. (1986) Large deformation shell analysis based on the degeneration concept, in T.J.R. Hughes and E. Hinton (eds.), *Finite Element Methods for Plate and Shell Structures*, Pineridge Press, Swansea, UK, pp. 365–393.
[33] Ramm, E., Schweizerhof, K. and Stegmüller, H. (1986) Ultimate load analysis of thin shells under pressure loads, in K.J. Bathe, P.G. Bergan and W. Wunderlich (eds.), *Proc. Europe-U.S. Symp. Finite Element Methods for Nonlinear Problems*, Springer-Verlag, pp. 339–357.
[34] Runesson, K., Samuelsson, A. and Bernspang, L. (1986) Numerical techniques in plasticity including solution advancement control, *Int. J. Num. Meth. Engg*, **22**, 769–788.
[35] Forde, B.W.R. and Stiemer, S.F. (1987) Improved arc length orthogonality methods for nonlinear finite element analysis, *Comput. Struct.*, **27**, 625–630.
[36] Gelin, J.G. and Picart, P. (1988) Use of quasi-Newton methods for large strain elastic-plastic finite element computations, *Comm. Appl. Num. Meth.*, **4**, 457–469.
[37] Lee, S.H. (1989) Rudimentary considerations for effective quasi-Newton updates in nonlinear finite element analysis, *Comput. Struct.*, **33**, 463–476.
[38] Schweizerhof, K. (1989) *Quasi-Newton Verfahren und Kurvenverfolgungsalgorithmen für die Lösung nichtlinearer Gleichungssysteme in der Strukturmechanik*, Report, Institüt für Baustatik, Universität Fridericiana Karlsruhe (TH), Germany.
[39] Nazareth, L. (1979) A relationship between the BFGS and conjugate gradient algorithms

and its implications for new algorithms, *SIAM J. Num. Anal.*, **16**, 794–800.
[40] Nocedal, J. (1980) Updating quasi-Newton matrices with limited storage, *Math. Comput.*, **35**, 773–782.
[41] Buckley, A. and LeNir, A. (1983) QN-like variable storage conjugate gradients, *Math. Program.*, **27**, 155–175.
[42] Papadrakakis, M. and Pantazopoulos, G. (1992) *Quasi-Newton Methods with Reduced Storage for Large-Scale Computation*, Report, Institute of Structural Analysis & Aseismic Research, National Technical University of Athens, Greece.
[43] Brodlie, K., Gourlay, A. and Greenstadt, J. (1973) Rank-one and rank-two corrections to positive definite matrices expressed in product form, *J. Inst. Math. Appl.*, **11**, 73–82.
[44] Gill, P.E. and Murray, W. (1979) *Conjugate Gradient Methods for Large Scale Nonlinear Optimization*, Report, Dept. of Operations Research, SOL 79 15, Stanford University.
[45] Perry, A. (1978) A self-correcting conjugate gradient algorithm, *Int. J. Comput. Math.*, **6**, 327–333.
[46] Shanno, D.F. (1978) Conjugate gradient methods with inexact searches, *Math. Oper. Res.*, **3**, 244–256.
[47] Buckley, A.G. (1978) A combined conjugate-gradient quasi-Newton minimization algorithm, *Math. Program.*, **15**, 200–210.
[48] Crisfield, M.A. (1979) A faster modified Newton-Raphson iteration, *Comput. Meth. Appl. Mech. Engg*, **20**, 267–278.
[49] Crisfield, M.A. (1985) New solution procedures for linear and nonlinear finite element analysis, in J. Whiteman (ed.), *The Mathematics of Finite Elements and Applications V*, Academic Press, pp. 49–81.
[50] Papadrakakis, M. and Gantes, C.J. (1989) Preconditioned conjugate and Secant-Newton methods for non-linear problems, *Int. J. Num. Meth. Engg*, **28**, 1299–1316.
[51] Powell, M.J.D. (1976) A view of unconstrained optimization, in L.C.W. Dixon (ed.), Academic Press, London, pp. 117–152.
[52] Toint, P.L. (1977) On sparse and symmetric matrix updating subject to a linear equation, *Math. Comput.*, **31**, 954–961.
[53] Kamat, M.P., Watson, L.T. and Vanden Brink, D.J. (1981) An assessment of quasi-Newton sparse update techniques for nonlinear structural analysis, *Comput. Meth. Appl. Mech. Engg*, **26**, 363–375.
[54] Curtis, A.R., Powell, M.J.D. and Reid, J.K. (1974) On the estimation of sparse Jacobian matrices, *J. Inst. Math. Appl.*, **13**, 117–119.
[55] Powell, M.J.D. and Toint, P.L. (1979) On the estimation of sparse Hessian matrices, *SIAM J. Num. Anal.*, **11**, 1060–1074.
[56] Dennis, J.E. and Marwil, E.S. (1982) Direct secant updates of matrix factorizations, *Math. Comput.*, **38**, 459–474.
[57] Tewarson, R.P. and Zhang, Y. (1987) Sparse quasi-Newton LDU updates, *Int. J. Num. Meth. Engg*, **24**, 1093–1100.
[58] Powell, M.J.D. (1977) Restart procedures for the conjugate gradient method, *Math. Program.*, **12**, 241–254.
[59] Papadrakakis, M. and Balopoulos, V. (1991) Improved quasi-Newton methods for large nonlinear problems, *J. Eng. Mech.*, ASCE, **117**, 1201–1219.
[60] Dembo, R.S. and Steihaug, T. (1983) Truncated-Newton algorithms for large scale unconstrained optimization, *Math. Program.*, **26**, 190–212.
[61] Nour-Omid, B., Parlett, B.N. and Taylor, R.L. (1983) A Newton-Lanczos method for solution of nonlinear finite element equations, *Comput. Struct.*, **16**, 241–252.
[62] Sherman, A.H. (1978) On Newton-iterative methods for the solution of systems of nonlinear equations, *SIAM J. Num. Anal.*, **15**, 755–771.
[63] Garg, N.K. and Tapia, R.A. (1978) *QDN: A Variable Storage Algorithm for Unconstrained*

Optimization, Report, Dept. of Mathematical Sciences, Rice University, Houston.
[64] O'Leary, D.P. (1982) A discrete Newton algorithm for minimizing a function of many variables, *Math. Program.*, **23**, 20–33.
[65] Nash, S.G. (1985) Preconditioning of truncated-Newton methods, *SIAM J. Sci. Stat. Comput.*, **6**, 599–616.
[66] Nash, S.G. (1984) Newton-type minimization via the Lanczos method, *SIAM J. Num. Anal.*, **21**, 770–788.
[67] Cervera, M., Liu, Y.C. and Hinton, E. (1986) Preconditioned conjugate gradient method for the non-linear finite element analysis with particular reference to 3D reinforced concrete structures, *Eng. Comput.*, **3**, 235–242.
[68] Borja, R.I. (1991) Composite Newton-PCG and quasi-Newton iterations for nonlinear consolidation, *Comput. Meth. Appl. Mech. Engg*, **86**, 27–60.
[69] Papadrakakis, M. (1990) A truncated Newton-Lanczos method for overcoming limit and bifurcation points, *Int. J. Num. Meth. Engg*, **29**, 1065–1077.
[70] Papadrakakis, M. and Gantes, C.J. (1988) Truncated Newton methods for nonlinear finite element analysis, *Comput. Struct.*, **30**, 705–714.
[71] Papadrakakis, M. and Nomikos, N. (1990) Automatic nonlinear solution with arc-length and Newton-Lanczos method, *Eng. Comp.*, **7**, 48–56.
[72] Papadrakakis, M. (1991) Computational strategies for tracing post-limit-point paths, *Int. J. Space Structures*, **6**, 116–131.
[73] Gill, P.E. and Murray, W. (1974) Newton-type methods of unconstrained and linearly constrained optimization, *Math. Program.*, **17**, 311–350.
[74] Batoz, J.L. and Dhatt, G. (1979) Incremental displacement algorithms for nonlinear problems, *Int. J. Num. Meth. Engg*, **14**, 1262–1267.
[75] Crisfield, M.A. (1981) A fast incremental/iterative solution procedure that handles snap-through, *Comput. Struct.*, **13**, 55–62.
[76] Ramm, E. (1981) Strategies for tracing nonlinear responses near limit points, in K.J. Bathe, E. Stein and W. Wunderlich (eds.), *Nonlinear Finite Element Analysis in Structural Mechanics*, Springer-Verlag, pp. 63–89.
[77] Watson, L.T., Kamat, M.P. and Reaser, M.H. (1985) A robust hybrid algorithm for computing multiple equilibrium solutions, *Eng. Comput.*, **2**, 30–34.
[78] Padovan, J. and Moscarello, R. (1986) Locally bound constrained Newton–Raphson solution algorithms, *Comput. Struct.*, **23**, 181–197.
[79] Crisfield, M.A. (1982) Accelerated solution techniques and concrete cracking, *Comput. Meth. Appl. Mech. Engg*, **33**, 585–607.
[80] Crisfield, M.A. and Wills, J. (1988) Solution strategies and softening materials, *Comput. Meth. Appl. Mech. Engg*, **66**, 267–289.
[81] Papadrakakis, M. and Theoharis, A.P. (1991) Tracing post-limit-point paths with incomplete or without factorization of the stiffness matrix, *Comput. Meth. Appl. Mech. Engg*, **88**, 165–187.
[82] Riks, E. (1972) The application of Newton's method to the problem of elastic stability, *J. Appl. Mech.*, ASME, **39**, 1060–1066.
[83] Wempner, G.A. (1971) Discrete approximations related to nonlinear theories of solids, *Int. J. Solids Structs.*, **7**, 1581–1599.
[84] Park, K.C. (1982) A family of solution algorithms for nonlinear structural analysis based on relaxation equations, *Int. J. Num. Meth. Engg*, **18**, 1337–1347.
[85] Bergan, P. (1980) Solution algorithms for nonlinear structural problems, *Comput. Struct.*, **12**, 497–509.
[86] Crisfield, M.A. (1984) Overcoming limit points with material softening and strain localisation, in E. Hinton, D.R.J. Owen and E. Onate (eds.), *Numerical Methods for Nonlinear Problems*, Vol. 2, pp. 244–277.

[87] Bathe, K.J. and Dvorkin, E.N. (1983) On the automatic solution of nonlinear finite element equations, *Comput. Struct.*, **17**, 871–879.

[88] Padovan, J. and Tovichakchaikul, S. (1983) On the solution of elastic-plastic, static and dynamic postbuckling collapse of general structure, *Comput. Struct.*, **16**, 199–205.

[89] Schweizerhof, K.H. and Wriggers, P. (1986) Consistent linearization for path following methods in nonlinear FE analysis, *Comput. Meth. Appl. Mech. Engg*, **59**, 261–279.

[90] Padovan, J. and Tovichakchaikul, S. (1982) Self-adaptive predictor-corrector algorithms for static nonlinear structural analysis, *Comput. Struct.*, **15**, 365–377.

[91] Felippa, C.A. (1984) Dynamic relaxation under general increment control, in W.K. Liu, T. Belytschko and K.C. Park (eds.), *Innovative Methods for Nonlinear Problems*, Pineridge Press, Swansea, UK, pp. 103–133.

[92] Skeie, G. and Felippa, C.A. (1990) A local hyperelliptic constraint for nonlinear analysis, *Proc. NUMETA '90 Conference*, Swansea, UK, **1**, 13–28.

[93] Fried, I. (1984) Orthogonal trajectory accession to the nonlinear equilibrium curve, *Comput. Meth. Appl. Mech. Engg*, **47**, 283–297.

[94] Yang, Y.B. and McGuire, W. (1985) A work control method for geometrically nonlinear analysis, *Proc. NUMETA '85 Conference*, Swansea, UK, pp. 913–921.

[95] Schweizerhof, K.H. and Wriggers, P. (1986) Consistent linearization for path following methods in nonlinear FE analysis, *Comput. Meth. Appl. Mech. Engg*, **59**, 261–279.

[96] Bellini, P.X. and Chulya, A. (1987) An improved automatic incremental algorithm for the efficient solution of nonlinear finite element equations, *Comput. Struct.*, **26**, 99–110.

[97] Clarke, M.J. and Hancock, G.J. (1990) A study of incremental-iterative strategies for nonlinear analysis, *Int. J. Num. Meth. Engg*, **29**, 1365–1391.

[98] Kouhia, R. and Mikkola, M. (1989) Tracing the equilibrium path beyond simple critical points, *Int. J. Num. Meth. Engg*, **28**, 2923–2941.

[99] Crisfield, M.A. (1991) *Non-linear Finite Element Analysis of Solids and Structures* — Vol. 1: *Essentials*, Wiley, Chichester.

[100] Bergan, P.G., Horrigmoe, G., Krakeland, B. and Soreide, T.H. (1978) Solution techniques for non-linear finite element problems, *Int. J. Num. Meth. Engg*, **12**, 1677–1696.

[101] Bergan, P.G. and Soreide, T. (1973) A comparative study of different solution techniques as applied to a nonlinear structural problem, *Comput. Meth. Appl. Mech. Engg*, **2**, 185–201.

[102] Jeusette, J.P., Lachet, G. and Idelsohn, S. (1989) An effective automatic incremental/iterative method for static nonlinear structural analysis, *Comput. Struct.*, **32**, 125–135.

[103] Riks, E. (1984) Some computational aspects of the stability analysis of nonlinear structures, *Comput. Meth. Appl. Mech. Engg*, **47**, 219–259.

[104] See, T. and McConnel, R.E. (1986) Large displacement elastic buckling of space structures, *J. Struct. Engg*, ASCE, **112**, 1052–1069.

[105] Cani, I.M. and McConnel, R.E. (1987) Collapse of shallow lattice domes, *J. Struct. Div.*, ASCE, **112**, 1806–1819.

[106] Crisfield, M.A. and Wills, J. (1988) Solution strategies and softening materials, *Comput. Meth. Appl. Mech. Engg*, **66**, 267–289.

[107] Wagner, W. and Wriggers, P. (1988) A simple method for the calculation of postcritical branches, *Eng. Comput.*, **5**, 103–109.

[108] Wriggers, P., Wagner, W. and Miehe, C. (1988) A quadratically convergent procedure for the calculation of stability points in finite element analysis, *Comput. Meth. Appl. Mech. Engg*, **70**, 329–347.

[109] Wriggers, P. and Simo, J.C. (1990) A general procedure for the direct computation of turning and bifurcation points, *Int. J. Num. Meth. Engg*, **30**, 155–176.

[110] Papadrakakis, M. and Yakoumidakis, M. (1987) A partial preconditioned conjugate gradient method for large eigenproblems, *Comput. Meth. Appl. Mech. Engg*, **62**, 195–207.

[111] Papadrakakis, M. and Yakoumidakis, M. (1987) On the preconditioned conjugate

gradient method for solving $(A - \lambda B)x = 0$, *Int. J. Num. Meth. Engg*, **24**, 1355–1366.
[112] Papadrakakis, M. and Kalathas, N. (1992) On the Newton-Lanczos approach for solving nonlinear finite element problems, *Comput. Struct.*, **44**, 389–397.
[113] Paige, C.C. and Saunders, M.A. (1975) Solution of sparse indefinite systems of linear equations, *SIAM J. Num. Anal.*, **12**, 617–629.
[114] Luenberger, D.G. (1969) Hyperbolic pairs in the method of conjugate gradient, *SIAM J. Appl. Math.*, **17**, 1263–1267.
[115] Luenberger, D.G. (1970) The conjugate residual method for constrained minimization problem, *SIAM J. Num. Anal.*, **7**, 390–398.
[116] Saad, Y. (1984) Practical use of some Krylov subspace methods for solving indefinite and nonsymmetric linear systems, *SIAM J. Sci. Stat. Comput.*, **5**, 203–228.
[117] Ashby, S.F., Manteuffel, T.A. and Saylor, P.E. (1990) A taxonomy for conjugate gradient methods, *SIAM J. Num. Anal.*, **27**, 1542–1568.
[118] Axelsson, O. and Munksgaard, N. (1979) A class of preconditioned conjugate gradient methods for the solution of a mixed finite element discretization of the biharmonic operator, *Int. J. Num. Meth. Engg*, **14**, 1001–1019.
[119] Bunch, J.R. and Parlett, B.N. (1971) Direct methods for solving symmetric indefinite systems of linear equations, *SIAM J. Num. Anal.*, **8**, 639–655.
[120] Duff, I.S., Munksgaard, N., Nielsen, H.B. and Reid, J.K. (1977) *Direct Solution of Sets of Linear Equations Whose Matrix is Sparse, Symmetric and Indefinite*, Harwell Report CSS 44.
[121] Schnabel, R.B. and Eskow, E. (1990) A new modified Cholesky factorization, *SIAM J. Sci. Stat. Comput.*, **11**, 1136–1158.
[122] Parlett, B.N. (1989) *The Symmetric Eigenvalue Problem*, Prentice-Hall, Englewood Cliffs, New Jersey, p. 10.
[123] Smith, L.A., Stanton, E.L., Gibson, W. and Goble, G.G. (1968) *Developments in Discrete Element Finite Deflection Structural Analysis by Function Minimization*, Report, AFFDL-TR-68-126, Wright-Patterson Air Force Base, Dayton, Ohio.
[124] Luenberger, D.G. (1973) *Introduction to Linear and Nonlinear Programming*, Addison-Wesley, Reading, Massachusetts.
[125] Lee, S.H. (1989) Rudimentary considerations for effective line search method in nonlinear finite element analysis, *Comput. Struct.*, **32**, 1287–1301.
[126] Crisfield, M.A. (1983) An arc-length method including line searches and accelerations, *Int. J. Num. Meth. Engg*, **19**, 1269–1289.
[127] Goldstein, A.A. (1968) On steepest descent, *SIAM J. Control Optim.*, **3**, 147–151.
[128] Armijo, L. (1968) Minimization of functions having Lipschitz continuous first partial derivatives, *Pacific J. Math.*, **16**, 149–154.
[129] Dixon, L.C.W. and Rice, R.C. (1988) Numerical experience with the truncated Newton method for unconstrained optimization, *J. Opt. Th. Applics.*, **56**, 245–255.
[130] Biegler-König, F. (1985) Quasi-Newton methods for saddle points, *J. Opt. Th. Applics.*, **47**, 393–399.
[131] Meek, J.L. and Tan, H.S. (1984) Geometrically nonlinear analysis of space frames by an incremental iterative technique, *Comput. Meth. Appl. Mech. Engg*, **47**, 261–282.
[132] Hinton, E. and Owen, D.J.R. (1984) *Finite Element Software for Plates and Shells*, Pineridge Press, Swansea, UK.

M. Papadrakakis
Institute of Structural Analysis and Aseismic Research
National Technical University of Athens
Zografou Campus
15773 Athens
GREECE

8
Mode Superposition Methods

P. Léger

École Polytechnique, Montreal, Canada

8.1 INTRODUCTION

Dynamic analysis of complex structural systems by the finite element method (FEM) requires detailed structural models with a large number of static/dynamic degrees-of-freedom (DOF) for accurate local stress recovery. To reduce the numerical effort, coordinate reduction procedures such as kinematic and static constraints (condensation) are often used to establish the system matrices. The response calculation is then carried out in modal coordinates because (1) a good approximation of the response can generally be obtained using a reduced system with a few coordinates, and (2) the reduced system usually becomes an independent set of equations that can be integrated one by one.

The use of eigenvectors, to reduce the size of structural systems, and to represent the structural behaviour by a few generalized coordinates, requires in its traditional formulation the solution of large eigenvalue problems [1]. The difficulties associated with the eigensolution are numerous:

1. the truncated basis does not span the complete problem space;
2. the generation of eigenvectors for large systems is very expensive and time consuming;
3. the automatic selection of the required number of eigenmodes for satisfactory convergence is difficult to estimate *a priori*; and
4. the eigenbasis ignores important information related to the specified loading characteristics such that computed eigenvectors can be nearly orthogonal to the applied loading and therefore will not participate significantly in the solution.

Solving Large-scale Problems in Mechanics, Edited by M. Papadrakakis
© 1993 John Wiley & Sons Ltd

A new method of dynamic analysis for structural systems subjected to fixed spatial distribution of the dynamic load was introduced by Wilson et al. [2, 3] as an economic alternative to classical mode superposition techniques. The solution is based on a transformation to a reduced system of generalized Ritz coordinates using load-dependent transformation vectors generated from the specified spatial distribution of the dynamic loads. These vectors can be generated at a fraction of the cost of eigenvectors, and directly include in the basis the static correction effects for the truncation of higher modes. The algorithm is particularly well suited for the analysis of large structural systems [4–7], and vectorization/parallelization [8–9] on more comprehensive computer systems. Computational variants and extensions have been presented to:

1. improve the efficiency of this solution technique and express the reduced systems in tridiagonal form in analogy with the Lanczos method [10–23];
2. compute the response to arbitrary loading that are a function of space and time [24–6];
3. compute the response of non-classically damped systems [27–31];
4. carry out dynamic substructure analysis [12, 32–7]; and
5. perform non-linear response analysis [38–41].

The purpose of this chapter is to present a review of recent developments and applications of eigen and load-dependent vector bases to solve structural dynamic problems. Classical modal summation techniques [42–9], and solution techniques using load-dependent vectors for various forms of loadings are first presented. The effectiveness of mode-superposition methods is then studied in the framework of elasto-plastic seismic analysis of MDOF structures.

8.2 DYNAMIC RESPONSE ANALYSIS BY VECTOR SUPERPOSITION METHODS

The first step in a structural analysis using the FEM is to discretize the structure to obtain the stiffness, $[K]$, the mass, $[M]$, and the damping, $[C]$, matrices needed for the formulation of the equations of dynamic equilibrium of order n that can be written as

$$[M]\{\ddot{U}(t)\} + [C]\{\dot{U}(t)\} + [K]\{U(t)\} = \{F(s,t)\} \qquad (n \times n) \qquad (8.1)$$

where $\{F(s,t)\}$ is the imposed dynamic load, which is a function of space and time. Usually the geometry of the structure does not permit the discretization in a few finite elements, but the behaviour may be accurately characterized by a few generalized DOF. A new discretization can then be carried out using the summation or the 'superposition' of a small number of global shape functions, the transformation vectors or 'modes', derived from the FE model.

Figure 8.1 summarizes the possible computational solution methods that can be employed for the dynamic analysis of linear systems by mode superposition. Depending on the system matrices considered, different types of transformation vectors can be generated. It can be expected that if more information about the characteristics of the structural system is included in the generation of the transformation vectors, a more accurate solution can be obtained. The key steps of the mode superposition methods presented in Figure 8.1 will be discussed in more detail in the following sections. These are:

1. the generation of the transformation vectors;
2. the formulation of the reduced system of equations in generalized coordinates; and
3. the subsequent summation techniques to obtain the final response in FE geometric coordinates.

Detailed descriptions of structural dynamic methods for time history or response spectra analyses of the reduced (or unreduced) system of equations can be found elsewhere [50–3].

8.2.1 Formation and solution of reduced systems of equations

The vector of nodal displacement, $\{U(t)\}$, can be approximated by a linear combination of r linearly independent transformation vectors, $[X_r]$, with r much less than n, as

$$\{U(t)\} = \sum_{i=1}^{r} \{X_i\} y_i(t) = [X_r]\{Y_r\}. \tag{8.2}$$

In equation (8.2), $[X_r]$ can be chosen to be $[M]$-orthonormal without loss of generality, and $y_i(t)$ are unknown parameters, the generalized coordinates, obtained by solving a reduced system of r equations written as

$$[M_r]^* \{\ddot{Y}_r(t)\} + [C_r]^* \{\dot{Y}_r(t)\} + [K_r]^* \{Y_r(t)\} = \{F_r(s,t)\}^* \qquad (r \times r) \tag{8.3}$$

where

$$[M_r]^* = [X_r]^T [M][X_r] = [I_r] \tag{8.4}$$

$$[C_r]^* = [X_r]^T [C][X_r] \tag{8.5}$$

$$[K_r]^* = [X_r]^T [K][X_r] \tag{8.6}$$

$$\{F_r(s,t)\}^* = [X_r]^T \{F_r(s,t)\}. \tag{8.7}$$

The objective of the transformation is to obtain new system mass, damping and

1. Equation of dynamic equilbrum

$$[M]\{\ddot{U}(t)\} + [C]\{\dot{U}(t)\} + [K]\{U(t)\} = \{F(s,t)\} \quad (n \times n)$$

2. System matrices

$[M], [K]$
$[M], [C], [K]$
$[M], [K], \{F(s,t)\}$
$[M], [K], [C], \{F(s,t)\}$

3. Transformation vectors: $[X_r] \quad (n \times r)$

Undamped free-vibration eigenvectors
Damped free-vibration eigenvectors
Undamped load-dependent vectors
Damped load-dependent vectors

4. Reduced system: $\{U(t)\} = [X_r]\{Y(t)\}$

$[M_r]\{\ddot{Y}(t)\} + [C_r]\{\dot{Y}(t)\} + [K_r]\{Y(t)\} = \{F_r(s,t)\} \quad (r \times r)$
$[M_r] = [X_r]^T[M][X_r]$
$[C_r] = [X_r]^T[C][X_r]$
$[K_r] = [X_r]^T[K][X_r]$
$\{F_r(s,t)\} = [X_r]^T\{F(s,t)\}$

5. Dynamic response analysis

$\{Y_r(t)\}$; (a) Direct Step-by-Step method
(b) Duhamel integral
(c) Frequency domain solution
(d) Piecewise exact method
$\{Y_{max}\}$; (e) Response Spectra Analysis

6(a) Modal summation methods: $\{U(t)\} = [X_r]\{Y_r(t)\} + \{U^c(t)\}$

Truncated Correction for higher
dynamic response mode truncation

MDM–LDM : $\{U^c(t)\} = 0$
SCM : $\{U^c(t)\} = [K]^{-1}([I] - [M][X_r][X_r]^T)\{F(s,t)\}$
MAM : $\{U^c(t)\} = ([K]^{-1} - [X_r][\omega_r^2]^{-1}[X_r]^T)\{F(s,t)\}$
DCM : $\{U^c(t)\} = \{X^c_{r+1}\}y^c_{r+1}(t)$; using $\{F(s,t)\} = \{f(s)\}g(t)$; $\{X^c_{r+1}\} = K^{-1}([I] - [M][X_r][X_r]^T)\{f(s)\}$
Orthonormalize $\{X^c_{r+1}\}$ use as 'Dynamic' vector $r + 1$ in Step 3.

6(b) Modal combination methods for response spectra: $\{Y_{max}\}$

CQC, SRSS, SABS... Using any set of transformation vectors, correction for higher-mode truncation can be added to the dynamic response.

Figure 8.1 Computational solution methods for mode superposition.

stiffness matrices that are reduced in size ($r \times r$) and have smaller bandwidths than the original system matrices while maintaining a good accuracy for the response quantities of interest. The solution of the transformed set of equations can then be carried out by any standard numerical methods used in structural dynamics such as direct step-by-step integration, mode superposition (if coupled), frequency-domain analysis, response spectra method, etc. as shown in Figure 8.1 (Step 5).

The best-known reduction method used in linear dynamic problems consists in choosing r free undamped vibration modes, $[\phi_r]$, coming from the solution of the eigenvalue problem $[K][\phi] = [M][\phi][\omega^2]$, as basis vectors, $[X_r]$. With this choice, it is easy to show that the reduced system matrices, expressed in generalized modal coordinates, become diagonal if proportional damping is assumed. Damped or complex vibration modes can be used to uncouple systems with non-proportional damping. However, it is not a necessary condition for reduction methods by vector superposition that the final system of differential equation be uncoupled.

8.2.2 Modal summation methods

In general, the amplitude of each vector component, as given by the corresponding generalized coordinates in equation (8.2), depends on the representation of the spatial distribution of the loading achieved by the truncated basis, and the frequency content of the loading as compared to the structural frequencies retained in the basis. Assuming for simplicity that the applied loads can be written as the product of a fixed spatial distribution, $\{f(s)\}$, and a prescribed time function, $g(t)$, as

$$\{F(s,t)\} = \{f(s)\} g(t). \tag{8.8}$$

Equation (8.7) then indicates that the generalized force

$$\{f_i(s)\}^* = (\{X_i\}^T \{f(s)\}) \tag{8.9}$$

will be insignificant if the spatial distribution of the applied load is totally dissimilar to the vector shape $\{X_i\}$. This vector can thus be omitted from the response without loss of accuracy. An important example of this type of behaviour is found in earthquake loading, where the load is distributed over the entire structures and interacts effectively only with its lower modes.

The frequency effect can be quantified by considering the relative contribution of the elastic and the inertia forces while resisting the applied dynamic loads. It can generally be recognized that for modes with structural frequencies, ω_s, about three times higher than the frequency of the applied loading, ω_1, the resistance is essentially elastic since the inertial and damping effects can be ignored. This suggests that higher modes resistance can be calculated as a static problem.

For most analysis using modal coordinates, the higher modes are generally of local character and have a negligible contribution to the overall or global structural response. However, they can be important for accurate local stress recovery if the load vector has significant components in its expansion into modal coordinates. The exact response to the residual force vector, $\{R_r(s)\}$, can be written as

$$[M]\{\ddot{U}^c(t)\} + [C]\{\dot{U}^c(t)\} + [K]\{U^c(t)\} = \{R_r(s)\}g(t) = (\{f(s)\} - \{f_r(s)\})g(t) \quad (8.10)$$

where $\{f_r(s)\}$ is the representation of the load in terms of the truncated vector basis given as

$$\{f_r(s)\} = \sum_{i=1}^{r} p_i[M]\{X_i\} = ([M][X_r][X_r]^T)\{f(s)\} \quad (8.11)$$

and $\{U^c(t)\}$ is the response of the system to the force component neglected by the vector basis representation of the loading. This displacement should be added to the response obtained from the truncated vector superposition to obtain exact results.

The classical approach to consider the complete spatial distribution of the dynamic load in mode superposition has been to calculate exact mode shapes for the structural system up to the point where $\omega_s/\omega_1 \geq 3$. The usual mode displacement summation method (MDM), as given by equation (8.2), is then modified by adding static correction terms to account for the flexibility of the higher modes not retained in the summation. In the static correction method (SCM), the acceleration and velocity in equation (8.10) are set to zero for all time to obtain

$$\{U^c(t)\} = [K]^{-1}([I] - [M][X_r][X_r]^T)\{f(s)\}g(t). \quad (8.12)$$

The total response of the system using static correction for higher modes thus becomes [42, 43, 48]

$$\{U(t)\} = [X_r]\{Y_r(t)\} + [K]^{-1}([I] - [M][X_r][X_r]^T)\{f(s)\}g(t). \quad (8.13)$$

A computational variant of the static correction method known as the mode acceleration method (MAM) is also widely used to include higher modes effects in the superposition. In this case $\{U(t)\}$ is written as [42–6]

$$\{U(t)\} = [X_r]\{Y_r(t)\} + ([K]^{-1} - [X_r][\omega_r^2]^{-1}[X_r]^T)\{f(s)\}g(t) \quad (8.14)$$

where the expression $[X_r][\omega_r^2]^{-1}[X_r]^T$ is the truncated expansion of the flexibility matrix using a reduced set of vectors, $[X_r]$. It has been shown that the MAM is exactly equivalent to the SCM [42, 43].

Dynamic correction methods (DCM) that are similar to the SCM except that the correction methods have dynamic effects, have also been proposed [37, 44]. Here, the truncated vector basis, $[X_r]$, is augmented by considering an $r+1$ Ritz vector, $[X_{r+1}^c]$, corresponding to the residual deflected shape representing the effect of higher modes. This vector, which is theoretically orthogonal to the $[X_r]$ set, is computed from equation (8.12) with $g(t) = 1$. It is however recommended to perform Gram–Schmidt orthogonalization to ensure the numerical accuracy of the orthogonality condition. A Ritz frequency, ω_{DCM}, representing an average of truncated eigenvalues, is then computed as

$$\omega_{\text{DCM}} = \left[\frac{\{X^c_{r+1}\}^T [K] \{X^c_{r+1}\}}{\{X^c_{r+1}\}^T [M] \{X^c_{r+1}\}} \right]^{1/2}. \tag{8.15}$$

The reduced system of equations, which is decoupled, is then augmented by considering one additional generalized coordinate, $y^c_{r+1}(t)$. The solution is obtained by applying the MDM on the augmented set of transformation vectors and generalized coordinates. The advantage of this approach is that except for the calculation of ω_{DCM}, standard programs based on mode superposition can be used without any change to carry out the dynamic analysis. For example, this approach is used in the commercial program STARDYNE [54] to account for the so-called 'missing mass' effect. A static analysis is first performed for the specified forcing function. Then using the computed modes, a steady-state solution at zero frequency is obtained. The difference between the results of these two solutions is used to compute the 'missing mass' mode for this forcing function. Additional computational variants to apply static corrections in mode superposition analyses can be found in the literature [42, 44, 45, 47].

The SCM, MAM, and DCM are modal summation methods that have been developed to include the system matrices $[K]$, $[M]$ and $\{F(s,t)\}$ in the vector basis instead of only $[K]$ and $[M]$ in the traditional eigensolution. Wilson et al. [2, 3] have presented a simple numerical algorithm, based on an inverse iteration type of scheme and using the spatial distribution of the dynamic load to generate a set of load-dependent transformation vectors. Ritz type of analyses based on load-dependent vectors, identified here as the load-dependent method (LDM), can maintain the high expected accuracy of modern computer analyses, and significantly reduce the execution time over traditional eigensolution procedures [2, 3, 5–41]. By using the LDM, the correction terms corresponding to the static correction components of the SCM or MAM are included directly in the basis as the vectors are generated.

8.3 LOAD-DEPENDENT TRANSFORMATION VECTORS

8.3.1 Generation of load-dependent vectors

The computational algorithm presented in Table 8.1 is used to generate load-dependent transformation vectors [16, 19]. This variant of the original algorithm proposed by Wilson et al. [2] is more effective to deal with systems that possess massless DOF. An initial vector, $\{U_0\}$, corresponding to the static response of the structure subjected to the spatial distribution of the specified loads, is first generated. As new vectors are computed, this initial static vector is updated using Gram–Schmidt orthogonalization to remove components common with the 'dynamic' vectors that are obtained by repeated multiplication with the mass matrix. The static residual is then added to the basis as a static correction term in analogy with the SCM, MAM and DCM. The main advantage of this procedure over the SCM and the MAM is that the static correction terms are directly included in the transformation basis such that the response can be obtained by the MDM.

Table 8.1 Algorithm for generation of load-dependent vectors using single vector iterations

1.	Dynamic equilibrium equation:	$[M]\{\ddot{U}\} + [C]\{\dot{U}\} + [K]\{U\} = \{F(s)\}g(t)$
2.	Triangularize stiffness matrix:	$[K] = [L]^T[D][L]$
3.	Solve for initial static deflected shape, $\{U_o\}$:	$[K]\{U_o\} = \{F(s)\}$
4.	Solve for Ritz vectors $i = 1, \ldots, r-1$:	$[K]\{\bar{X}_i\} = [M]\{U_{i-1}\}$
		$c_j = \{X_j\}^T[M]\{\bar{X}_i\}$
	[M]-orthogonalization (skip for $i = 1$)	$\{\tilde{X}_i\} = \{\bar{X}_i\} - \Sigma_{j=1}^{i-1} c_j \cdot \{X_j\}$
		$v = (\{\tilde{X}_i\}^T[M]\{\tilde{X}_i\})^{1/2}$
	[M]-normalization	$\{X_i\} = \{\tilde{X}_i\} \cdot 1/v$
	update static vector	$c_{u_i} = \{U_{i-1}\}^T[M]\{X_i\}$
		$\{U_i\} = \{U_{i-1}\} - c_{u_i} \cdot \{X_i\}$
5.	Add static residual, $\{U_{r-1}\}$, as static correction vector, $\{X_r\}$:	$b_r = (\{U_{r-1}\}^T[M]\{U_{r-1}\})^{1/2}$
		$\{X_r\} = \{U_{r-1}\} \cdot 1/b_r$
6.	Orthogonalize vectors with respect to stiffness matrix:	$[K_r] = [X_r]^T[K][X_r]$
	Solve $r \times r$ eigenvalue problem calculate orthogonal Ritz vectors	$([K_r] - [\omega_r^2][I])[Z_r] = [0]$
		$[X_r^0] = [X_r][Z_r]$

8.3.2 Evaluation of error norms

For linear earthquake response analysis based on vector superposition, an effective mass corresponding to the part of the total mass responding to the earthquake in each vector, is commonly used as an indicator of the relative contribution of a particular vector to the global structural response. The cumulative effective 'modal' mass, for a truncated set of r [M]-orthonormal eigen- or load-dependent vectors is

$$e_r = \frac{\sum_{i=1}^{r} p_i^2}{\{r\}^T[M]\{r\}} \times 100\% \tag{8.16}$$

where

$$p_i = \{X_i\}^T[M]\{r\} \tag{8.17}$$

and $\{r\}$ is the earthquake influence vector obtained from unit base rigid displacement. The value of e_r can be monitored during the vector computation process. An appropriate value, corresponding to 90–95% of the total mass, can be used to define a cut-off criterion to stop generating new vectors when a good representation of the spatial distribution of the earthquake load has been achieved. A generalization of this error measure to external loading can be obtained from the Euclidean norm of the residual loading as

$$e_r^* = \left| \frac{1 - \|\{f(s)\} - \{f_r(s)\}\|_2}{\|\{f(s)\}\|_2} \right| \times 100\%. \tag{8.18}$$

Additional computational variants of error norms can be found in the literature [11, 16, 19].

8.3.3 Extension to arbitrary loading

In the form presented in Table 8.1, the load-dependent vector generation algorithm is restricted to dynamic loads that can be defined as the product of one spatial load vector and one time function. An important case of this category is the earthquake problem. However, the most general form of dynamic loading can be expressed as

$$\{F(s,t)\} = \sum_{i=1}^{N} \{f_i(s)\} g_i(t) \tag{8.19}$$

where $\{f_i(s)\}$ represents the ith spatial distribution pattern and $g_i(t)$ the ith time variation function. Computational algorithms have been developed to extend the load-dependent vector generation algorithm to a recurrence sequence that uses a block of vectors corresponding initially to the multiload patterns [6, 24–6]. A block version of the load-dependent vector generation algorithm is presented in Table 8.2 and an extension of the single vectors iteration algorithm presented in Table 8.1. The solution of the reduced subspace eigenproblem as the vector generation proceeds (Step 4(b)) is optional. It provides greater numerical stability to the algorithm, especially for multi-spatial load patterns that will produce moderately to strongly linearly dependent starting deflected shapes, and for structures that have closely spaced modes [5, 25]. It also permits monitoring of the progress of the algorithms in terms of the structural frequencies spanned by the basis. This is important if the number of vectors that are iterated simultaneously is made fairly large. Then, the limited number of steps used in the algorithm to reach an acceptable number of vectors may lead to a poorer approximation of the dynamic effects of the inertial characteristics that are obtained by repeated multiplications by the mass matrix. It is therefore desirable to keep the number of initial loading patterns small. Note that the Rayleigh quotient can also be included in the single vector iteration algorithm to monitor the evolution of the frequency range spanned by the vector basis as new vectors are generated [11].

In complex structual systems, the selection of the initial deformation vectors, the starting static deflected shapes, should be made with the help of an automated linear independency criterion to reduce the number of repetitive load patterns to a minimum and to favour the generation of an orthogonal basis. For example, the cosine of the angle between different loading patterns can be used for this purpose. It will vary

Table 8.2 Algorithm for generation of load-dependent vectors using block vector iterations

1. Dynamic equilibrium equation: $[M]\{\ddot{U}\} + [C]\{\dot{U}\} + [K]\{U\} = [F(s)]\{g(t)\}$
2. Triangularize stiffness matrix: $[K] = [L]^T[D][L]$
3. Solve for initial static deflected shape, $\{U_0\}$: $[K][U]_0 = [F(s)]$
   $[M]$-orthonormalize $[U]_0$ (Gram–Schmidt)
4. Generate Ritz vectors $[X]_i$ $(i = 1, \ldots, n-1)$
   (a) Solve for $[X]_i$
   $$[F]_i = [M][U]_{i-1}$$
   $$[K][\bar{X}]_i = [F]_i$$
   (b) Form and solve reduced subspace eigenproblem (optional)
   reduced stiffness $\quad [\bar{K}] = [\bar{X}]_i^T[F]_i$
   reduced mass $\quad [\bar{M}] = [\bar{X}]_i^T[M][\bar{X}]_i$
   eigenproblem $\quad [\bar{K}][Z] = [\bar{M}][Z][V]$
   Ritz vectors $\quad [X]_i = [\bar{X}]_i[Z]$
   (c) $[M]$-orthogonalize $[X]_i$ against previous blocks (Gram–Schmidt)
   (d) $[M]$-orthonormalize block $[X]_i$
   (e) Remove new Ritz block $[X]_i$ from static block $[U]_{i-1}$
   $$[U]_i = [U]_{i-1} - [X]_i([X]_i^T[M][U]_{i-1})$$
5. Add static block residual $[U]_{n-1}$ as static correction vector, $\{X\}_n$
6. Make Ritz vectors, $[X]$, stiffness orthogonal (optional – uncouples equation of motion)
   reduce stiffness $\quad [K]^* = [X]^T[K][X]$
   eigenproblem $\quad ([K]^* - \bar{\omega}^2[I])[Y] = [0]$
   final Ritz vectors $\quad [X]^\circ = [X][Y]$

between 0 and 1, indicating the degree of linear independence of any vector pair [25].

8.4 APPLICATION EXAMPLES

Load-dependent vectors are automatically generated with a fraction of the numerical effort required for the calculation of the exact eigenvectors of the original system. They represent an efficient approach to the reduction of large three-dimensional structural systems such as soil structure, dam–foundation–reservoir, and offshore or space structures in which classical solution techniques are costly due to the large numerical effort required to solve the eigenvalue problem. Another important advantage of the load-dependent reduction methods is the possibility to carry out dynamic analysis of relatively large structures on inexpensive microcomputers. This section presents some application examples to illustrate the relative performance of load-dependent solutions (LDM) and eigensolutions (MDM, MAM/SCM) of structural dynamic problems.

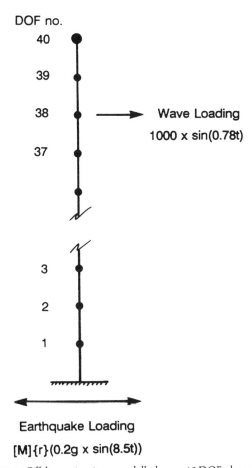

Figure 8.2 Offshore structure modelled as a 40 DOF shear beam.

Example 8.1: *40 DOF shear beam structure* [16, 19]. To illustrate the performance of the LDM a fictitious 40 DOF offshore platform modelled as a shear beam structure with a fundamental period of vibration equal to 5.93 s is first considered as shown in Figure 8.2. The steady-state responses to wave loading and earthquake loading idealized as sinusoidal forcing functions are examined. The wave loading is represented by a concentrated load near the top of the structure with a period of 8 s. Here, the participation of higher modes is important since a single concentrated load is used. The wave load is of low frequency such that the MAM should be very effective to compute the structural response with a few mode shapes. The earthquake loading has a period of 0.74 s and an amplitude of 0.2g. The MAM will not be as effective for this loading since the forcing frequency, w_1, is near important structural frequencies.

Figure 8.3 shows the maximum relative error in beam shear forces for wave loading with respect to the exact solution. A converged solution is assumed when a maximum relative error in beam shear force less than 1% is obtained. Convergence is obtained from 3 vectors for the LDM, 2 eigenvectors supplemented by the static

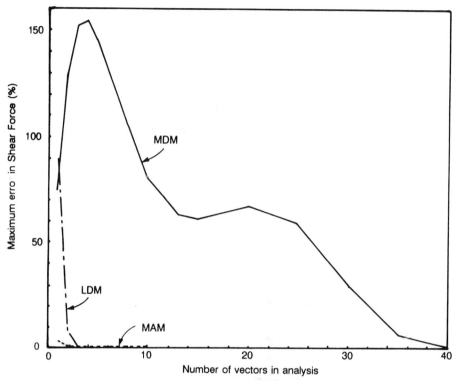

Figure 8.3 Relative error in beam shear forces for 40 DOF offshore structure subjected to wave loading.

correction vector for the MAM, and a complete basis of 40 eigenvectors for the MDM. Figure 8.4 shows the maximum relative error in beam shear force for earthquake loading. Convergence is obtained from 10 vectors for the LDM, 15 vectors for the MAM, and 30 vectors for the MDM.

For both analyses the MDM is not very effective to obtain accurate local forces (stresses) recovery. The LDM possesses convergence characteristics that are similar to the MAM in terms of the number of vectors to be included in the superposition to obtain accurate results. However, as will be shown in the next example dealing with large structural systems, the load-dependent vectors are much faster to generate that the 'exact' eigenvectors such that the LDM is more efficient than the MAM.

Example 8.2: *1944 DOF dam–foundation system* [5]. To illustrate the difference in computer execution time to generate load-dependent vectors and eigenvectors, vector bases are computed for the seismic analysis of dam–foundation systems, progressing from a two-dimensional model of 40 dynamic DOF (DDOF) to a three-dimensional model of nearly 2000 DDOF as shown in Figure 8.5. The characteristics of the models are summarized in Table 8.3. The number 2 or 3 in the model's name indicates a two- or three-dimensional model, the letter A or S is used to indicate a symmetric or antisymmetric model, and the expression F0 or FM is used to indicate a model with a massless foundation block or with a mass foundation block. The

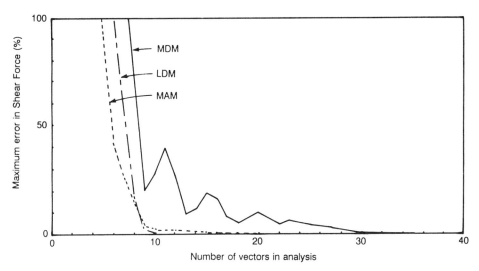

Figure 8.4 Relative error in beam shear forces for 40 DOF offshore structure subjected to earthquake loading.

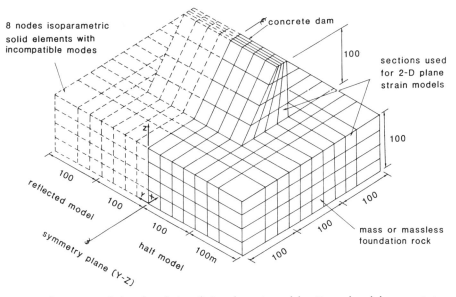

Figure 8.5 Geometry of dam-foundation finite element models. (Reproduced by permission of Elsevier plc.)

stiffness factorization times, working with double precision FORTRAN subroutines on an 8086/8087-2 (8 MHz) microcomputer, are also shown in the table.

The load-dependent vectors are generated in block form using the effective earthquake loads applied in the X-, Y-, and Z-directions as starting vectors. The computer execution time to evaluate the first ten transformation vectors (five for a symmetric or antisymmetric model) using the LDM and the subspace iteration

Table 8.3 Characteristics of dam-foundation models analysed. (Reproduced by permission of Elsevier plc.)

Model	Stiffness DOF	Mass DOF	Avge HBdw	Max. HBdw	No. of blocks in |K|	Factorization time (h:m:s)
DAM2	40	40	11	20	1	00:00:49
DAM2F0	112	50	16	28	1	00:01:47
DAM2FM	112	112	16	28	1	00:01:47
DAM3A	260	260	67	134	1	00:02:47
DAM3S	280	280	71	142	1	00:03:00
DAM3	540	540	78	156	2	00:06:35
DAM3AF0	944	325	116	321	6	00:27:56
DAM3SF0	1000	350	122	339	7	00:32:00
DAM3AFM	944	944	116	321	6	00:27:56
DAM3SFM	1000	1000	122	339	7	00:32:00
DAM3F0	1944	675	264	738	34	05:35:25
DAM3FM	1944	1944	264	738	34	05:35:25

Table 8.4 Computer execution times to generate transformation vectors for dam-foundation models analysed. (Reproduced by permission of Elsevier plc.)

Model	No. of vectors	Subspace iteration						Load-dependent algorithm			
		N_s	It	Time (h:m:s)	E_X (%)	E_Y (%)	E_Z (%)	Time (h:m:s)	E_X (%)	E_Y (%)	E_Z (%)
DAM2	10	14	13	00:03:45	—	100	100	00:00:41	—	100	100
DAM2F0	10	14	12	00:05:35	—	100	99	00:01:00	—	100	100
DAM2FM	10	14	15	00:07:02	—	93	89	00:01:01	—	100	99
DAM3A	5	9	10	00:05:52	—	84	—	00:01:30	—	93	—
DAM3S	5	9	10	00:06:37	89	—	73	00:01:37	90	—	81
DAM3	10	14	13	00:54:20	89	84	73	00:09:38	99	91	90
DAM3AF0	5	9	10	01:14:04	—	96	—	00:16:17	—	98	—
DAM3SF0	5	9	10	01:07:26	95	—	95	00:18:45	96	—	97
DAM3AFM	5	9	17	02:07:56	—	64	—	00:16:15	—	88	—
DAM3SFM	5	9	18	02:39:51	71	—	44	00:18:43	80	—	86
DAM3F0*	10	7	25	14:31:40	95	96	95	02:22:43	99	97	98
DAM3FM*	10	7	36	21:00:18	71	64	44	02:22:51	96	88	91

*N_s is less than number of vectors due to in-core memory limitations.

method are presented in Table 8.4 with the corresponding directional representation of the seismic loads, E_x, E_y and E_z expressed as a percentage of the total mass (the 'effective' modal mass). The size of the subspace, N_s, and the number of iterations, It, required to generate the eigenvectors by the subspace iteration method are also given. The results show that the LDM has a significant advantage over subspace iterations solutions in terms of a much reduced computational effort to generate the transformation vectors and improved convergence rate of critical error norm parameters specially in the vertical 'Z' direction.

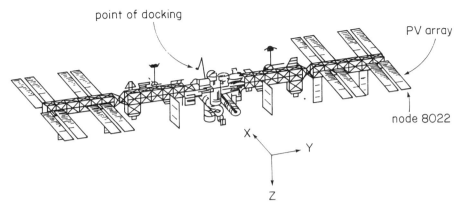

Figure 8.6 Space Station Freedom. (Adapted from Ricles [6].)

For example, the use of the LDM for the largest model that involved 1944 DDOF reduced the execution time by a factor of 9 as compared with the subspace iteration solution. The average error in load representation was also reduced from 40% for the subspace iteration to 8% for the LDM. The better representation of the vertical load is because the initial spatial distribution of the loading is producing significant deformations in the vertical direction, favouring the early appearance of stiff axial deformation mode. This is a major advantage of the LDM over 'exact' eigensolutions if vertical excitation is to be included in the analysis.

Example 8.3: *1000/2800 DOF space structures* [6, 7]. FE models of space structures are characterized by a large number of DOF. The dynamic response analysis of space structures is generally carried out using modal methods. The vector basis is typically truncated because of the enormous computational effort required to obtain a complete eigenbasis. It is very important to maintain reasonable computer costs for any analysis such that inexpensive reanalysis become possible. Low computer cost of a typical analysis cycle allows some basic assumptions used in selecting models and loads to be studied to study the sensitivity of the results, modify the original design and conduct reliability evaluations. In that respect, the LDM presents an attractive solution strategy. Arnold *et al.* [7] were the first to report the use of the LDM for the dynamic analysis of large space structures. The original load-dependent vector generation algorithm proposed by Wilson *et al.* [2] has been implemented in the MSC/NASTRAN computer code using the DMAP matrix manipulation capabilities. Numerical applications for the dynamic analysis of 1000 DOF structural models of optical laser tracking systems indicated that the LDM 'modal' extraction time was approximately one tenth of the NASTRAN eigenvalue extraction procedure.

Ricles [6] used the block form of the LDM algorithm to investigate the transient response of the Space Station Freedom for a simulated docking with the Space Shuttle as shown in Figure 8.6. The model of the Station had 2803 DOF. The block form of the LDM was extended to the analyses of semi-positive definite systems to generate the transformation vectors. To handle the rigid body modes, a shifted stiffness matrix was used. The rigid body modes were then removed from the static block $[U]_0$ to initiate the vector generation process. Vector bases with 30, 60 and

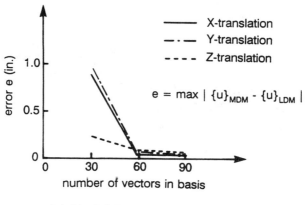

(a) Nodal Displacement {u}

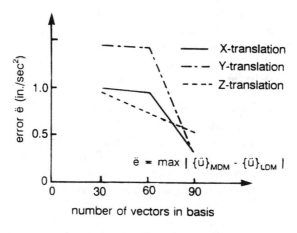

(b) Nodal Acceleration {ü}

Figure 8.7 Error between MDM and LDM solutions for (a) displacements u, and (b) accelerations $\ddot{u}$ at node 8022. (Adapted from Ricles [6].)

90 vectors were computed using a block size of 6 vectors corresponding to the static response of six individual docking loads acting on the structure. Comparisons were made between the LDM and the MDM using a basis of 210 exact mode shapes. Figure 8.7 shows the maximum discrepancies in the transient translational elastic displacements and total acceleration responses at selected DOF at node 8022. For the displacement response (Figure 8.7(a)), the LDM with 60 vectors was able to achieve the same result as the MDM with 210 mode shapes. The LDM solution for acceleration showed greater discrepancies with the MAM (Figure 8.7(b)). In this case, more vectors are required to achieve the level of accuracy found for the displacement response.

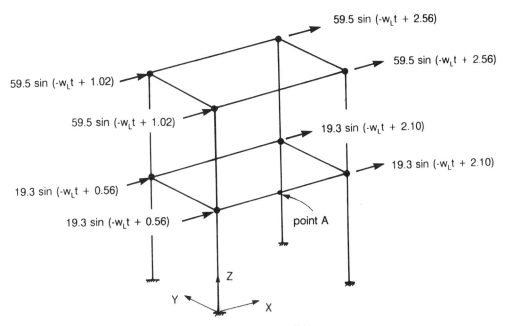

Figure 8.8 Flexible 48 DOF offshore structure.

Example 8.4: *48 DOF offshore structure* [19, 25]. Figure 8.8 shows a 48 DOF offshore structure subjected to sinusoidal wave loading with a period of 9 s. This example is selected to illustrate the typical response of a structural system where the spatial distribution of the load is a function of time. The loading was discretized using a time increment of 0.5 s. A study of the degree of linear independence of the resulting static deflection patterns indicated that only two initial deformation patterns (for $t = 7$ s and $t = 9$ s) were needed to initiate the generation of load-dependent vectors in block form. Dynamic stress responses, S_{comp}, were computed from the MDM, MAM, LDM using single vector iteration (starting vector chosen at $t = 0$ (or $t = 9$) s), and the LDM using block vector iteration. Six vectors were included in the summations and comparisons were established with the exact results, S_{exact}. Figure 8.9 shows the transient response of total longitudinal fibre stress at the bottom chord of the horizontal bracing member (point A in Figure 8.8). The blocked LDM and the MAM produced results that were virtually identical with the exact solution. The results from the LDM with single vector iteration were completely erroneous except at times $t = 0, 4.5, 9$ s where the spatial distribution of the loading coincided with the single loading pattern selected to initiate the vector generation algorithm. The MDM using 6 vectors produced completely erroneous results.

It should be noted that for dynamic loadings where spatial distributions vary with time, the MAM requires a significant computational effort as compared to that MDM since new pseudo-static solutions, $\{U^c(t)\}$, have to be calculated at each time step. The block or 'subspace' LDM is thus found very efficient as compared to the other vector superposition methods for loadings that are function of space and time, since no pseudo-static corrections needs to be computed at each step of the solution.

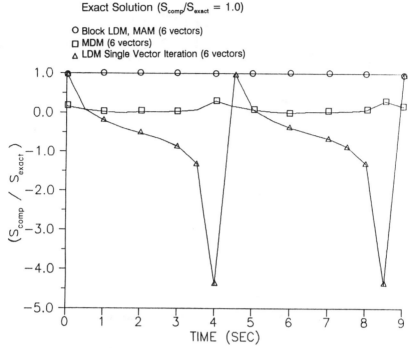

Figure 8.9 Convergence of stress response at point A from different vector bases. (From Leger [25], reproduced by permission of Pergamon Press plc.)

8.5 MODE SUPERPOSITION METHODS FOR NON-LINEAR DYNAMIC ANALYSIS

Step-by-step integration of the incremental form of the equations of dynamic equilibrium expressed in geometric coordinates is generally used to investigate the non-linear behaviour of the MDOF structures. The integration procedure mathematically corresponds to the simultaneous integration of all instantaneous modes of vibration using the same time step. The use of mode-superposition methods in non-linear analysis consists in performing a change of basis to a more effective system of equations. The effectiveness of vector superposition techniques in non-linear dynamic problems depends on:

1. the number of vectors required to simulate accurately the response;
2. the frequency of updating and recalculating the basis vectors, which is a function of the rate of change of these vectors with time; and
3. the efficiency of the algorithm used to calculate the initial vectors and updating them [55].

Vector superposition methods in non-linear structural dynamics can be based on either the tangent spectrum method [41, 56–62] (TSM) or the pseudo-force method

[38–40, 63–76] (PFM). In the TSM, the change of the basis is performed at each time step using vector shapes and frequencies corresponding to the instantaneous system matrices. In the PFM a single set of vectors, based on linear system matrices, is used throughout the analysis. The linear system matrices are employed during the complete response calculation, and the nonlinearities are taken as pseudo-forces on the right hand side of the equations of motion. The pseudo-force approach avoids the solution of the tangent eigenproblem at each time step during non-linear behaviour, and was shown competitive in terms of computational effort with direct integration operators in geometric coordinates in numbers of simple non-linear structural dynamic problems [38–40, 63–76]. For large problems the geometric displacements and velocities should be evaluated to compute the internal forces only at the DOF where non-linear effects occur. If a large portion of the system is known to remain linear, and a few modes are required to describe the motion of the system, the PFM will require significantly less computer resources as compared to a direct integration of the equation of motion [54].

Tangent methods provide a knowledge of the spectrum of frequencies of the dominating modes throughout the inelastic response that is not available in the PFM. It presents several advantages such as:

1. a rationalization of non-linear behaviour in an elastic format to evaluate period elongation, and Rayleigh damping matrices based on tangent modal properties;
2. the definition of new damage indices based on the evolution of tangent modal properties; and
3. a direct control of the participation of higher modes that can be explicitly excluded of the solution instead of relying on numerical damping.

Tangent methods have not been developed in the past due to the high computational cost required in the repetitive solution of the eigenproblem at each time step. But recent development in computer hardware and vectorization/parallelization of solution algorithms can now bring modal tangent solution methods to an economic possibility and technical feasibility without making compromises in the modelling to the point of rendering questionable any results obtained. Idelsohn and Cardona [41, 58] studied the dynamic response of simple geometrically non-linear structures subjected to harmonic loads by the TSM using truncated vector bases. Besides the usual modal truncation error, a new source of error was identified each time a change of basis was performed. Incompatibility between the ability of the old basis and the new basis to represent the initial displacement, velocity and acceleration at the beginning of a new time step introduced a continuously growing lack of equilibrium that produced an unstable solution. The proposed remedy to this problem was to improve the vector basis by using load-dependent Ritz vectors [41] and by the addition of modal derivatives to avoid the need to update the basis thus using a strategy similar to the PFM.

To investigate the use of mode superposition methods for non-linear structural dynamic problems, the following sections present the elasto-plastic seismic response of a simple MDOF structure using the PFM and the TSM. The emphasis is put on the effect of modal truncation on the non-linear response. Particular attention is given to the ductility demand and the hysteretic and damping energy dissipated during the seismic response. The effectiveness of the LDM, which directly provides a static correction for the truncation of higher modes, is studied in the framework of elasto-plastic analysis. An equilibrium correction method to stabilize tangent solutions when truncated vectors bases are updated is presented.

8.5.1 Solution of the non-linear equations of dynamic equilibrium

The equations of dynamic equilibrium for seismic response analysis of an elasto-plastic system in geometric coordinates are:

$$[M]\{\ddot{U}(t)\} + [C]_t\{\dot{U}(t)\} + \{R(t)\} = -[M]\{r\}\ddot{U}_g(t) = \{F(s,t)\} \tag{8.20}$$

where $[M]$ is the mass matrix, $[C]_t$ is the tangent damping matrix, $\{R(t)\}$ is the non-linear restoring force vector, $\{r\}$ is the influence vector from unit base displacement, and $\ddot{U}_g(t)$ is the specified ground acceleration. The restoring force vector, $\{R(t)\}$, can be written in terms of the tangent stiffness matrix, $[K]_t$, as

$$\{R(t)\} = [K]_t\{U(t)\} = ([K]_l + [K]_n)\{U(t)\} \tag{8.21}$$

where $[K]_l$ is the linear stiffness corresponding to the reference state of the structure, and $[K]_n$ is the stiffness component dependent on displacements. In non-linear seismic analysis, viscous damping is generally modelled by a Rayleigh-type representation given as [60]

$$[C]_t = a[M] + b[K]_t + b_0[K]_l = (a[M] + (b + b_0)[K]_l) + b[K]_n = [C]_l + [C]_n. \tag{8.22}$$

The pseudo-force formulation of the equation of equilibrium is obtained by substituting equations (8.21) and (8.22) in equation (8.20):

$$[M]\{\ddot{U}(t)\} + [C]_l\{\dot{U}(t)\} + [K]_l\{U(t)\} = \{F(s,t)\} - [K]_n\{U(t)\} - [C]_n\{\dot{U}(t)\}$$
$$= \{F(s,t)\} - \{F(s,t)\}_n. \tag{8.23}$$

Note that $[C]_n$ can represent also the non-proportional portion of the damping matrix [29, 77, 78]. To solve equation (8.20) or equation (8.23) by vector superposition methods, the nodal displacements, $\{U(t)\}$, can be approximated by a linear combination of a set of linearly independent free-vibration eigenvectors in a classical MDM or load-dependent vectors in the LDM, according to equation (8.2).

In the TSM, the transformation vectors are updated at each time step to reflect the evolution of non-linear behaviour in the solution. Equilibrium iterations can be optionally performed in the system of equations expressed in generalized coordinates. The equilibrium unbalance can be compensated by computing the acceleration vector from the original equations of dynamic equilibrium. This strategy will provide a stable tangent solution; however, the acceleration and velocity may then differ significantly from the exact solution using a complete vector basis. It should be noted that restoring equilibrium after each time step also provides some form of correction for the lack of equilibrium introduced by the truncation of higher modes.

The PFM and TSM solution algorithms developed in this study are based on the Newmark–Beta average acceleration method as shown in Table 8.5. The same algorithms can be used to compute the solution in geometric coordinates if the transformation to a reduced system of equations is ignored. Many computational variants are possible depending whether or not the coordinate transformation decouples the PFM or TSM reduced systems. This is a function of the type of damping model retained, and of the $[K]$-orthogonality condition of the $[M]$-orthonormal vector basis that is not a mandatory requirement for the validity of the proposed solution strategies. The Newmark–Beta method has therefore been selected for its ability to solve either coupled or uncoupled reduced systems, and to maintain a high degree of compatibility between the PFM, TSM, and the solution in geometric coordinates.

8.5.2 System analysed

Figure 8.10 shows the MDOF shear beam structure considered for numerical applications. A 25 DOF system is designed as a ductile system assuming a linear variation of strength and stiffness over the height. The system properties are calibrated such that the fundamental period of vibration, T_1, is 2.5 s. Figure 8.11 shows the bilinear hysteretic action-deformation model that is considered for the structural elements. A value of 10% is used for the strain hardening parameter α. Rayleigh damping based on initial elastic properties using 5% critical in the first mode and the mode required to obtain 95% effective modal mass is used in all analyses. The effect of damping matrices computed from tangent modal properties on non-linear response of MDOF structures has also been recently investigated [56, 79]. The first 20 s of the S00E component of the 1940 E1 Centro earthquake accelerogram multiplied by a 0.95 scaling factor is used as input motion.

To interpret the non-linear seismic response, peak values of ductility demand and energy response indicators are used. The energy calculations are performed using the 'absolute' input energy, $E_I(t)$, given as:

$$E_K(t) + E_D(t) + E_S(t) + E_H(t) = E_I(t) \tag{8.24}$$

where $E_K(t)$ is the 'absolute' kinetic energy, which includes the effects of the rigid-body translation of the structures, $E_D(t)$ is the viscous damping energy, $E_H(t)$ is the irrecoverable hysteretic energy, and $E_S(t)$ is the recoverable elastic strain energy at time t. Detailed expressions for each term of equation (8.24) can be found in Uang and Bertero [80]. The error in energy balance can then be used as a global indicator of the equilibrium achieved by a solution strategy. The normalized error in energy balance, EEB(t), can be defined as:

$$\text{EEB}(t) = \frac{|E_I(t) - E_K(t) - E_D(t) - E_H(t) - E_S(t)|}{|E_I(t)|} \times 100\%. \tag{8.25}$$

Table 8.5 Tangent spectrum method (TSM) and pseudo-force method (PFM) algorithms

(A) Initial calculations

Perform operations described in block C-Tangent properties with initial elastic properties.

(B) For each time-step

1. Form effective load vector:

$$\{\bar{F}(t,s)\} = \{\Delta F(t,s)\} + [M]\left(\frac{4}{\Delta t}\{\dot{U}(t)\} + 2\{\ddot{U}(t)\}\right) + 2[C]_i\{\dot{U}(t)\}$$

where $\{\Delta F(t,s)\} = \{F(t+\Delta t, s)\} - \{F(t,s)\}$

initialize $i = 1$, $\{\Delta u_0\} = 0$

2. (skip for PFM) If $i = 2$ and a change of basis is required, perform block (C) Tangent properties.

3. Reduce load vector:

$$\{\bar{F}(t,s)\}^* = [X]_i^T \{\bar{F}(t,s)\} \quad ; \quad \text{PFM: } [X]_i = [X]_i$$

4. Solve for incremental displacements:

$$[\bar{K}]_i^* \{\Delta Y_i\} = \{\bar{F}(t,s)\}^* \quad ; \quad \text{PFM: } [\bar{K}]_i^* = [\bar{K}]_i^*$$
$$\{\Delta U_i\} = [X]_i \{\Delta Y_i\} \qquad\qquad [X]_i = [X]_i$$

5. Accumulate incremental displacements:

$$\{\Delta U_i(t)\} = \{\Delta U_{i-1}(t)\} + \{\Delta U_i\}$$

6. Compute increment velocities and accelerations:

$$\{\Delta \dot{U}(t)\} = \frac{2}{\Delta t}\{\Delta U_i(t)\} - 2\{\dot{U}(t)\}$$

$$\{\Delta \ddot{U}(t)\} = \frac{4}{\Delta t^2}\{\Delta U_i(t)\} - \frac{4}{\Delta t}\{\dot{U}(t)\} - 2\{\ddot{U}(t)\}$$

7. Update motion:

$$\{U(t+\Delta t)\} = \{U(t)\} + \{\Delta U_i(t)\} \quad \{\dot{U}(t+\Delta t)\} = \{\dot{U}(t)\} + \{\Delta \dot{U}(t)\}$$
$$\{\ddot{U}(t+\Delta t)\} = \{\ddot{U}(t)\} + \{\Delta \ddot{U}(t)\}$$

Continued

8.5 SUPERPOSITION METHODS FOR NON-LINEAR ANALYSIS

Table 8.5 *continued*

8. State determination: (skip for PFM) if $i = 1$ and a member changed state, an update of the basis is required.	$R(t + \Delta t) = [K]_l \{U(t + \Delta t)\}$; PFM: $[K]_l = [K]_l + [K]_n$
9. Equilibrium check:	$\{RES\} = \{F(t + \Delta t, s)\} - \{R(t + \Delta t)\} - [C]_l[\dot{U}(t + \Delta t)\} - [M]\{\ddot{U}(t + \Delta t)\}$ PFM: $[C]_l = [C]_l + [C]_n$ IF $\|\{RES\}\|_2 >$ TOL $i = i + 1$ GOTO 2 with $\{\tilde{F}(t, s)\} = \{RES\}$ $\leq$ TOL CONTINUE
10. (skip for PFM) reestablish equilibrium:	$\{\ddot{U}(t + \Delta t)\} = [M]^{-1}(\{F(t + \Delta t, s)\} - \{R(t + \Delta t)\} - [C]_l\{\dot{U}(t + \Delta t)\})$
11. Proceed to next time step.	
(C) Tangent properties	
1. Compute tangent vector shapes.	$[K]_l[X]_l = [\omega^2]_l[M][X]_l$ for eigenvectors Table 8.1 or 8.2 for LD vectors
2. Compute damping matrix: (if required)	$[C]_l = a_l[M] + b_l[K]_l + b_o[K]_l$; PFM: $[C]_l = [C]_l + [C]_n$
3. Reduce the system of equation:	$[K]_l^* = [X]_l^T[K]_l[X]_l, \quad [C]_l^* = [X]_l^T[C]_l[X]_l,$ $[M]^* = [X]_l^T[M][X]_l = [I]$
4. Form effective stiffness:	$[\tilde{K}]_l^* = [K]_l^* + \dfrac{4}{\Delta t^2}[M]^* + \dfrac{2}{\Delta t}[C]_l^*$

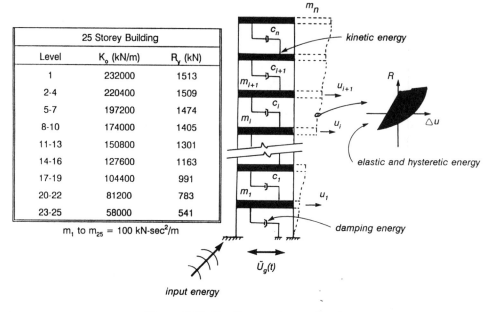

Figure 8.10 Non-linear system analysed.

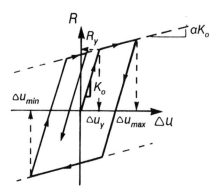

Figure 8.11 Action–deformation model of structural elements.

The maximum value of EEB(t) should be small to ensure the validity of the seismic response produced by a specific computational strategy.

8.5.3 *Analysis of structural response*

Figure 8.12(a, b) show the effect of equilibrium iterations on the ductility demand considering the TSM–MDM strategy with different time steps. A basis of 17 vectors is used since this number of vectors is required for energy convergence as shown in Figure 8.14(d). The solution without iteration (Figure 8.12(a)) is sensitive to the selected time step, especially in the bottom storeys. When equilibrium iterations are

8.5 SUPERPOSITION METHODS FOR NON-LINEAR ANALYSIS

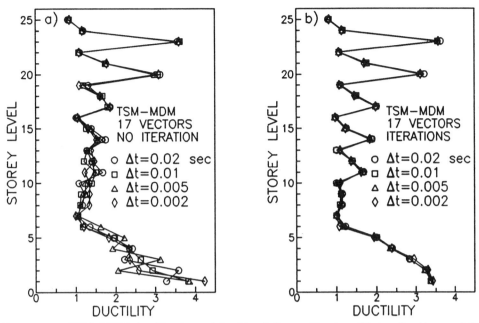

Figure 8.12 TSM–MDM ductility demand for (a) solutions without equilibrium iterations and (b) solutions with equilibrium iterations.

Table 8.6 Number of basis updates for different solution strategies for non-linear systems

Method	Time step Δt (s)	Number of basis updates	
		With iterations	Without iterations
TSM–MDM	0.02	102	142
	0.01	140	255
	0.005	164	500
	0.002	201	927
TSM–LDM	0.02	99	152
	0.01	133	258
	0.005	164	500
	0.002	216	1033

used (Figure 8.12(b)), the response is almost independent of the time step. An accurate response is obtained using Δt as large as 0.02 s corresponding to the interval used to describe the input accelerations. The TSM with equilibrium iterations is therefore very advantageous to limit the number of basis updates during the solution as shown in Table 8.6.

250 MODE SUPERPOSITION METHODS

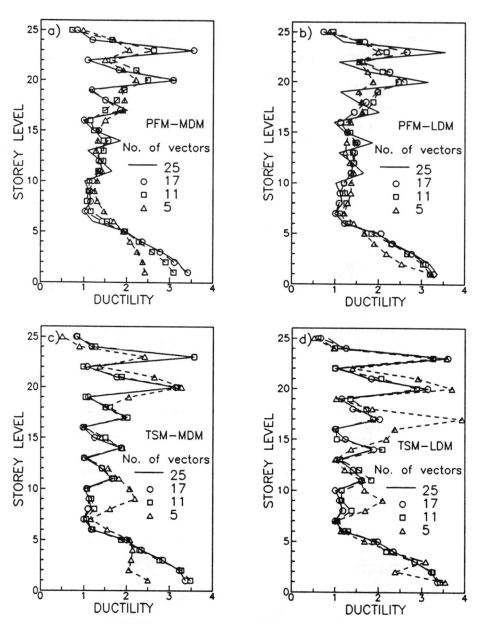

Figure 8.13 Effect of the number of transformation vectors on the peak ductility demand in each storey for different solution strategies: (a) PFM–MDM; (b) PFM–LDM; (c) TSM–MDM; and (d) TSM–LDM.

Figure 8.13 describes the effect of the number of transformation vectors on the peak ductility demand for different solution strategies. The time step is 0.01 s and the TSM is used with equilibrium iterations. The results shown in Figures 8.13(a) and 8.13(b) indicate that the peak ductility response is quite sensitive to vector basis

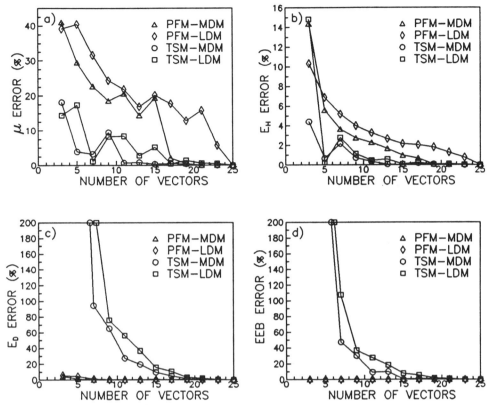

Figure 8.14 Convergence of various non-linear response indicators for different solution strategies. (a) ductility demand level 20; (b) hysteretic energy, E_H; (c) damping energy, E_D; and (d) error in energy balance, EEB.

truncation when the PFM is used. The solutions computed using eigenvectors, the PFM–MDM, and the load-dependent vectors, the PFM–LDM, yield comparable global performance. The truncated PFM–MDM ductility response is more accurate in the top storeys, while the PFM–LDM response is more accurate in the bottom storeys. Figures 8.13(c, d) show that the peak ductility response based on the TSM is much less sensitive to basis truncation than the solution obtained from the PFM. Excluding the solution carried out with 5 vectors, it is shown that the MDM and the LDM yield almost identical results.

Figure 8.14 shows the convergence of non-linear response indicators for different solution strategies. For the ductility, μ, the hysteretic energy, E_H, and damping energy, E_D, dissipated after the earthquake, the results are given in terms of the relative error with respect to the solution using a complete basis. For the error in energy balance, EEB, the maximum value of equation (8.25) is used. Figure 8.14(a) shows the convergence of the ductility demand in level 20 that represents the response in the upper region of the building. For solutions using from 5 to 17 vectors, the type of vectors considered does not influence significantly the results. For the PFM–MDM, 17 eigenvectors produce an error of 1%, while 25 vectors are required to get the same accuracy by the PFM–LDM. For a storey level representative of the bottom part of the structure, opposite results are observed, as shown in Figures 8.13(a, b).

Figure 8.14(b) indicates that E_H is not sensitive to modal truncation if the TSM is used. The E_H error is larger when the PFM is used instead of the TSM, although the values remain below 10% for any solution considering more than 5 vectors. Figure 8.14(c) shows the error in energy dissipated by viscous damping. The behaviours of the PFM and TSM solutions are very different. For the TSM, the computation of the acceleration vector from equilibrium condition introduces a significant error in the acceleration and velocity when a relatively small number of vectors is used. A solution using 17 vectors provides an E_D error below 10% using either vector basis. When the PFM is used, the E_D error is not sensitive to basis truncation. Figure 8.14(d) shows that the EEB error follows the convergence pattern of the E_D error. A value smaller than 5% should be obtained to ensure a reliable global performance of a particular solution strategy. However, this does not guarantee accurate results for local response indicators such as storey ductility demand. Contrary to elastic analyses, the criterion of 95% effective modal mass to obtain reliable global results has been found inapplicable to elasto-plastic systems.

The implementation of the non-linear solution algorithms used in this study have not been optimized to minimize the CPU execution time. The use of the TSM–LDM considering 17 vectors, required for energy convergence, decreases the CPU execution time by a factor of about two with respect to the TSM–MDM. The CPU time for the TSM–MDM was obtained by using the Lanczos method as the eigensolver. For a large structural system this difference would be more significant. For this system, which possesses a tridiagonal stiffness matrix, no significant CPU time differences are observed between the PFM–MDM, PFM–LDM and the direct integration in geometric coordinates since the vector basis is computed only once. The PFM and the direct itegration solutions require approximately 15% and 30% of the TSM–MDM and TSM–LDM execution time, respectively.

It should be noted that in an inelastic seismic analysis of a 19-storey shear beam type of structure, Chang and Mohraz [63] reported a reduction of the order of 40% using an optimized PFM including all modes as compared to a direct integration method. Mohraz et al. [56] have also shown in a recent paper that incremental mode superposition methods are more economical than direct integration methods if damping matrices based on instantaneous modal properties are used.

8.5.4 *Final remarks on non-linear mode superposition methods*

The use of vector superposition methods for non-linear dynamic analysis can present advantageous solution strategies as compared to the classical step-by-step integration of the fully coupled set of equations, particularly in terms of the knowledge of the evolution of the modal properties during inelastic response if tangent methods are used as shown in Figure 8.15. The use of load-dependent vectors can improve the effectiveness of the modal methods since it provides a more economical way to generate vector bases to transform the equations of the system while maintaining accuracy in the response quantities of interest. The TSM might emerge as a very valuable tool in earthquake engineering, and other fields where non-linear structural

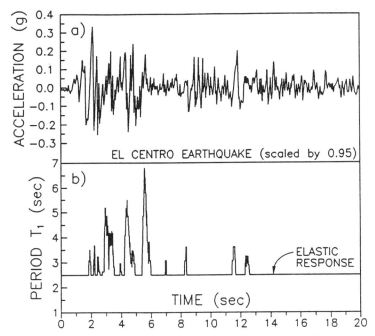

Figure 8.15 Input motion and evolution of tangent model properties for the system analysed; (a) El Centro accelerogram, (b) evolution of fundamental period T_1.

dynamic problems are frequently encountered, to improve further the understanding, and rationalize in a linear format, complex non-linear response mechanisms.

8.6 CONCLUSIONS

This chapter has presented computational solution methods that are very efficient for the dynamic analysis of large and/or non-linear structural systems. The numerical techniques are essentially based on the direct superposition of a special class of Ritz vectors generated from the spatial distribution of dynamic loads. The method first introduced by Wilson *et al.* [2] as an economic alternative to classical mode superposition, recognized the special nature of structural dynamic problems, mainly that a dynamic analysis can be interpreted as a static analysis that considers inertia forces.

This method has recently been incorporated into commercially available general purpose computer programs [81, 82] that are actively used to solve industrial problems. Recent research efforts, as reported in the numerous 'Ritz–Lanczos' related papers found in the literature, have clearly indicated that solution methods based on the direct superposition of load-dependent Ritz (Lanczos) vectors can be developed as complete numerical tools, able to improve the convergence characteristics and numerical efficiency of any classical dynamic analysis techniques that are using eigenvectors as bases for response computations.

BIBLIOGRAPHY

[1] Sehmi, N.S. (1989) *Large Order Structural Eigenanalysis Techniques*, Ellis Horwood, Chichester.
[2] Wilson, E.L., Yuan, M.W. and Dickens, J.M. (1982) Dynamic analysis by direct superposition of Ritz vectors, *Earthquake Engg and Struct. Dyn.*, **10**, 813–824.
[3] Wilson, E.L. (1985) A new method of dynamic analysis for linear and nonlinear systems, *Finite Elements in Analysis and Design*, **1**, 21–23.
[4] Nappi, A. and Perego, U. (1985) Solution of large eigenproblems on a microcomputer, *Avd. Eng. Software*, **7**, 15–20.
[5] Léger, P. (1989) Microcomputer solution of large structural dynamic problems, *Microcomputers in Civil Engineering*, **4**, 127–132.
[6] Ricles, J.M. (1990) *Development of Load-dependent Ritz Vector Method for Structural Dynamic Analysis of Large Space Structures*, Final Report NGT-44-005-803, NASA/ASEE, Johnson Space Center.
[7] Arnold, R.R., Citerley, R.L., Chargin, M. and Galant, D. (1985) Application of Ritz vectors for dynamic analysis of large structures, *Comput. Struct.*, **21**, 461–467.
[8] Farhat, C. and Wilson, E.L. (1986) Modal superposition dynamic analysis on concurrent multiprocessors, *Engg Comput.*, **3**, 305–311.
[9] Wilson, E.L. and Farhat, C. (1988) Linear and nonlinear finite element analysis on multiprocessor computer systems, *Commun. Appl. Num. Meth.*, **4**, 425–434.
[10] Ibrahimbegovic, A. and Wilson, E.L. (1990) Automated truncation of Ritz vector basis in modal transformation, *J. Engg Mech. ASCE*, **116**, 2506–2520.
[11] Joo, K.J., Wilson, E.L. and Léger, P. (1989) Ritz vectors and generation criteria for mode superposition analysis, *Earthquake Engg and Struct. Dyn.*, **18**, 149–167.
[12] Yuan, M., Chen, P., Xiong, S., Li, Y. and Wilson, E.L. (1989) The WYD method in large eigenvalue problems, *Engg Comput.*, **6**, 49–57.
[13] Chen, H.C. and Taylor, R.L. (1989) Using Lanczos vectors and Ritz vectors for computing dynamic responses, *Engg Comput.*, **6**, 151–157.
[14] Coutinho, A.L.G.A., Landau, L., Wrobel, L.C. and Ebecken, N.F.F. (1989) Modal solution of transient heat conduction utilizing Lanczos algorithm, *Int. J. Num. Meth. Engg*, **28**, 13–25.
[15] Alvarez, R., Molina, J. and Cerrolaza, M. (1988) Practical seismic analysis using Ritz basis, *Software for Engineering Workstations*, **4**, 186–192.
[16] Léger, P. and Wilson, E.L. (1987) Generation of load dependent Ritz transformation vectors in structural dynamics, *Engg Comput.*, **4**, 309–318.
[17] Coutinho, A.L.G.A., Landau, L., Lima, E.C.P. and Ebecken, N.F.F. (1987) The application of the Lanczos mode superposition method in dynamic analysis of offshore structures, *Comput. Struct.*, **25**, 615–625.
[18] Kline, K.A. (1986) Dynamic analysis using a reduced basis of exact modes and Ritz vectors, *AIAA J.*, **24**, 2022–2029.
[19] Léger, P., Wilson, E.L. and Clough, R.W. (1986) *The Use of Load Dependent Vectors for Dynamic and Earthquake Analysis*, Report No. UBC/EERC-85/04, Earthquake Engineering Research Center, University of California, Berkeley.
[20] Bayo, E.P. and Wilson, E.L. (1984) Use of Ritz vectors in wave propagation and foundation responses, *Earthquake Engg and Struct. Dyn.*, **12**, 499–505.
[21] Bayo, E.P. and Wilson, E.L. (1984) Finite element and Ritz vector techniques for solution to three-dimensional soil-structure interaction problems in the time domain, *Engg Comput.*, **1**, 298–311.
[22] Nour-Omid, B. and Clough, R.W. (1984) Dynamic analysis of structures using Lanczos coordinates, *Earthquake Eng. Struct. Dyn.*, **12**, 565–577.

[23] Chowdhury, P.C. (1975) An alternative to the normal mode method, *Comput. Struct.*, **5**, 315.
[24] Léger, P., Ricles, J.M. and Robayo, L.J. (1990) Reducing modal truncation error in the wave response of offshore structures, *Commun. Appl. Num. Meth.*, **6**, 7–16.
[25] Léger, P. (1988) Load dependent subspace reduction methods for structural dynamic computations, *Comput. Struct.*, **29**, 993–999.
[26] Nour-Omid, B. and Clough, R.W. (1985) Block Lanczos method for dynamic analysis of structures, *Earthquake Engg and Struct. Dyn.*, **13**, 271–275.
[27] Chen, H.C. and Taylor, R.L. (1990) Solution of viscously damped linear systems using a set of load-dependent vectors, *Earthquake Engg and Struct. Dyn.*, **19**, 653–665.
[28] Yiu, Y.-C. (1990) *Reduced Vector Basis for Dynamic Analysis of Large Damped Structures*, SEMM Report No. 90-6, Division of Structural Eng. and Struct. Mech., University of California, Berkeley.
[29] Ibrahimbegovic, A. and Wilson, E.L. (1989) Simple numerical algorithms for the mode superposition analysis of linear structural systems with non proportional damping, *Comput. Struct.*, **33**, 523–531.
[30] Léger, P. (1989) Coordinate reduction procedures for seismic analysis of dam-foundation-reservoir systems, *Finite Elements in Analysis and Design*, **5**, 73–85.
[31] Chen, H.C. and Taylor, R.L. (1988) Solution of eigenproblems for damped structural systems by the Lanczos algorithm, *Comput. Struct.*, **30**, 151–161.
[32] Balasubramanian, P., Jagadeesh, J.G., Suhas, H.K. and Ramamurti, V. (1991) Free-vibration analysis of cyclic symmetric structures, *Commun. Appl. Num. Meth.*, **7**, 131–139.
[33] Léger, P. (1990) Application of load-dependent vectors bases for dynamic substructure analysis, *AIAA J.*, **28**, 177–179.
[34] Abdallah, A.A. and Huckelbridge, A.A. (1990) Boundary flexibility method of component mode synthesis using static Ritz vectors, *Comput. Struct.*, **35**, 51–61.
[35] Leung, A.Y.T. (1988) A simple dynamic substructure method, *Earthquake Engg and Struct. Dyn.*, **16**, 827–837.
[36] Wilson, E.L. and Bayo, E.P. (1986) Use of special Ritz vectors in dynamic substructure analysis, *J. Struct. Eng. ASCE*, **112**, 1944–1954.
[37] Dickens, J.M. and Wilson, E.L. (1980) *Numerical Methods for Dynamic Substructure Analysis*, Report No. UBC/EERC-80/120, Earthquake Eng. Res. Center, University of California, Berkeley.
[38] Ibrahimbegovic, A. and Wilson, E.L. (1990) A methodology for dynamic analysis of linear structure-foundation systems with local non-linearities, *Earthquake Engg and Struct. Dyn.*, **19**, 1197–1208.
[39] Lima, E.C.P., Coutinho, A.L.G.A. and Alves, J.L.D. (1989) Non-linear dynamic substructure analysis using direct integration of steady state solution, *Commun. Appl. Num. Meth.*, **5**, 405–413.
[40] Venâncio Filho, F., Coutinho, A.L.G.A., Landau, L., Lima, E.C.P. and Ebecken, N.F.F. (1988) Nonlinear dynamic analysis using the pseudo-force method and the Lanczos algorithm, *Comput. Struct.*, **30**, 979–983.
[41] Idelsohn, S.R. and Cardona, A. (1985) A load-dependent basis for reduced nonlinear structural dynamics, *Comput. Struct.*, **20**, 203–210.
[42] Léger, P. and Wilson, E.L. (1988) Modal summation methods for structural dynamic computations, *Earthquake Engg and Struct. Dyn.*, **16**, 23–27.
[43] Soriano, L.H. and Venâncio-Filho, F. (1988) On the modal acceleration method in structural dynamics, Mode truncation and static correction, *Comput. Struct.*, **29**, 777–782.
[44] Borino, G. and Muscolino, G. (1986) Mode-superposition methods in dynamic analysis of classically and non-classically damped linear systems, *Earthquake Engg and Struct.*

Dyn., **14**, 705–717.

[45] Cornwell, R.E., Craig, R.R. and Johnson, C.P. (1983) On the application of the mode-acceleration method to structural engineering problems, *Earthquake Engg and Struct. Dyn.*, **11**, 679–688.

[46] Anagnostopoulos, S.A. (1982) Wave and earthquake response of offshore structures: evaluation of modal solutions, *J. Struct. Div. ASCE*, **108**, 2175–2191.

[47] Traill-Nash, R.W. (1981) Modal methods in the dynamics superposition analysis in structural dynamics, *Earthquake Engg and Struct. Dyn.*, **9**, 153–169.

[48] Hansteen, O.E. and Bell, K. (1979) On the accuracy of mode superposition analysis in structural dynamics, *Earthquake Engg and Struct. Dyn.*, **7**, 405–411.

[49] Clough, R.W. and Mojtahedi, S. (1976) Earthquake response analysis considering non-proportional damping, *Earthquake Engg and Struct. Dyn.*, **4**, 489–496.

[50] Humar, J.L. (1990) *Dynamics of Structures*, Prentice-Hall, Englewood Cliffs, New Jersey.

[51] Waburton, G.B. (1979) The influence of the finite element method on developments in structural dynamics, in R. Glowinski, E.Y. Rodin and O.C. Zienkienwicz (eds.), *Energy Methods in Finite Element Analysis*, Wiley, Chichester, pp. 59–80.

[52] Wilson, E.L. (1978) Numerical methods for dynamic analysis, in O.C. Zienkienwicz, R.W. Lewin and K.G. Stagg (eds.), *Numerical Methods in Offshore Engineering*, Wiley, Chichester.

[53] Clough, R.W. and Penzien, J. (1975) *Dynamics of Structures*, McGraw-Hill, New York.

[54] *STARDYNE User Information Manual*, General Micro Electronic Corporation, San Diego, California.

[55] Noor, A.K. (1981) Recent advances in reduction methods for nonlinear problems, *Comput. Struct.*, **13**, 31–44.

[56] Mohraz, B., Elghadamsi, F.E. and Chang, C.-J. (1991) An incremental mode superposition for nonlinear dynamic analysis, *Earthquake Engg and Struct. Dyn.*, **20**, 471–481.

[57] Maison, B.F. and Kasai, K. (1990) Analysis for type of structural pounding, *J. Struct. Engg*, ASCE, **116**, 957–977.

[58] Idelsohn, S.R. and Cardona, A. (1985) A reduction method for nonlinear structural dynamic analysis, *Comput. Meth. App. Mech. Engg*, **49**, 253–279.

[59] Idelsohn, S.R. and Cardona, A. (1984) Reduction methods and explicit time integration technique in structural dynamics, *Avd. Eng. Software*, **6**, 36–44.

[60] Gillies, A.G. and Shepherd, R. (1983) Prediction of post-elastic seismic response of structures by a mode superposition technique, *Bulletin of the New Zealand National Society for Earthquake Engineering*, **16**, 222–233.

[61] Remseth, S.N. (1979) Nonlinear static and dynamic analysis of frames structures, *Comput. Struct.*, **10**, 879–897.

[62] Nickell, R.E. (1976) Nonlinear dynamics by mode superposition, *Comput. Meth. Appl. Mech. and Engg*, **7**, 107–129.

[63] Chang, C.-J. and Mohraz, B. (1990) Modal analysis of nonlinear systems with classical and non-classical damping, *Comput. Struct.*, **36**, 1067–1080.

[64] Muscolino, G. (1990) Mode-superposition methods for elastoplastic systems, *J. Engg Mech. ASCE*, **115**, 2199–2215.

[65] Hanna, M.N. (1989) *An Efficient Mode Superposition Method for the Numerical Dynamic Analysis of Bilinear Systems*, Ph.D Dissertation, University of California, Irvine.

[66] Knight, N. (1985) Nonlinear structural dynamic analysis using a modified modal method, *AIAA*, **23**, 1594–1601.

[67] Dungar, R. (1982) An imposed force summation method for non-linear dynamic analysis, *Earthquake Engg and Struct. Dyn.*, **10**, 165–170.

[68] Bathe, K.J. and Gracewski, S. (1981) On nonlinear dynamic analysis using substructuring and mode superposition, *Comput. Struct.*, **13**, 699–707.

[69] Geschwindner, L.F. (1981) Nonlinear dynamic analysis by modal superposition, *J. Struct. Div. ASCE*, **107**, 2325–2336.
[70] Lukkunaprasit, P., Widartawan, S. and Karasudhi, P. (1980) Dynamic response of an elast-viscoplastic system in modal co-ordinates, *Earthquake Engg and Struct. Dyn.*, **8**, 237–250.
[71] Shah, V.N., Bohm, G.J. and Nahavandi, A.N. (1979) Modal superposition method for computationally economical nonlinear structural analysis, *ASME J. Pressure Vessel Technology*, **101**, 134–141.
[72] Clough, R.W. and Wilson, E.L. (1979) Dynamic analysis of large structural systems with local nonlinearities, *Comput. Meth. Appl. Mech. Engg*, **17/18**, 107–129.
[73] Hofmeister, L.D. (1978) Dynamic analysis of structures containing nonlinear springs, *Comput. Struct.*, **8**, 609–614.
[74] Stricklin, J.A. and Haisler, W.E. (1977) Formulations and solution procedures for nonlinear structural analysis, *Comput. Struct.*, **7**, 125–136.
[75] Morris, N.F. (1977) The use of modal superposition in nonlinear dynamics, *Comput. Struct.*, **7**, 65–72.
[76] Molnar, A.J., Vashi, K.M. and Say, C.W. (1976) Application of normal mode theory and pseudo-force methods to solve problems with nonlinearities, *ASME J. Pressure Vessel Technology*, 151–156.
[77] Claret, A.M. and Venâncio-Filho, F. (1991) A modal superposition pseudo-force method for dynamic analysis of structural systems with non-proportional damping, *Earthquake Engg and Struct. Dyn.*, **20**, 303–315.
[78] Udwadia, F.E. and Esfandiari, R.S. (1990) Nonclassically damped dynamic systems: an iterative approach, *J. Appl. Mech. ASME*, **57**, 423–433.
[79] Léger, P. and Dussault, S. (1992) Seismic energy dissipation in MDOF structures, *J. Struct. Engg ASCE*, **118**, 1251–1269.
[80] Uang, C.M. and Bertero, V.V. (1990) Evaluation of seismic energy in structures, *Earthquake Engg and Struct. Dyn.*, **19**, 77–90.
[81] Wilson, E.L. and Habibullah, A. (1989) *SAP 90 User Manual*, Computers and Structures Inc., Berkeley, CA, USA.
[82] PFrame *Users manual*, Softek Services Ltd., Vancouver, Canada (1988).

P. Léger
Department of Civil Engineering
École Polytechnique
Montreal University Campus
P.O. Box 6079, Station A
Montreal
QC H3C 3A7
CANADA

9
Recent Developments in Time Integration Methods for Structural and Interaction System Dynamics

K. C. Park

University of Colorado at Boulder, USA

9.1 INTRODUCTION

This survey is a follow-up on earlier ones (Park [1]; Park and Felippa [2]; Felippa and Park [3]; Park [4]) on direct time integration methods. The algorithmic characterization of the integration formulae offered therein, namely, stability, accuracy and implementation aspects, remains largely intact. Readers wishing to familiarize the algorithmic characterization may refer to the references cited above plus Hughes and Belytschko [5].

What we are about to cover herein reflects a steady shift of research thrusts in computational dynamics since the mid-1980s, from discipline-oriented dynamics to system-oriented dynamics, from sequential computations to parallel computations, and from efficiency/accuracy concern to system model improvements/refinements. The specific topics we survey in this chapter thus reflect their maturing stages; these developments do not fit into a coherent theory or categorization at the present time.

Since time integration algorithms have been presented within the context of linear structural dynamics for most instances, first we report on computational methods for non-linear multibody dynamics. Major challenges in the development of computational

---

Solving Large-scale Problems in Mechanics, Edited by M. Papadrakakis
© 1993 John Wiley & Sons Ltd

methods for multibody dynamics analysis have been the conservation of both energy and momentum, system constraint violations, and simulation speed. We will address some of these issues.

The second topic we will report is methods for the solution of coupled-field problems, primarily methods for control-structure interaction (CSI) problems. The design, modelling, analysis and real-time operation of CSI systems are one of the most intensely researched activities in recent years with applications ranging from aeroelastic tailoring, vibration control of reflectors deployed in low-earth orbits, to active vibration control of suspension systems. The third topic we will present is a computational method for transient thermal–structure interaction problems. This technique is relevant to the analysis of the thermal response of high-speed transport plane as well as integrated electronic chip thermal management problem.

9.2 SOLUTION TECHNIQUES FOR MULTIBODY DYNAMICS

The equations of motion for multibodies are characterized by two key features: highly nonlinear kinematical relations and complex constraints. It is *not* the purpose of this chapter to make an exhaustive survey of available solution techniques. Rather, we will examine selected techniques that meet our needs: computer implementability, adaptation to large-scale simulation, robustness and efficiency, in that order.

There are three aspects of solution techniques for multibody dynamics (MBD) analysis. First, we must have at hand an efficient and accurate algorithm for updating the kinematical quantities such as angular orientations, angular velocities. Second, direct time integration of the equations of motion that correspond to the unconstrained states of multibodies must be performed. Third, an accurate and efficient treatment of constraints is essential if the numerical solution is to maintain the given holonomic and nonholonomic constraints. In practice, the three aspects are intertwined so that one must achieve a careful balance in the employment of three computational aspects. As computer implementation of the three require different strategies, we will discuss them separately first and bring them together in the solution procedures.

The numerical solution procedure for MBD systems which we describe herein is termed a *staggered MBD solution procedure* that solves the generalized coordinates in a separate module from that for the constraint force (Park and Chiou [6]; Park *et al.* [7, 8]). A major advantage of such a partitioned solution procedure is that additional analysis capabilities such as active controller and design optimization modules can be easily interfaced without embedding them into a monolithic program. The solution of the equations of motion for constrained multibody systems, unlike typical structural dynamics problems, must satisfy at each time step the system constraints, whether holonomic or non-holonomic or time-specified manoeuvres. Because of this distinctive requirement, the reliability and cost of a multibody analysis package can be strongly affected by how efficiently and accurately the constraints are preserved during the numerical solution stage.

The system constraint forces can be either eliminated or retained depending upon the complexities associated with the elimination process. In general, it is preferable to eliminate the constraint degrees of freedom if they are associated only with open rigid links. On the other hand, if external torque or active control devices are attached to those joints, it is computationally more advantageous to solve the constraint forces (or Lagrange multipliers) simultaneously together with the generalized coordinates. Unfortunately, a straightforward way of computing the Lagrange multipliers can often lead to an unacceptable level of errors. The task of minimizing the propagation error due to violations of the constraint conditions is known as *stabilization*. We will describe a particular constraint stabilization which recasts the algebraic constraint conditions to a set of parabolic differential equations such that the constraint forces can also be integrated in time.

To solve for the generalized coordinates of the multibody system, the equations of motion are partitioned according to the translational and the rotational coordinates. This sets the stage for an efficient treatment of the rotational motions via the singularity-free Euler parameters. The translational part of the equations of motion is integrated via a standard central difference algorithm. The rotational part is treated by a modified central difference algorithm in order to preserve the discrete angular momentum. Once the angular velocities are obtained, the angular orientations are updated via the mid-point implicit formula employing the Euler parameters.

When the two algorithms, namely, the modified central difference algorithm for the rotational coordinates and the implicit staggered procedure for the constraint Lagrange multipliers, are brought together in a staggered manner, they constitute a staggered explicit-implicit procedure as detailed below.

9.2.1 Equations of motion for multibody systems

To motivate ourselves for the development of solution procedures for the multibody dynamics problems, let us introduce the following equations of motion:

$$\mathbf{M}\ddot{\mathbf{d}} = \mathbf{Q} - \mathbf{B}^T\boldsymbol{\lambda}, \qquad \dot{\mathbf{d}} = \begin{Bmatrix} \dot{\mathbf{u}} \\ \boldsymbol{\omega} \end{Bmatrix} \tag{9.1}$$

$$\Phi(\mathbf{d}, \dot{\mathbf{d}}) = 0 \tag{9.2}$$

where $\mathbf{M}$ is the system mass matrix, $\dot{\mathbf{d}}$ is the generalized velocity vector, $\mathbf{u}$ is the translational degrees of freedom, $\boldsymbol{\omega}$ is the angular velocity vector, $\mathbf{B} = \partial\Phi/\partial\mathbf{d}$ is the constraint projection matrix, $\boldsymbol{\lambda}$ is the constraint force vector, Φ are the constraint conditions that are imposed either on the subsystem boundaries or on the kinematical relations among the generalized coordinates, t is the time, (˙) denotes time differentiation, and $\mathbf{Q}$ is the generalized applied force plus non-linear inertia forces.

We observe from (9.1) and (9.2) that the task for solving the governing multibody dynamical equations constitutes three computational procedures: accurate computations of the constraint force, $\boldsymbol{\lambda}$, or their elimination from the equations of motion (9.1), updates of the angular orientations, and the direct integration of the translational displacement $\mathbf{u}$. To this end, we will first examine two distinctive ways of handling $\boldsymbol{\lambda}$.

9.2.2 Techniques for handling constraint conditions

In principle, it is better to eliminate the constraint conditions, if possible, if the corresponding forces are not needed for design or interface with other analysis modules. For example, if the system consists of open-tree configurations and no active controller is applied, then it is best to eliminate the joint constraint attributes. On the other hand, when the system includes multiple closed-loop configurations or active controllers are present on several joints, then it becomes important to compute the Lagrange multipliers as accurately as possible.

First, one can easily eliminate the system constraint forces via a coordinate partitioning strategy whenever any or all of the system components possess an open-tree topology. In the second procedure, we present a stabilization procedure for solving the Lagrange multipliers. A distinct feature of this stabilization procedure is that it can be implemented in a stand-alone module, thus can be interfaced not only with the equation solver for rigid-body systems but with that for flexible-body systems as well.

Parallel implementation of coordinate partitioning technique

In this technique, a projection matrix that spans the null space of the constraint Jacobian matrix Φ_u is first constructed (see, e.g., Wehage and Haug [9]). A parallel methodology (Chiou [10]; Chiou *et al.* [11]) based on an arrowhead algorithm then can be applied to the resulting complementary set of equations of motion. We will present the procedure for open-tree systems. For a system that contains closed loops, a cut-joint technique can be used so that the present scheme can be equally applied.

Let us introduce a projection matrix $\mathbf{A}$ such that, when its transposed matrix acts on the constraint force $\mathbf{B}^T \lambda$, it gives

$$\mathbf{A}^T \mathbf{B}^T \lambda = 0. \tag{9.3}$$

This projection matrix can be obtained by expressing the total generalized velocity $\dot{\mathbf{d}}^T = \langle \dot{\mathbf{u}}^T \, \omega^T \rangle$ in terms of the independent velocities $\dot{\mathbf{d}}^f$ and their time derivatives as

$$\dot{\mathbf{d}} = \mathbf{A}\dot{\mathbf{d}}^f, \qquad \ddot{\mathbf{d}} = \mathbf{A}\ddot{\mathbf{d}}^f + \dot{\mathbf{A}}\dot{\mathbf{d}}^f. \tag{9.4}$$

Due to the property of (9.3), premultiplying the equations of motion (9.1) by $\mathbf{A}^T$ yields

$$\mathbf{A}^T \mathbf{M} \ddot{\mathbf{d}} = \mathbf{A}^T \mathbf{Q}. \tag{9.5}$$

In conventional procedure, $\ddot{\mathbf{d}}$ in the above equation is replaced by (9.4b) and $\ddot{\mathbf{d}}^f$ is then solved from the reduced equations of motion. In the solution to be described below, instead of solving the reduced equations of motion, we augment (9.4b) to (9.5) to form an arrowhead matrix equation:

$$\begin{bmatrix} \mathbf{M} & -\mathbf{MA} \\ -\mathbf{A}^T \mathbf{M} & \mathbf{0} \end{bmatrix} \begin{Bmatrix} \ddot{\mathbf{d}} \\ \ddot{\mathbf{d}}^f \end{Bmatrix} = \begin{Bmatrix} \mathbf{M}\dot{\mathbf{A}}\dot{\mathbf{d}}^f \\ -\mathbf{A}^T \mathbf{Q} \end{Bmatrix} \tag{9.6}$$

which can be partitioned as

$$\begin{bmatrix} \mathbf{M}_{(1)} & 0 & 0 & 0 & \cdots & \mathbf{D}_{(1, n+1)} \\ 0 & \mathbf{M}_{(2)} & 0 & 0 & \cdots & \mathbf{D}_{(2, n+1)} \\ 0 & 0 & \mathbf{M}_{(3)} & 0 & \cdots & \mathbf{D}_{(3, n+1)} \\ 0 & 0 & 0 & \vdots & \cdots & \mathbf{D}_{(4, n+1)} \\ \vdots & \vdots & \vdots & \vdots & \mathbf{M}_{(n)} & \vdots \\ \mathbf{D}_{(n+1, 1)} & \mathbf{D}_{(n+1, 2)} & \mathbf{D}_{(n+1, 3)} & \cdots & \cdots & 0 \end{bmatrix} \begin{Bmatrix} \ddot{\mathbf{d}}_1 \\ \ddot{\mathbf{d}}_2 \\ \ddot{\mathbf{d}}_3 \\ \vdots \\ \ddot{\mathbf{d}}_n \\ \ddot{\mathbf{d}}^f \end{Bmatrix} = \begin{Bmatrix} \mathbf{g}_1 \\ \mathbf{g}_2 \\ \mathbf{g}_3 \\ \vdots \\ \mathbf{g}_n \\ \mathbf{g}^f \end{Bmatrix} \quad (9.7)$$

where n is the total number of bodies in the system. Decomposed in a manner convenient for parallel computations, one obtains

$$\mathbf{M}_j \ddot{\mathbf{d}}_j + \mathbf{D}_{(j, n+1)} \ddot{\mathbf{d}}^f = \mathbf{g}_j, \quad j = 1, \ldots, n$$

$$\sum_{j=1}^{n} \mathbf{D}_{(n+1, j)} \ddot{\mathbf{d}}_j = \mathbf{g}^f \quad (9.8)$$

where

$$\sum_{j=1}^{n} \mathbf{D}_{(n+1, j)} = -\sum_{j=1}^{n} \mathbf{A}_j^T \mathbf{M}_j, \quad \mathbf{D}_{(j, n+1)} = -\mathbf{M}_j \mathbf{A}_j, \quad j = 1, \ldots, n$$

$$\mathbf{g}_j = (\mathbf{M} \dot{\mathbf{A}} \dot{\mathbf{d}}^f)_j, \quad j = 1, \ldots, n, \quad \mathbf{g}^f = -\sum_{j=1}^{n} \mathbf{A}^T \mathbf{Q}_j.$$

Each diagonal submatrix $\mathbf{M}_j$ represents the local mass matrix which is decoupled and can be factorized concurrently. An off-diagonal submatrix $\mathbf{D}_{(j,n+1)}$ denotes the coupling between connecting bodies in the system. Since $\mathbf{M}$ is a constant matrix, (9.8) becomes

$$\ddot{\mathbf{d}}_j = \mathbf{M}_j^{-1} (\mathbf{D}_{(j, n+1)} \ddot{\mathbf{d}}^f - \mathbf{g}_j). \quad (9.9)$$

Substituting (9.9) into (9.8b) gives a form of *Schur complement*:

$$\sum_{j=1}^{n} \mathbf{D}_{(n+1, j)} \mathbf{M}_j^{-1} \mathbf{D}_{(j, n+1)} \ddot{\mathbf{d}}^f = \sum_{j=1}^{n} \mathbf{D}_{(n+1, j)} \mathbf{M}_j^{-1} \mathbf{Q}_j - \sum_{j=1}^{n} \mathbf{D}_{(n+1, j)} \dot{\mathbf{A}} \dot{\mathbf{d}}^f. \quad (9.10)$$

The preceding treatment of the reduced equations of motion provides several parallel computational features. First, the parallelism can be exploited by mapping each processor onto a group of bodies so that independent computations such as the left-hand side of (9.10) can be performed concurrently. Second, since $\mathbf{M}$ is a constant mass matrix, it needs to be factored only once. Third, to solve for $\ddot{\mathbf{d}}^f$, a parallel sparse solver may be utilized. Finally, once $\ddot{\mathbf{d}}^f$ is obtained, computations of $\mathbf{d}$ from (9.4) is also easily parallelizable.

Stabilization of constraint violations

When the Lagrange multipliers cannot be eliminated or are to be retained for other purposes, one must solve for them. It has been known for some time that a straightforward direct time integration of the governing differential equation (9.1), augmented with a linearized form of the system constraints (9.2), often incur unacceptably high errors in the numerical solution. Of several techniques proposed to date, perhaps the method proposed by Baumgarte [12, 13] is the earliest known stabilization technique for computing the constraint forces.

While the method due to Baumgarte works relatively well, it requires an *a priori* determination of stabilization parameters and the method breaks down when the number of independent system constraints change due to varying configurations. To cope with the varying system constraints without experiencing singularities, a penalty-based stabilization has been developed in Park and Chiou [6]. The penalty procedure recasts the constraint equation in the form

$$\lambda = \frac{1}{\varepsilon}\Phi \tag{9.11}$$

as the basic constraint equations instead of (9.2) for both the holonomic and non-holonomic constraint conditions. We then time-differentiate, for the holonomic case, the above penalty-based equation to obtain:

$$\dot{\lambda} = \frac{1}{\varepsilon}\mathbf{B}\,\dot{\mathbf{d}}, \qquad \dot{\mathbf{d}} = \begin{Bmatrix} \dot{\mathbf{u}} \\ \omega \end{Bmatrix}. \tag{9.12}$$

The numerical solution to the above companion differential equation is effected as follows.

The constrained equation of motion (9.1) is integrated once using the implicit integration rule

$$\dot{\mathbf{d}}^{n+1/2} = \dot{\mathbf{d}}^n + \delta\,\ddot{\mathbf{d}}^{n+1/2}, \qquad \delta = \frac{h}{2}$$

to yield

$$\dot{\mathbf{d}}^{n+1/2} = \delta\,\mathbf{M}^{-1}(\mathbf{Q}^{n+1/2} - \mathbf{B}^T \lambda^{n+1/2}) + \dot{\mathbf{d}}^n. \tag{9.13}$$

This expression is substituted into (9.12) to obtain the stabilized differential equation for the Lagrange multipliers

$$\varepsilon \dot{\lambda}^{n+1/2} + \delta \mathbf{B} \mathbf{M}^{-1} \mathbf{B}^{\mathrm{T}} \lambda^{n+1/2} = \delta \mathbf{B} \mathbf{M}^{-1} \mathbf{Q}^{n+1/2} + \mathbf{B} \dot{\mathbf{d}}^{n}.$$

When the above equation is integrated once more with the trapezoidal rule, we obtain the following discrete update for λ:

$$(\varepsilon I + \delta^{2} \mathbf{B} \mathbf{M}^{-1} \mathbf{B}^{\mathrm{T}}) \lambda^{n+1/2} = \varepsilon \lambda^{n} + \mathbf{r}_{\lambda}^{n+1/2} \tag{9.14}$$

$$\mathbf{r}_{\lambda}^{n+1/2} = \delta^{2} \mathbf{B} \mathbf{M}^{-1} \mathbf{Q}^{n+1/2} + \delta \mathbf{B} \dot{\mathbf{d}}^{n}. \tag{9.15}$$

The solution procedure for λ presented above can now be invoked in a staggered manner in conjuction with the generalized coordinate solver to be described below.

9.2.3 Time integration of MBD equations of motion

Once λ is computed by the procedure in Section 9.2.2 or $\ddot{\mathbf{d}}$ when using the partitioning algorithm, one still needs to compute $\dot{\mathbf{d}}$, u and the angular orientations and their parameters at each time step. This task is carried out by employing an explicit–implicit transient analysis algorithm that exploits the special kinematical relationships of the generalized rotational coordinates vs. the angular velocity, namely, the Euler parameters. First, the integration of the translational coordinates and the angular velocity is accomplished by the central difference formula. It should be mentioned that the use of the central difference formula does impose a step-size restriction due to its stability limit ($\omega_{\max} h \leq 2$) where $\omega_{\max}$ is the highest angular velocity of the system components for rigid-body systems or the highest frequency of the entire flexible members for flexible-body systems. The simplicity of its programming effort and robustness of its solution results can often become compelling enough to adopt an explicit formula, which is the view taken here.

Explicit translational coordinate integrator

In the conventional structural dynamics analysis, explicit time integration of the equations of motion by the central difference formula involves the following two updates per step:

$$\begin{aligned} \dot{\mathbf{u}}^{n+1/2} &= \dot{\mathbf{u}}^{n-1/2} + h \ddot{\mathbf{u}}^{n} \\ \mathbf{u}^{n+1} &= \mathbf{u}^{n} + h \dot{\mathbf{u}}^{n+1/2}. \end{aligned} \tag{9.16}$$

Unfortunately, the same integrator is not directly applicable to the rotational part of the equations of motion since ω is not directly integrable to yield the total rotational quantities except for some special kinematic configurations. This motivates us to partition $\dot{\mathbf{d}}$ into the translational velocity vector, $\dot{\mathbf{u}}$, which is directly integrable and the angular velocity vector, ω, which is not, and treat them differently, viz.:

$$\ddot{\mathbf{d}} = \begin{Bmatrix} \ddot{\mathbf{u}} \\ \dot{\boldsymbol{\omega}} \end{Bmatrix}, \quad \dot{\mathbf{d}} = \begin{Bmatrix} \dot{\mathbf{u}} \\ \boldsymbol{\omega} \end{Bmatrix}. \qquad (9.17)$$

The equations of motion can be written according to the above partitioning as

$$\begin{bmatrix} \mathbf{M}_u & 0 \\ 0 & \mathbf{M}_\omega \end{bmatrix} \begin{Bmatrix} \ddot{\mathbf{u}} \\ \dot{\boldsymbol{\omega}} \end{Bmatrix} = \begin{Bmatrix} \mathbf{Q}_u \\ \mathbf{Q}_\omega \end{Bmatrix} \qquad (9.18)$$

where

$$\begin{Bmatrix} \mathbf{Q}_u \\ \mathbf{Q}_\omega \end{Bmatrix} = \begin{Bmatrix} \mathbf{f}_u - \mathbf{D}_u(\dot{\mathbf{u}}) - \mathbf{S}_u(\mathbf{u}, \mathbf{q}) - \mathbf{B}_u^T \boldsymbol{\lambda} \\ \mathbf{f}_\omega - \mathbf{D}_\omega(\boldsymbol{\omega}) - \mathbf{S}_\omega(\mathbf{u}, \mathbf{q}) - \mathbf{B}_\omega^T \boldsymbol{\lambda} \end{Bmatrix} \qquad (9.19)$$

in which the subscripts $(\mathbf{u}, \boldsymbol{\omega})$ refer to the translational and the rotational motions, respectively, $\mathbf{f}$ is the external force vector, $\mathbf{D}$ is the generalized damping force including the centrifugal force, $\mathbf{S}$ is the internal force vector including member flexibility, $\mathbf{q}$ is the angular orientation parameters, $\mathbf{B}_u$ and $\mathbf{B}_\omega$ are the partition of the combined gradient matrices of the constraint conditions (9.2).

First, assume that $\mathbf{u}^{n+1/2}$ and $\mathbf{q}^{n+1/2}$ are already computed so that we can compute $\ddot{\mathbf{u}}^{n+1/2}$ and $\dot{\boldsymbol{\omega}}^{n+1/2}$:

$$\begin{Bmatrix} \ddot{\mathbf{u}}^{n+1/2} \\ \dot{\boldsymbol{\omega}}^{n+1/2} \end{Bmatrix} = \mathbf{M}^{-1} \begin{Bmatrix} \mathbf{Q}_u \\ \mathbf{Q}_\omega \end{Bmatrix}. \qquad (9.20)$$

Second, update the translational velocity at the step $(n+1)$ by

$$\dot{\mathbf{u}}^{n+1} = \dot{\mathbf{u}}^n + h\ddot{\mathbf{u}}^{n+1/2}. \qquad (9.21)$$

Third, we update the translational displacement, $\mathbf{u}$, by

$$\mathbf{u}^{n+3/2} = \mathbf{u}^{n+1/2} + h\dot{\mathbf{u}}^{n+1}. \qquad (9.22)$$

The updating of the angular orientations must be treated with care, and is described below.

Updating of angular velocity via discrete angular momentum conservation

In order to update the angular velocity and angular orientations, we combine judiciously a momentum-conserving form of the central difference algorithm and the mid-point implicit rule for computing the Euler parameters as follows (Park and Chiou [14]).

First, we retain the basic central difference formula for computing the angular velocity at the half steps:

$$\omega^{n+1/2} = \omega^{n-1/2} + h\dot{\omega}^n \tag{9.23}$$

where $\dot{\omega}$ is computed from the equations of motion, utilizing the angular velocity obtained from the discrete angular conservation law, as described shortly.

Second, $\omega^{n+1/2}$ is used to integrate the Euler parameters by

$$\dot{\mathbf{q}} = \frac{1}{2}\begin{bmatrix} 0 & -\omega \\ \omega & -\tilde{\omega} \end{bmatrix}\mathbf{q} = \mathbf{A}(\omega)\mathbf{q} \tag{9.24}$$

where $\mathbf{q} = (q_0\, q_1\, q_2\, q_3)^T$ are the Euler parameters expressed in the body-fixed frame and $\tilde{\omega}$ denotes the skew-symmetric angular velocity tensor given by

$$\tilde{\omega} = \begin{bmatrix} 0 & -\omega_3 & \omega_2 \\ \omega_3 & 0 & -\omega \\ -\omega_2 & \omega_2 & 0 \end{bmatrix}. \tag{9.25}$$

Implicit integration (9.24) by the mid-point rule yields

$$\mathbf{q}^{n+1} - \mathbf{q}^n = h\dot{\mathbf{q}}^{n+1/2} = h\mathbf{A}(\omega^{n+1/2})\mathbf{q}^{n+1/2}$$
$$= \frac{h}{2}\mathbf{A}(\omega^{n+1/2})(\mathbf{q}^{n+1} + \mathbf{q}^n) \tag{9.26}$$

where $\mathbf{A}(\omega^{n+1/2})$ can be viewed as the tangent matrix at the mid-configuration whereas $\mathbf{q}^{n+1/2} = (\mathbf{q}^{n+1} + \mathbf{q}^n)/2$ is the mid-point average value. It was shown in Park et al. [7] that $\mathbf{q}^{n+1}$ can be expressed as

$$\mathbf{q}^{n+1} = \frac{1}{\Delta}\left[\mathbf{I} + \frac{h}{2}\mathbf{A}(\omega^{n+1/2})\right]^2 \mathbf{q}^n \tag{9.27}$$

where $\Delta = 1 + h^2(\omega_1^2 + \omega_2^2 + \omega_3^2)/4$.

The updated Euler parameters $\mathbf{q}^{n+1}$ are then normalized to satisfy the constraint condition

$$\mathbf{q}^{n+1} \cdot \mathbf{q}^{n+1} = 1. \tag{9.28}$$

Once $\mathbf{q}^{n+1}$ is available, the angular orientation at t^{n+1} is then obtained by

$$\mathbf{R}^{n+1} = (2q_0^2 - 1)\mathbf{I} + 2\bar{\mathbf{q}}\bar{\mathbf{q}}^T - 2q_0\tilde{\bar{\mathbf{q}}}, \quad \bar{\mathbf{q}} = (q_1\, q_2\, q_3)^T \tag{9.29}$$

where the superscript $(n+1)$ is omitted on $\mathbf{q}$.

Third, the angular velocity ω^{n+1} that is needed to compute $\dot{\omega}$ for the next step integration is obtained via a discrete version of the angular momentum conservation law:

$$\mathbf{M}_\omega \omega^{n+1} - \mathbf{M}_\omega \omega^n = h\tau^{n+1/2} \tag{9.30}$$

where $\mathbf{M}_\omega$ is the moment of inertia, ω is the angular velocity vector, and τ is the applied moment, all expressed in body-fixed frame at the configuration t^k, $0 \le k \le n$. For computational simplicity, we will choose $k = 0$, i.e. the initial configuration so that (9.30) becomes

$$(\mathbf{R}^{n+1})^T \mathbf{M}_\omega \omega^{n+1} - (\mathbf{R}^n)^T \mathbf{M}_\omega \omega^n = h(\mathbf{R}^{n+1/2})^T \tau^{n+1/2} \tag{9.31}$$

where the matrix $\mathbf{R}$ is the rotation transformation matrix from the inertial basis $\mathbf{e}$ to the body fixed configuration $\mathbf{b}$ according to

$$\mathbf{b}^n = \mathbf{R}^n \mathbf{e} \tag{9.32}$$

and $\mathbf{M}_\omega$, ω and τ are now expressed in the superscript-indexed discrete $\mathbf{b}$-bases. Therefore, the discrete angular momentum equation (9.30) becomes

$$\omega^{n+1} = \mathbf{M}_\omega^{-1} \mathbf{R}^{n+1} \left((\mathbf{R}^n)^T \mathbf{M}_\omega \omega^n + h(\mathbf{R}^{n+1/2})^T \mathbf{Q}_\omega^{n+1/2} \right) \tag{9.33}$$

where τ is replaced by $\mathbf{Q}_\omega$ from the lefthand side of (9.18).

It is noted that, whereas the standard difference formula (9.16) satisfies the linear momentum conservation for constant and linearly varying $\mathbf{Q}_\omega$, (9.33) indicates that the use of a common basis is essential for the conservation of angular momentum. A similar approach was successfully utilized by Simo and Wong [15] in their development of a family of implicit algorithms.

Fourth, the angular acceleration needed for the next time step $(n + 1)$ is then computed for each rigid body by

$$\dot{\omega}^{n+1} = \mathbf{M}_\omega^{-1}(\mathbf{Q}^{n+1} - \tilde{\omega}^{n+1} \mathbf{M}_\omega \tilde{\omega}^{n+1}). \tag{9.34}$$

Equations (9.23)–(9.34) constitute the present modified central difference algorithm for integrating the rotational equations of motion in multibody systems. However, as many engineering multibody systems involve both holonomic and nonholonomic constraints, the computation of $\mathbf{Q}^{n+1/2}$ is not as straightforward as (9.18) implies.

For a momentum-conserving implicit algorithm, the reader may consult Simo and Wong [15]. For applications of the preceding MBD procedures to flexible multibody dynamics, one may refer to Downer [16], Downer *et al.* [17] and Downer and Park [18], who solve the flexible appendage deployment problem, among others. Finally, other MBD recent approaches can be found in Haug and Deyo [19].

9.3 ALGORITHMS FOR CONTROL-STRUCTURE INTERACTION SIMULATION

A second topic we should like to report in this survey is computational methods for the simulation of dynamic response of structures that are subject to active control forces. A general case of structural response under active control forces involves both large-angle rigid motions as well as transient flexible vibrations. Engineering examples include the manoeuvring of robotic arms, satellite attitude changes, deployment and vibration control of large space structures, and active vibration suppression of rotating machinery and vehicle suspension systems.

When relatively small size models are adequate for describing the predominant motions and vibrations, the resulting active control strategies also can consist of a small number of actuators and sensors. However, as the structural model needs to be large due to the physical nature of the problem or due to the high-precision requirement, so must be the size of the actuator/sensor numbers. It is for such large-scale control-structure problems that the following simulation methodology has been developed.

Specifically, simulation tasks for control-structure interaction (CSI) problems involve several computational elements and discipline-oriented models such as structural dynamics, control law synthesis, state estimation, actuator and sensor dynamics, thermal analysis, liquid sloshing and swirling, environmental disturbances, and manoeuvring thrusts and torques. Because each of these computational elements can be large, it is usually not practical to assemble these computational elements into a single set of equations of motion and perform the analysis in its totality, which will be referred to as the *simultaneous solution approach*. First, the equation size of the total system can be simply too large for many existing computers. Second, the simultaneous solution by treating the coupled interaction equations as one system may destroy the sparsity of the attendant matrices, thus requiring excessive computations and storage space. Most important of all, any changes in the model or in the computational procedures will engender significant modifications of the required analysis software modules and hence require a painstaking software verification effort.

In order to alleviate the aforementioned difficulties that exist in the *simultaneous solution approach*, a partitioned solution procedure that takes the following considerations into account has been developed. First, software development of any new capability is costly and time-consuming; thus, if at all possible, it is preferable to utilize existing single-field analysis modules to conduct the coupled-field interaction analysis. Second, the tasks for model generation and methods development of each field are best accomplished by relying on the experts of each single-field discipline. In order to accommodate both the software considerations and the single-field expertise, a *partitioned (or divide-and-conquer) analysis procedure* has been developed for control-structure interaction analysis for direct output feedback systems (Belvin [20]; Park and Belvin [21]). The procedure abandons the conventional way of treating the CSI problems as one entity. Instead, it treats the structure (or plant), the observer, and the controller/observer interaction terms as separate entities. Thus, the CSI problem is recognized as a coupled-field problem and a divide-and-conquer strategy adopted for the development of a real-time computational procedure. It should be mentioned that a similar concept has been successfully applied to other interaction analyses such as fluid-structure interactions (Park et al. [22]), multi-structural interaction systems (Park [23]; Felippa and Park [3]; Park and Felippa [2]), earth dam and pore-fluid interactions (Park [23]; Zienkiewicz et al. [24] and multibody systems with constraints (Park et al. [7]; Chiou [10]; Downer [16]).

The *partitioned analysis procedure* hinges on two software and computational aspects. First, at each discrete time increment, the equations of motion for each discipline are solved separately by considering the interaction terms as external disturbances or applied forces. Second, when necessary, computational stabilization and accuracy improvements are introduced through augmentations and/or equation modifications. It is important to note that such partitioned solutions of each discipline equation can be carried out on either a sequential or a parallel machine if certain message passing and memory-conflict issues are handled appropriately.

9.3.1 Equations of motion for control-structure interaction systems

The discrete equations of motion for control-structure interaction systems may be described by [25]

$$\left.\begin{array}{lll}\text{Structure:} & \text{(a)} & M\ddot{q} + D\dot{q} + Kq = f + Bu + Gw \\ & & q(0) = q_0, \quad \dot{q}(0) = \dot{q}_0 \\ \text{Sensor output:} & \text{(b)} & z = Hx + v \\ \text{Estimator:} & \text{(c)} & \dot{\bar{x}} = A\bar{x} + Ef + \bar{B}u + L\gamma \\ & & \bar{x}(0) = 0 \\ \text{Control force:} & \text{(d)} & u = -F\bar{x} \\ \text{Estimation error:} & \text{(e)} & \gamma = z - (H_d\bar{q} + H_v\dot{\bar{q}}) \end{array}\right\} \quad (9.35)$$

where

$$x = \begin{Bmatrix} q \\ \dot{q} \end{Bmatrix}, \quad \bar{x} = \begin{Bmatrix} \bar{q} \\ \dot{\bar{q}} \end{Bmatrix}$$

and

$$H = [H_d \ H_v], \quad L = \begin{bmatrix} L_1 \\ L_2 \end{bmatrix}, \quad E = \begin{bmatrix} 0 \\ M^{-1} \end{bmatrix}$$

$$A = \begin{bmatrix} 0 & I \\ -M^{-1}K & -M^{-1}D \end{bmatrix}, \quad \bar{B} = \begin{bmatrix} 0 \\ M^{-1}B \end{bmatrix}, \quad F = [F_1 \ F_2].$$

In the preceding equations, M is the mass matrix, D is the damping matrix, K is the stiffness matrix, $f(t)$ is the applied force, B is the actuator location matrix, G is the disturbance location matrix, q is the generalized displacement vector, w is a disturbance vector and the superscript dot denotes time differentiation. In (9.35b), z is the measured sensor output. The matrix H_d is the matrix of displacement sensor locations and H_v is the matrix of velocity sensor locations. The vector v is measurement noise. The state estimator in (9.35c) is assumed to be based on either the Kalman filter (Kalman and Bucy [26]) or a Luenberger observer [27] if the system is deterministic. The superscript ˜ denotes the estimated states. The actuator output, u, is a function of the state estimator variables, $\bar{q}$ and $\dot{\bar{q}}$, and F_1 and F_2 are control gains determined for example by pole–zero placement or from the solution of an optimal control problem. The observer is governed by L, the filter gain matrix. For the special case where L_1 is the null matrix (i.e. $\dot{\bar{q}} = \dot{\tilde{q}}$), a second-order state estimator can be expressed as

$$M\ddot{\bar{q}} + D\dot{\bar{q}} + K\bar{q} = f + Bu + ML_2\gamma. \tag{9.36}$$

The effect of the above simplification on the observer stability and convergence is discussed in detail in Belvin [20] and Belvin and Park [28].

9.3.2 Simultaneous solution approach

The numerical solution of (9.35) by the simultaneous solution approach begins with appropriate initial conditions, the feedback gain F and the filter gain L. The structure equation is written in first-order form

$$\dot{x} = Ax + Ef + \bar{B}u + \bar{G}w \tag{9.37}$$

where

$$\bar{G} = \begin{bmatrix} 0 \\ M^{-1}G \end{bmatrix}.$$

The control gains and observer gains can be synthesized independently by noting that the stability of the structural system and the observer error stability are uncoupled. Introducing the error equation by the deterministic form (9.35) as

$$\bar{e} = x - \tilde{x} = \begin{Bmatrix} q - \bar{q} \\ \dot{q} - \dot{\bar{q}} \end{Bmatrix} \tag{9.38}$$

and eliminating **u** yields

$$\begin{Bmatrix} \dot{x} \\ \dot{e} \end{Bmatrix} = \begin{bmatrix} A - \bar{B}F & \bar{B}F \\ 0 & A - LH \end{bmatrix} \begin{Bmatrix} x \\ e \end{Bmatrix} + \begin{bmatrix} E \\ 0 \end{bmatrix} f + \begin{bmatrix} \bar{G} \\ 0 \end{bmatrix} w. \qquad (9.39)$$

The stability of (9.39) is governed by the stability of $[A - \bar{B}F]$ and $[A - LH]$. Thus, the control gain **F** is suitably chosen from the matrix $[A - \bar{B}F]$ and the observer gain **L** from the matrix $[A - LH]$.

Subsequently, the simultaneous solution approach eliminates **u** and **z** from (9.35a, c) and then solves the observer based closed-loop equations

$$\begin{Bmatrix} \dot{x} \\ \dot{\bar{x}} \end{Bmatrix} = \begin{bmatrix} A & -\bar{B}F \\ LH & A - \bar{B}F - LH \end{bmatrix} \begin{Bmatrix} x \\ \bar{x} \end{Bmatrix} + \begin{Bmatrix} E \\ E \end{Bmatrix} f + \begin{Bmatrix} \bar{G} \\ 0 \end{Bmatrix} w. \qquad (9.40)$$

The embedding effects of both the controller and the state observer result in an unsymmetric and non-sparse system matrix of dimension ($4N$ by $4N$), where N is the number of structural degrees of freedom. Solution of (9.40) would require considerable software modifications of existing structural dynamics analysis programs for large-scale CSI simulation purposes. In addition to losing the computational advantages associated with the finite element based CSI equation, the simultaneous solution approach requires the control law to be embedded into the observer model. If the control law includes actuator, sensor and/or controller dynamics, additional states must be added to the observer. This greatly complicates the observer model and requires significant software development for each class of control law dynamics. The difficulties associated with the *simultaneous solution approach* have prompted development of *a partitioned solution approach* for the CSI equations as described below.

9.3.3 Stabilization for computations of control force and estimation error

The partitioned solution procedure numerically integrates the structural equations of motion (9.35a) and the observer equation (9.35c) by treating the control force **u** and the estimation error y as if they were applied terms in the right-hand sides. In this way, simulation of control-structure interaction systems using the partitioned solution procedure can be carried out by a judicious employment of three software modules: the structural analyser to obtain **q**, the state estimator to obtain $\bar{q}$, and the stabilized solver for the control force **u** and the state estimation error y. Thus the partitioned procedure becomes computationally efficient and can preserve software modularity by exploiting the symmetric matrix form on the left-hand sides of (9.35a) and (9.35c).

However, computations of the control force **u** and the state estimation error y by (9.35d) and (9.35e), respectively, can not only lead to an accumulation of errors but often can give rise to numerical instability. Hence, in order to make the partitioned

9.3 CONTROL-STRUCTURE INTERACTION SIMULATION

solution procedure robust, it is imperative to stabilize the partitioned solution process and/or numerically filter the solution errors in computing $\mathbf{u}$ and γ. This is addressed below.

First, we time-differentiate (9.35c) to obtain

$$\dot{\mathbf{u}} = \mathbf{F}_1 \dot{\mathbf{q}} - \mathbf{F}_2 \ddot{\mathbf{q}}. \tag{9.41}$$

Substituting $\ddot{\mathbf{q}}$ from (9.36) into the above equation, one obtains

$$\dot{\mathbf{u}} + \mathbf{F}_2 \mathbf{M}^{-1} \mathbf{B} \mathbf{u} = -\mathbf{F}_2 (\mathbf{M}^{-1} \dot{\bar{\mathbf{p}}} + \mathbf{L}_2 \gamma) - \mathbf{F}_1 \dot{\mathbf{q}} \tag{9.42}$$

where the generalized rate of momentum $\bar{\mathbf{p}}$ is given by

$$\dot{\bar{\mathbf{p}}} = (\mathbf{f} - \mathbf{D}\dot{\mathbf{q}} - \mathbf{K}\bar{\mathbf{q}}). \tag{9.43}$$

The parabolic stabilization that led to equation (9.42) for computing the control law is sometimes called an equation augmentation procedure as it has not altered any part of the basic governing equation (9.35) except one time-differentiation of $\mathbf{u}$ assuming $\dot{\mathbf{u}}$ exists. However, this assumption is later removed through time discretization as will be shown later in the chapter.

It is noted that the homogenous part of (9.42) has the filtering effect of the form $(s\mathbf{I} + \mathbf{F}_2 \mathbf{M}^{-1} \mathbf{B})^{-1}$ in parlance of classical control theory, where s is the Laplace transform operator, thus achieving the required stabilization. From the computational viewpoint, although $\mathbf{F}_2 \mathbf{M}^{-1} \mathbf{B}$ is in general a full matrix, its size is relatively small, as the size of $\mathbf{u}$ is proportional to the number of actuators placed on the structure.

Similarly, for the observer estimation error γ one can stabilize its computation first by time-differentiating it

$$\dot{\gamma} + \mathbf{H}_v \mathbf{L}_2 \gamma = \dot{\mathbf{z}} - (\mathbf{H}_d \dot{\mathbf{p}} + \mathbf{H}_v \ddot{\mathbf{q}}). \tag{9.44}$$

and substituting the observer equation into the above to obtain an augmented form of the observer error equation:

$$\dot{\gamma} \mathbf{H}_v \mathbf{L}_2 \gamma = \dot{\mathbf{z}} - \mathbf{H}_v \mathbf{M}^{-1} (\dot{\bar{\mathbf{p}}} + \mathbf{B}\mathbf{u}) - \mathbf{H}_d \dot{\mathbf{q}}. \tag{9.45}$$

9.3.4 Stabilized partitioned equations and solution process

The adoption of the second-order observer and the preceding stabilization thus replaces (9.35c), (9.35d) and (9.35e) by (9.36), (9.43) and (9.45), respectively, as summarized below.

Structure: (a) $\mathbf{M\ddot{q}} + \mathbf{D\dot{q}} + \mathbf{Kq} = \mathbf{f} + \mathbf{Bu} + \mathbf{Gw}$
$\mathbf{q}(0) = \mathbf{q}_0, \quad \mathbf{\dot{q}}(0) = \mathbf{\dot{q}}_0$

Sensor output: (b) $\mathbf{z} = \mathbf{Hx} + \mathbf{v}$

Estimator: (c) $\mathbf{M\ddot{\bar{q}}} + \mathbf{D\dot{\bar{q}}} + \mathbf{K\bar{q}} = \mathbf{f} + \mathbf{Bu} + \mathbf{ML}_2\gamma$ \hfill (9.46)
$\mathbf{\bar{q}}(0) = 0, \quad \mathbf{\dot{\bar{q}}}(0) = 0$

Control force: (d) $\mathbf{\dot{u}} + \mathbf{F}_2\mathbf{M}^{-1}\mathbf{Bu} = -\mathbf{F}_2(\mathbf{M}^{-1}\mathbf{\dot{p}} + \mathbf{L}_2\gamma) - \mathbf{F}_1\mathbf{\dot{q}}$

Estimation error: (e) $\mathbf{\dot{\gamma}} + \mathbf{H}_v\mathbf{L}_2\gamma = \mathbf{\dot{z}} - \mathbf{H}_v\mathbf{M}^{-1}(\mathbf{\dot{p}} + \mathbf{Bu}) - \mathbf{H}_d\mathbf{\dot{q}}$

Note that the difference between the original governing equation set (9.35) and the above stabilized set (9.46) is an obstacle to computation of the control forces and the state estimation error vector.

9.3.5 Stability and accuracy of partitioned solution procedure

Computational stability analysis of partitioned procedures for a general coupled system is still in an evolving stage. Hence, the analysis herein applies the relevant results from (Belvin [20]; Park and Belvin [21]) in the present stability analysis of the partitioned CSI solution procedure. The partitioned CSI solution procedure presented in (9.46), even when discretized by unconditionally stable implicit time integration formulae, may still suffer from computational instability as it involves extrapolations to obtain $\mathbf{u}^{n+1/2}$ and $\gamma^{n+1/2}$. A complete stability analysis of the partitioned solution procedure for the coupled structural dynamics, observer and controller equations is difficult to perform unless the observer characteristics $\mathbf{H}, \mathbf{L}$ and the controller characteristics $\mathbf{B}, \mathbf{F}$ are specified. Hence, the analysis that follows is restricted to an ideal observer, i.e. $\gamma = 0$. In what follows, it is assumed that all of the stabilized equation set is time-discretized by a mid-point version of the trapezoidal rule.

In order to assess the computational stability of the present partitioned solution procedure, we construct a model single degree-of-freedom interaction equation as follows. First, neglecting structural damping a modal structural equation of motion can be expressed as

$$\ddot{y} + \omega^2 y = -u \tag{9.47}$$

where y is a generalized coordinate and ω is its associated frequency.

Second, the model controller is assumed to consist of both the position and velocity feedback with appropriate weights given by

$$u = \eta\omega_c^2 y + \zeta\omega_c\dot{y}, \qquad \omega_{min} \leq \omega_c \leq \omega_{max} \tag{9.48}$$

where ω_c is the feedback frequency, which ranges from the minimum to the maximum of the structural frequency contents, and η and ζ are positive scalar coefficients that signify the strength of the position and the velocity feedback, respectively.

Combining (9.47) with the stabilized form of (9.48) we have the model interaction equation as

$$\ddot{y} + \omega^2 y = -u$$
$$\dot{u} + \zeta\omega_c u = \eta\omega_c^2 \dot{y} - \zeta\omega_c\omega^2 y. \tag{9.49}$$

Thus, the model interaction equations given by (9.49) represent the case of full state feedback. They do not, however, reflect the mode-to-mode coupling that can occur in reduced-order feedback controller. Nevertheless, an analysis of the computational stability using the above model interaction equations should shed insight on the overall stability of the present partitioned solution procedure.

Time integration of the above model problem (9.49) by the mid-point rule

$$\begin{aligned}
x^{n+1/2} &= x^n + \delta \dot{x}^{n+1/2} \\
\dot{x}^{n+1/2} &= \dot{x}^n + \delta \ddot{x}^{n+1/2} \\
x^{n+1} &= 2x^{n+1/2} - x^n
\end{aligned} \tag{9.50}$$

with $\gamma = 0$ yields

$$\begin{aligned}
y_p^{n+1/2} &= y^n + \delta \dot{y}^n \\
(1 + \delta\zeta\omega_c) u_p^{n+1/2} &= (\eta\omega_c^2 - \delta\zeta\omega_c\omega^2) y_p^{n+1/2} + \zeta\omega_c \dot{y}^n \\
(1 + \delta^2\omega^2) y^{n+1/2} &= -\delta^2 u^{n+1/2} + y^n + \delta \dot{y}^n \\
y^{n+1} &= 2y^{n+1/2} - y^n, \quad \dot{y}^{n+1/2} = (y^{n+1/2} - y^n)/\delta, \quad \dot{y}^{n+1} = 2\dot{y}^{n+1/2} - \dot{y}^n \\
u^{n+1} &= \eta\omega_c^2 y^{n+1} + \zeta\omega_c \dot{y}^{n+1}
\end{aligned} \tag{9.51}$$

where $y_p^{n+1/2}$ is a stable predictor that is needed to initiate the staggered solution.

Computational stability of the above difference equation can be assessed by seeking a non-trivial solution in the form

$$\begin{Bmatrix} u^{n+1} \\ y^{n+1} \end{Bmatrix} = \lambda \begin{Bmatrix} u^n \\ y^n \end{Bmatrix} \tag{9.52}$$

such that

$$|\lambda| \leq 1 \tag{9.53}$$

for stability.

Substituting (9.52) into (9.51) and eliminating $\dot{y}$, one obtains

$$\mathbf{J} \begin{Bmatrix} u \\ y \end{Bmatrix}^n = 0 \tag{9.54}$$

where

$$\mathbf{J} = \begin{bmatrix} \delta(1 + \delta\zeta\omega_c)(\lambda + 1)^2 & 4\lambda(\delta^2\zeta\omega_c\omega^2 - \delta\eta\omega_c^2) + 2\zeta\omega_c(1 - \lambda) \\ \delta^2(\lambda + 1)^2 & (1 + \delta^2\omega^2)(\lambda + 1)^2 - 4\lambda \end{bmatrix}.$$

In order to test the stability requirement (9.53) on the characteristic equation, i.e., $\det|\mathbf{J}| = 0$, one transforms $|\lambda| \leq 1$ into the entire left-hand plane of the z-plane by

$$\lambda = \frac{1+z}{1-z}, \qquad |\lambda| \leq 1 \Leftrightarrow \text{Re}(z) \leq 0. \tag{9.55}$$

Carrying out the necessary algebra we have from $\det|\mathbf{J}(z)| = 0$ the following z-polynomial equation:

$$(\delta^3 \zeta \omega_c \omega^2 - \delta^2 \eta \omega_c^2 + 1)z^2 + (\delta \zeta \omega_c)z + \delta^2(\eta \omega_c^2 + \omega^2) = 0. \tag{9.56}$$

A test of the polynomial equation (9.56) for possible positive real roots by the Routh–Hurwitz criterion (Ganthmacher [29]) indicates that the partitioned procedure as applied to the model coupled equations (9.47) and (9.48) give a stable solution provided

$$(\delta^3 \zeta \omega_c \omega^2 - \delta^2 \eta \omega_c^2 + 1) \geq 0. \tag{9.57}$$

Note that, if there is no position feedback (i.e., $\eta = 0$), the model interaction equations solved by the present partitioned solution procedure (9.46) yields unconditionally stable solutions as (9.57) is automatically satisfied. Hence, a more critical stability assessment can be made by assuming no velocity feedback (i.e. $\zeta = 0$) for which we have for stability from (9.57)

$$h \leq \frac{2}{\sqrt{\eta \omega_c}}. \tag{9.58}$$

The preceding stability analysis on the model interaction equations permits us to make the following observations. First, equation (9.58) indicates that feedback frequency (ω_c) and the strength of the position feedback (η) dictate the computational stability and not the structural frequency (ω). In other words, the position feedback dictates the allowable step size for stability. Thus the highest frequency of the controller governs stability, not the highest frequency of the structure. Since most controllers are designed with reduced order structure models that ignore high frequency dynamics, the present solution procedure is not unduly restricted by stability. Second, if velocity feedback is present, the allowable step size for stability increases until $\zeta \geq \sqrt{(4\eta^3/27)}$, at which point the solution becomes unconditionally stable.

It should be noted that, instead of the stabilized form of control force equation (9.46d) or (9.49b), if the scalar form of (9.48) is used in the preceding stability analysis, the resulting stability limit is given by

$$h \leq \min\left(\frac{2}{\zeta \omega_c}, \frac{2\zeta}{\eta \omega_c}\right). \tag{9.59}$$

Assuming $\zeta \ll 1$, the first term in the above condition allows a sufficiently large step size. However, since $\zeta/\eta \approx 1$ for a balanced control law, it imposes a step-size restriction $h \approx 2/\omega_c$, which approaches the limit imposed on by a typical explicit integration formula. This proves the advantage of the present stabilized partitioned solution equation (9.46) solely from the computational stability viewpoint.

Although not elaborated herein, a stability analysis that includes an observer model and the state estimation error equation has been conducted with the following parameter choices:

$$\mathbf{L}_2 = [l_{21} \ l_{22}], \quad \mathbf{H} = \begin{bmatrix} 1 & 0 \\ 0 & 1 \end{bmatrix} \tag{9.60}$$

in conjunction with the structural model and the controller model already used in (9.49). The analysis result yields the following step-size restriction:

$$h \le \min\left(\frac{2}{\zeta\omega_c}, \frac{2}{\sqrt{\eta\omega_c}}, \frac{2}{\sqrt{l_{21}}}\right). \tag{9.61}$$

It should be noted that l_{21} corresponds to the Kalman filter gain magnitude which can be adjusted to be sufficiently small compared with ω^2 as can be assessed from equation (9.39b). Hence, provided $l_{21} < \omega_c$, the condition given by (9.58) is seen to govern the maximum stable step size by the present partitioned solution procedure.

For the general multidimensional case governed by (9.46), one observes that the stiffness proportional control force in practice reaches only a fraction of the total internal force ($\mathbf{u} = \eta\mathbf{Kq}, \ \eta \ll 1$). Hence, even for a distributed stiffness proportional control configuration where $\omega_c \to \omega_{max}$, the stable step size given by (9.58) should be much larger than the maximum stable step size of a typical explicit integration algorithm (say, $h_{max} \le 2/\omega_{max}$). Therefore, the computational efficiency of the present partitioned solution procedure is established.

9.4 SOLUTION METHODS FOR COUPLED THERMAL–STRUCTURAL ANALYSIS

Coupled thermal–structural problems are becoming a major challenge in many engineering disciplines such as supersonic planes, satellites, superelectronic chips, and jet and combustion engines. Following the finite element formulations proposed by Wilson and Nickell [30], Nickell and Sackman [31], Oden [32], and Oden and Armstrong [33], among others, the semidiscrete coupled thermal–structural governing equations can be written as

$$\begin{aligned} \mathbf{M}\ddot{\mathbf{u}} + \mathbf{D}\dot{\mathbf{u}} + \mathbf{K}\mathbf{u} - \mathbf{C}\theta &= \mathbf{f} \\ \mathbf{Q}\dot{\theta} + \mathbf{H}\theta + \theta^o \mathbf{C}^T\dot{\mathbf{u}} &= \mathbf{r} \end{aligned} \tag{9.62}$$

where $\mathbf{M}$, $\mathbf{D}$ and $\mathbf{K}$ are the mass, damping and stiffness matrices, and $\mathbf{f}$ is the prescribed structural loading vector; $\mathbf{Q}$, $\mathbf{H}$ and $\mathbf{C}$ are the heat capacitance, heat diffusion and thermal expansion coupling matrices, and $\mathbf{r}$ is the external heat source, respectively; and θ_0 is the reference temperature.

9.4.1 Conventional implicit solution procedures

Suppose we are given two software modules, a structural analyser and a thermal conduction transient analysis module and are tasked to perform the coupled response analysis given by (9.62). The simplest way is then to move the coupling terms $\mathbf{C}\theta$ and $\mathbf{C}^T\dot{\mathbf{u}}$ in the above equation to the right-hand sides and treat them as if they are an applied force and an additional source term, respectively. This will permit the use of two single disciplined-oriented software modules for the analysis of coupled problems. Computationally, this amounts to employing the following *staggered solution procedure*:

$$\mathbf{M}\ddot{\mathbf{u}}^{n+1/2} + \mathbf{D}\dot{\mathbf{u}}^{n+1/2} + \mathbf{K}\mathbf{u}^{n+1/2} = \mathbf{f}^{n+1/2} + \mathbf{C}\theta_p^{n+1/2}$$

$$\mathbf{Q}\dot{\theta}^{n+1/2} + \mathbf{H}\theta^{n+1/2} = \mathbf{r}^{n+1/2} - \theta_0 \mathbf{C}^T\dot{\mathbf{u}}^{n+1/2}$$

(9.63)

where $\theta_p^{n+1/2}$ is the predicted temperature. It turns out that if $\mathbf{C}^T\dot{\mathbf{u}}^{n+1/2}$ is predicted instead of $\theta^{n+1/2}$, one ends up with the same accuracy and stability limits (Park et al. [22]).

While the above implicit–implicit staggered procedure is simple to implement, it can be shown that it is only conditionally stable, even though the implicit integrators used to integrate the left-hand sides of (9.62) are *algorithmically* unconditionally stable. The stabilization procedure that we will describe is a mid-point rule modification of Farhat et al. [34]).

9.4.2 Stabilization of implicit–implicit staggered solution procedure

Stabilization of a general staggered solution procedure for coupled-field problems can be accomplished either by a differential-level stabilization and algebraic-level stabilization. In the past both stabilization strategies have been employed for fluid–structure problems, coupled pore fluid–soil interactions, and structure–structure interaction problems (Park et al. [22]; Park [23]; Felippa and Park [3]; Park and Felippa [2]).

In general one can stabilize the implicit–implicit procedure by modifying both or just one of the two field equations. A successful stabilization is the one that minimizes the impact of stabilization in terms of software modification and computational overhead. Of several stabilization strategies studied, a concurrent adaptation of both differential and algebraic-level augmentations was found to yield the most attractive staggered procedure. We now outline the stabilization process.

First, we employ a mid-point version of the trapezoidal rule as

9.4 COUPLED THERMAL–STRUCTURAL ANALYSIS

$$\begin{aligned}
\dot{\mathbf{y}}^{n+1/2} &= \dot{\mathbf{y}}^n + \delta\ddot{\mathbf{y}}^{n+1/2} \\
\mathbf{y}^{n+1/2} &= \mathbf{y}^n + \delta\dot{\mathbf{y}}^{n+1/2} \\
\mathbf{y}^{n+1} &= 2\mathbf{y}^{n+1/2} - \mathbf{y}^n \\
\dot{\mathbf{y}}^{n+1} &= 2\dot{\mathbf{y}}^{n+1/2} - \dot{\mathbf{y}}^n
\end{aligned} \quad (9.64)$$

where $\mathbf{y}$ can be either the displacement or the temperature vector in (9.62) and δ is one-half of the step size, $\delta = 1/2\Delta t$, and Δt is the time step size.

Second, time discretization of the thermal coupled equations (9.62b) to obtain

$$(\mathbf{Q} + \delta\mathbf{H})\boldsymbol{\theta}^{n+1/2} = \delta\mathbf{r}^{n+1/2} + \mathbf{Q}\boldsymbol{\theta}^n - \delta\theta_0\mathbf{C}^T\dot{\mathbf{u}}^{n+1/2}. \quad (9.65)$$

Note that in the above difference equation, the unknown structural coupling term is associated with the velocity $\dot{\mathbf{u}}^{n+1/2}$. It is this vector that has been found to play a key role in stabilization of the present procedure. In order to stabilize the extrapolation of the coupling term, we utilize an integrated form of $\dot{\mathbf{u}}^{n+1/2}$ from the structural equation (9.61a):

$$\begin{aligned}
\dot{\mathbf{u}}^{n+1/2} &= \mathbf{B}[\mathbf{M}\dot{\mathbf{u}}^n + \delta(\mathbf{f}^{n+1/2} - \mathbf{K}\mathbf{u}^{n+1/2} + \mathbf{C}\boldsymbol{\theta}^{n+1/2})] \\
\mathbf{B} &= (\mathbf{M} + \delta\mathbf{D})^{-1}.
\end{aligned} \quad (9.66)$$

Upon substituting the above expression into (9.65), we obtain

$$\begin{aligned}
\mathbf{G}\boldsymbol{\theta}^{n+1/2} &= \mathbf{R}^{n+1/2} + \delta^2\theta_0\mathbf{C}^T\mathbf{B}\mathbf{K}\mathbf{u}_p^{n+1/2} \\
\mathbf{R}^{n+1/2} &= \delta\mathbf{r}^{n+1/2} + \mathbf{Q}\boldsymbol{\theta}^n - \delta\theta_0\mathbf{C}^T\mathbf{B}(\mathbf{M}\dot{\mathbf{u}}^n + \delta\mathbf{f}^{n+1/2}) \\
\mathbf{G} &= \mathbf{Q} + \delta\mathbf{H} + \delta^2\theta_0\mathbf{C}^T\mathbf{B}\mathbf{C} \\
\mathbf{u}_p^{n+1/2} &= \mathbf{u}^n.
\end{aligned} \quad (9.67)$$

It is observed that the solution matrix $\mathbf{G}$ for the thermal equation is augmented with the additional matrix $\delta^2\theta_0\mathbf{C}^T\mathbf{B}\mathbf{C}$ and the prediction of $\dot{\mathbf{u}}_p^{n+1/2}$ is replaced by $\delta\mathbf{B}\mathbf{K}\mathbf{u}_p^{n+1/2}$. Also, note that the predictor for $\mathbf{u}_p^{n+1/2}$ is simply the previous step solution which has been found the most stable predictor when used in conjunction with the trapezoidal rule (Park [23]).

Once the thermal equation is stabilized as described above, the structural equation (9.62a) can be integrated in an existing structural analysis program as if the term $\mathbf{C}\boldsymbol{\theta}$ is an external force at each integration step. Time discretization of (9.62a) by the mid-point rule (9.64) yields

$$\begin{aligned}
\mathbf{E}\mathbf{u}^{n+1/2} &= \mathbf{F}^{n+1/2} + \delta^2\mathbf{C}\boldsymbol{\theta}^{n+1/2} \\
\mathbf{E} &= \mathbf{M} + \delta\mathbf{D} + \delta^2\mathbf{K} \\
\mathbf{F}^{n+1/2} &= \delta^2\mathbf{f}^{n+1/2} + \mathbf{M}(\mathbf{u}^n + \delta\dot{\mathbf{u}}^n) + \delta\mathbf{D}\mathbf{u}^n.
\end{aligned} \quad (9.68)$$

The updating procedure for states at time step $(n + 1)$ is achieved as follows:

$$\begin{aligned}
\mathbf{u}_t^{n+1} &= 2\mathbf{u}^{n+1/2} - \mathbf{u}^n \\
\dot{\mathbf{u}}_t^{n+1} &= 2(\mathbf{u}^{n+1/2} - \mathbf{u}^n)/\delta - \dot{\mathbf{u}}^n \\
\boldsymbol{\theta}_t^{n+1} &= 2\boldsymbol{\theta}^{n+1/2} - \boldsymbol{\theta}^n \\
\ddot{\mathbf{u}}^{n+1} &= \mathbf{M}^{-1}(\mathbf{f}^{n+1} + \mathbf{C}\boldsymbol{\theta}_t^{n+1} - \mathbf{D}\dot{\mathbf{u}}_t^{n+1} - \mathbf{K}\mathbf{u}_r^{n+1}) \\
\dot{\mathbf{u}}^{n+1} &= \dot{\mathbf{u}}^n + \delta(\ddot{\mathbf{u}}^{n+1} + \ddot{\mathbf{u}}^n) \\
\mathbf{u}^{n+1} &= \mathbf{u}^n + \delta(\dot{\mathbf{u}}^{n+1} + \dot{\mathbf{u}}^n) \\
\dot{\boldsymbol{\theta}}^{n+1} &= \mathbf{Q}^{-1}(\mathbf{r}^{n+1} - \boldsymbol{\theta}_0 \mathbf{C}^T \dot{\mathbf{u}}^{n+1} - \mathbf{H}\boldsymbol{\theta}_t^{n+1}) \\
\boldsymbol{\theta}^{n+1} &= \boldsymbol{\theta}^n + \delta(\dot{\boldsymbol{\theta}}^{n+1} + \dot{\boldsymbol{\theta}}^n).
\end{aligned} \quad (9.69)$$

9.4.3 An analysis of stability and accuracy of stabilized procedure

Stability of the staggered procedure presented in (9.67)–(9.69) can be assessed by adopting an analysis procedure, for example, outlined in Park [4]. First, we assume that the step-by-step numerical solution for a uniform step integration can be characterized by

$$\mathbf{y}^{n+1} = \lambda \mathbf{y}^n. \qquad (9.70)$$

Hence, computational stability is maintained if

$$|\lambda| \leq 1. \qquad (9.71)$$

In order to invoke the well-known Routh–Hurwitz criterion, we map the stable zone, $|\lambda| \leq 1$, onto the left-hand side of a z-plane by the following idempotent transformation:

$$\begin{bmatrix} \mathbf{u}^{n+1} \\ \dot{\mathbf{u}}^{n+1} \\ \ddot{\mathbf{u}}^{n+1} \\ \boldsymbol{\theta}^{n+1} \\ \dot{\boldsymbol{\theta}}^{n+1} \end{bmatrix} = \frac{1+z}{1-z} \begin{bmatrix} \mathbf{u}^n \\ \dot{\mathbf{u}}^n \\ \ddot{\mathbf{u}}^n \\ \boldsymbol{\theta}^n \\ \dot{\boldsymbol{\theta}}^n \end{bmatrix}. \qquad (9.72)$$

Substitution of (9.70) and (9.72) into (9.67)–(9.69) with $\mathbf{D} = 0$ yields

$$\begin{bmatrix} z^2 \mathbf{M} + \delta^2 \mathbf{K} & -\delta^2 \mathbf{C} \\ -(1-z^2)\delta^2 \boldsymbol{\theta}_0 \mathbf{C}^T \mathbf{M}^{-1} \mathbf{K} & z^2 \mathbf{Q} + z\delta \mathbf{H} + \delta^2 \boldsymbol{\theta}_0 \mathbf{C}^T \mathbf{M}^{-1} \mathbf{C} \end{bmatrix} \begin{bmatrix} \mathbf{u}^n \\ \boldsymbol{\theta}^n \end{bmatrix} = \begin{bmatrix} 0 \\ 0 \end{bmatrix} \quad (9.73)$$

whose characteristic equation is obtained from

$$\text{Det}| \mathbf{M}z^3 + \mathbf{VM}\delta z^2 + \delta^2(\mathbf{K} + \theta_0 \mathbf{CQ}^{-1}\mathbf{C}^T + \delta^2\theta_0 \mathbf{CQ}^{-1}\mathbf{C}^T\mathbf{M}^{-1}\mathbf{K})z$$
$$+ \delta^3 \mathbf{VK}| = 0 \tag{9.74}$$

where

$$\mathbf{V} = \mathbf{CUC}^T$$
$$\mathbf{U} = \mathbf{Q}^{-1}\mathbf{H}(\mathbf{C}^T\mathbf{C})^{-1}.$$

The reader may find a complete stability analysis in Farhat et al. [34]. Hence, we offer the following synopsis. First, for a two-degree-of freedom problem, we have

$$\mathbf{M} = 1, \quad \mathbf{K} = \omega^2, \quad \mathbf{Q} = q, \quad \mathbf{H} = h, \quad \mathbf{C} = c \tag{9.75}$$

which, when substituted into the above characteristic equation, gives

$$a_3 z^3 + a_2 z^2 + a_1 z + a_0 = 0 \tag{9.76}$$

where

$$a_3 = 1, \quad a_2 = \delta q, \quad a_1 = \delta^2[\omega^2 + \frac{\theta_0 c^2}{qm}(1 + \delta^2\omega^2)], \quad a_0 = \delta^3 q\omega^2.$$

Since δ, h, q, ω^2, θ_0, c^2, and $m \geq 0$, all the coefficients of the polynomial (9.76) in z are positive. Moreover, the quantity

$$a_1 a_2 - a_0 a_3 = \theta_0 hc^2 \delta^3 mq^2(1 + \delta^2\omega^2)$$

is also positive, which demonstrates that the stabilized staggered solution procedure is unconditionally stable for the 2-d.o.f. model problem.

For multi-dimensional cases, the limiting case of $\mathbf{K} = 0$, which gives rise to a quadratically growing structural response due to thermal coupling, can be used as a pathological test:

$$|\mathbf{M}z^2 + \delta \mathbf{VM}z + \delta^2\theta_0 \mathbf{CQ}^{-1}\mathbf{C}^T| = 0. \tag{9.77}$$

Since $\mathbf{M}$ is positive definite and the other two matrices are at least semi-definite, the stabilized staggered procedure for this limiting case is unconditionally stable via Bellman's theorem [35] as successfully utilized in Park [23]. Hence, we conclude that the procedure given by (9.67)–(9.69) is unconditionally stable.

As for accuracy, it can be shown that the stabilized staggered procedure is second-order accurate. This can be done first by expanding the difference equations (9.67)–(9.69) and using the governing coupled semidiscrete equations (9.62) where needed.

9.4.4 Computational sequence

When we ignore structural damping, diagonal structural mass and diagonal capacitance matrix, the stabilized computational procedure can be summarized as follows:

$$\mathbf{R}^{n+1/2} = \delta \mathbf{r}^{n+1/2} + \mathbf{Q}\boldsymbol{\theta}^n - \delta\theta_0 \mathbf{C}^T \mathbf{B}[\dot{\mathbf{u}}^n + \delta \mathbf{M}^{-1}(\mathbf{f}^{n+1/2} - \mathbf{K}\mathbf{u}^n)] \tag{9.78}$$

$$(\mathbf{Q} + \delta \mathbf{H} + \delta^2 \theta_0 \mathbf{C}^T \mathbf{M}^{-1}\mathbf{C})\boldsymbol{\theta}^{n+1/2} = \mathbf{R}^{n+1/2} \tag{9.79}$$

$$\mathbf{F}^{n+1/2} = \delta^2 \mathbf{f}^{n+1/2} + \mathbf{M}(\mathbf{u}^n + \delta \dot{\mathbf{u}}^n) + \delta \mathbf{D}\mathbf{u}^n + \delta^2 \mathbf{C}\boldsymbol{\theta}^{n+1/2} \tag{9.80}$$

$$(\mathbf{M} + \delta^2 \mathbf{K})\mathbf{u}^{n+1/2} = \mathbf{F}^{n+1/2} \tag{9.81}$$

$$\begin{aligned}
\mathbf{u}_t^{n+1} &= 2\mathbf{u}^{n+1/2} - \mathbf{u}^n \\
\boldsymbol{\theta}_t^{n+1} &= 2\boldsymbol{\theta}^{n+1/2} - \boldsymbol{\theta}^n \\
\ddot{\mathbf{u}}^{n+1} &= \mathbf{M}^{-1}(\mathbf{f}^{n+1} + \mathbf{C}\boldsymbol{\theta}_t^{n+1} - \mathbf{K}\mathbf{u}_t^{n+1}) \\
\dot{\mathbf{u}}^{n+1} &= \dot{\mathbf{u}}^n + \delta(\ddot{\mathbf{u}}^{n+1} + \ddot{\mathbf{u}}^n) \\
\mathbf{u}^{n+1} &= \mathbf{u}^n + \delta(\dot{\mathbf{u}}^{n+1} + \dot{\mathbf{u}}^n) \\
\dot{\boldsymbol{\theta}}^{n+1} &= \mathbf{Q}^{-1}(\mathbf{r}^{n+1} - \theta_0 \mathbf{C}^T \dot{\mathbf{u}}^{n+1} - \mathbf{H}\dot{\boldsymbol{\theta}}^{n+1}) \\
\boldsymbol{\theta}^{n+1} &= \boldsymbol{\theta}^n + \delta(\dot{\boldsymbol{\theta}}^{n+1} + \dot{\boldsymbol{\theta}}^n).
\end{aligned} \tag{9.82}$$

The key for the efficiency of the above procedure compared with other candidate procedures is to utilize the matrix $\mathbf{C}^T \mathbf{M}^{-1} \mathbf{C}$, which appears in the left-hand side of (9.79) and is a symmetric banded matrix. Other possible stabilization involves $\mathbf{C}\mathbf{Q}^{-1}\mathbf{C}^T$ into the left-hand side of (9.79), which has much larger bandwidth than the former.

Some two-dimensional solutions of the thermal–structural interaction problems based on the above procedure are reported in Farhat et al. [34].

9.5 APPLICATION EXAMPLES

In the preceding sections three computational methods for performing coupled-field dynamics analyses have been surveyed. It is anticipated that as the analyst demands more realistic models, all the single-field components, viz., structures, control, thermal and multibody systems, may have to be included in a typical analysis. An example would be a satellite undergoing solar panel deployment as well as attitude stabilization and vibration control. Two examples that we include are a scenario of shuttle-based assembly of the space station for one-cargo segment, and a vibration control of a generic Earth-observing platform when it is subject to a reboosting thrust.

MB-1   MB-2

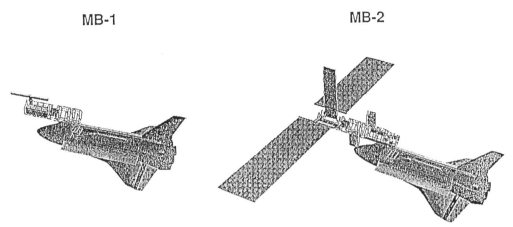

Figure 9.1 Incremental construction of space station. (Courtesy McDonnell Douglas Space Systems Co.)

9.5.1 *Manoeuvring of the SRMS with prescribed motion constraints*

Figure 9.1 illustrates the assembly of the first and second modules of the space station to be deployed and assembled. Each module is lifted by the shuttle remote manipulator system (SRMS) from the shuttle cargo bay and deployed for the first module and subsequently assembled into the partially assembled space station. In order to simulate the assembly process, first, we have studied the effect of the SRMS dynamics due to the required manoeuvring constraints. This incremental in-space construction of the space station must meet stringent geometry, weight and stiffness requirements as shown in Figure 9.2. The arm boom assemblies comprise two thin-walled graphite-epoxy circular sections called the upper arm, lower arm and end effector. These arms are connected by a shoulder joint (modelled by a universal joint), an elbow joint (modelled by a revolute joint) and a wrist joint (modelled by a spherical joint). The properties of the joints and arms are shown as follows (Hunter et al. [36]):

(1) Upper arm:
Young's modulus:
$E_u = 1.27 \times 10^{11}$ Pa
Shear modulus:
$G_u = 3.18 \times 10^{10}$ Pa
Length: $L_u = 6.38$ m
Cross-section area:
$A_u = 0.0022$ m^2
Moment of intertia:
$I_u = 3.16 \times 10^{-5}$ m^4
Weight: $W_u = 24.97$ kg

(2) Lower arm:
Young's modulus:
$E_l = 1.09 \times 10^{11}$ Pa
Shear modulus:
$G_l = 3.30 \times 10^{10}$ Pa
Length: $L_l = 7.06$ m
Cross-section area:
$A_l = 0.0015$ m^2
Moments of inertia:
$I_l = 2.19 \times 10^{-5}$ m^4
Weight: $W_l = 24.06$ kg

Properties of SRMS:
- Weight = 410 Kg
- Length = 15 m
- Cross Section Area = 0.0022 m²
- Young's Module = 1.27 X 10¹¹ Pa
- Shear Module = 3.18 X 10¹⁰ Pa
- Density = 1.2 X 10⁴ Kg/m³
- Tip Maneuvering Speed (without payload) = 0.6 m/s

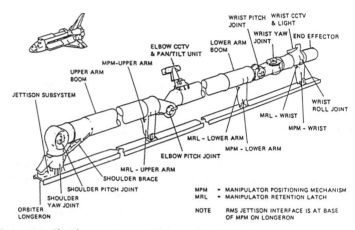

Figure 9.2 Shuttle remote manipulator system [34].

(3) End effector:
Length: $L_e = 1.82$ m
End effector weight:
$W_e = 107.14$ kg

(4) Joint weights:
Shoulder joint weight:
$W_s = 117.13$ kg
Elbow joint weight:
$W_{el} = 53.12$ kg
Wrist joint weight:
$W_w = 84.44$ kg.

The effect of the motion constraints to the orbital motion stability can be assessed by modelling (1) both the space shuttle and the SRMS modeled to be rigid, (2) the shuttle to be rigid and the SRMS as a flexible beam (discretized into 4 elements and 5 nodal points). By imposing angular velocity (a cubic type) at the tip of the SRMS (Figure 9.3), the manipulator will slew through 90° with respect to the space shuttle. Figures 9.4 and 9.5 illustrate the pitching angles of the rigid and flexible SRMS, and Figure 9.6 shows the angular velocity of the rigid and flexible SRMS. Note that the terminal velocities of the flexible case are non-zero, implying that the SRMS manoeuvring would trigger vibrations on the space shuttle modules after assembly. To overcome this difficulty, a more refined SRMS manoeuvring motion is necessary

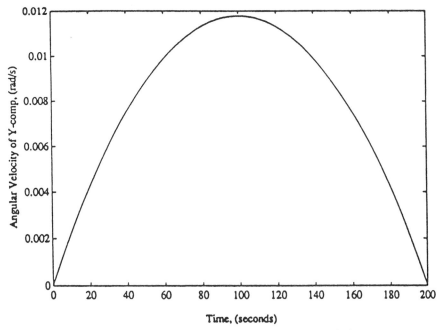

Figure 9.3 Imposed angular motions of arm and the platform.

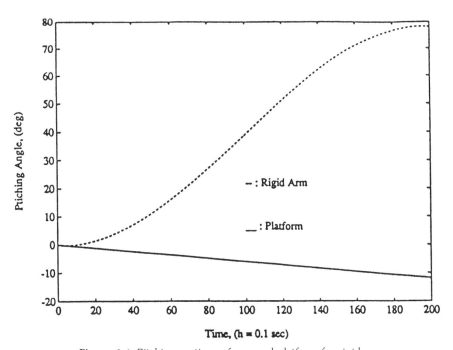

Figure 9.4 Pitching motions of arm and platform for rigid case.

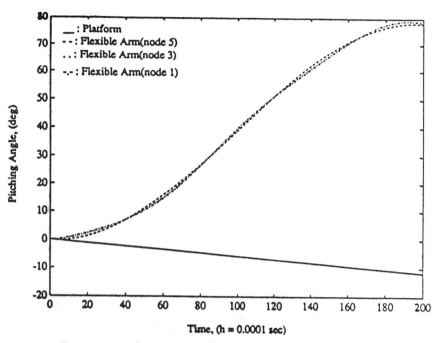

Figure 9.5 Pitching motions of arm and platform for flexible case.

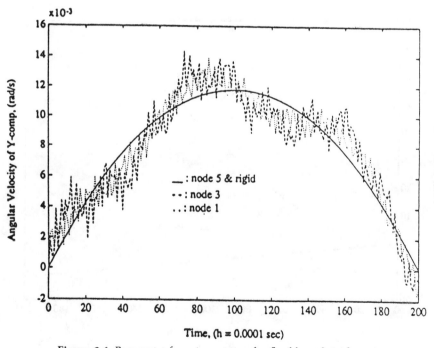

Figure 9.6 Response of yawing motion for flexible and rigid cases.

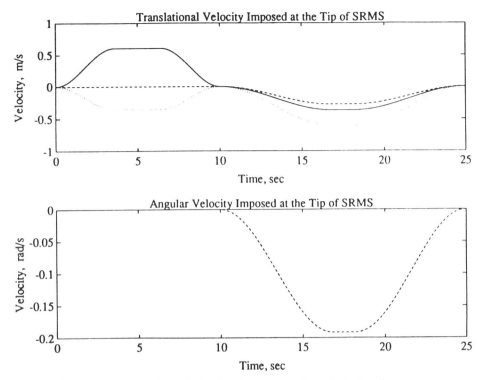

Figure 9.7 Imposed translational and angular motions at the tip of SRMS.

as described in Chiou et al. [37]. By adopting the refined starting and stopping conditions, the in-space construction of the space station can be divided into the following stages:

1. manoeuvring of the SRMS to the position where its end effector is ready to attach the space structure which is lying in the shuttle cargo bay;
2. contact/impact when the end effector of the SRMS collides with the space structure;
3. manoeuvring of the SRMS with the space structure to attach to another space structure which is floating in space;
4. contact/impact when the SRMS with the space structure collides with another space structure in space.

For the first stage, the motion constraints for the tip of the SRMS are given by Figure 9.7 where 25 seconds of manoeuvring time is used to place the end effector of the SRMS to the position where the space structure/payload is located. As indicated in Figures 9.8 and 9.9, the angular velocity vectors for the upper arm and lower arm of the rigid and flexible SRMS experience almost the same behaviour in terms of trends and magnitudes which prove that the present motion constraints are valid in manoeuvring the rigid and flexible SRMS. At the second stage, where the contact/impact has occurred, the end effector of the SRMS is approaching the

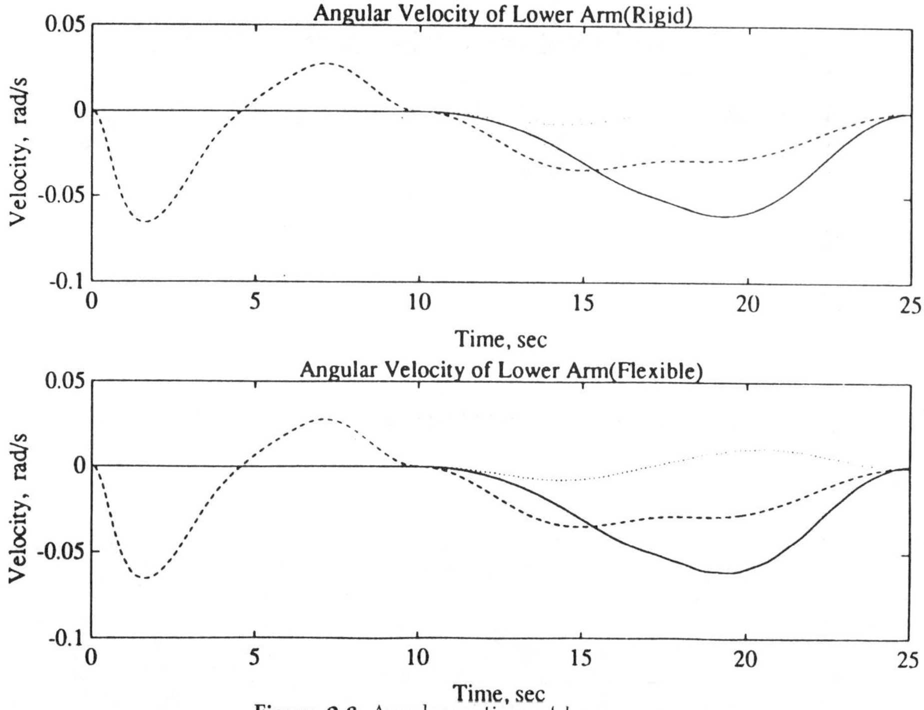

288 TIME INTEGRATION METHODS FOR SYSTEM DYNAMICS

Figure 9.8 Angular motions of lower arm.

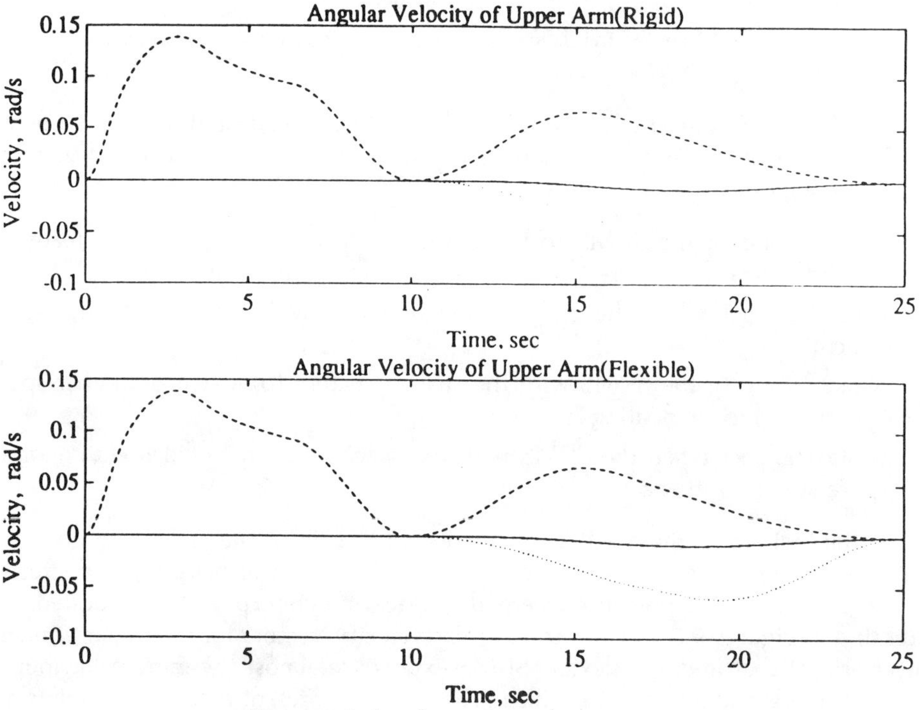

Figure 9.9 Angular motions of upper arms.

9.5 APPLICATION EXAMPLES   289

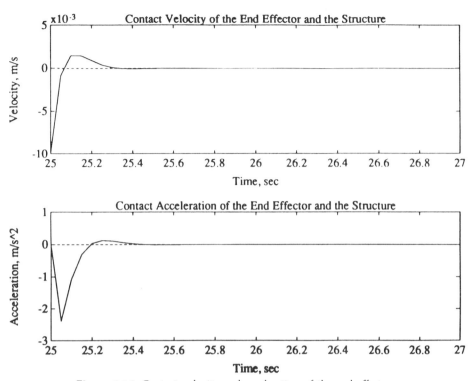

Figure 9.10 Contact velocity and acceleration of the end effector.

structure with velocity equal to −0.01 m/s (Figure 9.10(a)). When two bodies make contact at 25 s, the velocity of the end effector drops from −0.01 m/s to 0.0018 m/s to almost 0 m/s in less than one second of contact/impact time. From Figure 9.10(b), the contact/impact provides a peak acceleration (−2.4 m/s^2) on the end effector which eventually dies down because of the large mass ratio between the end effector and the structure. At the third stage, the SRMS lifts the structure with a motion constraint that is given by Figure 9.11. The purpose of this motion constraint is to manoeuvre the structure into the position where the previously existing structure is located so that the assembly of two structures can take place via contact/impact. From Figures 9.12 and 9.13, even though the angular velocities of the flexible SRMS still maintain the trends as in the rigid SRMS case, the high vibration modes can easily be seen as the stopping conditions of the flexible SRMS are applied. Consequently, due to these vibrations, the non-zero terminal velocities have occurred, which makes the assembly of the two structures very difficult to carry out. In conclusion, since the current motion constraints cannot provide the zero terminal velocities for the flexible SRMS, the control strategy in damping out these vibrations needs to be studied in order to proceed to the final stage of the present construction process. In Figure 9.14, the contact/impact of the fourth stage has been carried out by using the rigid SRMS model that the velocity of the approaching structure is −0.01 m/s which produces of accelerations for both assembling structures during 1.8 s of contact/impact time. Note that after two seconds of contact/impact, both structures are traveling with the same velocity as indicated in Figure 9.14(a).

290 TIME INTEGRATION METHODS FOR SYSTEM DYNAMICS

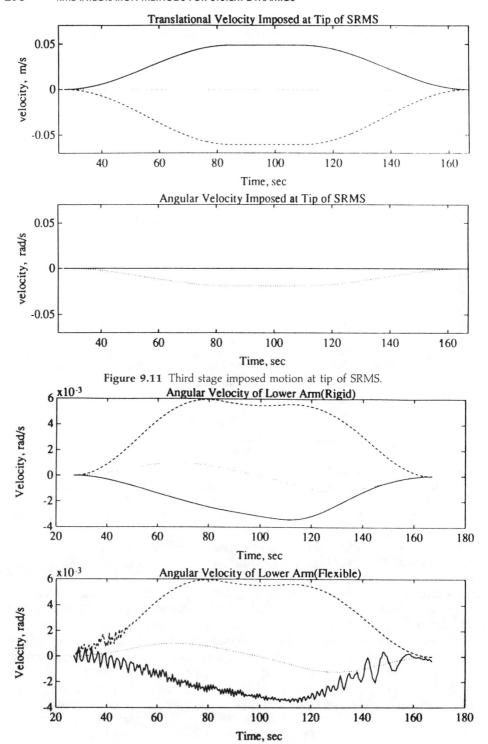

Figure 9.11 Third stage imposed motion at tip of SRMS.

Figure 9.12 Response of lower arm during third stage.

9.5 APPLICATION EXAMPLES

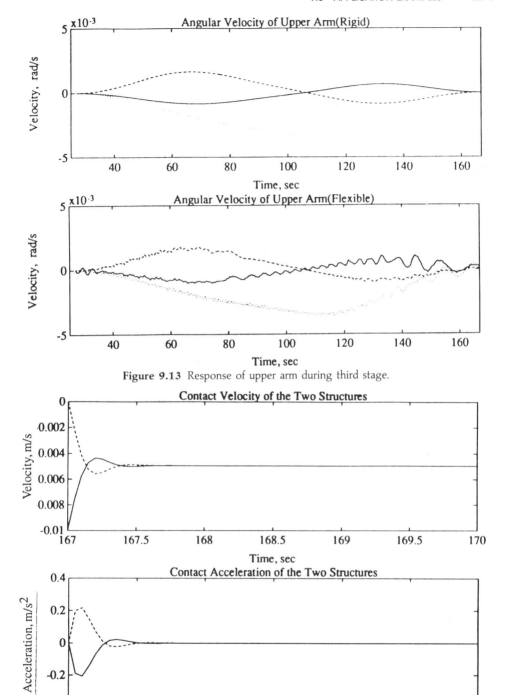

Figure 9.13 Response of upper arm during third stage.

Figure 9.14 Contact velocity and acceleration during third stage.

9.5.2 Control of earth-observing platform

The computational efficiency achieved by the partitioned solution procedure as compared with the conventional solution method is sketched out in Figure 9.15. Assumed in the construction of the chart are the ratio of the bandwidth and the stiffness matrix to be 0.2, the number of the actuator and that of the sensor to be the same, and the ratio of the structural degrees of freedom and those of the actuator to be 0.1. Note that, given a real-time processing computer that can perform a wall-clock rate of 200 samples/second command and control, the conventional method can at most handle the real time control of a simple beam articulation, whereas the partitioned method can handle the real-time control of complex truss-beam vibrations. For nonlinear problems, the advantages of the partitioned method is more pronounced, as can be seen from the chart.

The partitioned CSI simulation procedure as derived in (9.46) has been implemented as a stand-alone package (Park *et al.* [38]). The present software implementation emphasizes the use of the widely available sequential and parallel analysis modules specially developed for the solution of structural dynamics equations. Note that the solution algorithm for both the structural system and the state estimator is the same, hence the software module, provided the right hand terms are treated as applied forces. Although the stabilized form of the controller and the filtered measurements are solved in a coupled manner, their size in general is substantially smaller, typically a fraction of the size of the structural system for large-scale problems.

Figure 9.16 illustrates a test-bed evolutionary model of an Earth-pointing satellite. Eighteen actuators and 18 sensors are applied to the system (see Figure 9.16 for their locations) for vibration control and their locations are provided in Tables 9.1 and 9.2. Figures 9.17–9.19 are representative of the responses for open-loop, direct output feedback, and dynamically compensated case does drift away initially even though the settling time is about the same as that by the direct output feedback case. However, the sensor outputs are assumed to be noise-free in these two numerical experiments. Further simulations with the present procedure should shed light on the performance of dynamically compensated feedback systems or large-scale systems as they are computationally more feasible than heretofore possible.

Tables 9.3 and 9.4 illustrate the computational overhead associated with the direct output feedback vs. the use of a dynamic compensation scheme by the output present Kalman filtering equations, compared in those tables are for two simpler tests cases, viz., a 3-d.o.f. system and a truss beam model. In the numerical experiments herein, we have relied on the Matlab software package for the synthesis of both the control law gains and the discrete Kalman filter gain matrices. Results of the full state feedback (FSFB) utilizing a direct feedback vs a dynamically compensated feedback based on the Kamlan filter (K. Filter) indicate that they become competitive as the size of the model increases. Reported in those two tables are also the effects of various implementation versions, from a nominal code (version A1) to a fully parallelized version on a shared memory machine (allied with 8 processors). The present numerical results indicate that CPU requirements for dynamically compensated CSI simulations would in general require about three times that of a typical structure-only transient analysis. It should be mentioned that even though there is a slight

9.5 APPLICATION EXAMPLES

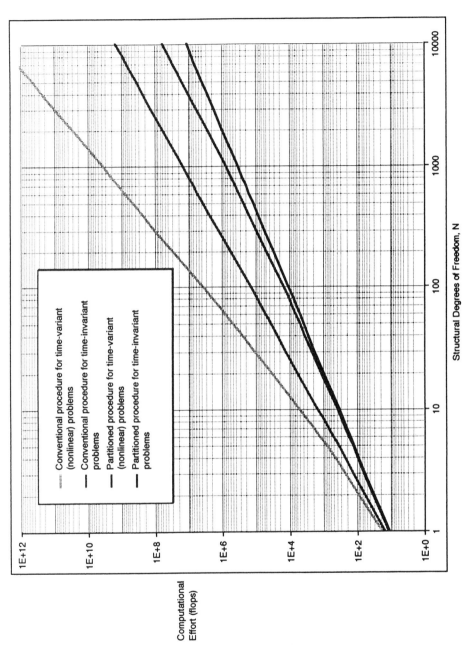

Figure 9.15 Efficiency of partitional CSI procedure.

EARTH POINTING SATELLITE DESIGN PROBLEM

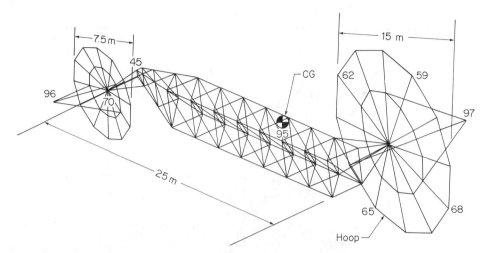

Figure 9.16 A generic Earth-pointing satellite.

Table 9.1 Actuator placement for EPS example problem.

Actuator	Node	Component
1	97	x
2	97	z
3	96	x
4	96	z
5	65	y
6	68	y
7	59	y
8	62	y
9	45	y
10	45	z
11	70	y
12	70	z
13	95	x
14	95	y
15	95	z
16	95	ϕ_x
17	95	ϕ_y
18	95	ϕ_z

increase of total CPU units from the compiler optimized sequential run to the parallel case, the actual run-clock time on the parallel machine is about one-eighth of the sequential case.

To conclude, it is seen that the use of the present second-order discrete Kalman

9.5 APPLICATION EXAMPLES

Table 9.2 Sensor placement for EPS example problem.

Sensor	Type	Node	Component
1	Rate	97	x
2	Rate	97	z
3	Rate	96	x
4	Rate	96	z
5	Rate	65	y
6	Rate	68	y
7	Rate	59	y
8	Rate	62	y
9	Rate	45	y
10	Rate	45	z
11	Rate	70	y
12	Rate	70	z
13	Position	95	x
14	Position	95	y
15	Position	95	z
16	Position	95	ϕ_z
17	Position	95	ϕ_y
18	Position	95	ϕ_z

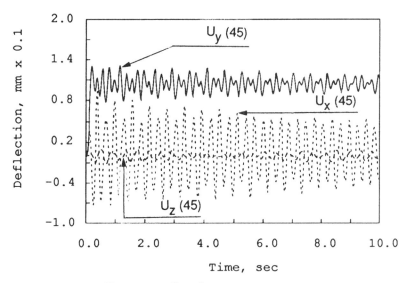

Figure 9.17 Open loop transient response.

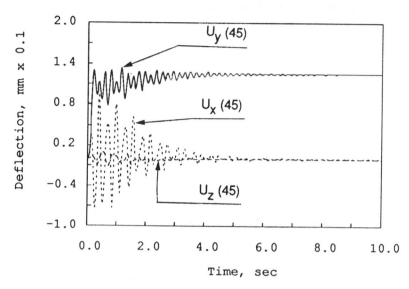

Figure 9.18 Full state feedback response.

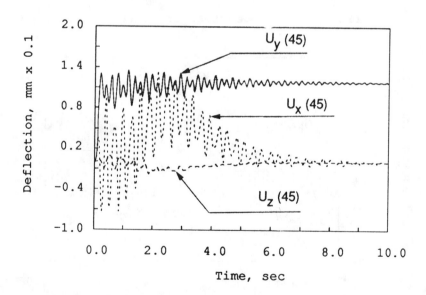

Figure 9.19 Dynamically compensated response via Kalman filter.

Table 9.3 CPU results for versions of ACSIS.

Model	Problem type	(A1) Nominal code	(A2) Compiler optimized	(A3) Parallel observer
3DOF Spring	Transient	6.6	2.1	2.1
	FSFB	8.0	3.3	3.3
	K. Filter	12.3	3.5	3.3
54 DOF Truss	Transient	78.2	5.7	5.6
	FSFB	97.1	9.4	10.2
	K. Filter	170.7	13.0	10.7
582 DOF EPS7	Transient	3506	98.6	100.3
	FSFB	7040	190.2	294.5
	K. Filter	n/a	284.2	312.5

Table 9.4 CPU results for ACSIS with EBE computations.

Model	Problem type	(A4) E-B-E computation	(A5) Parallel E-B-E	(A6) Parallel Obs. & EBE
3 DOF Spring	Transient	3.8	3.3	3.3
	FSFB	4.9	4.4	4.9
	K. Filter	6.6	5.6	5.0
54 DOF Truss	Transient	31.7	13.0	13.0
	FSFB	35.5	16.9	35.6
	K. Filter	62.6	27.3	36.2
582 DOF EPS7	Transient	391.7	153.9	n/a
	FSFB	485.9	245.9	n/a
	K. Filter	n/a	n/a	n/a

filtering equations for constructing dynamically compensated control laws add computational overhead, but is only the equivalent of open-loop transient analysis of symmetric sparse systems of order N instead of $2N \times 2N$ dense systems.

9.6 CLOSING REMARKS

The present survey have focused mostly on the research activities on the computational methods for multibody dynamics, control-structure interactions, and coupled thermal–structural transients undertaken by the researchers at the Center for Space Structures and Controls, University of Colorado. No attempt has thus been made to include many important advances made by researchers around the world. A notable omission in this survey is computational methods for parallel computations, which is intensely pursued by many researchers. We hope to report on the progress on parallel computational methods at a later occasion.

ACKNOWLEDGEMENTS

This survey consists of work performed by the author's colleagues, in alphabetical order, K. F. Alvin, W. K. Belvin, J. C. Chiou, J. D. Downer, C. Farhat and C. A. Felippa. Without having had the privilege of working with them over the past six years, it would not have been possible for the author to put together the present survey. He is deeply indebted to them for this privilege. However, any misconception or imbalance in the coverage of the materials solely rests on the author and he hopes they will forgive him for not presenting their work as they conceive to be the case if any misrepresentation has been made. This survey was prepared while the author was visiting the Department of Aeronautics and Astronautics, Massachusetts Institute of Technology, during Spring 1992. He thanks Prof. E. F. Crawley and his staff for providing a nurturing environment for his sabbatical stay at the Space Engineering Research Center, a NASA-sponsored space engineering center at MIT.

BIBLIOGRAPHY

[1] Park, K.C. (1988) *Transient Analysis Methods in Computational Dynamics*, in D.L. Dwoyer, M.Y. Hussaini and R.G. Voigt (eds.), *Finite Elements: Theory and Applications*, Springer-Verlag, pp. 240–267.
[2] Park, K.C. and Felippa, C.A. (1983) Partitioned analysis of coupled systems, in T. Belytschko and T.J.R. Hughes (eds.), *Computational Methods for Transient Analysis*, Elsevier, pp. 157–219.
[3] Felippa, C.A. and Park, K.C. (1980) Staggered transient analysis procedures for coupled-field mechanical systems: formulation, *Comput. Meth. Appl. Mech. Engg*, **24**, 61–111.
[4] Park, K.C. (1977) Practical aspects of numerical time integration, *Comput. Struct.*, **7**, 343–353.
[5] Hughes, T.J.R. and Belytschko, T. (1985) A précis of developments in computational methods for transient analysis, *J. Appl. Mech.*, **50**, 1033–1041.
[6] Park, K.C. and Chiou, J.C. (1988) Stabilization of computational procedures for constrained dynamical systems, *J. Guidance, Control and Dynamics*, **11**, 365–370.
[7] Park, K.C., Chiou, J.C. and Downer, J.D. (1990) Explicit–implicit staggered procedure for multibody dynamics analysis, *J. Guidance, Control, and Dynamics*, **13**, 562–570.
[8] Park, K.C., Downer, J.D., Chiou, J.C. and Farhat, C. (1991) A modular multibody analysis capability for high-precision, active control and real-time applications, *Int. J. Num. Meth. Engg*, **32**, 1767–1798.
[9] Wehage, R.A. and Haug, E.J. (1982) Generalized coordinate partitioning for dimension reduction in analysis of constrained dynamic systems, *ASME J. Mech. Design*, **104**, 247–255.
[10] Chiou, J.C. (1990) *Constraint Treatment Techniques and Parallel Algorithms for Multibody Dynamic Analysis*, Ph.D. Thesis, University of Colorado.
[11] Chiou, J.C., Park, K.C. and Farhat, C. (1991) A natural partitioning scheme for parallel simulation of multibody systems, *1991 AIAA Structures, Dynamics and Materials Conference*, Paper No. AIAA-91-1111, Baltimore, Maryland, 8–10 April 1991.
[12] Baumgarte, J.W. (1972) Stabilization of constraints and integrals of motion in dynamical systems, *Comput. Meth. Appl. Mech. Engg*, **1**, 1–16.
[13] Baumgarte, J.W. (1983) A new method of stabilization for holonomic constraints, *J. Appl. Mech.*, **50**, 869–870.

[14] Park, K.C. and Chiou, J.C. (1992) A Discrete Momentum-Conserving Explicit Algorithm for Multibody Dynamic Analysis, to appear in *Int. J. Num. Meth. Engg.*

[15] Simo, J.C. and Wong, K.K. (1991) Unconditionally stable algorithms for rigid body dynamics that exactly preserve energy and momentum, Int. J. Num. Meth. Engg, **31**(1), 19–52.

[16] Downer, J.D. (1990) *A Computational Procedure for the Dynamics of Flexible Beams within Multibody Systems*, Ph.D. Thesis, University of Colorado.

[17] Downer, J.D., Park, K.C. and Chiou, J.C. (1991) A Computational Procedure for Multibody Systems Including Flexible Beam Dynamics, to appear in *Comput. Meth. Appl. Mech. Engg, Proc. the 1990 AIAA Dynamics Specialist Conference*, Paper No. AIAA-90-1237, Long Beach, California, 5–6 April, 1990.

[18] Downer, J.D. and Park, K.C. (1991) Dynamics of Spacecraft with Deploying Flexible Appendages, *Comput. Meth. Appl. Mech. Engg,* 1991; *Proc. the 1992 AIAA Dynamics Specialist Conference*, Paper No. AIAA-92-2087, Dallas, Texas, 16–17, April 1992.

[19] Haug, E.J. and Deyo, R.C. (eds.) (1991) *Real-Time Integration Methods for Mechanical Systems Simulation*, NATO, ASI Series, Springer-Verlag, Berlin.

[20] Belvin, W.K. (1989) *Simulation and Interdisciplinary Design Methodology for Control-Structure Interaction Systems*, Ph.D. Thesis, Center for Space Structures and Controls, University of Colorado, Report No. CU-CSSC-89-10, July.

[21] Park, K.C. and Belvin, W.K. (1991) A partitioned solution procedure for control-structure interaction simulations, *J. Guidance, Control and Dynamics*, **14**(1), January–February, 59–67.

[22] Park, K.C., Felippa, C.A. and DeRuntz, J.A. (1977) Stabilization of staggered solution procedures for fluid-structure interaction analysis, in Belytschko, T. and Geers, T.L. (eds.) *Comput. Meth. Fluid-Structure Interaction Problems*, ASME, AMD Vol. 26, New York, N.Y., pp. 95–124.

[23] Park, K.C. (1980) Partitioned analysis procedures for coupled-field problems: stability analysis, *J. Appl. Mech.*, **47**, 370–378.

[24] Zienkiewicz, O.C., Paul, D.K. and Chan, A.H.C. (1988) Unconditionally stable staggered solution procedure for soil-pore fluid interaction problems, *Int. J. Num. Meth. Engg*, **26**, 1039–1055.

[25] Kwarkernaak, H. and Sivan, R. (1974) *Linear Optimal Control Systems*, Wiley-Interscience, New York.

[26] Kalman, R.E. and Bucy, R.S. (1961) New results in linear filtering and prediction theory, *ASME J. Basic Engg*, **83**, 95–108.

[27] Luenberger, D.G. (1974) Observing the state of a linear system, *IEEE Trans. Military Electronics*, **8**, 74–80.

[28] Belvin, W.K. and Park, K.C. (1989) On the state estimation of structures with second order observers, *Proc. 30th Structures, Structural Dynamics and Materials Conference*, AIAA Paper No. 89-1241, April 3–5.

[29] Gantmacher, F.R. (1959) *The Theory of Matrices*, 2, Chelsea, New York, pp. 190–196.

[30] Wilson, E.L. and Nickell, R.E. (1966) Application of the finite element method to heat conduction analysis, *Nucl. Eng. Des.*, **4**, 276–286.

[31] Nickell, R.E. and Sackman, J.L. (1968) The extended Ritz method applied to transient coupled thermoelastic boundary-value problems, *J. Appl. Mech.*, **35**, 255–266.

[32] Oden, J.T. (1969) Finite element analysis of nonlinear problems in the dynamical theory of coupled thermoelasticity, *Nucl. Eng. Des.*, **10**, 465–475.

[33] Oden, J.T. and Armstrong, W.H. (1971) Analysis of nonlinear dynamic coupled thermoviscoelasticity problems by the finite element method, *Comput. Struct.*, **1**, 603–621.

[34] Farhat, C., Park, K.C. and Dubois-Pelerin, Y.D. (1991) An unconditionally stable

staggered algorithm for transient finite-element analysis of coupled thermoelastic problems, *Comput. Meth. Appl. Mech. Engg,* **85**, 349–365.
[35] Bellman, R. (1970) *Introduction to Matrix Analysis* (2nd edn), McGraw-Hill, New York, pp. 249–262.
[36] Hunter, J.A., Ussher, T.H. and Gossain, D.M. (1982) Structural dynamic design consideration of the shuttle remote manipulator system, *Proc. the 1982 AIAA Structures, Structural Dynamics and Materials Conf.,* Paper No. AIAA-82-0762, pp. 499–505.
[37] Chiou, J.C., Downer, J.D., Natori, M.C. and Park, K.C. (1992) Interaction dynamics of an orbiter and a flexible space structure undergoing incremental in-space construction, to be presented at the 32nd SDM Conf., Dallas, TX, April 1992.
[38] Park, K.C., Belvin, W.K. and Alvin, K.F. (1992) Second-order discrete Kalman filtering equations for control-structure interaction simulations, to appear in *Journal of Guidance, Control, and Dynamics.*

K.C. Park
Department of Aerospace Engineering Sciences and Center for Space Structures and Controls
University of Colorado at Boulder
Campus Box 429
Boulder
CO 80309-0429
USA

10

Coupled Field Problems — Solution Techniques for Sequential and Parallel Processing

I. St. Doltsinis

University of Stuttgart, Germany

10.1 INTRODUCTION

In field problems of mechanics and physics, the variables usually interact in the sense that changes at a certain location affect several parts of the domain occupied by the medium. Marked differences in the intensity of the interaction or in the local response may distinguish subdomains within an otherwise homogeneous medium on the one hand, as for instance in elastoplasticity [1]. On the other hand, such a distinction is provided by the equipresence of heterogeneous media, e.g. in fluid-structure interaction [2]. Furthermore, a number of problems are characterized by the interaction of essentially different physical phenomena within a common material domain; for instance thermal and mechanical [3]. Under the above circumstances, problems may be considered as coupled in the sense that they are composed of distinct sub-problems, each exhibiting a certain coherence. The latter characteristic may suggest or necessitate the application of individual solution techniques to the partial problems which have to be coupled in order to establish a solution to the original problem. The separation into reasonable subtasks is a particular requirement in large scale applications, in order to facilitate the numerical treatment in sequential and parallel computations.

Solving Large-scale Problems in Mechanics, Edited by M. Papadrakakis
© 1993 John Wiley & Sons Ltd

A variety of numerical solutions to coupled problems may be found in [4] together with a certain classification, cf. also [5]. Some applications treated in [6] include the dynamic interaction between fluid flow and flexible solid systems as well as the coupling of thermal phenomena to the motion of fluids on the one hand and to the large deformation of non-linear solids on the other hand. Here, it is interesting to notice the concerted action of distinct branches of a single software system — cf. [7] for an early documentation — towards the computation of different coupled problems.

The present treatise mainly summarizes and extends previous work by the author in the field. It deals in particular with iteration techniques coupling partial solutions of the complete problem. Although most of the results are applicable to several classes of coupled problems, reference is made exclusively to the interaction of different physical fields within a common material domain; this should be beneficial to the conciseness of the presentation. In this context, a statement of the pertinent tasks is given in Section 10.2 for stationary and transient applications. An elementary consideration on the time integration of coupled evolution processes is included for completeness, and discusses the possible use of a distinct time incrementation for the interacting phenomena; for further details on transient procedures for coupled problems the reader may consult [5, 8, 9]. Ultimately, a system of vector equations which are usually non-linear in the variables has to be treated at several stages of the evolution process in association with an incremental continuation algorithm.

The solution of the equation system is the subject of Sections 10.3 and 10.4. In particular, Section 10.3 refers to the treatment of the coupled problem as a whole. To this purpose, the formal indication of a customary iteration procedure for the solution of the compound non-linear system is followed by the presentation of a substitution algorithm for the coupled variables, which replaces the original procedure by a number of steps with reduced problem dimensions [6]. The processing of coupling matrices is, however, still required. Section 10.4, on the other hand, deals with the iterative coupling of partial solutions. Thereby, the individual fields may be treated either in parallel by a Jacobi-type algorithm or in sequence by a Gauss–Seidel procedure [10]. In both cases the values of the variables may be exchanged either after a complete solution of the partial problem or after each iteration cycle. An assessment of the proposed execution modes is given for a system of two non-linear scalar equations and is interpreted for the simple model problem of the adiabatic expansion of a tensile specimen at elevated temperatures.

Section 10.5 is concerned with the implementation of the iterative coupling algorithms on parallel computers with distributed memory [11]. Within the conceptual basis of the spatial decomposition, the domain of the problem is distributed among all available processors. In this scheme the variables of the interacting physical phenomena are treated in sequence as by the iterative Gauss–Seidel algorithm. The concept of the physical decomposition, on the other hand, deals with the distinct field problems concurrently along the lines of a Jacobi iteration for the coupling; combination with the spatial decomposition of the partial physical problems increases the granularity of the parallel computation. The techniques are investigated by means of a test example.

Section 10.6 touches upon the subject of the finite element discretization in connection with mesh generation, adaptive modification, and decomposition for parallel computations. A particular aspect of coupled processes is the use of individual discretizations for each constituent, and is discussed here.

Despite the essentially general scope of the study, the theory is uniquely illustrated with reference to thermomechanically coupled large deformation processes. The equations governing the finite element representation of this class of problems are specified in Section 10.7 and provide the basis for the numerical simulation of industrial metal forming. A first application deals with the hot rolling of a rail where the initial unsteady phase of the process is followed up to stationary conditions in the deformation zone. In conclusion, the process of the hot rolling of a steel wire is considered as a demonstration of the parallel processing for coupled field problems.

10.2 FORMULATION OF THE COUPLED FIELD PROBLEM

We consider a physical domain in which two distinct sets of field variables appear and are represented in a discretized manner by the vector arrays v and w. The length of the above vector arrays is equal to the product: number of points representing the discretized field × number of components of the respective field variable. Let the interacting physical phenomena be governed by the system of equations

$$\hat{f}(v, w) = 0, \qquad \hat{g}(w, v) = 0 \qquad (10.1)$$

which may be linear or non-linear in the variables v and w. Each of the particular sets of equations in (10.1) expresses a unique condition governing the single physical phenomenon under consideration. In particular, the set $\hat{f} = 0$ governs the field variables v whilst $\hat{g} = 0$ is responsible for the variables w; this associates the number of single equations in each set with the dimension of the respective vector variables.

The system (10.1) does not contain any time rates of the variables v and w, and is therefore referred to time-independent problems. Nevertheless, its solution may be required at several time instants in connection with an incremental continuation procedure if the history of the field variables is requested for timely variable boundary conditions and/or in order to facilitate the numerical analysis of strongly non-linear problems. Transient processes, on the other hand, are characterized by the appearance of the time rates of the variables v and w in the governing equations. For instance first order systems may be described by

$$\tilde{f}(v, \dot{v}; w, \dot{w}) = 0, \qquad \tilde{g}(w, \dot{w}; v, \dot{v}) = 0. \qquad (10.2)$$

Both the variables, v, w and their time rates $\dot{v}, \dot{w}$ are unknown quantities in (10.2) and are linked via the integration in time

$$v = \int \dot{v} \, dt, \qquad w = \int \dot{w} \, dt \qquad (10.3)$$

which is usually performed by approximation following an incremental scheme. The construction of different approximate integration formulae is outlined in [12] from a mathematical point of view; the time integration of distinct physical processes is addressed in [13]. For the purpose of the present discussion, the single step formula

$$u = {}^a u + (1-\zeta)\tau {}^a \dot{u} + \zeta \tau {}^b \dot{u} = {}^b u$$
$$u = v = w \tag{10.4}$$

is considered. In (10.4), time is advanced from $t = {}^a t$ to $t = {}^a t + \tau = {}^b t$ and $0 \le \zeta \le 1$ is the collocation parameter taken as $\zeta = 0$ in an explicit integration scheme, and often as $\zeta = \frac{1}{2}$ in an implicit integration scheme which is associated with $\zeta > 0$.

In the explicit case, the system (10.2) is used in the form

$$\tilde{f}({}^a v, {}^a \dot{v}; {}^a w, {}^a \dot{w}) = 0, \qquad \tilde{g}({}^a w, {}^a \dot{w}; {}^a v, {}^a \dot{v}) = 0 \tag{10.5}$$

which is a system of equations for the rates ${}^a \dot{v}, {}^a \dot{w}$ at the beginning of the time increment under consideration. Whilst the values of the variables v, w at the beginning of the increment have to be known, those at the end of the increment are determined by (10.4) using the solution of (10.5). A special case arises if at least one of the vector equations in (10.5) does not depend on the time rate of the other set of variables, as for instance in

$$\tilde{f}({}^a v, {}^a \dot{v}; {}^a w, -) = 0, \qquad \tilde{g}({}^a w, {}^a \dot{w}; {}^a v, {}^a \dot{v}) = 0 \tag{10.6}$$

Here, the solution of the problem is in fact uncoupled since ${}^a \dot{v}$ can be obtained directly from $\tilde{f} = o$ and then enables us to solve $\tilde{g} = 0$ for ${}^a \dot{w}$. The same procedure can be applied to the coupled case as well if the actual time rate of the coupling variables in one of the vector equations is replaced by that of the preceding time increment, as for instance in

$$\tilde{f}({}^a v, {}^a \dot{v}; {}^a w, {}^a \dot{w}_-) = 0, \qquad \tilde{g}({}^a w, {}^a \dot{w}; {}^a v, {}^a \dot{v}) = 0 \tag{10.7}$$

where ${}^a \dot{w}_-$ is known from the previous incremental computation. By this algorithm coupling is effected during the course of time in a staggered manner which may either take advantage of the weakest interaction between the physical phenomena under consideration or be applied alternately to both sets of equations. For a discussion of the staggered coupling procedure and some variance for implicit integration the reader is referred to [8] and [9]; see also [5].

In the case of an implicit integration by (10.4), where $\zeta > 0$, both the field variables and their time rates must be known at the beginning of the time increment whilst their values at the end of the time increment satisfy the system of equations (10.2) in the form

$$\tilde{f}({}^b v, {}^b \dot{v}; {}^b w, {}^b \dot{w}) = 0, \qquad \tilde{g}({}^b w, {}^b \dot{w}; {}^b v, {}^b \dot{v}) = 0. \tag{10.8}$$

As the field variables and their time rates appearing in (10.8) are linked by the approximate integration scheme (10.4), the coupling between the individual vector equations is effected here by both types of unknowns. Therefore, the remarks to (10.6) based on the time rates in the explicit integration are no longer applicable.

The time step of the approximate integration is limited by the requirement of stability of the numerical scheme. In this context we refer for convenience, in conjunction with a unique time integration of both sets of equations in (10.2), to the compound array of the variables

$$W = \begin{bmatrix} v \\ w \end{bmatrix}.\qquad(10.9)$$

Accordingly, the single system

$$\tilde{F}(W, \dot{W}) = 0 \qquad (10.10)$$

comprises the individual vector equations.

For a stability analysis we consider the propagation by the integration scheme of deviations from the solution W of (10.10). Let the deviations be collected in the vector array,

$$D = \begin{bmatrix} d \\ e \end{bmatrix} \qquad (10.11)$$

where d and e are associated with the individual sets of variables. They are propagated by (10.4) in accordance to

$$^bD - \zeta\tau^b\dot{D} = {}^aD + (1-\zeta)\tau^a\dot{D} \qquad (10.12)$$

where D and its time rate $\dot{D}$ may be related via an incrementation of (10.10) as by

$$\dot{D} = -\left[\frac{\partial \tilde{F}}{\partial \dot{W}}\right]^{-1} \frac{\partial \tilde{F}}{\partial W} D = -ND \qquad (10.13)$$

with the matrix

$$N = \left[\frac{\partial \tilde{F}}{\partial \dot{W}}\right]^{-1} \frac{\partial \tilde{F}}{\partial W} \qquad (10.14)$$

introduced here as a generalization of the respective considerations in [14]. Applying (10.13) in (10.12) and with the identity matrix I we obtain for the deviation

$$^bD = [I + \zeta\tau^bN]^{-1}[I - (1-\zeta)\tau^aN]^aD \qquad (10.15)$$

which must not be amplified, to ensure local stability. Hence a necessary condition states that the spectral norm of the amplification matrix in (10.15) must be less than unity. This ultimately may be expressed by the condition

$$\left| \frac{1 - (1 - \zeta)\tau^a\rho}{1 + \zeta\tau^b\rho} \right| < 1 \tag{10.16}$$

where

$$\rho = \rho(N) \tag{10.17}$$

denotes the spectral norm of the matrix product in (10.14). As a result, the spectral norm $\rho(N)$ limits the time step for conditional stability to

$$\tau < \frac{2}{{}^a\rho_N - ({}^a\rho_N + {}^b\rho_N)\zeta} \quad \text{and} \quad 0 \leq \zeta < \frac{{}^a\rho_N}{{}^a\rho_N + {}^b\rho_N} \tag{10.18}$$

whereas unconditional stability is achieved when

$$\tau \to \infty \quad \text{and} \quad \frac{{}^a\rho_N}{{}^a\rho_N + {}^b\rho_N} \leq \zeta \leq 1. \tag{10.19}$$

The stability condition for the explicit integration follows from (10.18) for $\zeta = 0$; linear problems are characterized by ${}^b\rho_N = {}^a\rho_N = \rho_N$.

Instead of a unique integration, it may be suitable to adapt the numerical scheme to the individual evolution properties of the sub-problems. For a discussion on this subject we detail the matrix N according to the individual sets of variables v and w as follows:

$$N = \begin{bmatrix} N_{vv} & N_{vw} \\ N_{wv} & N_{ww} \end{bmatrix}. \tag{10.20}$$

Uncoupled problems are characterized by the absence of the off-diagonal submatrices in (10.20), in which case the foregoing stability considerations may be applied to the individual integration of each of the two processes. For an elementary discussion we refer in the following to linear problems for which the stability criterion (10.18) suggests the relation

$$\frac{\tau_w}{\tau_v} = \frac{\rho_v}{\rho_w} \tag{10.21}$$

between the individual time increments in conjunction with a common integration scheme. In (10.21), ρ_v and ρ_w denote the spectral norm of the matrices N_{vv} and N_{ww}

in (10.20) respectively. For a synchronization of the two problems the quotient of the time steps should be viewed as an integer number ($\tau_w > \tau_v$ by convention).

Let the time period under consideration be subdivided into $k_v \tau_v$ or $k_w \tau_w$ incremental units. Common integration of both evolution processes with $\tau = \tau_v$ implies that each problem has to be evaluated at k_v stages, whilst individual incrementation requires $k_v + k_w$ solutions in total. A comparison of the respective number of incremental solutions leads to the expression

$$\frac{k_v + k_w}{2k_v} = \frac{1 + \tau_v/\tau_w}{2} \tag{10.22}$$

which indicates that individual incrementation may imply a reduction up to 1/2 if the time scales of the two processes are substantially different. Alternatively, the slower process might be integrated explicitly with the time increment $\tau = \tau_w$, which is used also for the faster process in conjunction with an implicit integration.

In actually coupled problems the off-diagonal submatrices do not vanish in (10.20), and this complicates the assessment of stability for individual integration of the interacting phenomena. We therefore consider a system of two scalar evolution equations and explore then the use of different values ζ_v and ζ_w of the collocation parameter and of different time increments, τ_v and $\tau_w = n\tau_v$. For the propagation of perturbations of the solution via the integration scheme we then have

$$^b\begin{bmatrix} d \\ e \end{bmatrix} - \tau \begin{bmatrix} \zeta_v & 0 \\ 0 & n\zeta_w \end{bmatrix} {}^b\begin{bmatrix} \dot{d} \\ \dot{e} \end{bmatrix} = {}^a\begin{bmatrix} d \\ e \end{bmatrix} + \tau \begin{bmatrix} 1 - \zeta_v & 0 \\ 0 & n(1 - \zeta_w) \end{bmatrix} {}^a\begin{bmatrix} \dot{d} \\ \dot{e} \end{bmatrix} \tag{10.23}$$

instead of (10.12).

Bearing in mind that the time rates are related to the perturbations by means of (10.13), equation (10.23) may be written as

$$[I + \tau^b A]^b D = [I - \tau^a A]^a D. \tag{10.24}$$

The matrix $^b A$ is defined by

$$^b A = \begin{bmatrix} \zeta_v & 0 \\ 0 & n\zeta_w \end{bmatrix} \begin{bmatrix} N_{vv} & N_{vw} \\ N_{wv} & N_{ww} \end{bmatrix} \tag{10.25}$$

and the matrix $^a A$ by

$$^a A = \begin{bmatrix} (1 - \zeta_v) & 0 \\ 0 & n(1 - \zeta_w) \end{bmatrix} \begin{bmatrix} N_{vv} & N_{vw} \\ N_{wv} & N_{ww} \end{bmatrix}. \tag{10.26}$$

In the above expressions the matrix N from (10.13) is considered constant as appertaining to linear evolution processes; in the non-linear case, the respective entries in (10.25) and (10.26) are different.

Following the reasoning leading to (10.18), we obtain here the condition for stability in the generalized form,

$$\tau < \frac{2}{{}^a\rho_A - {}^b\rho_A} \tag{10.27}$$

where ${}^a\rho_A$ and ${}^b\rho_A$ denote the spectral norm of the matrices ${}^aA$ and ${}^bA$ respectively. For the purpose of comparison, we first evaluate (10.27) for a unique integration with $\zeta_w = \zeta_v = \zeta$ and $\tau_w = \tau_v = \tau$ and obtain the stability limit as

$$\tau = \frac{2}{(1 - 2\zeta)\rho_N}. \tag{10.28}$$

The eigenvalues λ_N of the present 2×2 matrix N are given by the expression

$$\frac{2\lambda_N}{N_{vv}} = (1 + q) \pm [(1 - q)^2 + 4\bar{q}q]^{1/2} \tag{10.29}$$

and with the simplifying assumptions

$$q = \frac{N_{ww}}{N_{vv}} \ll 1, \qquad 0 \leq \bar{q} = \frac{N_{vw}}{N_{vv}} \frac{N_{wv}}{N_{ww}} \leq 1 \tag{10.30}$$

the spectral norm ρ_N follows as from

$$\frac{\rho_N}{N_{vv}} = \frac{\max|\lambda_N|}{N_{vv}} = 1 + \bar{q}q. \tag{10.31}$$

For a discussion of the individual incrementation, we consider the case $\zeta_w = \zeta_v = \zeta_n$ and $\tau_v = \tau_n, \tau_w = n\tau_n$. Here, the stability limit $\bar{\tau}_n$ may formally be obtained by (10.28) but the eigenvalues of the modified matrix N read now

$$\frac{2\lambda_n}{N_{vv}} = (1 + nq) \pm [(1 - nq)^2 + 4n\bar{q}q]^{1/2} \tag{10.32}$$

and for

$$n = \frac{N_{vv}}{N_{ww}} = \frac{1}{q} \tag{10.33}$$

and with the second condition in (10.30), the spectral norm is obtained from (10.32) as

$$\frac{\rho_n}{N_{vv}} = \frac{\max|\lambda_n|}{N_{vv}} = 1 + \bar{q}^{1/2}. \tag{10.34}$$

10.2 FORMULATION OF THE COUPLED FIELD PROBLEM

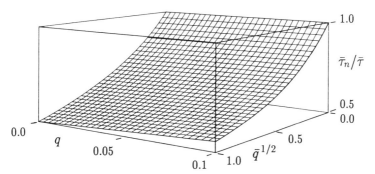

Figure 10.1 Stability limit $\bar{\tau}_n$ of individual incrementation related to stability limit $\bar{\tau}$ of unique incrementation ($\zeta_n = \zeta$).

From (10.28) we obtain for the stability limits of the two integration methods with (10.31) and (10.34) the relation

$$\frac{\bar{\tau}_n}{\bar{\tau}} = \frac{2}{(1 - 2\zeta)\rho_n} \frac{(1 - 2\zeta)\rho_N}{2} = \frac{1 - 2\zeta}{1 - 2\zeta_n} \frac{1 + \bar{q}q}{1 + \bar{q}^{1/2}}. \tag{10.35}$$

For $\zeta_n = \zeta$ and with reference to the expressions in (10.30), the range of variation of the quotient in (10.35) is found to be

$$\frac{1}{2} < \frac{\bar{\tau}_n}{\bar{\tau}} \leq 1. \tag{10.36}$$

The upper limit in (10.36) associated with $\bar{q} = 0$, is applicable when one of the processes does not depend on the other (one-sided coupling); the lower limit corresponds to the fully coupled case as defined by $\bar{q} = 1$. The gradual transition between the above limits for $\zeta_n = \zeta$ is depicted in Figure 10.1 for a varying q.

In the present coupled case, the ratio of the incremental solutions for the individual and the unique time incrementation of the two processes is not given by (10.22) but by the expression

$$\frac{k_v + k_w}{2k} = \frac{1 + \tau_v/\tau_w}{2} \frac{\tau}{\tau_w}. \tag{10.37a}$$

The first term on the right-hand side of (10.37a) reproduces the uncoupled case of (10.22) where the number of individual solutions may be reduced up to 1/2 if $\tau_v/\tau_w = q$ becomes negligible with respect to unity. The second term accounts for the effect of coupling on the stability limit as controlled by (10.35). Accordingly, an increasing intensity of the coupling diminishes this benefit and compensates it completely for $\bar{q} = 1$. On the other hand, the number of coupled solutions is reduced in accordance with the relation

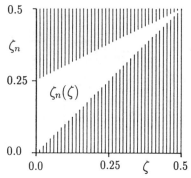

Figure 10.2 Domain of dependence $\zeta_n(\zeta)$ for same stability limit of individual and unique incrementation.

$$\frac{k_w}{k} = \frac{\tau}{n\tau_v} = q\frac{\tau}{\tau_v}. \tag{10.37b}$$

The above discussion of (10.35) compares the two integration methods for a unique value of the collocation parameter ζ. Alternatively, the stability limits in (10.35) may be determined with ζ_n for the individual incrementation and ζ for the unique incrementation; the two limits are equal if

$$2\zeta_n = 1 - (1 - 2\zeta)\frac{1 + \bar{q}q}{1 + \bar{q}^{1/2}}. \tag{10.38}$$

Also, with reference to the restriction by (10.36) of the variation of the quotient on the right-hand side of (10.38) it may be found that

$$\zeta \leq \zeta_n < \frac{1 + 2\zeta}{4}. \tag{10.39}$$

In the uncoupled case a unique stability limit is obtained with $\zeta_n = \zeta$, whilst $\zeta_n > \zeta$ for coupled problems, the difference diminishing as unconditional stability is approached, with $\zeta_n = \zeta = 1/2$, cf. Figure 10.2.

We next consider an implicit integration of the fast process ($\zeta_v > 0$) in conjunction with an explicit integration of the slow process ($\zeta_w = 0$); the time increment is unique ($\tau_v = \tau_w = \tau_\zeta$). For a discussion of the stability condition (10.27) the determination of the eigenvalues of the matrices ^bA and ^aA in (10.25) and (10.26) for the above conditions yields

$$\frac{2^b\lambda_A}{N_{vv}} = \zeta_v(1 \pm 1) \tag{10.40}$$

and

10.2 FORMULATION OF THE COUPLED FIELD PROBLEM

$$\frac{2\,{}^a\lambda_A}{N_{vv}} = (1 - \zeta_v + q) \pm [(1 - \zeta_v + q)^2 + 4(1 - \zeta_v)\bar{q}q]^{1/2} \tag{10.41}$$

respectively. The corresponding spectral norms are given by

$$\frac{{}^b\rho_A}{N_{vv}} = \frac{\max |{}^b\lambda_A|}{N_{vv}} = \zeta_v \tag{10.42}$$

and for $0 \leq q \ll (1 - \zeta_v)$, $0 \leq \bar{q} \leq 1$, by

$$\frac{{}^a\rho_A}{N_{vv}} = \frac{\max |{}^a\lambda_A|}{N_{vv}} = 1 - \zeta_v + \bar{q}q. \tag{10.43}$$

The stability limit $\bar{\tau}_\zeta$ may now be obtained from (10.27) and is compared to the stability limit (10.28) of the unique integration by means of the quotient

$$\frac{\bar{\tau}_\zeta}{\bar{\tau}} = \frac{2}{{}^a\rho_A - {}^b\rho_A} \frac{(1 - 2\zeta)\rho_N}{2} = \frac{(1 - 2\zeta)(1 + \bar{q}q)}{1 - 2\zeta_v + \bar{q}q} \tag{10.44}$$

which is indicative of the number of increments required by the alternative schemes.

Considering a unique value of the collocation parameter $\zeta_v = \zeta$, the quotient in (10.44) reduces to

$$\frac{\bar{\tau}_\zeta}{\bar{\tau}} = 1 - \frac{2\zeta \bar{q}q}{1 - 2\zeta + \bar{q}q}. \tag{10.45}$$

For $\zeta = 0$, the above two methods are of course identical, whilst for $\zeta = 1/2$ the unique integration is unconditionally stable ($\bar{\tau} = \infty$) but the mixed integration not; this causes the quotient in (10.45) to formally vanish. The variation of the quotient with the collocation parameter ζ and the product $\bar{q}q$ is depicted in Figure 10.3. As a consequence of the assumption to (10.43), in the absence of coupling ($\bar{q}q = 0$) the stability limit is controlled solely by the implicit process and is therefore identical in both cases ($\bar{\tau}_\zeta/\bar{\tau} = 1$). With increasing coupling intensity the stability limit $\bar{\tau}_\zeta$ of the mixed integration decreases in comparison to $\bar{\tau}$ of the unique integration; this effect is enhanced as the collocation parameter is increased towards $\zeta = 1/2$.

Whilst the above discussion refers to a unique value of the collocation parameter ζ, it might be of interest to explore also conditions which conserve the stability limit in either scheme. To this end, the quotient in (10.44) is equated to unity and it may be deduced that for $\bar{\tau}_\zeta = \bar{\tau}$,

$$\zeta_v = (1 + \bar{q}q)\zeta \tag{10.46}$$

which means that for $\bar{q}q \ll 1$ a slight modification of the collocation parameter in the mixed integration establishes the stability limit of the unique integration of the coupled problem.

312 COUPLED FIELD PROBLEMS

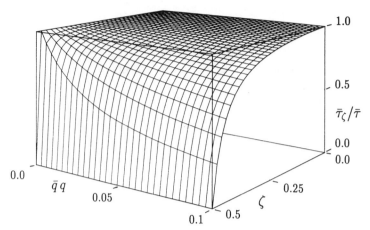

Figure 10.3 Stability limit $\bar{\tau}_\zeta$ of mixed integration related to stability limit $\bar{\tau}$ of unique integration ($\zeta_v = \zeta$).

In summary, equations (10.2) governing the unsteady coupled process provide in conjunction with the suitable approximate integration scheme a system for the computation of either the state variables or their time rates at each time instant. In this context, the essential problem is considered to be the solution of a system of equations

$$f(x, y) = 0, \qquad g(y, x) = 0 \tag{10.47}$$

at distinct time instants during the course of an incremental continuation procedure, and will be discussed in the following. In (10.47), the vector arrays x, y comprise either the field variables or their time rates as unknown quantities, and f, g are usually non-linear functions of their arguments. As a matter of fact, the system (10.47) encompasses the time-independent situation represented by (10.1).

10.3 SOLUTION OF THE COUPLED PROBLEM TAKEN AS A WHOLE

10.3.1 Iteration procedure for the non-linear system

The equation system (10.47) may be solved by a coupled method operating in the entire domain of the variables x and y at the same time. For this purpose we define the compound arrays

$$Y = \begin{bmatrix} x \\ y \end{bmatrix}, \qquad F = \begin{bmatrix} f(x, y) \\ g(y, x) \end{bmatrix} \tag{10.48}$$

and solve the vector equation

$$F(Y) = 0 \tag{10.49}$$

for the unknown variables Y. The above collective representation of the coupled field problem may either be composed from individual discretizations of the interacting phenomena (cf. for instance [3]) or may directly result from a concurrent discretization of the two fields (cf. for instance [15]). In any case, the solution of a non-linear equation (10.49) requires the application of an iterative procedure. A suitable scheme is

$$Y_{i+1} = Y_i + H_i F(Y_i). \tag{10.50}$$

Following the recurrence formula in (10.50), the ith iteration cycle starts with an approximation to the solution and supplies a new estimate with the aid of the iteration matrix H_i. The matrix H may be considered an auxiliary one, usually chosen with respect to best numerical properties and computational convenience. It can be seen from (10.50) that convergence of the iteration process implies that (10.49) is fulfilled. For an assessment of the convergence properties of the iteration procedure, we deduce from (10.50) for the difference D_{i+1}, D_i between the temporary solution Y_{i+1}, Y_i in consecutive iterations and the ultimate solution Y the expression

$$D_{i+1} = [I + HG]D_i \tag{10.51}$$

where I denotes the identity matrix and

$$G = \frac{dF}{dY} \tag{10.52}$$

the gradient of the compound vector function F with respect to the coupled variables Y used in the linearization. The requirement for convergence as based on the spectral norm of the amplification matrix in (10.51) reads

$$\|I + HG\| < 1 \tag{10.53}$$

and consequently convergence depends on the relation of the iteration matrix H to the gradient matrix G of the system (10.48). From (10.51), the best choice for the iteration matrix obviously is

$$H = -G^{-1} \tag{10.54}$$

and is associated with the Newton procedure. In this case we arrive at the recursive scheme of (10.50) by formally solving the linearised form of the equation (10.49)

$$F(Y_{i+1}) = F(Y_i) + G_i[Y_{i+1} - Y_i] = 0 \tag{10.55}$$

for Y_{i+1} in the vicinity of the estimate Y_i. Usually, the exact gradient matrix G is replaced by suitable approximations in order to facilitate the computation, which brings us back to the general iteration scheme of (10.50).

Apart from the above technique which is based on standard procedures for the solution of non-linear problems, we aim in the following at the development of an alternative algorithm characterized by the reduction of the bulk dimensions of the problem. For this purpose we take advantage of the inherent block structure of the system as shown in (10.48) by the distinct sets of variables coupled via their physical interaction.

10.3.2 Substitution algorithm for the coupled variables

We consider once more the Newton technique and point out the physical block structure of the problem by detailing (10.55) with reference to (10.48) as

$$\begin{bmatrix} f(x_i, y_i) \\ g(y_i, x_i) \end{bmatrix} + \begin{bmatrix} F_x & F_y \\ G_x & G_y \end{bmatrix} \begin{bmatrix} x_{i+1} - x_i \\ y_{i+1} - y_i \end{bmatrix} = \begin{bmatrix} 0 \\ 0 \end{bmatrix}. \tag{10.56}$$

The partial gradient matrices appearing in (10.56) are defined as

$$F_x = \frac{\partial f}{\partial x}, \quad F_y = \frac{\partial f}{\partial y}; \quad G_x = \frac{\partial g}{\partial x}, \quad G_y = \frac{\partial g}{\partial y}. \tag{10.57}$$

In order to solve the system (10.56) within the physical domain of each sub-problem while maintaining the full influence of the coupling terms, we first separate the two subsets of the equation and obtain in place of (10.56)

$$f_i + F_x[x_{i+1} - x_i] + F_y[y_{i+1} - y_i] = 0$$
$$g_i + G_y[y_{i+1} - y_i] + G_x[x_{i+1} - x_i] = 0. \tag{10.58}$$

As a matter of fact (10.58) represents the standard Taylor-series expansion to the first order of the sets of equation (10.47). Secondly, we introduce an auxiliary variable

$$z = [x_{i+1} - x_i] + F_x^{-1} F_y [y_{i+1} - y_i] \tag{10.59}$$

and modify the equation system (10.58) to

$$f_i + F_x z = 0$$
$$g_i + G_x z + [G_y - G_x F_x^{-1} F_y][y_{i+1} - y_i] = 0 \tag{10.60}$$

the first equation in (10.60) yields a solution for the auxiliary variable z which enables us to solve the second equation for y_{i+1}. Subsequently, the new estimate x_{i+1} may be established via (10.59)

$$x_{i+1} - x_i = z - F_x^{-1} F_y [y_{i+1} - y_i]. \tag{10.61}$$

The above treatment simply presents a solution of the equation system (10.58) by substitution as given in [6] for non-linear coupled field problems. The technique is equally applicable in conjunction with simplified approximations instead of the exact gradient matrices involved in the Newton iteration procedure. In the particular case that the mixed gradients are disregarded in the approximation of the gradient matrix of the system, i.e. if the iteration is based on

$$\begin{bmatrix} F_x & 0 \\ 0 & G_y \end{bmatrix} \Rightarrow \begin{bmatrix} F_x & F_y \\ G_x & G_y \end{bmatrix} = G \tag{10.62}$$

the solution of each of the equation sets in (10.58) can be performed independently in the current iteration cycle. The coupling is then essentially effected by the subsequent evaluation of the functions $f(x, y), g(y, x)$ with the solution x_{i+1}, y_{i+1}, which is used also for the determination of the direct gradients F_x, G_y required in the next iteration cycle.

10.4 THE COUPLED PROBLEM DIVIDED INTO PHYSICAL SUBDOMAINS

Towards a distribution of the computation among different software and hardware units it proves advantageous to handle each of the vector equations in (10.47) individually. A feasible technique involves then the consecutive solution of (10.47) in an iterative manner which assumes that one of the sets of variables x and y is temporary fixed. This procedure has been adopted as a standard algorithm in the solution of thermomechanically coupled problems in [3, 6, 16]. The separate treatment of the vector equations in the system (10.47) can be effected in different ways and depends on the intensity of the coupling [10].

10.4.1 Jacobi iteration

A Jacobi iteration [17, 18], for instance, applied to the vector variables in (10.47), starts with the estimates x_i, y_i to the solution and provides new estimates x_{i+1}, y_{i+1} by solving the equations

$$f(x_{i+1}, y_i) = 0, \qquad g(y_{i+1}, x_i) = 0. \tag{10.63}$$

Since the equations (10.63) are uncoupled with respect to the temporary unknowns x_{i+1}, y_{i+1}, they may be treated also concurrently. Once the two vector equations (10.63) have been solved for the current iteration cycle, the resulting x and y are exchanged between them, and a subsequent iteration cycle is started.

For an assessment of the convergence of the generalized Jacobi iteration (10.63) for the coupled problem, we consider the deviations d and e of the temporary solution to the ultimate solution x, y. We then have,

$$x_{i+1} = x + d_{i+1}, \qquad y_i = y + e_i$$
$$y_{i+1} = y + e_{i+1}, \qquad x_i = x + d_i \tag{10.64}$$

and may write (10.63) in the form,

$$f(x + d_{i+1}, y + e_i) \cong F_x d_{i+1} + F_y e_i = 0$$
$$g(y + e_{i+1}, x + d_i) \cong G_y e_{i+1} + G_x d_i = 0 \tag{10.65}$$

the coefficient matrices in (10.65) denoting the partial derivatives of the functions f and g as defined by (10.57). From (10.65),

$$\begin{bmatrix} d_{i+1} \\ e_{i+1} \end{bmatrix} = - \begin{bmatrix} O & F_x^{-1} F_y \\ G_y^{-1} G_x & O \end{bmatrix} \begin{bmatrix} d_i \\ e_i \end{bmatrix} \tag{10.66}$$

and in order for the deviations d, e to decrease during the iteration, the spectral norm of the amplification matrix of the Jacobi scheme

$$M_j = - \begin{bmatrix} O & F_x^{-1} F_y \\ G_y^{-1} G_x & O \end{bmatrix} \tag{10.67}$$

must be less than unity,

$$\|M_j\| < 1. \tag{10.68}$$

The above condition limits the applicability of the present method (10.63) to a restricted degree of coupling in the original equation system (10.47).

The non-linear vector equations in (10.63) are usually solved by the recurrence procedure

$$x_{j+1} = x_j + H_j f(x_j, y_i)$$
$$y_{k+1} = y_k + K_k g(y_k, x_i) \tag{10.69}$$

which furnishes x_{i+1}, y_{i+1} as a result of the iteration with the auxiliary matrices H, K, while x_i, y_i are held fixed. Considering next the deviations d, e of the iterates to the solution x_{i+1}, y_{i+1} such that

$$x_j = x_{i+1} + d_j, \qquad y_j = y_{i+1} + e_j \tag{10.70}$$

we obtain from (10.69)

$$d_{j+1} = [I + HF_x]d_j, \qquad e_{j+1} = [I + KG_y]e_j. \tag{10.71}$$

In analogy to (10.53), the condition for convergence of the iterative solution of the individual sets of equations as by (10.69) reads

$$\|I + HF_x\| < 1, \qquad \|I + KG_y\| < 1, \tag{10.72}$$

and accordingly, convergence depends on the relation of the iteration matrices H, K to the partial gradient matrices F_x, G_y.

The aforementioned two different types of iteration, one for the individual equations as by (10.69) and the other for the coupled system as by (10.63), may be replaced by a single iteration loop whereby the recurrence procedure of (10.69) is executed as

$$\begin{aligned} x_{i+1} &= x_i + H_i f(x_i, y_i) \\ y_{i+1} &= y_i + K_i g(y_i, x_i). \end{aligned} \tag{10.73}$$

In the above scheme, the ith iteration cycle is started with the estimates x_i, y_i to the solution and yields immediately the new estimates x_{i+1}, y_{i+1} with the aid of the iteration matrices H, K. It may be confirmed that convergence of (10.73) is equivalent to the solution of the original equations system (10.47). Also, since the two sets of equations can be treated simultaneously, the term Jacobi iteration is justified here as well.

Introducing the deviations d, e from the solution x, y in (10.73) we deduce the relation

$$\begin{bmatrix} d_{i+1} \\ e_{i+1} \end{bmatrix} = \begin{bmatrix} I + HF_x & HF_y \\ KG_x & I + KG_y \end{bmatrix} \begin{bmatrix} d_i \\ e_i \end{bmatrix} \tag{10.74}$$

between consecutive iterations. In (10.74), the amplification matrix for the deviations from the solution appears to be

$$\bar{M}_J = \begin{bmatrix} I + HF_x & HF_y \\ KG_x & I + KG_y \end{bmatrix} \tag{10.75}$$

and convergence requires its spectral norm to be less than unity

$$\|\bar{M}_J\| < 1. \tag{10.76}$$

The amplification matrix $\bar{M}_J$, (10.75), of the present scheme differs in general from the amplification matrix M_J, (10.67), of the former procedure. If, however, the recurrence formulae (10.73) are operated in conjunction with

$$H = -F_x^{-1}, \qquad K = -G_y^{-1} \tag{10.77}$$

which are the iteration matrices in the respective Newton procedures, then $\bar{M}_J$, and M_J, are equivalent.

10.4.2 Gauss–Seidel iteration

The solution of the equation system (10.47) may alternatively be effected by the application of a Gauss–Seidel procedure, cf. [17, 18] with respect to the vector variables x, y. We define this algorithm as

$$f(x_{i+1}, y_i) = 0, \qquad g(y_{i+1}, x_{i+1}) = 0 \tag{10.78}$$

where the iteration is started with an estimate to one of the sets of variables, say y_i. The first equation is then solved for the other set of variables, x_{i+1}, and this information is used in the second equation which provides the new estimate y_{i+1}.

The communication of the intermediate result necessitates here a sequential treatment of the two matrix equations in (10.78) in contrast to a possible parallel execution of the Jacobi algorithm (10.63).

For a study of the development of the deviations d, e from the solution during the course of the Gauss–Seidel iteration we first approximate (10.78) by

$$\begin{aligned} f(x + d_{i+1}, y + e_i) &\cong F_x d_{i+1} + F_y e_i = 0 \\ g(y + e_{i+1}, x + d_{i+1}) &\cong G_y e_{i+1} + G_x d_{i+1} = 0 \end{aligned} \tag{10.79}$$

and derive subsequently the relation

$$\begin{bmatrix} d_{i+1} \\ e_{i+1} \end{bmatrix} = \begin{bmatrix} O & -F_x^{-1} F_y \\ O & G_y^{-1} G_x F_x^{-1} F_y \end{bmatrix} \begin{bmatrix} d_i \\ e_i \end{bmatrix}. \tag{10.80}$$

The amplification matrix for the deviations from the solution in the Gauss–Seidel type of iteration is

$$M_{GS} = \begin{bmatrix} O & -F_x^{-1} F_y \\ O & G_y^{-1} G_x F_x^{-1} F_y \end{bmatrix} \tag{10.81}$$

and convergence is restricted to coupled problems where

$$\|M_{GS}\| < 1. \tag{10.82}$$

The above condition for the spectral norm has to be supplemented by the requirement for a solution of the two non-linear matrix equations in (10.78). For this purpose, the solution procedure of (10.69) may be interpreted here as

$$\begin{aligned} x_{j+1} &= x_j + H_j f(x_j, y_i) \\ y_{k+1} &= y_k + K_k g(y_k, x_{i+1}) \end{aligned} \tag{10.83}$$

where x_{i+1} in the second equation denotes the convergent solution of the first equation. The requirement for convergence of (10.83) is again given by (10.72) but the matrices are evaluated at a different state of the variables x, y.

A simplified version of the Gauss–Seidel type of the method of subsequent solutions evolves directly from (10.83). It may be symbolized as

$$x_{i+1} = x_i + H_i f(x_i, y_i)$$
$$y_{i+1} = y_i + K_i g(y_i, x_{i+1}). \tag{10.84}$$

The single iteration loop starts the ith iteration cycle with the values x_i, y_i as estimates. The new estimate x_{i+1} for the one set of variables is provided by the first vector equation in (10.84) and is used for the computation of the new estimate y_{i+1} for the other set of variables by the second equation.

A relation for the deviations d, e of consecutive iterates from the exact solution x, y may be derived from (10.84) and reads

$$\begin{bmatrix} d_{i+1} \\ e_{i+1} \end{bmatrix} = \begin{bmatrix} I + HF_x & HF_y \\ KG_x[I + HF_x] & I + KG_y + KG_x HF_y \end{bmatrix} \begin{bmatrix} d_i \\ e_i \end{bmatrix}. \tag{10.85}$$

The amplification matrix of the present Gauss–Seidel type of algorithm

$$\bar{M}_{GS} = \begin{bmatrix} I + HF_x & HF_y \\ KG_x[I + HF_x] & I + KG_y + KG_x HF_y \end{bmatrix} \tag{10.86}$$

is in general different from that of (10.81) appertaining to the previous Gauss–Seidel technique. If, however, the iteration matrices H and K are chosen in accordance with the Newton method, cf. (10.77), the amplification matrices of the two Gauss–Seidel version are equivalent.

The requirement for convergence is again based on the spectral norm of the amplification matrix and may be expressed here as

$$\|\bar{M}_{GS}\| < 1. \tag{10.87}$$

10.4.3 Comparison of solution procedures

For a closer examination of the convergence properties of the proposed iteration schemes for the solution of coupled problems we consider in the following a system of two scalar equations

$$f(x, y) = 0, \quad g(y, x) = 0 \tag{10.88}$$

instead of the coupled vector equations of (10.47). The partial derivatives of the non-linear functions f and g with respect to the variables x and y are denoted here by

$$f_x = \frac{\partial f}{\partial x}, \qquad f_y = \frac{\partial f}{\partial y}, \qquad g_y = \frac{\partial g}{\partial y}, \qquad g_x = \frac{\partial g}{\partial x}. \tag{10.89}$$

Application of the complete block Jacobi iteration of (10.63) to the solution of the scalar system (10.88) reveals the 2×2 amplification matrix of the process

$$M_J = -\begin{bmatrix} 0 & f_y/f_x \\ g_x/g_y & 0 \end{bmatrix} \tag{10.90}$$

which represents a special form of (10.67). The eigenvalues of the amplification matrix in (10.90) are given by

$$\lambda_J = \pm \left(\frac{g_x f_y}{g_y f_x} \right)^{1/2} \tag{10.91}$$

and therefore the requirement for convergence (10.68) for the coupling iteration may be expressed as

$$|\lambda_J|_{\max} = \left| \frac{g_x f_y}{g_y f_x} \right|^{1/2} < 1. \tag{10.92}$$

In addition to (10.92), convergence of the iterative solution in the subdomain of each equation as by (10.69) demands that the conditions (10.72) are fulfilled. Consequently, one obtains

$$-1 < 1 + hf_x < 1, \qquad -1 < 1 + kg_y < 1 \tag{10.93}$$

where the scalar quantities h and k replace in the present case the iteration matrices H and K respectively in (10.69). Following conditions (10.92) and (10.93) the numerical properties of the above Jacobi technique depend on the parameters

$$p_x = 1 + hf_x, \qquad p_y = 1 + kg_y, \qquad p_{xy} = \frac{g_x f_y}{g_y f_x}. \tag{10.94}$$

They limit the applicability of the solution scheme within the cube indicated in Figure 10.4. Obviously, p_x and p_y are characteristic of the numerical iterative solution for the individual equations in (10.88) for x and y respectively, while p_{xy} defines a physical coupling parameter.

Alternatively, the non-linear equation system (10.88) may be solved by the single step alternating Jacobi iteration defined in (10.73). Thereby the values of the variables are exchanged at the end of each iteration cycle at the subdomain level. The amplification matrix related to the evolution of the error in this process is given by (10.75) and reduces in the scalar case considered here to

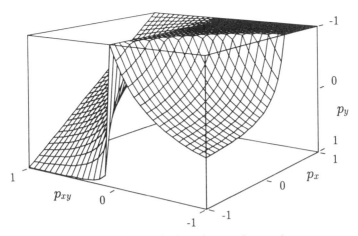

Figure 10.4 Domain of convergence for Jacobi algorithm. Single-step alternating iteration vs. complete block iteration (cube).

$$\bar{M}_j = \begin{bmatrix} 1 + hf_x & hf_y \\ kg_x & 1 + kg_y \end{bmatrix}. \quad (10.95)$$

In determining the eigenvalues of the amplification matrix (10.95) we make use of the abbreviations p_x, p_y of (10.94) and introduce in addition the parameter $\bar{p}_{xy}$ as

$$\bar{p}_{xy} = hf_y kg_x = (1 - p_x)(1 - p_y)p_{xy}. \quad (10.96)$$

This new parameter is related to the coupling between the equations (10.73) and is expressed in terms of the former parameters p_x, p_y and p_{xy} by simply exploiting the definitions in (10.94). The expression for the eigenvalues of the amplification matrix (10.95) may thus be written as

$$2\bar{\lambda}_j = (p_x + p_y) \pm [(p_x - p_y)^2 + 4\bar{p}_{xy}]^{1/2} \quad (10.97)$$

and in conformity with (10.76) the single condition for convergence of (10.73) reads

$$|\bar{\lambda}_j|_{\max} < 1. \quad (10.98)$$

On the basis of the requirement (10.98) in conjunction with expression (10.97) a surface can be defined in terms of the parameters p_x, p_y, p_{xy} which limits the applicability of the single step Jacobi method. The limiting surface is depicted in Figure 10.4, and may be compared with the cube appertaining to the complete iteration. It is seen that for the present procedure the domain of convergence is larger with respect to negative values of the coupling parameter $p_{xy} < 0$. On the other hand the method is more restrictive with respect to negative values of both parameters p_x and p_y. The latter effect appears along the negative as well as the positive p_{xy}-axis and is more pronounced for positive values of the coupling parameter $p_{xy} > 0$.

We next turn our attention to the complete block Gauss–Seidel procedure (10.78) as applied to the solution of the scalar equation system (10.88). The associated amplification matrix (10.81) assumes here the form

$$M_{GS} = \begin{bmatrix} 0 & -f_y/f_x \\ 0 & g_x f_y/g_y f_x \end{bmatrix} \qquad (10.99)$$

and possesses the eigenvalues zero and

$$\lambda_{GS} = g_x f_y / g_y f_x = p_{xy}. \qquad (10.100)$$

In accordance with the criterion (10.82), convergence is restricted to coupled problems where $|\lambda_{GS}|_{max} < 1$, i.e.

$$-1 < p_{xy} < 1 \qquad (10.101)$$

Besides, following (10.93), the conditions

$$-1 < p_x < 1, \qquad -1 < p_y < 1 \qquad (10.102)$$

ensure convergence of the iterative solution of the scalar equations (10.88) by the method of (10.83). The domain of applicability of the Gauss–Seidel procedure as limited by (10.101), (10.102) in the p_x, p_y, p_{xy}-space (cf. Figure 10.5) is the same cube as in the first Jacobi method, Figure 10.4. The maximum eigenvalue of the amplification matrix M_{GS}, (10.100), appears however to be less than the corresponding eigenvalue of M_J, (10.91), appertaining to the Jacobi method and convergence of the Gauss–Seidel iteration should therefore proceed faster.

The simplified Gauss–Seidel solution of the non-linear equation system (10.88) follows the recurrence formulae of (10.84). The evolution of the error in this scheme is given by (10.85). Interpretation for the present scalar case reveals in place of (10.86) the amplification matrix

$$\bar{M}_{GS} = \begin{bmatrix} 1 + hf_x & hf_y \\ kg_x(1 + hf_x) & 1 + kg_y + kg_x hf_y \end{bmatrix} \qquad (10.103)$$

for the error during iteration. The eigenvalues of the amplification matrix (10.103) are given by the expression

$$2\bar{\lambda}_{GS} = (p_x + p_y + \bar{p}_{xy}) \pm [(p_x + p_y + \bar{p}_{xy})^2 - 4p_x p_y]^{1/2} \qquad (10.104)$$

and according to (10.87) convergence of the single step Gauss–Seidel iteration scheme requires the condition

$$|\bar{\lambda}_{MS}|_{max} < 1 \qquad (10.105)$$

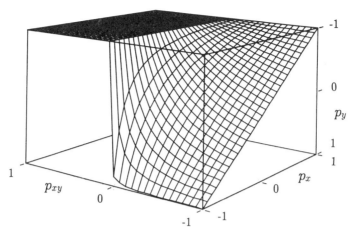

Figure 10.5 Domain of convergence for Gauss–Seidel algorithm. Single step alternating iteration vs. complete block iteration (cube).

to be fulfilled. The requirement (10.105) defines in conjunction with the expression (10.104) a surface limiting the applicability of the simplified Gauss–Seidel technique in the p_x, p_y, p_{xy}-space. This surface is shown in Figure 10.5 together with the cube limiting the applicability of the previous complete block Gauss–Seidel procedure. It may be concluded from the figure that the domain of convergence is enlarged by the present method with respect to the negative values of the coupling parameter $p_{xy} < 0$. On the other hand the method poses restrictions additional to the complete Gauss–Seidel iteration with respect to negative values of p_x, p_y along the negative p_{xy}-axis whilst for $p_{xy} > 0$ the domain of convergence remains unaltered.

In conclusion, single step alternating iteration reduces the domain of convergence of both the Jacobi and the Gauss–Seidel type of solution of the coupled non-linear equations. This reduction appears when the numerical iteration parameters p_x, p_y are negative. Thereby, the Gauss–Seidel procedure is affected only if the physical coupling parameter p_{xy} possesses negative values, while negative and positive values of p_{xy} may be concerned in the case of the Jacobi method of solution.

10.4.4 Thermomechanical coupling

The present discussion deals with the following model problem. A uniaxial specimen subject to prescribed stress is considered in conjunction with different constitutive laws relevant to metal forming. For simplicity, deformations are assumed here entirely irreversible and the process adiabatic. Under these conditions the local balance of energy reads

$$\rho c \dot{T} = \sigma \dot{\delta} \qquad (10.106)$$

where ρ, c denote the density and the specific heat capacity of the material respectively, T the absolute temperature, σ the uniaxial stress and δ the corresponding rate of deformation.

If the material can be considered viscous without strain hardening, the stress follows a functional dependence of the form

$$\sigma = \varphi(\delta, T) \tag{10.107}$$

and is considered prescribed in what follows while temperature changes are governed by (10.106).

The thermomechanical problem described by (10.107) and (10.106) is originally uncoupled with respect to the rate quantities δ and $\dot{T}$. Implicit integration for the temperature as by

$$T = {}^aT + (1 - \zeta)\tau{}^a\dot{T} + \zeta\tau\dot{T} \tag{10.108}$$

for instance, couples the equations. In (10.108) ${}^aT, {}^a\dot{T}$ denote the values at the beginning of the time interval τ and $T, \dot{T}$ those at the end, cf. (10.4).

The coupling parameter p_{xy} may be evaluated for the present variables $\delta, \dot{T}$ in accordance with the definition in (10.94). To this end, the explicit equations (10.107) and (10.106) have first to be brought into the implicit form of (10.88) for f and g. The expression for the coupling parameter is then derived for a prescribed stress as

$$p_{\delta\dot{T}} = \zeta\tau \frac{\sigma}{\rho c} \frac{-\partial\varphi/\partial T}{\partial\varphi/\partial\delta} > 0. \tag{10.109}$$

Since the stress usually increases with the rate of deformation but decreases with increasing temperature, the coupling parameter given by (10.109) is a positive quantity. It vanishes if the integration for the temperature is explicit ($\zeta = 0$) as pointed out before. For positive values of the coupling parameter, the domain of convergence is theoretically the same for both the complete and the simplified Gauss–Seidel method, Figure 10.5. The two versions of the Jacobi method, on the other hand, exhibit different properties, Figure 10.4. In the latter method, the single-step alternating iteration demands some more caution with respect to the solution of the individual equations.

We consider next the case of a rigid-plastic material exhibiting strain dependence. While equation (10.106) for the temperature rate is still valid, the stress reads here

$$\sigma = \varphi(\gamma, T). \tag{10.110}$$

The permanent strain is denoted by γ and may be obtained by an approximate integration in time of the rate of deformation δ, as

$$\gamma = {}^a\gamma + (1 - \zeta)\tau{}^a\delta + \zeta\tau\delta. \tag{10.111}$$

The temperature T and its time rate $\dot{T}$ are linked by the analogous expression in (10.108). With the aid of the integration schemes (10.111) and (10.108) the rate

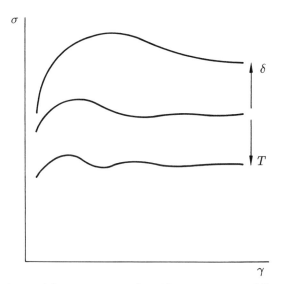

Figure 10.6 Dependence of flow stress σ on logarithmic strain γ at different temperature T and rates of deformation $\dot{\delta}$ (steel material at elevated temperature; schematic).

quantities $\dot{\delta}$ and $\dot{T}$ may be retained as the variables of the coupled problem. The coupling parameter is then given in the present case of a rigid plastic material under prescribed stress by the expression

$$p_{\delta \dot{T}} = \frac{\sigma}{\rho c} \frac{-\partial \varphi / \partial T}{\partial \varphi / \partial \gamma}. \qquad (10.112)$$

For a discussion of (10.112) we notice that the flow stress decreases if the temperature is raised. Consequently, the coupling parameter is a positive quantity as long as the material hardens with strain. Under hot-working conditions, however, isothermal stress–strain curves may exhibit a maximum indicating that during deformation strain hardening is balanced by dynamic softening phenomena (recrystallization), which are thermally activated, cf. Figure 10.6. As the stress approaches its maximum value, the denominator in (10.112) approaches zero and the coupling parameter becomes infinite. For negative values of the denominator appearing along the softening branch of the flow curve, the coupling parameter is negative. It may be seen from Figures 10.4 and 10.5 that for negative values of the coupling parameter simplification as by single step alternating iteration reduces the domain of convergence not only of the Jacobi method but for the Gauss–Seidel method as well.

Materials exhibiting a dependence on the strain γ as well as on the rate of deformation $\dot{\delta}$ may be described by the functional dependence

$$\sigma = \varphi(\gamma, \dot{\delta}, T) \qquad (10.113)$$

for the flow stress while (10.106) still applies for adiabatic deformation processes. The total quantities γ and T appearing in (10.113) may be linked to the rate quantities

δ and $\dot{T}$ of (10.106) by the approximate integration schemes of (10.111) and (10.108). Under these conditions the coupling parameter (10.47) evaluated for a prescribed stress σ and the variables δ and $\dot{T}$ in the case of the present material model, reads

$$p_{\delta\dot{T}} = \frac{\sigma}{\rho c} \frac{-\zeta\tau\partial\varphi/\partial T}{\zeta\tau\partial\varphi/\partial\gamma + \partial\varphi/\partial\delta} \tag{10.114}$$

and comprises the previously discussed constitutive models as special cases.

By inspection of the denominator in (10.114) it may be concluded that the effect of strain softening on the coupling parameter can here be controlled by the choice of the time increment τ in conjunction with the positive rate sensitivity $\partial\varphi/\partial\delta$.

A particular form of the constitutive function in (10.113) is given by the expression

$$\sigma = ke^{m_1(T-T_0)}\delta^{m_2}\gamma^{m_3}e^{m_4\gamma} = \varphi(\gamma, \delta, T) \tag{10.115}$$

which appertains to the hot rolling application described in Section 10.6. The associated coupling parameter (10.114) is

$$p_{\delta\dot{T}} = \frac{\sigma}{\rho c} \frac{\zeta\tau(-m_1)}{\zeta\tau(m_3/\gamma + m_4) + m_2/\delta} \tag{10.116}$$

and in the presence of a rate sensitivity does not approach infinity as the prescribed stress approaches the maximum value for $\gamma = -m_3/m_4$. Expression (10.115) may be used also as a guide for the specification of the functional dependences (10.107) and (10.110) appertaining to the viscous and the plastic material model respectively.

10.5 PARALLEL PROCESSING OF COUPLED FIELD PROBLEMS

The two iteration procedures outlined in the preceding section will be considered in the following with respect to their parallelization in the context of finite element applications. The execution mode of the Jacobi iteration is:

Estimate $x_0 = x^0, y_0 = y^0$
Loop $i = 1, i_{max}$
$f(x_{i+1}, y_i) = 0 \rightarrow x_{i+1}$ (10.117a)
$g(y_{i+1}, x_i) = 0 \rightarrow y_{i+1}$ (10.117b)
Exchange x_{i+1}, y_{i+1}
End loop

The Gauss–Seidel iteration, on the other hand, always uses the most recent results within each iteration:

Estimate $x_0 = x^0, y_0 = y^0$

Loop $i = 1, i_{max}$

$$f(x_{i+1}, y_i) = 0 \rightarrow x_{i+1} \qquad (10.118a)$$

Exchange x_{i+1}

$$g(y_{i+1}, x_{i+1}) = 0 \rightarrow y_{i+1} \qquad (10.118b)$$

Exchange y_{i+1}

End loop

The Gauss–Seidel scheme is a sequential algorithm, whereas the Jacobi iteration can be used directly in the parallel computation of the two sets of variables. In both procedures, the iterative treatment of the individual vector iterations is ultimately based on the repeated solution of a linear system with varying coefficients. As exposed previously two variants of the Jacobi algorithm (10.117) and the Gauss–Seidel algorithm (10.118) may be distinguished. In the one case, the field data are exchanged only after full convergence is achieved in the individual non-linear problems (complete block solution). Alternatively, data are exchanged after one iteration pass for each part of the coupled problem (single-step iteration).

Several approaches to the problem decomposition necessary for parallelization of finite element procedures are generally feasible, two of which will be presented and compared here, namely spatial decomposition of the problem domain and physical decomposition of the coupling algorithm [11]. The considerations refer to multiprocessor computers with distributed memory.

10.5.1 Spatial decomposition

The concept of spatial decomposition is based on the distribution of the discretized problem domain onto all available processors. Each processor stores all relevant problem data only for the subdomain associated with it, plus information regarding the connectivity to neighbouring subdomains. It is required that the decomposition of the finite element model guarantees good load balancing, the computational load being determined by a number of factors including the number of degrees of freedom and the number of elements within each subdomain. For an automatic decomposition algorithm the reader may consult [19].

The application of spatial decomposition to coupled processes leaves the stream of operations of the sequential algorithm essentially unchanged and allows us to retain the Gauss–Seidel iteration procedure. The individual fields of the coupled problem are parallelized independently of each other, both being distributed across all available processors, as indicated in Figure 10.7. If a unique discretization mesh is used for both fields, no message passing is necessary between the two problems since all relevant data are held in the same processor for each subdomain; use of different meshes complicates the communication pattern.

Spatial decomposition allows several important steps of the finite element computation, such as the formation of the element matrices, which is required in every iteration of each time increment, to be carried out in parallel with very little requirement for communication and with only minor modifications of the sequential

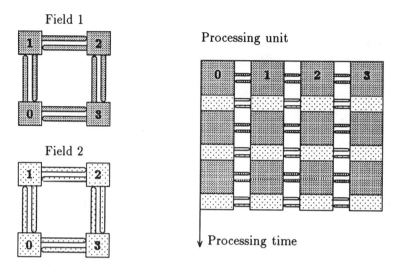

Figure 10.7 Spatial decomposition for coupled problems; fields in sequence.

code. The repeated solution of the linear system of equations, which tends to dominate the overall CPU-time for large scale applications, is more critical. For each discretized field of the coupled problem, the system assumes the general form

$$Ax = b \qquad (10.119)$$

where A is the symmetric banded matrix with different dimensions $N \times N$ for each system, x represents the vector of unknown degrees of freedom and b the vector of generalized forces.

The considerations are restricted here to the direct solution of (10.119) by two distinct algorithms, a parallel Cholesky decomposition and a substructuring procedure. In addition to the subsequent conceptual presentation, a detailed comparison and assessment of the algorithms *per se* may be found in [19].

Parallel Cholesky decomposition

A Cholesky decomposition is generally performed in three stages:

Factorization	$A = LL^T$	(10.120a)
Forward substitution	$Lu = b \to u$	(10.120b)
Backward substitution	$L^T x = u \to x.$	(10.120c)

The array L is a lower triangular matrix and is stored in the same location as A. In order to parallelize the algorithm the system matrix A has to be distributed among

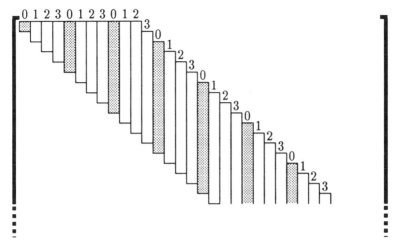

Figure 10.8 Column-wise distribution of the system matrix on four processors (0 to 3).

the processors. This is generally done in the column-wise manner shown in Figure 10.8, which ensures good load balancing during the factorization phase [20]. The assembly and distributed storage of the system matrix A requires contributions from element matrices computed in different processing units. In order to minimize the associated communication overhead, element contributions which have to be transferred are first stored in temporary buffers and sent only when these are full.

The factorization and forward substitution are performed in one step in order to avoid spurious communication. The arrays L and u are treated in a row-wise fashion, whereby for each row first the diagonal element is computed in a sequential step by the processor storing it, and then off-diagonal terms are processed in parallel. During the execution of the parallel parts of the algorithm, each processor works only on those elements it is storing and thus, as a consequence of the column-wise distribution of the data, good load balancing is achieved if the bandwidth of the matrix is large in relation to the number of processors. Also, it can be shown that the sequential portion of the procedure does not seriously affect the efficiency if the dimension N is sufficiently large in relation to the number of processors. The communication overhead, however, can by no means be neglected, as will be demonstrated in the numerical study below.

Parallel substructuring

Substructuring is a solution technique [21] which has been used for many years in large-scale finite element applications on conventional computers. It has attained new importance in the context of parallel processing, because it provides a systematic scheme for the distributed treatment of the problem domain as required by the spatial decomposition method.

An essential characteristic of substructuring is the reduction of the $N \times N$ system matrix to one comprising only degrees of freedom lying on the internal boundaries between subdomains by eliminating all other (interior) degrees of freedom. In the parallel computation, this elimination process can be executed concurrently without any need of communication. The reduced system matrix is then solved with a conventional solution procedure, for instance by the parallel Cholesky algorithm described previously.

The implementation of the method demands a separation of the interior (i) and the boundary (b) degrees of freedom, so that the system of equations appertaining to each individual subdomain assumes the form

$$\begin{bmatrix} A_{ii} & A_{ib} \\ A_{bi} & A_{bb} \end{bmatrix} \begin{bmatrix} x_i \\ x_b \end{bmatrix} = \begin{bmatrix} b_i \\ b_b \end{bmatrix}, \quad \text{where } A_{bi} = A_{ib}^{\mathrm{T}}. \tag{10.121}$$

Elimination of the interior part x_i from the system leaves for x_b the equation

$$\underbrace{[A_{bb} - A_{bi} A_{ii}^{-1} A_{ib}]}_{A_{bb}^*} x_b = \underbrace{[b_b - A_{bi} A_{ii}^{-1} b_i]}_{b_b^*} \tag{10.122}$$

which implies the formation of A_{bb}^* and b_b^*. To this end the following computation steps have to be performed in each processor:

$$A_{ii} = LL^{\mathrm{T}} \tag{10.123a}$$

$$LM_{ib} = A_{ib} \to M_{ib}, \qquad Lu_i = b_i, \quad L^{\mathrm{T}} z_i = u_i \tag{10.123b}$$

$$A_{bb}^* = A_{bb} - M_{ib}^{\mathrm{T}} M_{ib}, \qquad b_b^* = b_b - A_{bi} z_i \tag{10.123c}$$

where M_{ib} is an intermediate array of dimension (number of interior d.o.f.) × (number of boundary d.o.f.).

The individual matrices A_{bb}^* from each processor are then assembled into the reduced matrix of the global system, which at the same time is distributed periodically column by column to all processors in order that the solution for x_b at the connection boundaries is performed by the parallel Cholesky algorithm. Subsequently, the solution for x_i in the interior of each subdomain is obtained by

$$x_i = A_{ii}^{-1}[b_i - A_{ib} x_b] \tag{10.124}$$

which can again be executed independently by all processors.

10.5.2 Physical decomposition

Physical decomposition may be considered an alternative approach to the parallelization of coupled field problems. Here, the physical sub-problems are calculated concurrently on different processors (or different groups of processors) implying the use of a Jacobi iteration scheme. For instance, in the case of two individual phenomena:

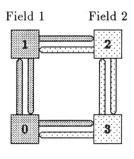 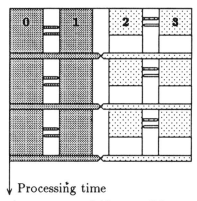

Figure 10.9 Physical decomposition on four processors; fields in parallel.

Estimate $x_0 = x^0$, $y_0 = y^0$
Loop $i = 1, i_{max}$
$f(x_{i+1}, y_i) = 0 \rightarrow x_{i+1}$ $g(y_{i+1}, x_i) = 0 \rightarrow y_{i+1}$ (10.125)
Send x_{i+1} Send y_{i+1}
Receive y_{i+1} Receive x_{i+1}
End Loop

Starting from an appropriate estimate in each time step of the evolution process, both units calculate independently a new estimate of the respective field data in each iteration cycle, as by (10.125). These data are then exchanged as required by the other part of the coupled problem.

In the most straightforward case, each of the coupled fields is assigned to exactly half of the available processing units. If a total of more than two processors is available, each field will be distributed on the group of processors assigned to it using the spatial decomposition methods described above. This is illustrated for a four-processor configuration in Figure 10.9. As indicated in the figure, load balancing problems can arise if the two distinct physical processes require different amounts of CPU-time. In some cases, this may coincide with the necessity to use a finer spatial discretization or smaller time steps for the subproblem requiring less computer resources, but in general this problem can only be alleviated by assigning a larger number of processors to the part requiring more CPU time. This requires, of course, a more complicated communication pattern.

The implied transition from a Gauss–Seidel to a Jacobi iteration procedure, leading one to expect an impaired convergence behaviour of the numerical process, may be considered another drawback of the physical decomposition. On the other hand, when applied in conjunction with spatial decomposition for the physical sub-problems it leads to a larger grain size, i.e. a larger size of the subdomains being assigned to one processor, than in pure spatial decomposition. A larger grain size, however, is

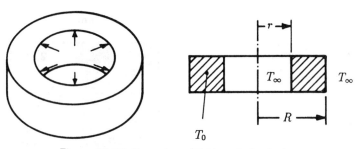

Figure 10.10 Expansion of thick-walled cylinder.

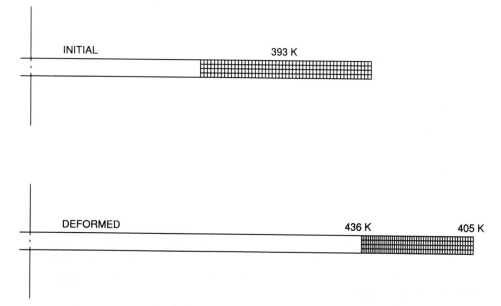

Figure 10.11 Temperature distribution at initial and deformed stage.

one of the most important factors towards good efficiency in parallel finite element computations.

10.5.3 Investigation of a test problem

The performance of the above parallel algorithms is studied on the Intel iPSC/2 hypercube with four nodes. The small test problem considered is a thick-walled cylinder (Figure 10.10) expanding under constant velocity of the inner radius [3]. The coupled problem consists of a mechanical and a thermal part, the coupling caused by a temperature dependence of the flow stress on the one hand, and deformation-dependent thermal effects, in particular the dissipation of mechanical work, on the other hand, cf. Section 10.7.

Figure 10.11 presents the temperature levels at the initial and the deformed

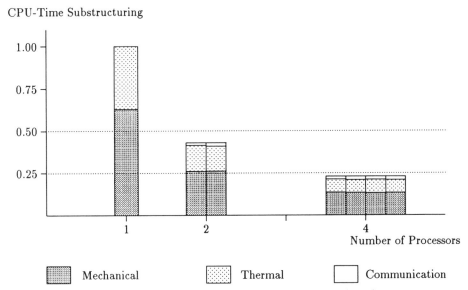

Figure 10.12 Performance of parallel substructuring algorithm.

geometry. The cylinder was discretized by 10 × 100 4-node axisymmetric elements in order to increase the size of the problem for the test. The simulation required 11 time increments in conjunction with an implicit integration ($\zeta = 1/2$) of the rate equations governing deformation and temperature. In each incremental step 2–5 sequential Gauss–Seidel iterations furnished a convergent solution (relative residuum $< 10^{-4}$).

Parallelization of the Gauss–Seidel algorithm is effected by the spatial decomposition method. Due to the regularity of the present discretization, the decomposition into subdomains is straightforward and good load balancing could be achieved by one subdivision along the axis and one along the radius of the cylinder. Figures 10.12 and 10.13 show the normalized run times on the iPSC/2 for different numbers of processors with a constant size of the discretized problem for parallel substructuring and parallel Cholesky, respectively. The diagrams are based on the overall run time of the simulation, including all steps of the algorithm except for data input and output of results. These operations were not taken into account because there was no parallel I/O facility available, and I/O performance depends therefore on a variety of factors not related to the algorithm.

The results indicate a superior performance for the parallel substructuring algorithm as compared to the parallel Cholesky solver. The parallel Cholesky solver incurs a large number of message-passing steps, which affect the efficiency of the algorithm; the associated communication time increases with the number of processors. Both parallel techniques are related to the same sequential algorithm based on an optimized band matrix solver. For this reason, the parallel substructuring algorithm — which is optimized to a degree that could not be achieved for the sequential solver — leads to apparent efficiencies (sequential execution time/number of processors × parallel execution time) greater than 1.0. Moreover, this algorithm requires minor communic-

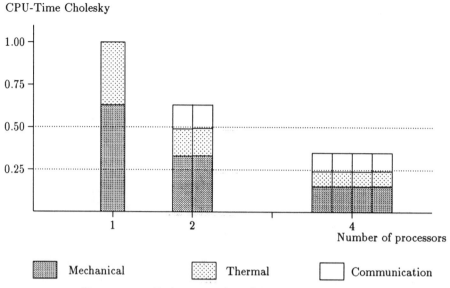

Figure 10.13 Performance of parallel Cholesky algorithm.

ation due to the assembly and solution of a reduced system matrix.

The test problem is alternatively computed by an application of the physical decomposition, which implies a transition from the Gauss–Seidel to the Jacobi iteration of the coupled fields. In order to investigate the effect of this transition on the convergence behaviour of the algorithm, the parallel calculations were performed on two processors, each assigned one field of the coupled problem. Figure 10.14 summarizes the convergence behaviour, both for complete block and for single step iterations. It is seen that in each increment the Jacobi algorithm requires 1–2 additional mechanical iterations and 1–3 additional thermal iterations. Over the 11 increments of the entire analysis the number of cycles is increased by 15% in the mechanical and 27% in the thermal problem.

The performance of the overall computation with physical decomposition on a two-processor configuration is illustrated in Figure 10.15. The efficiency is clearly reduced due to the fact that the thermal part of the analysis requires less than 50% of the CPU time of the mechanical part, and that consequently the respective processor is idle for large intervals of time. Complete block solution and single step iteration require approximately the same amount of CPU time, but the more frequent exchange of data increases the communication overhead by approximately 8% in the single step iteration method. In order to utilize the idle time of the thermal processor, the time increment of the thermal part which appears to be more highly non-linear, was reduced to one half of the increment length of the mechanical part, and data were exchanged only after every other thermal increment. The result is shown on the right-hand side of Figure 10.15 and confirms that the idle time of the thermal processor is reduced considerably to the benefit of the numerical solution. The improvement of the overall run time of the coupled problem is of course only marginal.

10.6 FINITE ELEMENT MODEL

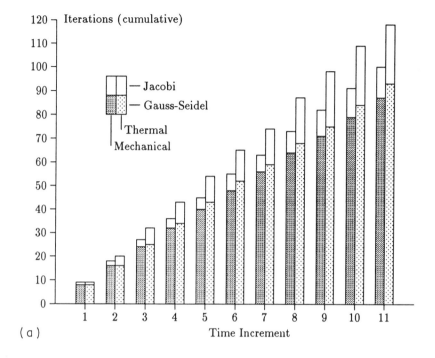

(a)

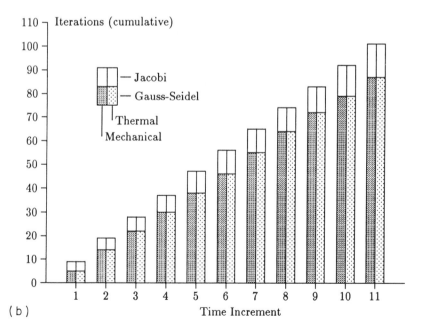

(b)

Figure 10.14 Convergence behaviour of Jacobi vs. Gauss–Seidel algorithm: (a) complete block solution, (b) single-step iteration.

336 COUPLED FIELD PROBLEMS

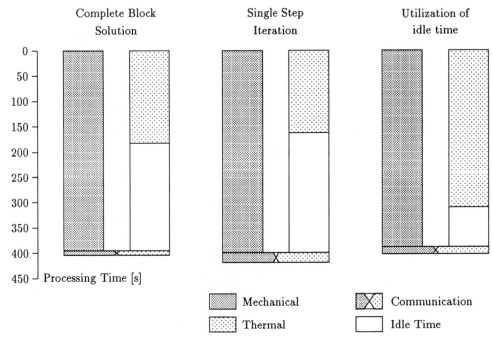

Figure 10.15 Utilization of processors during parallel computation of fields.

10.6 FINITE ELEMENT MODEL

10.6.1 *Preparation and handling of mesh*

In large scale simulations the description of the input for the computation may constitute a considerable part of the overall effort. With respect to this, the availability of a mesh generation procedure for arbitrary geometries appears particularly helpful. A systematic scheme for mesh generation developed on the basis of the multi-block concept was described in [22]. The structure of the original algorithm could be utilized in an extended version for parallel computing, for the balanced decomposition of the discretized domain on the block level [19]. This allows for concurrent generation of the ultimate finite element mesh on all available processors such that the subsequent finite element computations can be started immediately in parallel.

An important issue in process simulations is the modification of the discretization in compliance with the requirements of the evolving numerical solution. Adaptive mesh modification may be effected by simply displacing nodal points in conjunction with an error indicator, cf. [23, 24], or by the appropriate addition elements, cf. [25] for instance. A discussion of adaptivity in connection with the parallel algorithms presented here may be found in [19].

A specific task related to the discretization of coupled field problems will be addressed below and concerns the possible use of individual discretizations for each field. In the computer implementation this mainly affects the exchange of information between the interacting physical phenomena and consequently complicates the communication patterns of the parallel execution.

10.6.2 Modelling of coupled processes

In coupled processes, the intensity of the interaction between the physical phenomena may locally vary within the problem domain in the course of time. Accordingly, the interacting field variables may develop differently requiring individual discretizations, both in space and time in the computation. While the incrementation in time was considered before in Section 10.2, the subject here is the discretization of a domain by a finite element mesh. In this respect, some remarks may be appropriate in connection with thermomechanically coupled deformation processes, as an example. Thermomechanical coupling is most pronounced in the limiting case of adiabatic conditions. However, the local temperature is then a consequence of the deformation only and does not require the solution of a thermal conduction problem. Therefore initial discretization and adaptive improvement of the finite element mesh can be based on the mechanical solution. The deformation field is decisive also when temperature gradients are relaxed and the thermal process approaches isothermal conditions. Between the above two limiting cases, deformation and temperature may, however, develop completely differently requiring individual discretization meshes in the computation. This will be demonstrated in the following by means of an example.

An axisymmetric, initially cylindrical specimen with a central bore is subject to upsetting at elevated temperature. For the numerical analysis half of the symmetric problem is modelled, the action of the tool is simulated by the imposition of a unique vertical velocity on the upper surface of the specimen and the suppression of all radial velocities there. Thermal phenomena are activated by the contact with the cold tool, the heat exchange with the surrounding air and by the dissipation of mechanical work. The determination of the temperature distribution is required for the evaluation of the flow stress throughout the course of the deformation process.

The upsetting process is simulated up to a reduction of the initial height of the specimen by approximately one half. The shape of the specimen at this stage is shown in Figure 10.16 as a result of the computation, together with the distribution of the equivalent strain. It is seen that the most intense deformations of the material appear in the central region and at the edge. Consequently, if the finite element mesh is adapted to the deformation, nodal points are moved to the above regions (Figure 10.16, left). Temperature gradients, on the one hand, are more pronounced as the region contacting the tool, Figure 10.17, so that modification of the mesh for the requirements of the thermal solution attracts nodal points to this part of the specimen (Figure 10.17, right). Figures 10.16 and 10.17 demonstrate to which extent the different modifications of the mesh affect the mechanical and the thermal solution of the coupled problem. It may be concluded that in the present demonstration significant temperature inhomogeneities appear up to the final stage of the upsetting, which cannot be appropriately modelled by a mesh designed for the mechanical process.

10.6.3 Individual discretization of the physical sub-problems

An assessment under the above point of view of the solution methods presented in Sections 10.3 and 10.4 reveals that the treatment of the coupled process as a single

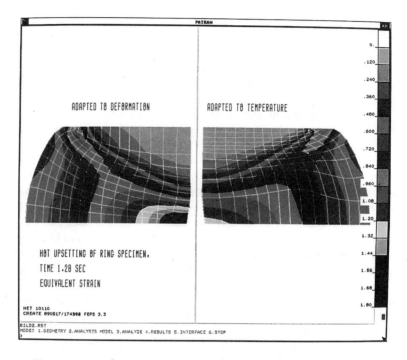

Figure 10.16 Hot upsetting of ring specimen. Equivalent strain for alternative modifications of the mesh.

problem along the lines of Section 10.3 does not favour the use of individual discretizations of the interacting fields. If the compound repesentation of the problem by (10.48) emanates from coupled finite elements uniquely responding to both variables [26], an individual discretization is not possible. On the other hand, the collective arrays in (10.48) may be associated with individual contributions from the interacting fields treated on separate finite element meshes in the following way

$$f(x_f, y_f) = 0, \qquad g(y_g, x_g) = 0. \tag{10.126}$$

Here the transition from one discretization mesh to the other requires an interpolation of the variables as by

$$y_f = E_{fg}(y_g), \qquad x_g = E_{gf}(x_f) \tag{10.127}$$

where E_{fg}, E_{gf} symbolize the transition operation from mesh f to mesh g and vice versa. In (10.49), we have then to consider the arrays

$$Y = \begin{bmatrix} x_f \\ y_g \end{bmatrix}, \qquad F = \begin{bmatrix} f(x_f, E_{fg}(y_g)) \\ g(y_g, E_{gf}(x_f)) \end{bmatrix} \tag{10.128}$$

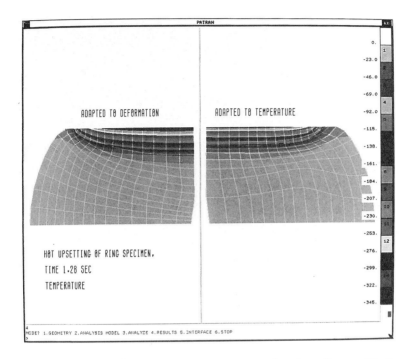

Figure 10.17 Hot upsetting of ring specimen. Temperature for alternative modifications of the mesh.

instead of those in (10.48). In the solution process of (10.49) for Y by the methods of Section 10.3, the relation between the individual meshes as established by the operators E_{fg}, E_{gf} is necessary for the evaluation of $F(Y)$ in the course of the iteration, cf. (10.50), and for the formation and approximation of the coupling entities F_y, G_x in the gradient matrix of the system, cf. (10.56).

An application of the physical subdomain solutions outlined in Section 10.4 in conjunction with individual discretization meshes modifies the Gauss–Seidel iteration algorithm of (10.118) for the sequential iteration of the interacting fields as follows:

$$\text{Loop } i = 1, i_{\max}$$
$$\text{Mesh } f: \ f(x_{i+1}, E_{fg}(y_i)) = 0 \rightarrow x_{i+1}; E_{gf}(x_{i+1})$$
$$\text{Exchange } E_{gf}(x_{i+1})$$
$$\text{Mesh } g: \ g(y_{i+1}, E_{gf}(x_{i+1})) = 0 \rightarrow y_{i+1}; E_{fg}(y_{i+1})$$
$$\text{Exchange } E_{fg}(y_{i+1})$$
$$\text{End Loop}$$

Analogously, the Jacobi algorithm of (10.125) for the iteration of the fields in parallel reads:

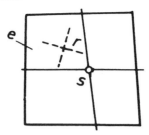

Figure 10.18 Interpolation technique. Location of nodal point r of mesh g in mesh f (nodal point s, element e).

Loop $i = 1, i_{max}$

Mesh f:
$f(x_{i+1}, E_{fg}(y_i)) = 0 \to x_{i+1}$
$E_{gf}(x_{i+1})$
Send $E_{gf}(x_{i+1})$
Receive $E_{fg}(y_{i+1})$
End Loop

Mesh g:
$g(y_{i+1}, E_{gf}(x_i)) = 0 \to y_{i+1}$
$E_{fg}(y_{i+1})$
Send $E_{fg}(y_{i+1})$
Receive $E_{gf}(x_{i+1})$

Both algorithms differ from their original version merely by the interpolation of the coupling variables for the transition from one mesh to the other which has to be carried out prior to the exchange between the equations governing the interacting physical phenomena. In order to refer the solution obtained by the one mesh to the other mesh, the procedure outlined in [23] may be applied. To this end let the field quantities x be known at the nodal points of mesh f. The position of the nodal points of mesh g is given by the respective coordinates. In addition, the transfer of the field quantities to mesh g requires a specification of the positions of its nodal points with respect to the structure of mesh f. To this purpose, a first step obtains, for each nodal point r in mesh g, the nodal point s in mesh f closest to r. Subsequently, the particular element e in mesh f connected with s is found, which contains the nodal point r of mesh g, Figure 10.18. The field quantities at r are then obtained in accordance with the finite element approximation

$$x_r = w_e(r_r)x_e. \qquad (10.129)$$

Here, the vector array x_e comprises the values at the nodal points of the element e in mesh f and w_e contains the interpolation function for this element. The value x_r at nodal point r in mesh g is obtained by an evaluation of the interpolation for the position r_r. As a matter of fact, equation (10.129) details for a single nodal point the symbolic operation on the right-hand side of equation (10.127) which refers to the entire mesh g.

In parallel computations, this step necessitated by the different discretization of the constituent fields of the coupled problem can lead to very complex communication patterns during the course of the numerical simulation [11].

10.7 APPLICATIONS

10.7.1 *Thermomechanically coupled deformation processes*

For a consideration of thermomechanical coupling in connection with applications to metal forming, the inelastic behaviour of the solid is described here within the conceptual basis of a non-linear viscous material. Such a constitutive approach is well suited for deformations at elevated temperatures [6, 14] and may adequately be used also in the context of inviscid, rigid-plastic models [27]. For alternative material descriptions considering elastic constituents, the reader may be referred to [3] and [28], for instance. The material law establishes here a relation between the Cauchy stress $\boldsymbol{\sigma}$ (6 × 1) vector and the corresponding rate of deformation $\boldsymbol{\delta}$ which is defined by the symmetric part of the velocity gradient in the solid.

Isotropic materials are characterized by the deviatoric response

$$\boldsymbol{\sigma}_D = 2\mu\boldsymbol{\delta}_D \qquad (10.130)$$

with the viscosity coefficient μ. From (10.130) a relation can be derived between the von Mises equivalent stress $\bar{\sigma}$ and rate of deformation $\bar{\delta}$ and is compared to the uniaxial flow stress of the material $\sigma = \varphi(\bar{\delta}, \gamma, T)$. This specifies the viscosity coefficient

$$\mu = \frac{\bar{\sigma}}{3\bar{\delta}} = \frac{\varphi(\bar{\delta}, \bar{\gamma}, \bar{T})}{3\bar{\delta}} \qquad (10.131)$$

as a function of the equivalent rate of deformation $\bar{\delta}$, the accumulated strain $\bar{\gamma}$ and the absolute temperature T.

Under isothermal conditions, the deformation process is considered isochoric. Temperature variations do not affect the deviatoric deformation in (10.130) but may cause volumetric changes

$$\delta_V = \beta \dot{T} \qquad (10.132)$$

in connection with β, the linear coefficient of differential thermal expansion. In the numerical analysis, the volumetric condition (10.132) must be approximately utilized for a determination of the hydrostatic stress which is not related to the deformation by a material law [6]. In connection with completely dissipative materials where the reversible deformation due to thermal expansion does not appear, it is convenient to apply a relaxed, penalty approach to the isochoric condition in the form

$$3\delta_V = \frac{1}{\bar{\kappa}}\sigma_H, \qquad \bar{\kappa} \to \infty \qquad (10.133)$$

which implies a relation for the hydrostatic stress σ_H based on the penalty factor $\bar{\kappa}$.

For a presentation of the thermodynamics of deformation processes the reader may be referred to [29]. The specific elements required for the thermomechanical

COUPLED FIELD PROBLEMS

description of the material under consideration follow the treatment in [30]. Accordingly, the local energy balance is written in the form

$$\dot{u} = \frac{1}{\rho}\boldsymbol{\sigma}^T\boldsymbol{\dot{\delta}} + \dot{q} \qquad (10.134)$$

where $u(T)$ denotes the specific internal energy of the material and $\dot{q}$ the heat supply per unit mass. The temperature is here the unique independent state variable because the irreversible rate of deformation in (10.130) is assumed entirely dissipative, whilst the reversible rate of deformation, the thermal expansion in (10.132), is a consequence of the temperature variation.

In the energy balance, the rate of mechanical work can be partitioned into hydrostatic and deviatoric contributions so that (10.134) is brought into the form

$$c\dot{T} = \frac{1}{\rho}\boldsymbol{\sigma}_D^T\boldsymbol{\dot{\delta}}_D + \dot{q} \qquad (10.135)$$

where the specific heat capacity is given by

$$c = \frac{du}{dT} - 3\beta\sigma_H = \left(\frac{\partial q}{\partial T}\right)_{\delta_D=0} \qquad (10.136)$$

For a numerical analysis of the deformation process the solid is represented by a finite element model. The geometry is then specified by the coordinates of the nodal points of the finite element mesh collected in the vector array $\boldsymbol{X}$, the applied forces accordingly form the vector $\boldsymbol{R}$. The forces resulting at the nodal points from stresses via the accumulation of elemental contributions are collected in the vector $\boldsymbol{S}$. With reference to the above definitions quasistatic deformation of the solid is governed by the vector equation

$$\boldsymbol{R}(t, \boldsymbol{X}) = \boldsymbol{S}(\boldsymbol{\sigma}, \boldsymbol{X}) = \bar{\boldsymbol{D}}(\boldsymbol{V}, \boldsymbol{X}, T)\boldsymbol{V}. \qquad (10.137)$$

In the finite element formulation of the problem, the rate of deformation in expressions (10.130) and (10.133) for the stress is a function of the nodal velocities in the vector $\boldsymbol{V}$ and of the actual geometry $\boldsymbol{X}$. This dependence transfers to the stress resultants in (10.137); their representation via the viscosity matrix $\bar{\boldsymbol{D}}$ of the system as based on the penalty approach to the isochoric condition is useful for the solution of the problem. A non-linear constitution of the viscous material, cf. (10.131), is reflected by the dependence on the velocity in (10.137), whilst the temperature sensitivity requires knowledge of the vector $\boldsymbol{T}$ comprising the temperature at the nodal points.

The temperature is governed by the energy balance (10.135) which in the finite element formulation of the thermal problem leads to the equation system

$$\boldsymbol{C}(\boldsymbol{T}, \boldsymbol{X})\dot{\boldsymbol{T}} + \boldsymbol{L}(\boldsymbol{T}, \boldsymbol{X})\boldsymbol{T} = \dot{\boldsymbol{Q}}(\boldsymbol{V}, \boldsymbol{T}, \boldsymbol{X}). \qquad (10.138)$$

In (10.138), $\boldsymbol{C}$ denotes the heat capacity matrix, $\boldsymbol{L}$ the heat conductivity matrix and the vector $\dot{Q}$ accounts for the externally applied thermal actions and dissipative mechanical contributions introducing a dependence on the velocity V. The thermal problem is stated at the deformed geometry X of the solid.

Unsteady processes call for an application of incremental integration techniques for the numerical treatment of the equilibrium equation (10.137) and the thermal equation (10.138) in the course of time. Thereby, starting with a known thermomechanical state of the solid at time $t = {}^a t$, the next state at time $t = {}^a t + \tau = {}^b t$ is obtained so that the mechanical equilibrium condition and the thermal equation are satisfied at the same instant for externally applied actions advanced in accordance with the time increment, and for the new stresses, temperature and geometry of the solid.

Geometry X and velocity V appear as unknown variables in unsteady problems and are linked via an approximate integration of the velocity within the time increment as by (10.4), which here reads

$$X = {}^a X + (1 - \zeta)\tau {}^a V + \zeta \tau {}^b V = {}^b X. \tag{10.139}$$

Analogously for the temperature T,

$$T = {}^a T + (1 - \zeta)\tau {}^a \dot{T} + \zeta \tau {}^b \dot{T} = {}^b T. \tag{10.140}$$

An explicit integration scheme ($\zeta = 0$) for geometry and temperature requires the rates at the beginning of the time increment $t = {}^a t$. In this case the equilibrium equation (10.137) is stated for $V = {}^a V$ with $X = {}^a X$ and $T = {}^a T$ known, and may be solved independently of the thermal equation (10.138). The latter receives the resulting velocity and provides then the temperature rate $\dot{T} = {}^a \dot{T}$ with all other quantities fixed. In the present case, the absence of the temperature rate in the equation for mechanical equilibrium decouples the problem as outlined in Section 10.2, cf. equation (10.6).

If the integration is implicit ($\zeta > 0$), the governing equations are considered at the end of the time increment $t = {}^b t$ with $X = {}^b X = X({}^b V)$ and $T = {}^b T = T({}^b \dot{T})$ in accordance to (10.139) and (10.140); the interacting variables are then $V = {}^b V$ and $\dot{T} = {}^b \dot{T}$. The coupled thermomechanical problem governed by the system of equations (10.137) and (10.138) may be solved by parts in conjunction with the iterative coupling algorithms of Section 10.4; in particular, the sequential Gauss–Seidel technique is frequently applied. The non-linear solution of the individual equations as based on the Newton-type scheme described before, makes use of the system matrices as approximation to the gradient matrix, whilst the associated terms appearing in the residual vectors are assembled directly from elemental contributions. To be specific, the iterative solution of (10.137) for the velocity $V = {}^b V$ is based on the viscosity matrix $\bar{D}$, the iteration of (10.138) for the temperature rate $\dot{T} = {}^b \dot{T}$ is based on a combination of the heat capacity matrix C and the heat conductivity matrix L, motivated by the introduction via (10.140) of the rate $\dot{T} = {}^b \dot{T}$ as an unknown variable instead of the temperature $T = {}^b T$ in the conduction term, cf. [6].

In certain cases, only the temperature requires an implicit integration, whilst the deformation problem can be treated explicitly. Within a common incremental step, the velocity $V = {}^a V$ may then be obtained from (10.137) without any reference to

(10.138) and is transferred to the thermal equation for the computation of the temperature rate $\dot{T} = {}^b\dot{T}$ in conformity with the implicit algorithm; an iterative coupling is not needed.

Stationary situations are characterized by the appearance of the velocity V and the temperature T as the interacting variables in the thermomechanical problem. The coupled equations (10.137) and (10.138) reduce then to the system

$$R - \bar{D}(V, T)V = 0$$

$$\dot{Q}(V, T) - L(T)T = 0 \qquad (10.141)$$

which is written in the residual form of (10.47) and may be solved by parts in conjunction with coupling by iteration as outlined in Section 10.4.

As a consequence of the Lagrangian description in (10.137) and (10.138), the equation system (10.141) refers to stationary conditions from the material point of view. Therefore, if stationary conditions are alternatively defined with respect to the space temporarily occupied by the solid, the conductivity matrix has to be supplemented by a convection term which is non-symmetric and introduces an additional dependence on the velocity [6].

10.7.2 Hot rolling of a rail

A first application of the computational methodology deals with the numerical simulation of the hot rolling of a rail from steel material originally reported in [31]. This rolling process necessitates a complete three-dimensional analysis accounting for the physical coupling between thermal and mechanical phenomena. In particular the initial unsteady phase of a rolling pass is followed up to the appearance of stationary conditions in the deformation zone.

The hot rolling process considered here has been the subject of extensive investigations from the experimental point of view in [32]. The laboratory model corresponds to a $1:3.2$ reduction in scale of the actual geometry of the cross-section of the rail. The initially quadratic cross section of a slab ($90 \times 90 \times 500$ mm) is formed into the profile of a rail by a sequence of rolling passes interrupting the process.

The present computer simulation of the hot rolling process deals with the first step in the final twin pass which leads to the ultimate cross-section. In the calculation both the rollers and the rail are discretized by finite elements. The rollers are assumed to behave rigidly and impose the geometric boundary conditions on the work-piece material [23]. The discretized description of the surface of the rollers by 1717 plane quadrilateral elements (1843 grid points) is depicted in Figure 10.19. The domain shown in the figure proved to be sufficient for an adequate consideration of the boundary conditions.

The shape of the rail prior to the rolling pass is indicated by the first of the three-

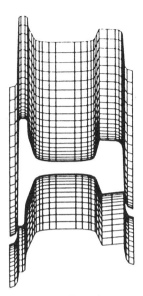

Figure 10.19 Hot rolling of a rail. Discretization of rollers.

dimensional views in Figure 10.20. At this stage, the cross-section is discretized by the application of the mesh generation procedure of [22]. The resulting mesh forms the basis for the twenty layers along 100 mm of the longitudinal axis of the rail. This mesh comprising 3340 hexahedral elements (8 nodes) connected at 4576 nodal points is used both for the simulation of the deformation process and for the thermal phenomena. Initially, the temperature distribution is homogeneous throughout the rail with $^{0}T = 1273$ K. It is cooled down in air at $T_{\infty} = 293$ K during a time period of $t = 6.59$ s. This phase of the process is governed solely by thermal phenomena. Heat exchange with the exterior takes place by convection and radiation. Both effects are suitably modelled in the computation by radiation with an emission coefficient $\varepsilon = 0.4$. The specific heat capacity of the steel material is assumed to depend on the temperature with the relation

$$c = 590 + 0.275(\theta - 800) \quad \text{J/kg K} \tag{10.142}$$

where θ denotes the temperature in °C. The thermal conductivity is taken as

$$\lambda = 2.24 \times 10^{-8}\theta^2 - 5.09 \times 10^{-5}\theta + 5.32 \times 10^{-2} \quad \text{W/mm K}. \tag{10.143}$$

A numerical analysis of the cooling process provides the temperature distribution in the rail prior to the rolling pass, which represents the initial thermal condition for the subsequent deformation process.

The computation of the actual hot rolling pass refers to the theoretical background of thermodynamically coupled deformation processes given in Section 10.7.1. In the present case the work-piece material may be described adequately within the conceptual basis of a rigid viscous solid exhibiting strain dependence. A specification

346 COUPLED FIELD PROBLEMS

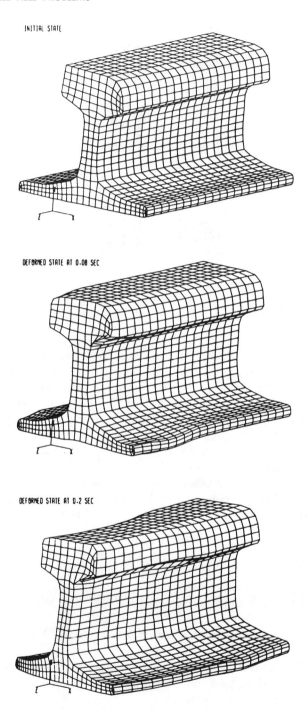

Figure 10.20 Stages of rolling process.

of the relevant material characteristics requires the flow stress of the material, which is given by the relation

$$\sigma = k e^{m_1 \theta} \delta^{m_2} \gamma^{m_3} e^{m_4 \gamma} \quad \text{N/mm}^2 \tag{10.144}$$

with $k = 4150$, $m_1 = -0.0031$, $m_2 = 0.1202$, $m_3 = 0.2889$, $m_4 = -0.5006$, and depends not only on the strain γ but also on the temperature θ and on the rate of deformation δ. While (10.144) is used for the evaluation of the deviatoric viscosity coefficient μ, a volumetric viscosity coefficient $\kappa = 10^5$ is taken in conjunction with the penalty approach to the isochoric condition for viscous deformations. In this context, the volumetric response of the finite element is evaluated by the application of a reduced integration rule [33].

The action of the rollers is transmitted to the work-piece material via the friction forces on the contact surface [23]. In this context, the Coulomb law is applied with a friction coefficient $c_f = 0.3$. Furthermore, a possible rigid motion downwards of the undeformed part of the rail is prohibited by appropriate kinematic restrictions. The process of rolling is started in the computation by pushing the rail with a longitudinal velocity $v_p = 385$ mm/s prescribed at the end cross-section. The angular speed of the rollers is $\omega = 3.226 \, \text{s}^{-1}$, the associated longitudinal velocity in the middle between the rotation axes is $\omega_m = 500$ mm/s. If the action of the rollers results in a faster motion of the rail than the prescribed one, the prescribed velocity is removed. This occurs when the end cross-section is displaced by 30 cm along the longitudinal axis after the initiation of the rolling process.

The thermal phenomena are enhanced in the rail by the contact with the cold rollers ($T_\infty = 293$ K). The thermal action of the tool has been substituted in the computation by a heat transfer coefficient $\alpha = 0.0044$ W/mm² K relevant for the part of the material surface actually contacting the rollers. The remainder of the rail surface still exchanges heat with the air as in the cooling phase prior to the rolling process. Local dissipation of mechanical work in the deforming material also contributes to the thermal phenomena.

The thermomechanical analysis of the hot rolling process was performed using 100 time increments of $\tau = 0.002$ s. An explicit integration of the deformation process ($\zeta = 0$) was applied in conjunction with an implicit integration of the thermal problem ($\zeta = 1/2$) which leads within each increment to the one-sided coupled situation mentioned in Section 10.7.1. Accordingly, coupling by iteration is not necessary; the iterative solution of the partial problems requires 3 to 5 cycles. The computation of the time dependent process — dealing at each stage with 13 728 unknown velocities and 4576 unknown temperatures — consumed approximately 15 hours of CPU time on the CRAY 2 system installed at the University of Stuttgart.

Results of the numerical simulation are presented in Figures 10.20–2. Figure 10.20 shows three-dimensional views of the deformed rail up to stationary conditions in the deformation zone ($t = 0.2$ s). The deformed configurations may be compared with the undeformed one in the figure at the beginning of the rolling pass. The temperature distribution on the surface of the rail at $t = 0.2$ s is demonstrated in Figure 10.21. The figure clearly shows the reduction of the temperature on the surface of the hot material contacting the cold rollers.

Figure 10.22 indicates the evolution of the thermomechanical process in a cross-

348 COUPLED FIELD PROBLEMS

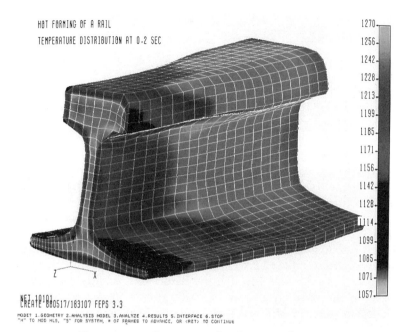

Figure 10.21 Temperature distribution in the rail.

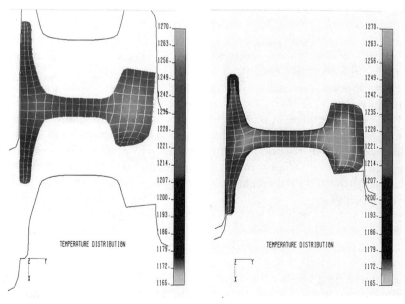

Figure 10.22 Cross-section at the beginning (left) and the end of the rolling pass (right). Distribution of temperature.

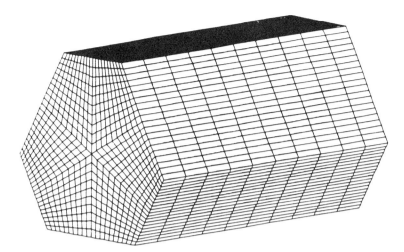

Figure 10.23 Hot rolling of a wire. Initial configuration.

section of the rail deforming during the rolling pass. In particular, the left part of the figure shows the position of the cross-section relative to the rollers at the initial stage of the process. At this stage, the distribution of the temperature in the cross-sections is a result of the cooling process prior to the rolling pass. It may be compared to the temperature distribution at the ultimate stage of the pass, on the right part of Figure 10.22. In this figure, cooling of outer regions due to the contact with the rollers is demonstrated, as also is heating in interior regions due to the dissipation of mechanical work during deformation.

10.7.3 Hot rolling of steel wire

The application of parallel processing to industrial metal forming is demonstrated by a three-dimensional process simulation on the iPSC/2 of the Institute for Computer Applications [11]. In particular, the treatment was based on the spatial decomposition technique outlined in Section 10.5.1; the solution of the system was performed by the parallel substructuring algorithm.

The process considered is the hot rolling of a steel wire with hexagonal cross-section [34]. It refers to the third rolling pass through the Kocks three-roll block[35] in use at the Acciaierie di Bolzano. This pass appears to be critical because the cross-section undergoes a drastic reduction and some surface defects are produced in the material.

The simulation deals also in the present case with the initial unsteady phase of the rolling pass up to the appearance of stationary conditions in the deformation zone. Figure 10.23 shows the initial section of the wire used in the analysis. Due to symmetry considerations only one-sixth of the cross-section was discretized. The three-dimensional mesh consists of 1331 nodal points and 1000 8-node hexahedral elements. Figure 10.24 shows the discretized structure after it was decomposed into four subdomains for the parallel computations. This decomposition was accomplished

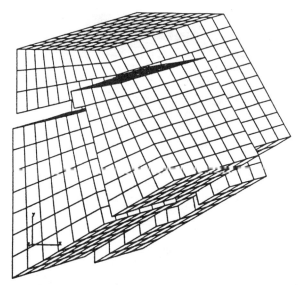

Figure 10.24 Decomposition of the discretized structure into four subdomains.

with an automatic, element-level domain decomposer [36]. Starting at an arbitrary or user-specified element, this scheme propagates through the finite element mesh by adding neighbouring elements until the required number of elements or degrees of freedom for one subdomain is reached, and then proceeds with the next subdomain in the same fashion.

The resulting decomposition for the three-dimensional section of the wire was well balanced in terms of the number of elements and the number of degrees of freedom per subdomain. However, due to the irregularity of the subdomains, the bandwidth of the local system matrices varied for the different subdomains, as did the number of internal boundary nodes, i.e. nodes that are common to at least two subdomains. These factors cannot be controlled effectively with the element-level domain decomposer that was used here, and the load balancing problems cannot be avoided in the substructuring solution. The above balancing problems have been resolved by the alternative decomposition method developed in [19]. By this method, the problem is decomposed during the course of the mesh generation procedure, which is based on a multiblock algorithm. As soon as the block structure is defined together with certain characteristics of the mesh, the computational load for the subsequent parallel finite element solution can be estimated prior to the generation of the ultimate mesh. Thus, a balanced decomposition is accomplished at this stage by recursively splitting the computational domain in half such that the work loads on both half-domains, as composed of the contained blocks, are equal.

Figure 10.25 depicts the set-up of the three rollers (diameter 486.4 mm, circumferential velocity 1.2 m/s), which are positioned at an angle of 120 degrees relative to each other. The thermomechanical material data are as in the previous example; heat transfer and friction are also taken into account with the coefficients specified in Section 10.7.2. The contact algorithm of [23] which uses a discretized representation of the rigid rollers, was employed to simulate the mechanical and thermal interactions

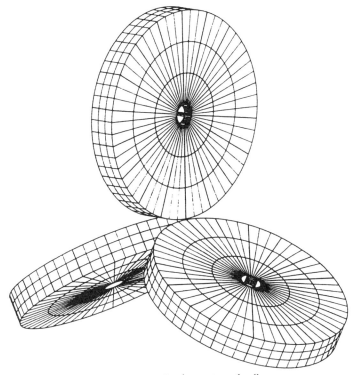

Figure 10.25 Configuration of rollers.

of wire and rollers. Starting from an initially homogeneous temperature (1050 °C), the top surface of the wire cools off after coming into contact with the cold rollers (20 °C). In the area of largest deformation, however, the temperature rises above the intial value due to dissipative effects. This fact is associated with severe temperature gradients which may affect the deformability of the material. Figure 10.26 depicts the deformation and temperature distribution on the decomposed configuration, as resulting upon the achievement of the stationary conditions in the deformation zone; the final shape of the complete wire section, reassembled from four subdomains, is presented in Figure 10.27.

The computation of the unsteady process was based on time increments $\tau = 0.001$ s; stationary conditions in the deformation zone were achieved after 60 incremental steps. As in the previous application, iterative coupling within each increment of time could be avoided by the combination of an explicit integration for the deformation process with an implicit scheme for the thermal problem. The processing of the 3993 unknown velocities and 1331 unknown temperatures on the four units of the iPSC/2 required 15 CPU hours in total. This may be compared with 2 hours CPU time on the CRAY 2 for the sequential computation. Also, the use of single-precision arithmetic in conjunction with the iPSC/2 increased the number of the iteration cycles for the individual solutions.

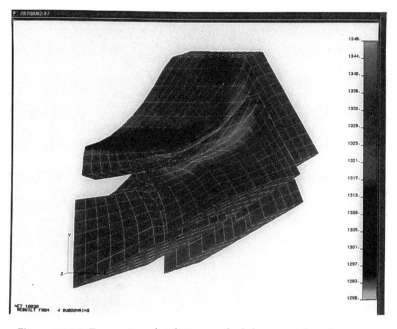

Figure 10.26 Temperature distribution on final decomposed configuration.

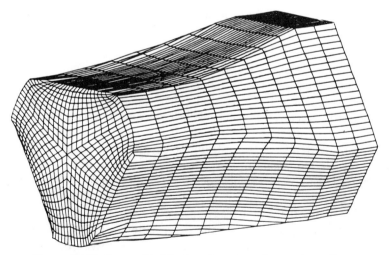

Figure 10.27 Reassembled final configuration of wire after rolling.

10.7.4 Remarks

For the purpose of demonstration of the interaction of distinct physical fields within a common material domain, Section 10.7 focuses on thermomechanically coupled deformation processes. The selected examples refer to metal forming and indicate the large dimensions that problems of some industrial relevance may assume. A consideration of the physics underlying the coupled processes in question, helps to simplify the numerical treatment. To be specific, the different time scales in which deformation and temperature changes occur allow the application of distinct integration techniques avoiding the iterative coupling within each increment of time. This, however, is not always possible and therefore the performance of fully coupled numerical solution algorithms was examined by means of a test case in Section 10.5.3.

Despite the aforementioned simplification, the sequential computation remains highly time consuming even on a CRAY 2 system. The utilization of distributed processing aims at a significant reduction of the computation time and the transition to alternative computer configurations. Although the application reported in the preceding Section 10.73 refers to four units, aspects of massive parallelization are discussed in [37].

Both examples presented here in Sections 10.7.2 and 10.7.3 would benefit from an adaptive modification of the discretization mesh during the course of the computation. The equipresence of the deformation velocity and the temperature, however, makes the decision on how to steer a modification of the unique finite element discretization of the two differently developing fields difficult, cf. Section 10.6.2. For this reason, use of distinct discretizations is envisaged in forthcoming applications.

BIBLIOGRAPHY

[1] Balmer, H., Doltsinis, I.St. and König, M. (1974) Elastoplastic and creep analysis with the ASKA program system, *Comput. Meth. Appl. Mech. Engg*, **3**, 87–104.
[2] Donea, J., Giuliani, S. and Halleux, J.P. (1982) An arbitrary Lagrangian-Eulerian finite element method for transient dynamic fluid-structure interactions, *Comput. Meth. Appl. Mech. Engg*, **33**, 689–723.
[3] Argyris, J. and Doltsinis, I.St. (1981) On the natural formulation and analysis of large deformation coupled thermomechanical problems, *Comput. Meth. Appl. Mech. Engg*, **25**, 195–253.
[4] Lewis, R.W., Bettes, P. and Hinton, E. (eds.) (1984) *Numerical Methods in Coupled Systems*, Wiley, Chichester.
[5] Zienkiewicz, O.C. and Chan, A.H.C. (1989) Coupled problems and their numerical solution, in I.St. Doltsinis (ed.), *Advances in Computational Nonlinear Mechanics*, Springer Verlag.
[6] Argyris, J., Doltsinis, I.St., Fischer, H. and Wüstenberg, H. (1985) TA ΠΑΝΤΑ PEI, *Comput. Meth. Appl. Mech. Engg*, **51**, 289–362.
[7] Wüstenberg, H. (1986) FEPS 3.3, *Finite Element Programming System — User's Guide*, ICA Report No. 21, Stuttgart.
[8] Felippa, C.A. and Park, K.C. (1980) Staggered transient analysis procedures for coupled mechanical systems, *Comput. Meth. Appl. Mech. Engg*, **24**, 61–111.
[9] Park, K.C. and Felippa, C.A. (1984) Recent developments in coupled field analysis methods, in R.W. Lewis *et al.* (eds.), *Numerical Methods in Coupled Systems*, Wiley, Chichester.
[10] Doltsinis, I.St. (1990) Aspects of modelling and computation in the analysis of metal forming, *Engg Comput.*, **7**, 2–20.

[11] Doltsinis, I.St. and Nölting, S. (1991) Studies on parallel processing of coupled field problems, *Comput. Meth. Appl. Mech. Engg*, **89**, 497–521.
[12] Van der Houwen, P.J. (1977) *Construction of Integration Formulas for Initial Value Problems*, North-Holland, Amsterdam.
[13] Belytschko, T. and Hughes, T.J.R. (eds.) (1982) *Computational Methods in Transient Analysis*, North-Holland, Amsterdam.
[14] Argyris, J. and Doltsinis, I.St. (1984) A primer on superplasticity in natural formulation, *Comput. Meth. Appl. Mech. Engg*, **46**, 83–131.
[15] Oden, J.T. (1969) Finite element analysis of nonlinear problems in the dynamical theory of coupled thermoelasticity, *Nucl. Engg Des.*, **10**, 465–475.
[16] Argyris, J., Doltsinis, I.St., Pimenta, P.M. and Wüstenberg, H. (1982) Thermomechanical response of solids at high-strains — natural approach, *Comput. Meth. Appl. Mech. Engg*, **32**, 3–57.
[17] Jennings, A. (1977) *Matrix Computation for Engineers and Scientists*, Wiley, Chichester.
[18] Ortega, J.M. and Rheinboldt, W.C. (1970) *Iterative Solution of Nonlinear Equations in Several Variables*, Academic Press, New York.
[19] Doltsinis, I.St. and Nölting, S. (1992) Generation and decomposition of finite element models for parallel computations, *Comput. Syst. Engg*, **2**, 427–449.
[20] Farhat, C. and Wilson, E. (1988) A parallel active column equation solver, *Comput. Struct.*, **28**, 289–304.
[21] Argyris, J. (1960) *Energy Theorems and Structural Analysis*, Butterworths, London.
[22] Doltsinis, I.St. and Nölting, S. (1989) Netzgenierierung für die numerische Simulation von Umformprozessen, *Proc. Workshop Numerische Methoden der Plastomechanik*, Institut für Mechanik, Universität Hannover, May 29–31.
[23] Doltsinis, I.St., Luginsland, J. and Nölting, S. (1987) Some developments in the numerical simulation of metal forming processes, *Engg Comput.*, **4**, 266–280.
[24] Doltsinis, I.St. and Luginsland, J. (1988) Computer-aided modelling of metal forming processes, *Proc. Materials '88 Conf.*, London, May 9–13. The Institute of Metals, London, 1989.
[25] Argyris, J., Doltsinis, I.St. and Friz, H. (1989) Hermes space shuttle: Exploration of reentry aerodynamics, *Comput. Meth. Appl. Mech. Engg*, **73**, 1–51.
[26] Miehe, C. (1988) *Zur numerischen Behandlung thermomechanischer Prozesse*, Dr.-Ing. Dissertation, Universität Hannover.
[27] Zienkiewicz, O.C. (1984) Flow formulation for the simulation of forming processes, in J.F.T. Pittman *et al.* (eds.), *Numerical Analysis of Forming Processes*, Wiley, Chichester.
[28] Simo, J.C. and Miehe, C. (1992) Associative coupled thermoplasticity at finite strains: Formulation, numerical analysis and implementation, *Comput. Meth. Appl. Mech. Engg*, **98**, 41–104.
[29] Ziegler, H. (1977) *An Introduction to Thermomechanics*, North-Holland, Amsterdam.
[30] Doltsinis, I.St. (1983) *Thermodynamics of Deformation Processes*, Unpublished Notes, Stuttgart.
[31] Doltsinis, I.St., Luginsland, J., Dohmen, P.M., de Souza, M.M. and Kopp, R. (1988) Coupled thermal-mechanical simulation of three-dimensional hot-rolling, *Proc. 4th Aachener Stahlkolloquium*, Aachen, June 30–July 1.
[32] Kallabis, H.-P. (1987) Untersuchungen zur lokalen thermomechanischen Behandlung beim Profilwalzen, *Umformtechnische Schriften 2*, Stahleisen, Düsseldorf.
[33] Taylor, R.L. and Zienkiewicz, O.C. (1982) Mixed finite element solution of fluid flow problems, in R.H. Gallagher *et al.* (eds.), *Finite Elements in Fluids*, Vol. 4, Wiley, New York.
[34] Doltsinis, I.St., Manini, L., Jacucci, G., Contro, R., Pegoretti, A. and Juliani, F. (1990) Computer simulation of an industrial forming process: Three-dimensional hot rolling of

wire rods, *Proc. 2nd Int. Conf. Advanced Manufacturing Systems and Technology*, AMST'90, Trento, Italy. Also: Computer simulation of three-dimensional hot rolling of wire rods, *WIRE*, **42**, 91–97.

[35] Blos, O. (1986) New developments in rod mills, *Iron and Steel Engineering*, **63**, 35–46.
[36] Farhat, C. (1988) A simple and efficient automatic domain decomposer, *Comput. Struct.*, **28**, 579–602.
[37] Doltsinis, I.St. and Nölting, S. (1992) *Simulation von Umformprozessen auf Parallelrechnern mit verteiltem Speicher*, ICA Report No. 34, Stuttgart.

I. St. Doltsinis
Institute for Computer Applications
University of Stuttgart
Pfaffenwaldring 27
D-7000 Stuttgart 80
GERMANY

11
Direct Time-integration Methods: Stabilized Space–Time Finite Element Formulation of Incompressible Flows

S. Mittal and T. E. Tezduyar

University of Minnesota, MN, USA

11.1 INTRODUCTION

In this chapter the stabilized space–time finite element formulation of incompressible flows, including those involving moving boundaries and interfaces, is reviewed. Also reviewed are the efficient iteration techniques employed to solve the equation systems resulting from the space–time finite element discretization of these flow problems. Results are presented for applications to certain unsteady flows past a circular cylinder. In some of these flows the cylinder is moving, either with a prescribed motion or with an unknown motion which needs to be determined as part of the solution.

The space–time formulation reviewed in this chapter is a direct time-integration method and is used in conjunction with the Galerkin/least-squares (GLS) stabilization. The GLS stabilization prevents the numerical oscillations that might be produced by the presence of dominant advection terms in the governing equations or by not using an acceptable combination of interpolation functions to represent the velocity

---

Solving Large-scale Problems in Mechanics, Edited by M. Papadrakakis
© 1993 John Wiley & Sons Ltd

and pressure fields. In this kind of stabilization, a series of stabilizing terms are added to the Galerkin formulation of the problem. These terms can be obtained by minimizing the sum of the squared residual of the momentum equation integrated over each element domain. The GLS stabilization leads to a consistent formulation, in the sense that an exact solution still satisfies the stabilized formulation. Consequently, it introduces minimal numerical diffusion, and therefore results in solutions with minimal loss of accuracy. This approach has been successfully applied to Stokes flows by Hughes and Franca [1], to compressible flows by Hughes et al. [2] and Shakib [3], and to incompressible flows at finite Reynolds numbers by Tezduyar et al. [4, 5], Liou and Tezduyar [6] and Hansbo and Szepessy [7].

The space–time finite element formulation with the GLS stabilization has recently been used for various problems with fixed spatial domains. These authors are most familiar with the work of Shakib [3], Hansbo and Szepessy [7], Hughes et al. [8], and Hughes and Hulbert [9]. The fundamentals of the space–time formulation, its implementation, and the associated stability and accuracy analysis can be found in these references.

In the space–time formulation, the finite element discretization is applied not only spatially but also temporally. Consequently, the deformation of the spatial domain is taken into account automatically. This feature of the stabilized space–time formulation was first exploited by Tezduyar et al. [4, 5]. They introduced the deforming-spatial-domain/space–time (DSD/ST) procedure and applied it to several unsteady incompressible flow problems involving moving boundaries and interfaces, such as free-surface flows, liquid drops, two-liquid flows and flows with drifting cylinders. In the DSD/ST procedure the frequency of remeshing is minimized. We define remeshing as the process of generating a new mesh, and projecting the solution from the old mesh to the new one. Since remeshing, in general, involves projection errors, minimizing the frequency of remeshing results in minimizing the projection errors.

It is important to realize that the finite element interpolation functions are discontinuous in time so that the fully discrete equations are solved one space–time slab at a time, and this makes the computations feasible. Still, the computational cost associated with the space–time finite element formulations using piecewise linear functions in time is quite heavy. For large-scale problems it becomes imperative to employ efficient iteration methods to reduce the cost involved. This was achieved in Liou and Tezduyar [6] by using the generalized minimal residual (GMRES) [10] iteration algorithm with the clustered element-by-element (CEBE) preconditioners.

The CEBE preconditioning is a generalized version of the standard element-by-element (EBE) preconditioning. The EBE preconditioners, which were first introduced by Hughes et al. [11, 12], are defined as sequential products of element level matrices. The iterative computations with EBE preconditioners are performed in an element-by-element fashion and are highly vectorizable (see Hughes and Ferencz [13]). In the CEBE preconditioning the elements are partitioned into clusters of elements, with a desired number of elements in each cluster, and the iterations are performed in a cluster-by-cluster fashion. The number of clusters should be reviewed as an optimization parameter to minimize the computational cost (both memory and CPU time). By specifying the number of clusters, one can select an algorithm anywhere in the spectrum of algorithms ranging from the direct solution technique (when the number of clusters is one) to the standard element-by-element method (when the number of clusters is same as the number of elements). Parallel implementation of the CEBE preconditioning is very similar to that of the grouped element-by-element (GEBE) [14, 15] preconditioning. Therefore in this article we also review the GEBE method, a variation of the standard EBE method.

The GEBE preconditioners are sequential products of the element group matrices, with the condition that no two elements in the same group can be 'neighbours'. This is achieved by a simple element grouping algorithm which is applicable to arbitrary meshes. In this form, vectorization and parallel implementation of the EBE method becomes more clear. In parallel computations, before we can start with a new element group we have to finish the one that we have already started with. To minimize the overhead associated with this synchronization, we try to minimize the number of groups. Furthermore, to increase the vector efficiency of the computations performed, within each group the elements are processed in packets of 64 elements (or whatever the optimum packet size for a given vector environment is). When parallel processing is invoked, each of the available processors works on one packet at a time until all packets in a group are finished.

Computation of time-dependent incompressible flow problems on fixed spatial domains can also be performed by using the finite element discretization in space only, rather than in both space and time. In this case first the GLS stabilization for the steady-state equations of incompressible flows is considered. Then in the definition of the stabilizing terms, the residual of the steady-state equations is replaced with the ones for the time-dependent equations. These stabilizing terms are added to the Galerkin formulation of the time-dependent equations. Furthermore, if, at the element interiors, the contribution to the weighting function from the viscous terms is neglected (it is identically zero for linear velocity interpolation) one obtains a formulation with the combination of streamline-upwind/Petrov–Galerkin (SUPG) and pressure-stabilizing/Petrov–Galerkin (PSPG) stabilizations.

The SUPG formulation, which prevents the numerical oscillations caused by the presence of dominant advection terms, was introduced by Hughes and Brooks [16]. A comprehensive description of the formulation, together with various numerical examples, can be found in Brooks and Hughes [17]. For hyperbolic systems in general, and compressible Euler equations in particular, the SUPG stabilization was first reported by Tezduyar and Hughes [18]. The implementation of the SUPG formulation in Brooks and Hughes [17] was based on Q1P0 (bilinear velocity/constant pressure) elements and one-step time-integration of the semi-discrete equations obtained by using such elements. The SUPG stabilization for the vorticity-stream function formulation of incompressible flow problems, including those with multiply connected domains, was introduced by Tezduyar et al. [19].

It was shown that (see Brezzi and Pitkaranta [20], and Hughes et al. [21]), with proper stabilization, elements which do not satisfy the Brezzi condition can still be used for Stokes flow problems. The Petrov–Galerkin stabilization proposed in Hughes et al. [21] is achieved, just as in the SUPG stabilization, by adding to the Galerkin formulation a series of integrals over element domains. The PSPG stabilization proposed in Tezduyar et al. [22] is a generalization, to finite Reynolds number flows, of the Petrov–Galerkin stabilization proposed in Hughes et al. [21] for Stokes flows. In Tezduyar et al. [22], the SUPG and PSPG stabilizations are used together with both one-step (T1) and multi-step (T6) time-integration schemes. With the T1 scheme, the SUPG and PSPG stabilizations are applied simultaneously. With the T6 scheme, on the other hand, the SUPG stabilization is applied only to the steps involving the advective terms, while the PSPG stabilization is applied only to the steps involving the pressure terms. Both schemes were implemented in Tezduyar et al. [22] based on the Q1Q1 (bilinear velocity and pressure) and P1P1 (linear velocity and pressure) elements, and were successfully applied to a set of nearly standard test problems. Also, recently Lundgren and Mansour [23] applied this type of stabilization technique to Lagrangian finite element computation of viscous free-surface flows.

In the SUPG, PSPG and GLS stabilizations the stabilizing terms added involve the residual of the momentum equation as a factor. Consequently, when an exact solution is substituted into the stabilized formulation, these added terms vanish, and as a result the stabilized formulation is satisfied by the exact solution in the same way as the Galerkin formulation is satisfied. It is because of this property of the SUPG, PSPG and GLS stabilizations that numerical oscillations are prevented without introducing excessive numerical diffusion (i.e., without 'over-stabilizing') and therefore without compromising the accuracy of the solution.

11.2 THE GOVERNING EQUATIONS OF UNSTEADY INCOMPRESSIBLE FLOWS

Let $\Omega_t \in \mathbb{R}^{n_{sd}}$ be the spatial domain at time $t \in (0, T)$, where n_{sd} is the number of space dimensions. Let Γ_t denote the boundary of Ω_t. We consider the following velocity–pressure formulation of the Navier–Stokes equations governing unsteady incompressible flows:

$$\rho\left(\frac{\partial \mathbf{u}}{\partial t} + \mathbf{u} \cdot \nabla \mathbf{u}\right) - \nabla \cdot \boldsymbol{\sigma} = \mathbf{0} \quad \text{on} \quad \Omega_t \quad \forall t \in (0, T), \tag{11.1}$$

$$\nabla \cdot \mathbf{u} = 0 \quad \text{on} \quad \Omega_t \quad \forall t \in (0, T) \tag{11.2}$$

where ρ and $\mathbf{u}$ are the density and velocity, and $\boldsymbol{\sigma}$ is the stress tensor given as

$$\boldsymbol{\sigma}(p, \mathbf{u}) = -p\mathbf{I} + 2\mu\boldsymbol{\varepsilon}(\mathbf{u}) \tag{11.3}$$

with

$$\boldsymbol{\varepsilon}(\mathbf{u}) = \tfrac{1}{2}(\nabla \mathbf{u} + (\nabla \mathbf{u})^{\mathrm{T}}). \tag{11.4}$$

Here p and μ are the pressure and the dynamic viscosity, and $\mathbf{I}$ is the identity tensor. The part of the boundary at which the velocity is assumed to be specified is denoted by $(\Gamma_t)_g$:

$$\mathbf{u} = \mathbf{g} \quad \text{on} \quad (\Gamma_t)_g \quad \forall t \in (0, T). \tag{11.5}$$

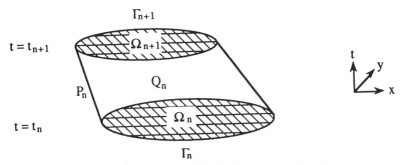

Figure 11.1 The space-time slab for the DSD/ST formulation.

The 'natural' boundary conditions associated with (11.1) are the conditions on the stress components, and these are conditions assumed to be imposed at the remaining part of the boundary:

$$\mathbf{n} \cdot \boldsymbol{\sigma} = \mathbf{h} \quad \text{on} \quad (\Gamma_t)_\mathbf{h} \quad \forall t \in (0, T). \tag{11.6}$$

The homogeneous version of (11.6), which corresponds to the 'traction-free' (i.e. zero normal and shear stress) conditions, is often imposed at the outflow boundaries. As initial condition, a divergence-free velocity field $\mathbf{u}_0(\mathbf{x})$ is specified over the domain Ω_t at $t = 0$:

$$\mathbf{u}(\mathbf{x}, 0) = \mathbf{u}_0(\mathbf{x}) \quad \text{on} \quad \Omega_0. \tag{11.7}$$

11.3 THE STABILIZED SPACE-TIME FINITE ELEMENT FORMULATION — A DIRECT TIME-INTEGRATION METHOD

In the space–time finite element formulation, the time interval $(0, T)$ is partitioned into subintervals $I_n = (t_n, t_{n+1})$, where t_n and t_{n+1} belong to an ordered series of time levels $0 = t_0 < t_1 < \cdots < t_N = T$. It was first shown in Tezduyar et al. [4, 5] that the stabilized space–time finite element formulation can be effectively applied to fluid dynamics computations involving moving boundaries and interfaces. In this formulation the spatial domains at various time levels are allowed to vary. We let $\Omega_n = \Omega_{t_n}$ and $\Gamma_n = \Gamma_{t_n}$, and define the space–time slab Q_n as the space–time domain enclosed by the surfaces Ω_n, Ω_{n+1} and P_n (see Figure 11.1). Here P_n, the lateral surface of Q_n, is the surface described by the boundary Γ, as t traverses I_n. Similarly to the way it was represented by equations (11.5) and (11.6), P_n is decomposed into $(P_n)_\mathrm{g}$ and $(P_n)_\mathrm{h}$ with respect to the type of boundary condition being imposed.

Finite element discretization of a space–time slab Q_n is achieved by dividing it into elements Q_n^e, $e = 1, 2, \ldots, (n_{el})_n$, where $(n_{el})_n$ is the number of elements in the space–time slab Q_n. Associated with this discretization, for each space–time slab we define the following finite element interpolation function spaces for the velocity and pressure:

$$(S_\mathbf{u}^h)_n = \{\mathbf{u}^h \mid \mathbf{u}^h \in [H^{1h}(Q_n)]^{n_\mathrm{sd}}, \mathbf{u}^h \doteq \mathbf{g}^h \quad \text{on} \quad (P_n)_\mathrm{g}\} \tag{11.8}$$

$$(V_{\mathbf{u}}^h)_n = \{\mathbf{w}^h \mid \mathbf{w}^h \in [H^{1h}(Q_n)]^{\text{nsd}}, \mathbf{w}^h \doteq \mathbf{0} \quad \text{on} \quad (P_n)_{\mathbf{g}}\} \tag{11.9}$$

$$(S_p^h)_n = (V_p^h)_n = \{q^h \mid q^h \in H^{1h}(Q_n)\}. \tag{11.10}$$

Here $H^{1h}(Q_n)$ represents the finite-dimensional function space over the space–time slab Q_n. This space is formed by using, over the parent (element) domains, first-order polynomials in space and time. It is also possible to use zeroth-order polynomials in time. In either case, globally, the interpolation functions are continuous in space but discontinuous in time.

The space–time formulation of (11.1)–(11.7) can be written as follows: start with

$$(\mathbf{u}^h)_0^- = (\mathbf{u}_0)^h \tag{11.11}$$

sequentially for $Q_1, Q_2, \ldots, Q_{N-1}$, given $(\mathbf{u}^h)_n^-$, find $\mathbf{u}^h \in (S_{\mathbf{u}}^h)_n$ and $p^h \in (S_p^h)_n$, such that

$$\forall\, \mathbf{w}^h \in (V_{\mathbf{u}}^h)_n \quad \text{and} \quad \forall\, q^h \in (V_p^h)_n$$

$$\int_{Q_n} \mathbf{w}^h \cdot \rho\left(\frac{\partial \mathbf{u}^h}{\partial t} + \mathbf{u}^h \cdot \nabla \mathbf{u}^h\right) dQ + \int_{Q_n} \varepsilon(\mathbf{w}^h) : \sigma(p^h, \mathbf{u}^h)\, dQ$$

$$- \int_{(P_n)_h} \mathbf{w}^h \cdot \mathbf{h}\, dP + \int_{Q_n} q^h \nabla \cdot \mathbf{u}^h\, dQ + \int_{\Omega} (\mathbf{w}^h)_n^+ \cdot ((\mathbf{u}^h)_n^+ - (\mathbf{u}^h)_n^-)\, dQ$$

$$+ \sum_{e=1}^{(n_{el})_n} \int_{Q_n^e} \tau \frac{1}{\rho}\left[\rho\left(\frac{\partial \mathbf{w}^h}{\partial t} + \mathbf{u}^h \cdot \nabla \mathbf{w}^h\right) - \nabla \cdot \sigma(q^h, \mathbf{w}^h)\right]$$

$$\times \left[\rho\left(\frac{\partial \mathbf{u}^h}{\partial t} + \mathbf{u}^h \cdot \nabla \mathbf{u}^h\right) - \nabla \cdot \sigma(p^h, \mathbf{u}^h)\right] dQ = 0. \tag{11.12}$$

In the variational formulation given by (11.12), the following notation is being used:

$$(\mathbf{u}^h)_n^\pm = \lim_{\delta \to 0} \mathbf{u}^h(t_n \pm \delta) \tag{11.13}$$

$$\int_{Q_n} (\ldots)\, dQ = \int_{I_n}\int_{\Omega} (\ldots)\, d\Omega\, dt \tag{11.14}$$

$$\int_{P_n} (\ldots)\, dP = \int_{I_n}\int_{\Gamma} (\ldots)\, d\Gamma\, dt. \tag{11.15}$$

Remarks

1. If we were to consider a spatial finite element discretization, rather than a space–time one, the Galerkin formulation of (11.1)–(11.7) would have consisted of the

first four integrals (their spatial versions of course) appearing in equation (11.12). In the space–time formulation, because the interpolation functions are discontinuous in time, the fifth integral in equation (11.12) enforces, weakly, the temporal continuity of the velocity field. The remaining series of integrals in equation (11.12) are the least-squares terms added to the Galerkin variational formulation to assure the numerical stability of the computations. The coefficient τ determines the weight of such added terms.

2. This kind of stabilization of the Galerkin formulation is referred to as the Galerkin/least-squares (GLS) procedure, and can be considered as a generalization of the stabilization based on the streamline-upwind/Petrov–Galerkin (SUPG) and the pressure-stabilizing/Petrov–Galerkin (PSPG) procedure employed for incompressible flows [22]. It is with such stabilization procedures that it is possible to use elements which have equal-order interpolation functions for velocity and pressure, and which are otherwise unstable.

3. It is important to realize that the stabilizing terms added involve the momentum equation as a factor. Therefore, despite these additional terms, an exact solution is still admissible to the variational formulation given by equation (11.12).

The coefficient τ used in this formulation is obtained by a simple multi-dimensional generalization of the optimal τ given in Shakib [3] for one-dimensional space–time formulation. The expression for the τ used in this formulation is

$$\tau = \left[\left(\frac{2 \|\mathbf{u}^h\|}{h} \right)^2 + \left(\frac{4\nu}{h^2} \right)^2 \right]^{-1/2} \tag{11.16}$$

where ν is the kinematic viscosity, and h is the spatial element length. For derivation of τ for higher-order elements see Franca et al. [24].

Remarks

4. Because the finite element interpolation functions are discontinuous in time, the fully discrete equations can be solved one space–time slab at a time. Still, the memory needed for the global matrices involved in this method is quite substantial. For example, in two dimensions, the memory needed for space–time formulation (with interpolation functions which are piecewise linear in time) of a problem is approximately four times more compared to using the finite element method only for spatial discretization. However, iteration methods can be employed to substantially reduce the cost involved in solving the linear equation systems arising from the space–time finite element discretization (see Sections 11.4 and 11.5).

5. In the DSD/ST procedure, to facilitate the motion of free surfaces, interfaces and solid boundaries, we need to move the boundary nodes with the normal component of the velocity at those nodes. Except for this restriction, we have the freedom to move all the nodes any way we would like to. With this freedom,

we can move the mesh in such a way that we only need to remesh when it becomes necessary to do so to prevent unacceptable degrees of mesh distortion and potential entanglements. By minimizing the frequency of remeshing we minimize the projection errors expected to be introduced by remeshing. In fact, for some computations, as a byproduct of moving the mesh, we may be able to get a limited degree of automatic mesh refinement, again with minimal projection errors. For example, a mesh moving scheme suitable for a single cylinder drifting in a bounded flow domain is described in Tezduyar et al. [5]. We use the same mesh moving scheme for all the results to be presented in this article.

11.4 THE GROUPED ELEMENT-BY-ELEMENT (GEBE) ITERATION METHOD AND VECTORIZATION/PARALLELIZATION CONCEPTS

It was pointed out before, in Remark 4, that the memory needed for the global matrices involved in the space–time method is quite substantial. Iteration techniques can be effectively used to reduce the associated cost significantly. In this section we review the grouped element-by-element (GEBE) iteration method. The description of the GEBE preconditioners and the results reported in this section have been extracted from Tezduyar et al. [14, 15].

After the linearization of the fully discretized equations, the following system needs to be solved for the nodal values of the unknowns:

$$\mathbf{A}\mathbf{x} = \mathbf{b}. \tag{11.17}$$

We rewrite (11.17) in a scaled form

$$\tilde{\mathbf{A}}\tilde{\mathbf{x}} = \tilde{\mathbf{b}} \tag{11.18}$$

where

$$\tilde{\mathbf{A}} = \mathbf{W}^{-1/2}\mathbf{A}\mathbf{W}^{-1/2} \tag{11.19}$$

$$\tilde{\mathbf{x}} = \mathbf{W}^{1/2}\mathbf{x} \tag{11.20}$$

$$\tilde{\mathbf{b}} = \mathbf{W}^{-1/2}\mathbf{b}. \tag{11.21}$$

The scaling matrix $\mathbf{W}$ is defined as

$$\mathbf{W} = \text{diag } \mathbf{A}. \tag{11.22}$$

With this definition of $\mathbf{W}$, diag $\tilde{\mathbf{A}}$ becomes an identity matrix.

For the formulations presented in this article, the matrix $\mathbf{A}$ is not in general symmetric and positive-definite. Therefore, the proposed GEBE preconditioner (and the CEBE preconditioner to be described in the next section) will be used in conjunction with the GMRES method [10]; an outline of the GMRES method used is given below.

(i) Set the iteration counter $m = 0$, and start with an initial guess $\tilde{\mathbf{x}}_0$,

(ii) Calculate the residual scaled with the preconditioner matrix $\tilde{\mathbf{P}}$:

$$\tilde{\mathbf{r}}_m = \tilde{\mathbf{P}}^{-1}(\tilde{\mathbf{A}}\tilde{\mathbf{x}}_m - \tilde{\mathbf{b}}). \tag{11.23}$$

(iii) Construct the Krylov vector space:

$$\mathbf{e}^{(1)} = \tilde{\mathbf{r}}_m / \|\tilde{\mathbf{r}}_m\| \tag{11.24}$$

$$\mathbf{f}^{(j)} = \tilde{\mathbf{P}}^{-1}\tilde{\mathbf{A}}\mathbf{e}^{(j-1)} - \sum_{i=1}^{j-1} (\tilde{\mathbf{P}}^{-1}\tilde{\mathbf{A}}\mathbf{e}^{(j-1)}, \mathbf{e}^{(i)})\mathbf{e}^{(i)}, \qquad 2 \leq j \leq k \tag{11.25}$$

$$\mathbf{e}^{(j)} = \mathbf{f}^{(j)} / \|\mathbf{f}^{(j)}\| \tag{11.26}$$

where k is the dimension of the Krylov space and $\mathbf{e}^{(i)}$, $i = 1, 2, \ldots, k$, are the basis vectors.

(iv) Update the unknown vector:

$$\tilde{\mathbf{x}}_{m+1} = \tilde{\mathbf{x}}_m + \sum_{j=1}^{k} s_j \mathbf{e}^{(j)} \tag{11.27}$$

where $\mathbf{s} = \{s_j\}$ is the solution of the equation system

$$\mathbf{Qs} = \mathbf{z} \tag{11.28}$$

with

$$\mathbf{Q} = [(\tilde{\mathbf{P}}^{-1}\tilde{\mathbf{A}}\mathbf{e}^{(i)}, \tilde{\mathbf{P}}^{-1}\tilde{\mathbf{A}}\mathbf{e}^{(j)})], \qquad 1 \leq i, j \leq k, \tag{11.29}$$

$$\mathbf{z} = \{(\tilde{\mathbf{P}}^{-1}\tilde{\mathbf{A}}\mathbf{e}^{(i)}, -\tilde{\mathbf{r}}_m)\}, \qquad 1 \leq i \leq k. \tag{11.30}$$

(v) For the next iteration, set $m \leftarrow m + 1$ and goto i:
The iterations continue until the ratio $\|\tilde{\mathbf{r}}_m\|/\|\tilde{\mathbf{r}}_0\|$ falls below a predetermined value. It should be noted that the matrix $\mathbf{Q}$ is symmetric and positive-definite.

Remark

6. The convergence rate of this algorithm depends on the condition number of the matrix $\tilde{\mathbf{P}}^{-1}\tilde{\mathbf{A}}$. Therefore one would like to select a preconditioner that involves minimal inversion cost, and provides, within cost limitations, an optimal representation of $\tilde{\mathbf{A}}$. For example, the Jacobi conjugate gradient method, in

which $\tilde{\mathbf{P}} = \text{diag } \tilde{\mathbf{A}} \ (= \mathbf{I})$, involves minimal cost for the inversion of $\tilde{\mathbf{P}}$; however, this representation of $\tilde{\mathbf{A}}$ is rather a poor one, and therefore it usually takes too many iterations to converge.

In the GEBE method, the set of elements $\mathbb{E}$ is partitioned into 'parallelizable' groups $\mathbb{E}_K$, $K = 1, 2, \ldots, N_{pg}$, where N_{pg} is the number of such groups. The element grouping is achieved in such a way that no two elements within a group can be neighbours. Here we consider two elements being neighbours if they have at least one common node. The global matrix $\tilde{\mathbf{A}}_K$ associated with the group K is defined as

$$\tilde{\mathbf{A}}_K = \sum_{e \in \mathbb{E}_K} \tilde{\mathbf{A}}^e \tag{11.31}$$

where $\tilde{\mathbf{A}}^e$ is the element level matrix associated with element e.

The matrix $\tilde{\mathbf{A}}$ can then be expressed as

$$\tilde{\mathbf{A}} = \mathbf{I} + \sum_{K=1}^{N_{pg}} \tilde{\mathbf{B}}_K \tag{11.32}$$

where

$$\tilde{\mathbf{B}}_K = \tilde{\mathbf{A}}_K - \tilde{\mathbf{W}}_K, \qquad K = 1, 2, \ldots, N_{pg}. \tag{11.33}$$

Remarks

7. Because there is no inter-element coupling within each group, computations performed in element-by-element fashion (such as operations performed on a group matrix) do not depend on the ordering of the elements.
8. In parallel computations, before we can start with a new element group we first have to finish with the one that we have already started with. To minimize the overhead associated with this synchronization, we try to minimize the number of groups. For example, in a two-dimensional problem with rectangular domain and 100×100 mesh, assuming that we use quadrilateral elements, there will be four groups with groups having 2500 elements each and arranged in a checkerboard style.

An element grouping algorithm for arbitrary meshes

Initialization

$e = 1$	(start with the first element)
$N_{pg} = 1$	(create the first group)
$e \in \mathbb{E}_{N_{pg}}$	(put the element in that group)

Next element

$e = e + 1$	(take the next element)
if $e > n_{e1}$ then return	(if there is no such element than return)
$K = 0$	(otherwise get ready to search for an eligible group among the existing groups, starting with the first existing group)

Next existing group

$K = K + 1$	(take the next existing group)
if e has no neighbours in $\mathbb{E}_K$ then	(check if the element can be put in that group)
$\quad e = \mathbb{E}_K$	(if so, do it)
$\quad$ go to the next element	
else if $K < N_{pg}$	(otherwise check if there is a next existing group)
then	
$\quad$ go to next existing group	(if so, try that next group)
end if	

Create a new group

$N_{pg} = N_{pg} + 1$	(otherwise create a new group)
$e = \mathbb{E}_{N_{pg}}$	(and put the element in that group)
go to the next element	

Remark

9. To increase the vector efficiency of the computations performed, within each group the elements are processed in packets of n_{ep} elements, where n_{ep} is the optimum packet size for a given vector environment (e.g. 64). For the example of Remark 8, with an assumed packet size of 64 elements, each group would have 40 packets.

The GEBE preconditioning is based on the approximation of (11.32) by a sequential product of group level matrices. Given below are the definitions for three different GEBE preconditioners.

Crout factorization based GEBE preconditioner:

$$\tilde{\mathbf{P}} = \prod_{K=1}^{N_{pg}} \hat{\mathbf{L}}_K \prod_{K=N_{pg}}^{1} \hat{\mathbf{U}}_K \qquad (11.34)$$

where $\hat{\mathbf{L}}_K$ and $\hat{\mathbf{U}}_K$ are the matrices resulting from the following Crout factorization:

$$(\mathbf{I} + \tilde{\mathbf{B}}_K) = \hat{\mathbf{L}}_K \hat{\mathbf{U}}_K, \qquad K = 1, 2, \ldots, N_{pg}. \tag{11.35}$$

2-Pass GEBE preconditioner:

$$\tilde{\mathbf{P}} = \prod_{K=1}^{N_{pg}} \left(\mathbf{I} + \tfrac{1}{2}\tilde{\mathbf{B}}_K \right) \prod_{K=N_{pg}}^{1} \left(\mathbf{I} + \tfrac{1}{2}\tilde{\mathbf{B}}_K \right). \tag{11.36}$$

Gauss–Seidel factorization based GEBE preconditioner:

$$\tilde{\mathbf{P}} = \prod_{K=1}^{N_{pg}} (\mathbf{I} + \tilde{\mathbf{A}}_K^L) \prod_{K=N_{pg}}^{1} (\mathbf{I} + \tilde{\mathbf{A}}_K^U) \tag{11.37}$$

where $\tilde{\mathbf{A}}_K^L$ and $\tilde{\mathbf{A}}_K^U$ are the matrices resulting from the following lower-triangular–diagonal–upper-triangular decomposition:

$$\tilde{\mathbf{A}}_K = \tilde{\mathbf{A}}_K^L + \tilde{\mathbf{A}}_K^D + \tilde{\mathbf{A}}_K^U, \qquad K = 1, 2, \ldots, N_{pg}. \tag{11.38}$$

In a multi-processor vector machine each processor is an individual vector processor. The single vector processor can gain high performance through overlapping multiple operations using segmented functional units. When a program is restructured or vectorized to facilitate this overlapping, the speed can be improved by an order of magnitude. Performance improvements from vectorization can be augmented through the use of parallel processing in a multi-processor machine. Parallel processing allows a single program to use more than one CPU to do the required work. This results in a reduction of the elapsed time needed to do the job. The amount of reduction is proportional to the percentage of the program that can be executed independently. In the implementation in Tezduyar et al. [15], the focus is on the vectorization and parallel processing of element level matrix–vector calculations, factorization and back substitution, and global residual calculations.

The structure of the vectorized code contains an inner loop over elements that belong to the particular packets of elements with identical calculations. An important implementation point is the restructuring of the code to enable compiler optimization. A specific example is the partitioning of the element level matrix–vector calculations into multiple loops. Of course this introduces additional storage requirements as we need temporary arrays to hold the immediate computational results across loops. Another example is the deployment of CVMxx, CRAY compiler extensions for vectorizing IF constructs.

Additionally, all calculations are parallel processed using Microtasking, a compiler-directive-driven preprocesser for parallel processing on the CRAY. Microtasking is implemented with comment card directives. Before compiling the code, a preprocessor expands the directives into FORTRAN code, which includes calls to the Microtasking library. It is important to note that if the preprocessor is not used, the code is portable to other machines since the Microtasking directives appear as Fortran comment cards.

If there are n_{CPU} CPUs available for parallel processing, we would like to distribute the work among all the CPUs in such a way that large task granularity, good load balance and synchronization are satisfactorily achieved. The parallel performance gain must be considered against the deterioration of the vectorization performance.

In the implementation of the GEBE procedure, within each group, elements are subgrouped into n_{pa} packets with each packet containing n_{ep} (e.g. 64) elements. The last packet in each group may contain less than n_{ep} elements. The choice of n_{pa} and n_{ep} is made in such a way that n_{pa} is a multiple of n_{CPU}. When parallel processing is invoked (with a compiler directive in the outer loop over n_{pa}), each of the available n_{CPU} CPUs works on one packet at a time until all packets are finished. Because each packet contains the same amount of work (except for the last packet) and the overall number of packets can evenly be distributed to n_{CPU} CPUs, load balancing is ensured.

The tests for the GEBE computations reported in Tezduyar et al. [15] were performed on a CRAY-XMP/416 : 4 CPUs, 8.5 ns clock, 128 megabytes of memory, COS 1.16 operating system and products (Cray Research Mendota Heights data center). The test problem chosen was the unsteady flow past a circular cylinder at Reynolds number 100. The finite element mesh used consists of 1940 elements and 2037 nodal points. The dimensions of the computational domain, normalized by the cylinder diameter, are 16 and 8 in the flow and the cross flow directions, respectively. These computations were performed using the vorticity–stream function formulation of the incompressible Navier–Stokes equations. The benchmark test lasted for 10 time steps with several non-linear iterations per time step. The vector performance reached 100 Mflops for the element level matrix–vector computations, 70 Mflops for factorization and back substitution, and 25 Mflops for the global residual calculation. The overall program performance was 48 Mflops. Higher performance rates are expected if the program is vectorized further and if the initial set-up/overhead cost is filtered out. Parallel processing on the element level matrix–vector computations yielded speed-ups of 3.5–3.8. The speed up on the factorization/back substitution and global residual calculation is around 2–3. This is due to the fact that the task granularity is small because of the relatively small size of the test problem.

11.5 THE CLUSTERED ELEMENT-BY-ELEMENT (CEBE) ITERATION METHOD

The CEBE preconditioning was first proposed by Liou and Tezduyar [25] to be used with the conjugate gradient method for solving problems with symmetric spatial operators (e.g. for problems governed by the Poisson equation). It was later used by Liou and Tezduyar [6] in conjunction with the GMRES method to solve compressible and incompressible flow problems.

In CEBE preconditioning, the set of elements $\mathbb{E}$ is partitioned into clusters of elements $\mathbb{E}_J$, $J = 1, 2, \ldots, N_{cl}$, where N_{cl} is the number of the clusters. The global matrix $\tilde{\mathbf{A}}_J$ associated with the cluster J is defined as

$$\tilde{\mathbf{A}}_J = \sum_{e \in \mathbb{E}_J} \tilde{\mathbf{A}}^e. \tag{11.39}$$

The matrix $\tilde{\mathbf{A}}$ can then be expressed as

$$\tilde{\mathbf{A}} = \mathbf{I} + \sum_{J=1}^{N_{cl}} \tilde{\mathbf{B}}_J \tag{11.40}$$

where

$$\tilde{\mathbf{B}}_J = \tilde{\mathbf{A}}_J - \tilde{\mathbf{W}}_J, \qquad J = 1, 2, \ldots, N_{cl}. \tag{11.41}$$

The CEBE preconditioning is based on the approximation of (11.40) by a sequential product of cluster level matrices. The Crout CEBE preconditioner is defined as

$$\tilde{\mathbf{P}} = \prod_{J=1}^{N_{cl}} \hat{\mathbf{L}}_J \prod_{J=N_{cl}}^{1} \hat{\mathbf{U}}_J \tag{11.42}$$

where $\hat{\mathbf{L}}_J$ and $\hat{\mathbf{U}}_J$ are the matrices resulting from the following Crout factorization:

$$(\mathbf{I} + \tilde{\mathbf{B}}_J) = \hat{\mathbf{L}}_J \hat{\mathbf{U}}_J, \qquad J = 1, 2, \ldots, N_{cl}. \tag{11.43}$$

Remarks

10. The convergence of the algorithm depends on the numbering of the clusters but not on the numbering of the elements within each cluster. By treating each cluster as a 'super element', we can identify the CEBE preconditioning as a generalization of the standard element-by-element preconditioning.
11. We have the option of storing the cluster matrices and their inverses, or recomputing them as they are needed.
12. To facilitate vectorization and parallel processing, just as it is done in GEBE preconditioning, the clusters can be grouped in such a way that no two clusters in any group have shared nodes. Furthermore, depending on the cluster size (i.e. the number of elements in the cluster), elements within each cluster can again be grouped in the same way. Then in the limiting case when the number of element groups is the same as the number of elements, the CEBE preconditioning reduces to a GEBE preconditioning.

11.6 APPLICATIONS: UNSTEADY FLOWS PAST A CIRCULAR CYLINDER

All solutions presented here were obtained with linear-in-time interpolation functions. In all cases, the computational values for the cylinder radius and the free-stream velocity are, respectively, 1.0 and 0.125; a time step of 1.0 is used for the computations. The dimensions of the computational domain, normalized by the cylinder radius, are 61.0 and 32.0 in the flow and cross-flow directions, respectively.

The mesh employed consists of 4060 elements and 4209 nodes. Symmetry conditions are imposed at the upper and lower computational boundaries, and the traction-free condition is imposed at the outflow boundary. The periodic solution is obtained by introducing a short term perturbation to the symmetric solution. For all computations, we use the CEBE iteration method to solve the resulting equation system. At each time step about 25 000 equations are solved simultaneously. We chose a Krylov vector space of dimension 25 and an average cluster size of 23 elements. This cluster size, determined by numerical experimentation, is nearly optimal in minimizing the CPU time. For this set of problems, the CEBE technique takes less then one-sixth the CPU time and less then one-third the storage needed by the direct method. The nodal values of the stationary stream function (normalized with the free-stream velocity) and vorticity are obtained by the least-squares interpolation. All the flow-field pictures shown in this article display the part of the domain enclosed by a rectangular region, with the lower left and upper right co-ordinates (13,10) and (43,22) respectively, relative to the lower left corner of the domain.

11.6.1 A fixed cylinder at Reynolds number 100

In this problem the cylinder location is fixed at (16,16) relative to the lower left corner of the domain. Figure 11.2 shows time history of the lift, drag and torque coefficients for the fixed cylinder. The Strouhal number obtained is 0.167. The difference between this value and the ones reported in Tezduyar et al. [22], computed with different formulations and on a finer mesh, is less than 2%. Figures 11.3 and 11.4 show a sequence of frames for the vorticity and stationary stream function during one period of the lift coefficient. In both figures, the first, third and last frames correspond to zero lift coefficient; the second and fourth frames correspond to the trough and crest of the lift coefficient, respectively. As expected, in each of these figures, the first and the last frames are very similar, while the first and third and the second and fourth frames are mirror images of each other.

11.6.2 A cylinder with forced horizontal oscillations at Reynolds number 100

It is well known that at Reynolds number 100, the flow past a fixed circular cylinder leads to the classical unsymmetrical vortex shedding. In such a case the lift and torque coefficients oscillate with a frequency corresponding to the related Strouhal number, while the drag oscillates with twice that frequency.

The case in which the cylinder is subjected to forced horizontal oscillations shows some very interesting features. Depending on the amplitude and the frequency (f_f) of the forced oscillations of the cylinder, two modes of vortex shedding are possible. This phenomenon, for vortex-induced oscillations, has been discussed in the review papers by King [26] and Sarpkaya [27]. Oscillations with a low reduced frequency ($F_f = 2f_f a/U_\infty$, where a is the radius of the cylinder and U_∞ is the free-stream velocity) lead to unsymmetric modes of vortex shedding. For higher values of F_f, on the other hand, symmetric vortex shedding is observed. However, such a symmetric

372 DIRECT TIME-INTEGRATION METHODS

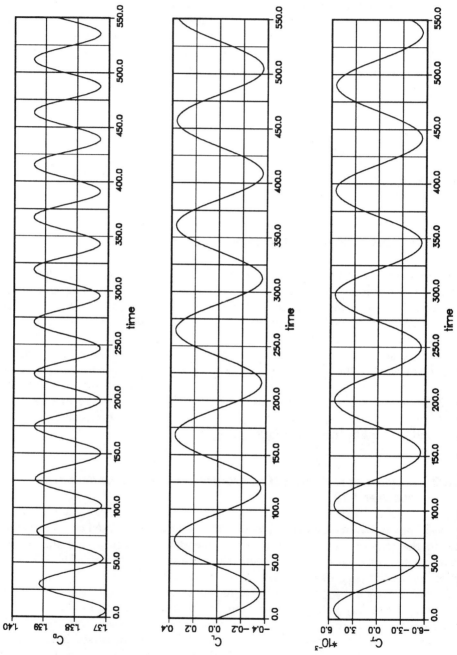

Figure 11.2 Flow past a fixed cylinder at Re = 100: time history of the lift, drag and torque coefficients.

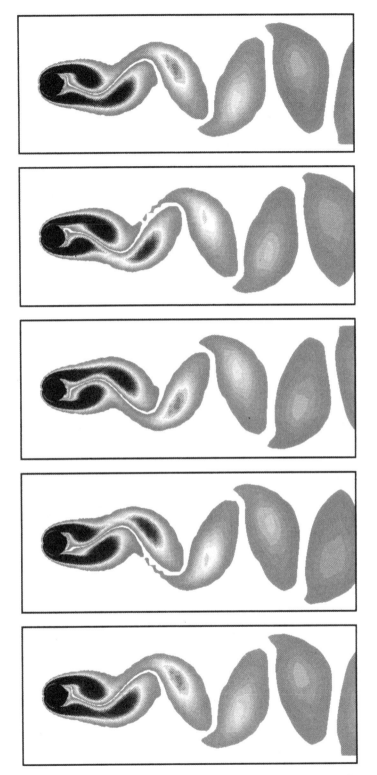

Figure 11.3 Flow past a fixed cylinder at Re = 100: vorticity at various instants during one period of the lift coefficient.

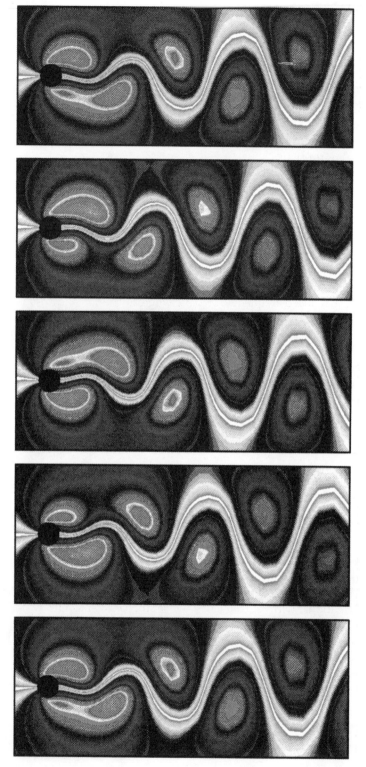

Figure 11.4 Flow past a fixed cylinder at Re = 100: stationary stream function at various instants during one period of the lift coefficient.

arrangement of the vortices is unstable, and consequently the vortices coalesce downstream.

We simulate the flow with symmetrical shedding by forcing the cylinder to oscillate horizontally with the following prescribed displacement (normalized by the cylinder radius):

$$X = 1 - \cos(\omega_f t) \tag{11.44}$$

where $\omega_f = 2\pi f_f$. For this case, the value of f_f corresponds to a reduced frequency of 0.35. The initial condition for this simulation is prescribed as the unsteady solution for flow past a fixed cylinder at Re = 100 (from the previous example). Figure 11.5 shows the time history of the drag, lift and torque coefficients and the normalized horizontal displacement and velocity (normalized by the free-stream velocity) of the cylinder. We observe that the drag coefficient for the horizontally oscillating cylinder is significantly larger than that for a fixed cylinder. Furthermore, the drag coefficient oscillates with a reduced frequency of 0.35 whereas the lift and torque coefficients approach zero. The fact that we start from an unsymmetric solution and still obtain a symmetric mode of shedding demonstrates that this mode is a stable one. Figures 11.6 and 11.7 show a sequence of frames for the vorticity and stationary stream function during one period of the cylinder motion. During each period two pairs of symmetric vortices are shed from the cylinder. In both figures, the first, third and last frames correspond to the mean cylinder location, while the second and fourth frames correspond, respectively, to the left and right extreme positions of the cylinder.

11.6.3 A cylinder with vortex-induced vertical oscillations at Reynolds number 324

In the first numerical example we observed that for sufficiently high Reynolds numbers (>40) flow past a fixed cylinder leads to unsymmetric vortex shedding. This causes the cylinder to experience alternating lift force at a frequency corresponding to the Strouhal number for that Reynolds number. Now, if the cylinder is mounted on a flexible support, then under certain conditions it can undergo sustained oscillations with a frequency close to, or coincident with, its natural frequency. These oscillations can alter the vortex shedding mechanism which in turn can change the cylinder response and so on. This leads to a complex non-linear fluid-structure interaction phenomenon and has been addressed by several researchers [26–9]. We simulate this phenomenon for a cylinder which is allowed to move only in the vertical direction. The motion of the cylinder is governed by the following equation:

$$\ddot{Y} + 2\pi F_n \zeta \dot{Y} + (\pi F_n)^2 Y = \frac{C_L}{M}. \tag{11.45}$$

Here $\ddot{Y}$, $\dot{Y}$ and Y are, respectively, the normalized vertical acceleration, velocity and

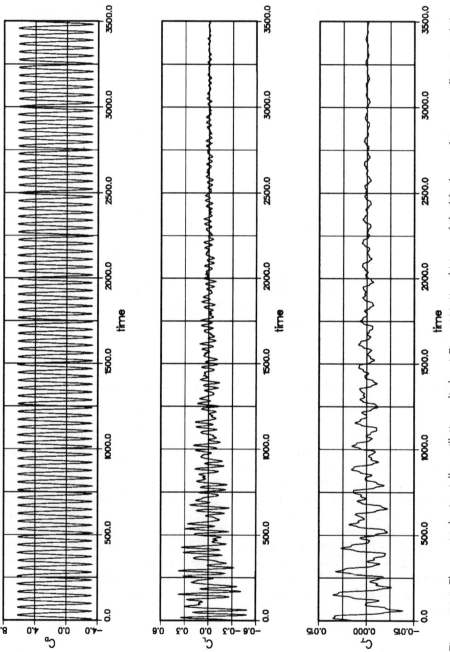

Figure 11.5 Flow past a horizontally oscillating cylinder at Re = 100: time history of the lift, drag and torque coefficients and the normalized displacement and velocity of the cylinder.

11.6 APPLICATIONS 377

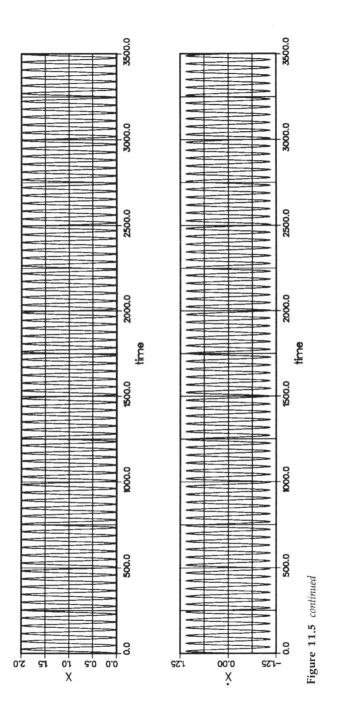

Figure 11.5 continued

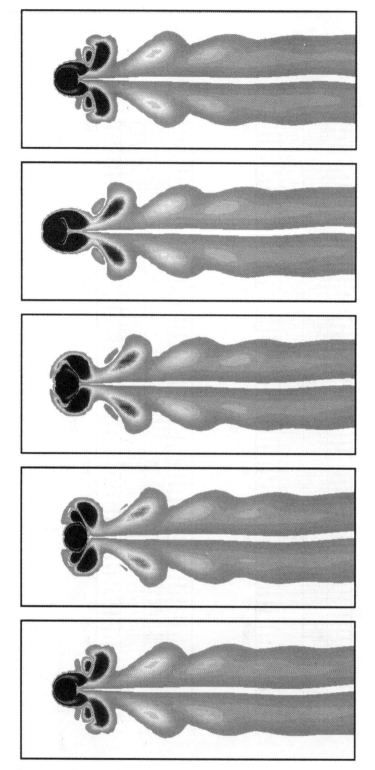

Figure 11.6 Flow past a horizontally oscillating cylinder at Re = 100: vorticity at various instants during one period of the cylinder motion.

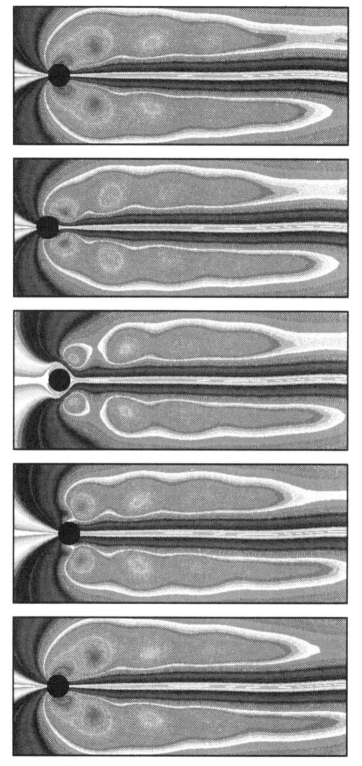

Figure 11.7 Flow past a horizontally oscillating cylinder at Re = 100: stationary stream function at various instants during one period of the cylinder motion.

380 DIRECT TIME-INTEGRATION METHODS

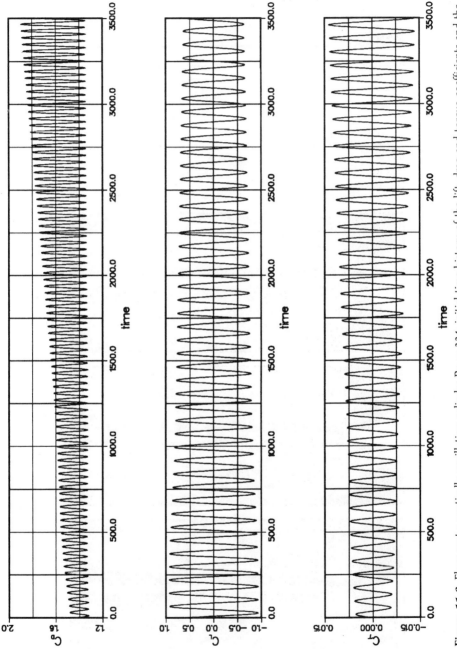

Figure 11.8 Flow past a vertically oscillating cylinder Re = 324: initial time history of the lift, drag and torque coefficients and the normalized displacement and velocity of the cylinder.

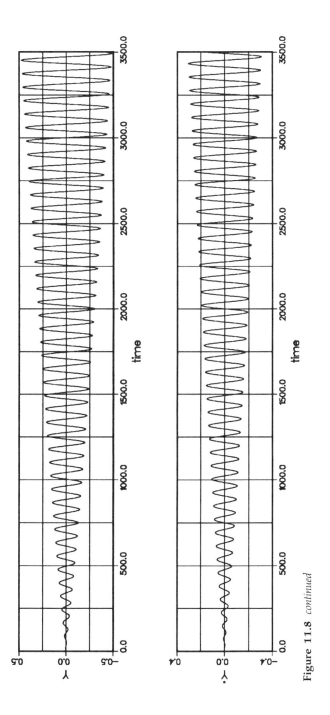

Figure 11.8 *continued*

382 DIRECT TIME-INTEGRATION METHODS

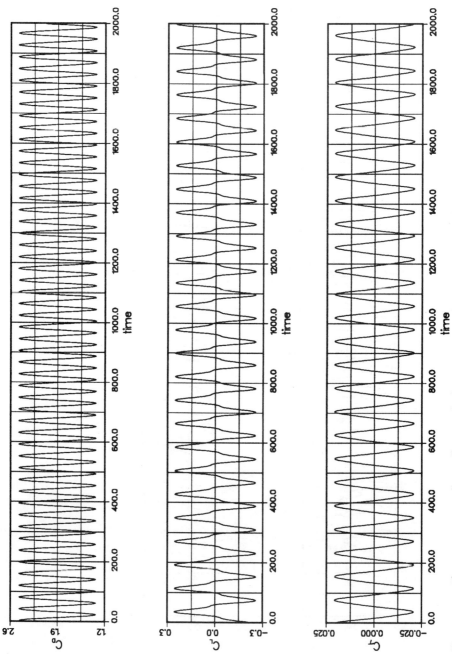

Figure 11.9 Flow past a vertically oscillating cylinder at $Re = 324$: later time history of the lift, drag and torque coefficients and the normalized displacement and velocity of the cylinder.

11.6 APPLICATIONS

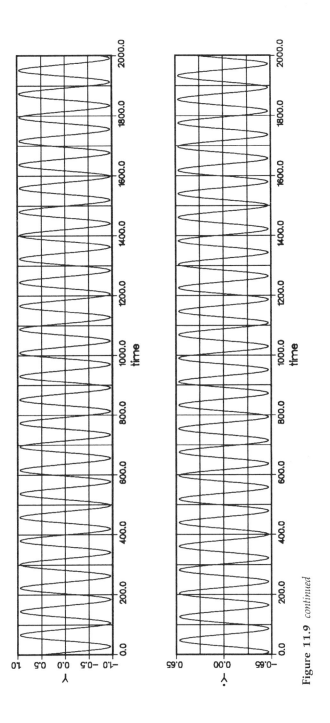

Figure 11.9 *continued*

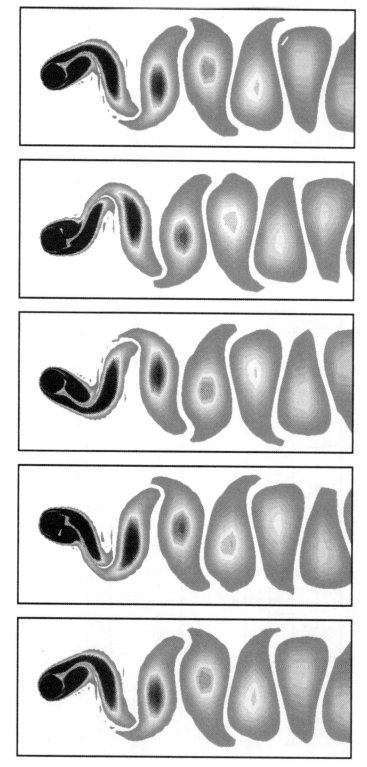

Figure 11.10 Flow past a vertically oscillating cylinder at Re = 324: vorticity at various instants during one period of the cylinder motion.

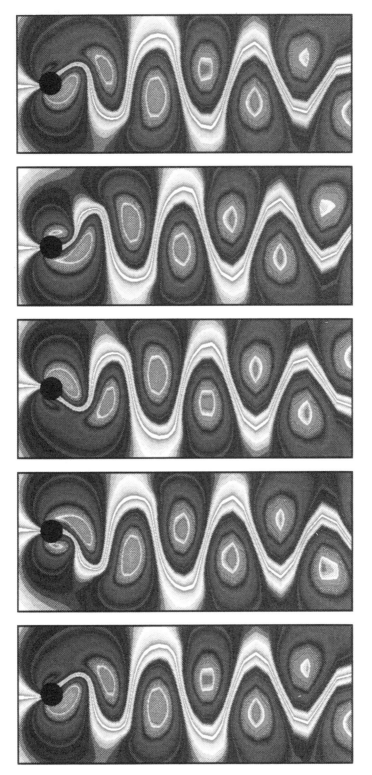

Figure 11.11 Flow past a vertically oscillating cylinder at Re = 324: stationary stream function at various instants during one period of the cylinder motion.

displacement of the cylinder. The displacement and velocity of the cylinder are normalized by its radius and the free-stream velocity, respectively. M is the non-dimensional mass/unit length of the cylinder, ζ is the structural damping coefficient associated with the system, and C_L denotes the lift coefficient for the cylinder. F_n, the reduced natural frequency of the spring–mass system, is defined as

$$F_n = \frac{2f_n a}{U_\infty} \qquad (11.46)$$

where f_n is the actual natural frequency of the system. For our problem $F_n = 0.204$, $M = 472.74$ and $\zeta = 3.3 \times 10^{-4}$.

At Reynolds number 324 the reduced natural frequency of the spring mass system and the Strouhal number for flow past a fixed cylinder have very close values. Therefore, we decided to carry out this simulation for Reynolds number 324. The periodic solution for flow past a fixed cylinder at the same Reynolds number is used as the initial condition. Figure 11.8 shows, for the initial stages of the simulation, time history of the lift, drag and torque coefficients and the normalized vertical displacement and velocity of the cylinder. We observe that the cylinder oscillates with an increasing amplitude. The drag and torque coefficients for the cylinder also increase while the lift coefficient shows a decreasing amplitude. It is interesting to note that both the mean and peak values of the drag coefficient increase with time, but the trough value remains almost constant. The quantities displayed in Figure 11.8 are shown in Figure 11.9 for a later stretch of time when the cylinder reaches a steady-state oscillation amplitude of about one radius. The cylinder oscillates with its natural frequency, and so does the torque coefficient; the drag coefficient oscillates with twice the natural frequency of the cylinder. The dominant frequency for the lift coefficient corresponds to the natural frequency of the cylinder. In addition, there is a very small component of the lift coefficient with thrice the frequency of the dominant one. Figures 11.10 and 11.11 show a sequence of frames for the vorticity and stationary stream function during one period of the cylinder motion. In both figures the first, third and last frames correspond to mean cylinder location, while the second and fourth frames correspond, respectively, to the lower and upper extreme positions of the cylinder.

11.7 CONCLUDING REMARKS

The space–time finite element formulation with the GLS stabilization, for incompressible flows, was reviewed. The GEBE and CEBE iteration methods employed to solve the equation systems resulting from these space–time finite element discretizations were also reviewed. In the space–time formulation the deformation of the spatial domain is automatically taken into account. Therefore this formulation is very suitable for flow problems involving moving boundaries and interfaces, such as free-surface flows, liquid drops, two-liquid flows, and flows with moving objects. The Galerkin/least-squares stabilization leads to a formulation which is consistent. That is, the stabilization terms added to the Galerkin formulation of the problem vanish when an exact solution is substituted into the stabilized formulation. Consequently, this stabilization method introduces minimal numerical diffusion, and therefore results in solutions with minimal loss of accuracy. By employing iteration techniques with GEBE and CEBE preconditioners we are able to substantially reduce the computational cost associated with solving the fully-discretized equations of the space–time formulation. These iteration techniques are highly vectorizable and parallelizable.

As applications of the stabilized space–time formulation and the iteration methods, computations were performed for unsteady incompressible flow problems involving fixed and oscillating cylinder. Some interesting physical phenomena were observed as a result of these computations. While for flow past a fixed cylinder, the usual, unsymmetric vortex shedding was observed, when the cylinder was subjected to horizontal oscillations with certain prescribed frequency and amplitude, symmetrical vortex shedding was observed instead. The case of vortex-induced vertical oscillations was also simulated. These oscillations result in an increase in the drag and torque coefficients and a decrease in the lift coefficient.

ACKNOWLEDGEMENTS

This work was sponsored by NASA-Johnson Space Center (under grant NAG 9-449), NSF (under grant MSM-8796352), University of Minnesota Army High Performance Computing Research Center, and the ALCOA Foundation.

BIBLIOGRAPHY

[1] Hughes, T.J.R. and Franca, L.P. (1987) A new finite element formulation for computational fluid dynamics: VII. The Stokes problem with various well-posed boundary conditions: symmetric formulations that converge for all velocity/pressure spaces, *Comput. Meth. Appl. Mech. Engg*, **65**, 85–96.

[2] Hughes, T.J.R., Franca, L.P. and Hulbert, G.M. (1989) A new finite element formulation for computational fluid dynamics: VIII. The Galerkin/least-squares method for advective-diffusive equations, *Comput. Meth. Appl. Mech. Engg*, **73**, 173–189.

[3] Shakib, F. (1988) *Finite Element Analysis of the Compressible Euler and Navier–Stokes Equations*, Ph.D. Thesis, Department of Mechanical Engineering, Stanford University, Stanford, California.

[4] Tezduyar, T.E., Behr, M. and Liou, J. (1992) A New Strategy for Finite Element Computations Involving Moving Boundaries and Interfaces — the DSD/ST Procedure: I. The Concept and the Preliminary Numerical Tests, *Comput. Meth. Appl. Mech. Engg.*, **94**, 339–351.

[5] Tezduyar, T.E., Behr, M., Mittal, S. and Liou, J. (1992) A New Strategy for Finite Element Computations Involving Moving Boundaries and Interfaces — the DSD/ST Procedure: II. Computation of Free-surface Flows, Two-liquid Flows, and Flows with Drifting Cylinders, *Comput. Meth. Appl. Mech. Engg.*, **94**, 353–371.

[6] Liou, J. and Tezduyar, T.E. (1990) *Computation of Compressible and Incompressible Flows with the Clustered Element-by-element Method*, University of Minnesota Supercomputer Institute Research Report, UMSI 90/215, October.

[7] Hansbo, P. and Szepessy, A. (1990) A velocity–pressure streamline diffusion finite

element method for the incompressible Navier–Stokes equations, *Comput. Meth. Appl. Mech. Engg,* **84**, 175–192.
[8] Hughes, T.J.R., Franca, L.P. and Mallet, M. (1987) A new finite element formulation for computational fluid dynamics: VI. Convergence analysis of the generalized SUPG formulation for linear time-dependent multi-dimensional advective–diffusive systems, *Comput. Meth. Appl. Mech. Engg,* **63**, 97–112.
[9] Hughes, T.J.R. and Hulbert, G.M. (1988) Space–time finite element methods for elastodynamics: formulations and error estimates, *Comput. Meth. Appl. Mech. Engg,* **66**, 339–363.
[10] Saad, Y. and Schultz, M.H. (1983) *GMRES: a Generalized Minimal Residual Algorithm for Solving Nonsymmetric Linear Systems,* Research Report, YALEU/DCS/RR-254.
[11] Hughes, T.J.R., Levit, I. and Winget, J. (1983) An element-by-element solution algorithm for problems of structural and solid mechanics, *Comput. Meth. Appl. Mech. Engg,* **36**, 241–254.
[12] Hughes, T.J.R., Winget, J., Levit, I. and Tezduyar, T.E. (1983) New alternating direction procedures in finite element analysis based upon EBE approximate factorizations, in S.N. Atluri and N. Perrone (eds.), *Recent Developments in Computer Methods for Nonlinear Solids and Mechanics,* AMD-Vol. 54, ASME, New York, pp. 75–109.
[13] Hughes, T.J.R. and Ferencz, R.M. (1988) in R. Glowinski, G.H. Golub, G.A. Meurant and J. Periaux (eds.), *First International Symposium on Domain Decomposition Methods for Partial Differential Equations,* SIAM, Philadelphia, p. 261.
[14] Tezduyar, T.E. and Liou, J. (1989) Grouped element-by-element iteration schemes for incompressible flow computations, *Comput. Phys. Commun.,* **53**, 441–453.
[15] Tezduyar, T.E., Liou, J., Nguyen, T. and Poole, S. (1988) Adaptive implicit–explicit and parallel element-by-element iteration schemes, *Proc. 2nd Int. Symp. on Domain Decomposition Methods* (eds. Chan, T.F., Glowinski, R., Periaux, J. and Widlund, O.B.), Los Angeles, January (SIAM, Philadelphia), pp. 443–463.
[16] Hughes, T.J.R. and Brooks, A.N. (1979) A multi-dimensional upwind scheme with no crosswind diffusion, in Hughes, T.J.R. (ed.), *Finite Element Methods for Convection Dominated Flows,* AMD-Vol. 34, ASME, New York, pp. 19–35.
[17] Brooks, A.N. and Hughes, T.J.R. (1982) Streamline-upwind/Petrov–Galerkin formulations for convection dominated flows with particular emphasis on incompressible Navier–Stokes equation, *Comput. Meth. Appl. Mech. Engg,* **32**, 199–259.
[18] Tezduyar, T.E. and Hughes, T.J.R. (1982) *Development of Time-accurate Finite Element Techniques for First-order Hyperbolic systems with Particular Emphasis on the Compressible Euler Equations,* Report prepared under NASA-Ames University Consortium Interchange No. NCA2-OR745-104.
[19] Tezduyar, T.E., Glowinski, R. and Liou, J. (1984) Petrov–Valerkin methods on multiply-connected domains for the vorticity-stream function formulation of the incompressible Navier–Stokes equations, *Int. J. Num. Meth. Fluids,* **8**, 1269–1290.
[20] Brezzi, F. and Pitkaranta, J. (1984) On the stabilization of finite element approximations of Stokes problem, *Efficient Solutions of Elliptic Systems, Notes on Numerical Fluid Mechanics,* Viewig, **10**, 11–19.
[21] Hughes, T.J.R., Franca, L.P. and Balestra, M. (1986) A new finite element formulation for computational fluid dynamics: V. Circumventing the Babuska–Brezzi condition: A stable Petrov–Galerkin formulation of the Stokes problem accommodating equal-order interpolations, *Comput. Meth. Appl. Mech. Engg,* **59**, 85–99.
[22] Tezduyar, T.E., Mittal, S., Ray, S.E. and Shih, R. (1992) Incompressible Flow Computations with Stabilized Bilinear and Linear Equal-order-interpolation Velocity–Pressure Elements, *Comput. Meth. Appl. Mech. Engg.,* **95**, 221–242.
[23] Lundgren, T. and Mansour, N.N. (1990) A stabilized Lagrangian finite element method

for viscous free surface flows, *in preparation*.
[24] Franca, L.P., Frey, S.L. and Hughes, T.J.R. (1992) Stabilized Finite Element Methods: I. Application to the Advective-Diffusive Model, *Comput. Meth. Appl. Mech. Engg.*, **95**, 253–276.
[25] Liou, J. and Tezduyar, T.E. (1990) *A Clustered Element-by-Element Method for Finite Element Computations*, University of Minnesota Supercomputer Institute Research Report UMSI 90/116, July.
[26] King, R. (1977) A review of vortex shedding research and its application, *Ocean Engineering*, **4**, 141–172.
[27] Sarpkaya, T. (1979) Vortex-induced oscillations, *J. Appl. Mech.*, **46**, 241–258.
[28] Blevins, R.D. (1990) *Flow-induced Vibration*, Van Nostrand Reinhold, New York (2nd edn).
[29] Griffin, O.M. and Koopmann, G.H. (1979) The vortex-excited lift and reaction forces on resonantly vibrating cylinders, *J. Sound and Vibration*, **54**(3), 435–448.

S. Mittal and T.E. Tezduyar
Department of Aerospace Engineering and Mechanics, Army High Performance Computing Research
Center and Minnesota Supercomputer Institute
University of Minnesota
1200 Washington Avenue South
Minneapolis
MN 55415
USA

12

Methods for Optimization of Large-scale Systems*

J. S. Arora

University of Iowa, IA, USA

12.1 INTRODUCTION

This chapter presents fundamental concepts of optimum design and numerical methods. Basic concepts common to most optimization algorithms are described. The most modern algorithms based on linearization and quadratic programming are presented and discussed. These algorithms need to be implemented in the computer in a robust way. Therefore, computational aspects of the algorithms as well as their computer implementation aspects are discussed. Several other optimization methods are also briefly presented. The problem of optimum design of structural systems modelled by the finite elements is discussed. Numerous other applications of the algorithms are mentioned; however, due to space restrictions, details of the algorithms and their applications cannot be included here. References are cited, however, for more details on these topics. The problem of optimum design of large-scale systems requires special attention, so it is also discussed. Most of the material for the chapter is taken from books and review articles, such as References [1–5].

12.2 DESIGN OPTIMIZATION MODEL

Formulation of the problem is an important step in the design optimization process. Problems of static and dynamic response, shape optimization of mechanical and

---

*Part of the material in this chapter is derived with permission from author's book, *Introduction to Optimum Design*, published by McGraw-Hill Book Co., New York, 1989.
Solving Large-scale Problems in Mechanics, Edited by M. Papadrakakis
© 1993 John Wiley & Sons Ltd

structural systems using finite element and other simulation methods, reliability-based design, optimal control of systems and many others can be formulated as *non-linear programming* (NLP) problems [1–20]. The problem formulation process requires identification of *design variables* (that describe the system), a *cost function or an objective function* (that needs to be optimized), and *constraints* for safe and proper performance of the system. Depending on the class of problems and needs, several types of design variables and cost functions can be identified. Constraints usually involve physical limitations, material failure, buckling load, vibration frequencies and other such response quantities. These problems can be transcribed into a *standard NLP model* defined as minimization of a function subject to equality constraints and inequality constraints expressed in a '$\leq$' form as

Problem P Find a design variable vector X of dimension n to minimize a cost function $f(X)$ with X in the *feasible set* (also called the *constraint set*) defined as

$$S = \{X: g_j(X) = 0, j = 1 \text{ to } p; g_j(X) \leq 0, j = p+1 \text{ to } m\} \tag{12.1}$$

where p is the number of equality constraints and m is the total number of constraints.

Note that most of the optimization problems can be transformed to the form of Problem P, i.e. *maximization* of $F(X)$ can be replaced by minimization of $f(X) = -F(X)$ and $G_i(X) \geq 0$ can be replaced by $g_i(X) = -G_i(X) \leq 0$. In the numerical optimization process, the *explicit lower and upper bonds* on the design variable are easy to impose. Therefore they are treated separately in computer implementation of the algorithms to improve efficiency. However, for sake of simplicity of presentation, such *simple bound constraints* are assumed to be included in the equality constraints $g_j(X) \leq 0$.

An inequality constraint $g_i(X) \leq 0$ is said to be *active (tight or binding)* at a design point X if it is satisfied as an equality at that point, i.e. $g_i(X) = 0$. It is said to be *inactive* if it has negative value at that point, and *violated* if it has positive value. Note that an *equality* constraint is always either *active or violated* for any design point.

In the numerical optimization process, it is important to use the idea of *ε-active constraints*. To define such constraints, we select a small positive number ε. Any inequality constraint $g_i(X) \leq 0$ is said to be *ε-active* at the point X if $g_i(X) < 0$ but $g_i(X) + \varepsilon \geq 0$. An ε-active constraint for a design point simply means that the design point is close to the constraint boundary on the feasible side (within an ε-band).

If cost and constraint functions of the problem are linear in terms of the variables, the Problem P is called a *linear programming* (LP) problem. If the cost function is quadratic and the constraints are linear, the problem is called a *quadratic programming* (QP) problem. There are many practical applications that can be formulated as LP or QP problems. In addition, most of the numerical algorithms for non-linear problems solve either an LP or QP subproblem in their iterative process. Therefore these problem areas are quite important, and it is not surprising that a substantial amount of research has been done and presented in the literature to develop special methods for their solution.

Basic concepts related to optimum design of systems have been recently summarized by Arora [3]. *Local and global minima* are defined and *existence of solution* for the design optimization model is discussed. *Optimality conditions (Karush–Kuhn–Tucker (KKT)) necessary conditions and sufficient conditions*) for the model are presented and discussed. *Convexity* of the design model is discussed and the question of a *global solution* for the problem is addressed. *Post optimality analyses* are briefly discussed. *Duality* results for nonlinear programming problems are also presented and discussed. These concepts are important in practical applications because they give insights for the problem if the optimization process fails to produce a useful solution. They are also important because most of the numerical methods for optimization are based on these concepts. For a review of these concepts, References [1–3, 21–23] should be consulted.

It is important to note the following points for the foregoing NLP model relative to the structural and mechanical system design problems:

1. The model is applicable to all problems with continuous design variables. *Multi-objective* [24] and *discrete variable* [25] problems can also be treated after certain extensions.

2. The functions of the problem are assumed to be *twice differentiable*. Problems having *non-differentiable functions* can be treated with additional computations effort [26]. Also, gradients of *active constraints* are assumed to be linearly independent at the optimum, so that Karush–Kuhn–Tucker optimality conditions hold at the minimum point [1].

3. The *cost and constraint functions are usually implicit* as well as explicit functions of the design variables. These functions, however, can be calculated using various analysis methods once a design is specified. There are methods that replace the problems having implicit functions with a series of approximate problems that have explicit dependence on the design variables.

4. *Gradients* of the functions are needed in numerical methods of optimization. Efficient methods to calculate them taking advantage of the structure of engineering design problems have been developed [27]. Two methods are presented later in this chapter for structural design problems modelled by finite elements.

5. Realistic analysis models can be quite large, requiring enormous calculation to evaluate implicit functions at a trial design. Thus, the *number of calls for function evaluations* is a measure of the efficiency of an optimization algorithm.

6. Engineering design problems usually have a large number of inequality constraints. However, a majority of them are not binding at the optimum point. Since this set of binding constraints is not known *a priori*, it must be determined during the solution process. Some algorithms differentiate all the inequalities at each iteration to define a direction-finding sub-problem while others differentiate only a subset of them. Since gradient evaluation for problems involving implicit functions needs large computation, the *number of gradient evaluations* at each iteration, therefore, is another measure of efficiency.

12.3 BASIC CONCEPTS RELATED TO NUMERICAL ALGORITHMS

Computational methods for design optimization of structural and mechanical systems have matured during the 1970s and 1980s and are beginning to be utilized in realistic engineering design applications. This field has become known as *computational design optimization* (CDO). Many papers have been published that show applications of the methods. It is not possible to include a detailed review of all the methods and their variations, or reference to all the published articles. Only a linearization method and sequential quadratic programming methods are described in some detail and a few others are only summarized. Numerous review articles, text books and conference proceedings have appeared from time to time to discuss various aspects of the CDO field. A recent review article [2] cites 11 review articles, 19 textbooks, 15 conference proceedings, and 60 other journal articles. This body of literature contains work of almost all the researchers in the field.

Engineering design problems are usually quite complex with each application having its own design requirements, simulation methods and constraints to meet. In addition, as more powerful computers become available, the desire to solve more complex and larger problems also grows. Further, since the CDO field has matured substantially during the last decade, more non-experts of optimization algorithms are beginning to use this new technology in their design work. All these factors dictate the *use of a theoretically sound and reliable algorithm*. Use of such an algorithm in routine design environment can remove uncertainty about the optimization algorithm behaviour, allowing the designers to concentrate on their design problems. Such theoretically sound algorithms, although computationally more expensive, can be *more cost-effective* in a general design environment. In reality they can save human resources and facilitate investigation of alternative design concepts in a relatively short time, thus resulting in better designs.

In this section, some basic concepts related to numerical algorithms for optimization of small as well as large systems are presented and discussed. The ideas of a descent function, constraint normalization, and potential constraint strategy are introduced. Convergence of an algorithm is discussed and attributes of a good algorithm are presented. It is important to note that all the algorithms discussed in this chapter converge only to a *local minimum* point for Problem P. Algorithms for finding a *global solution* require extensive numerical calculations [28] and are not discussed here.

12.3.1 A basic algorithm

Some general concepts are presented that form the basis for many computational algorithms for design optimization. Most of the algorithms are based on the general iterative prescription:

$$\mathbf{X}^{(k+1)} = \mathbf{X}^{(k)} + \alpha_k \mathbf{d}^{(k)}, \qquad k = 0, 1, 2, \ldots, \tag{12.2}$$

where the superscript k represents the iteration number, $\mathbf{X}^{(k)}$ is the current estimate of the optimum point, $\alpha_k \mathbf{d}^{(k)}$ is a change in design, α_k is a step size, $\mathbf{d}^{(k)}$ is a search direction calculated using function values and their gradients, and $X^{(0)}$ is an initial design estimate given by the designer.

Optimization algorithms are broadly classified as *primal and transformation methods*. In *primal methods* the direction vector $\mathbf{d}^{(k)}$ is calculated using the function values and their gradients at the point $X^{(k)}$. The step-size calculation needs only the function values. Different algorithms can be generated depending on how the direction $\mathbf{d}$ and step size α are calculated. In many algorithms, $\mathbf{d}$ is calculated by solving a linear or quadratic programming sub-problem. Several methods based on this idea are described later in this chapter.

The *transformation methods* convert Problem P to a sequence of unconstrained optimization problems whose solutions converge to solution of the original problem. They include barrier and penalty function methods as well as the *augmented Lagrangian or multiplier methods* [29–32]. In these methods, a transformed function is constructed by adding a penalty term for the constraint violations to the cost function, as $\Phi(X, \mathbf{r}) = f(X) + P(\mathbf{g}(X), \mathbf{r})$, where $\mathbf{r}$ is a scalar or vector of penalty parameters and P is a real-valued function whose action of imposing the penalty is controlled by $\mathbf{r}$. Details of the transformation methods can be found in the foregoing references and the literature cited there.

In *primal methods*, it is assumed that the optimum solution lies on the boundary of the feasible set S even if there are only inequality constraints in the problem. If the solution is unconstrained, then unconstrained methods can be used to locate it. Several philosophies have been used to develop various algorithms. For example, if an intermediate design or the starting design is infeasible, many methods will iterate through the infeasible domain to reach the final solution; many others correct the constraints to reach the feasible domain first and then move along the boundary to reach the solution point. Still others make special calculations not to violate constraints during the iterative process. Some algorithms generate and use second-order information about the problem as the iterations progress.

Several algorithms based on the strategies described in the foregoing have been developed are evaluated. Some algorithms are better for a certain class of problems than others. A few algorithms work well if the problem has only inequality constraints whereas others can treat both equality and inequality constraints simultaneously. We shall concentrate only on general algorithms that have no restriction on the form of the functions or the constraints. Such algorithms are usually based on the following *four basic steps*:

1. *Linearization of cost and constraint functions* about the current design. Some methods require linearization of all the constraints while others only for a subset of them.

2. *Definition of a search direction determination sub-problem* using the linearized cost and constraint functions.

3. *Solution of the sub-problem* which gives a search direction in the design space.

4. *Calculation of a step size* to minimize a descent function (defined later) in the search direction.

12.3.2 Constraint normalization

In numerical calculations, it is essential to normalize all the constraint functions. This way the constraints can be easily compared to each other, and one value for ε (say 0.10) can be used for all constraints in defining the ε-active set. If they are not normalized, it is not proper to use one ε for all the constraints. For example, consider a stress constraint $\sigma \leq \sigma_a$, or $\sigma - \sigma_a \leq 0$, and a displacement constraint $\delta \leq \delta_a$, or $\delta - \delta_a \leq 0$, where $\sigma =$ calculated stress, $\sigma_a =$ an allowable stress, $\delta =$ calculated deflection, and $\delta_a =$ an allowable deflection. Since the units for the two constraints are different their values are a widely differing orders of magnitude. If they are violated during the iterative solution process, it is difficult to judge the severity of their violation. However, if they are normalized as $R - 1.0 \leq 0$, where $R = \sigma/\sigma_a$ for the stress constraint, and $R = \delta/\delta_a$ for the deflection constraint (note that σ_a and δ_a are assumed to be positive; otherwise, the sense of the inequality will change), then it is easy to compare them and define an ε-active set using the same value of ε for all of them.

There are other constraints that must be normalized as $1.0 - R \leq 0$. For example, the fundamental vibration frequency ω of a structure must be above a given threshold value of ω_a, i.e. $\omega \geq \omega_a$. When the constraint is normalized and converted to the standard 'less than' form, it becomes $1.0 - R \leq 0$, where $R = \omega/\omega_a$.

It is *assumed* in the remaining chapter that all the constraints for the problem have been appropriately normalized.

12.3.3 Potential constraint strategy

To evaluate the search direction, one needs to know the cost and constraint functions and their gradients at the current design point. The algorithms can be classified based on whether gradients of all the constraints or only a subset of them are required at a design iteration. The algorithms that use gradients of only a subset of the constraints are said to use a *potential constraint strategy*. A potential constraint set can be defined in several different ways. In general, it comprises active, ε-active and violated constraints, such as the following index set I_k at the kth design point $\mathbf{X}^{(k)}$:

$$I_k = \begin{bmatrix} \{i : i = 1 \text{ to } p\} \text{ and} \\ \{i : g_i(\mathbf{X}^{(k)}) + \varepsilon \geq 0, i = (p+1) \text{ to } m\} \end{bmatrix}. \tag{12.3}$$

Note that the equality constraints are always included in the index set I_k by definition.

12.3.4 Descent function

A function used to monitor progress of the iterative optimization process towards the minimum is called the descent function or the merit function. The cost function is the obvious descent function for the unconstrained optimization problems. Use of the cost function, however, as a descent function for constrained optimization can be tedious because it does not account for the constraints of the problem. Therefore, many other descent functions have been proposed and used for constrained problems. These functions must include the effect of constraint violations. We shall present some of the functions later in this chapter. At this point, the purpose of the descent function should be well understood. The basic idea is to compute a step size along the search direction $\mathbf{d}^{(k)}$ such that the descent function is reduced at each iteration of the algorithm (note that the cost function for the problem may actually increase at some iterations). With this requirement proper progress towards the minimum point is maintained. The descent function also has the property that its minimum value is the same as that of the original cost function.

12.3.5 Convergence of an algorithm

An algorithm is said to be convergent if it reaches a minimum point starting from an arbitrary initial design. An algorithm that has been proved to converge starting from an arbitrary point is called a *globally convergent* method. Such an algorithm satisfies the following two requirements:

1. There is a descent function for the algorithm that is required to decrease at each iteration. This way, progress towards the minimum point can be monitored.

2. The direction of design changes $\mathbf{d}^{(k)}$ is a continuous function of the design variables. This requirement implies that there is no 'oscillation', or 'zigzagging' in the set of active constraints during the iterative design process (zigzagging implies that a constraint is coming in and out of the potential set in alternate iterations).

In addition to the preceding two requirements, *existence of a solution* for the problem must be guaranteed. This is possible if the feasible domain (feasible set S) for the problem is closed and bounded, and the cost function is continuous. These requirements are related to the formulation of the problem. Note that if $f(\mathbf{X})$ is continuous on a closed and bounded set S, it is a *bounded* function. If all the constraint functions $g_i(\mathbf{X})$ are continuous and there are no strict inequalities in the problem formulation (such as $g_i(\mathbf{X}) < 0$), then the set S is *closed*. Therefore this condition is not difficult to check. The set S is *non-empty* if there are feasible designs, i.e. there are no conflicting constraints and the problem is not over-constrained. This condition is difficult to check before initiating the solution process. If the solution process does not yield a feasible point, then one can suspect the set S to be empty. In that case, the problem formulation needs to be re-examined, so that there are feasible designs for the problem. The *boundedness* of S is difficult to check. It must be shown that there exists a number R such that for any $\mathbf{X}$ in S, $(\mathbf{X} \cdot \mathbf{X}) < R$, i.e. the elements of the set have lower and upper bounds. Note that $(\mathbf{x} \cdot \mathbf{y}) = \mathbf{x}^T\mathbf{y}$ represents *scalar product* of two vectors. This notation is used throughout this chapter.

12.3.6 Good algorithm: a definition

Based on the foregoing discussion a good algorithm has the following attributes listed in the order of their priority [1,2]:

1. *Robustness*: The algorithm must be reliable for general design applications and thus must be theoretically guaranteed to converge to a solution point starting from an initial design estimate (even bad designs). This must be possible with one execution of the software based on the algorithm without adjusting any parameters.
2. *Generality*: The algorithm must be general implying that it should be able to treat equality as well as inequality constraints and should not impose any restrictions on the form of the cost and constraint functions.
3. *Accuracy*: The ability of an algorithm to converge to the precise mathematical optimum point is important even though it may not be required in practice. An accurate optimization algorithm is likely to have a sound mathematical basis and thus higher reliability.
4. *Ease of use*: The algorithm must be easy to use by the experienced as well as inexperienced designers. An algorithm that requires selection of tuning parameters which are problem dependent is difficult to use.
5. *Efficiency*: The algorithm must be efficient for general engineering design applications. To be efficient, the number of repeated analyses of the system (such as finite element analyses) must be kept to a minimum. Thus, an efficient algorithm has (i) faster rate of convergence requiring fewer number of iterations and consequently, fewer system analyses, and (ii) least number of calculations within one design iteration.

It is important to note that there exists a trade-off between robustness, generality, accuracy and ease of use of an algorithm versus its efficiency. Some computational penalty must be paid in terms of CPU time to achieve robustness, generality, accuracy and ease of use of an algorithm [33].

Optimum design of *large-scale systems* presents special challenges for any numerical algorithm in terms of its efficiency. For such systems, the algorithm should be selected judiciously. For example, if an algorithm does not use the potential constraint strategy in defining the search direction determination subproblem, then it is not suitable for large scale problems. Also an algorithm that requires tuning of certain parameters for its proper performance will not be suitable because it may require several trials before proper values for the parameters are found. This is not only inconvenient but also inefficient. Such algorithms should be avoided for optimization of large-scale systems. It is important to note that a *good algorithm* must be robust (i.e. it must converge to a local minimum) for small as well as large-scale problems. The algorithm may be slightly less efficient for small scale problems (small problems require very little CPU time any way), but it must be more efficient for large-scale problems.

12.4 LINEARIZATION OF THE PROBLEM

At each iteration, most numerical methods for constrained optimization compute a design change by solving an approximate sub-problem which is obtained by writing linear Taylor's expansions for the cost and constraint functions. All search methods start with an initial design estimate and iteratively improve it. Let $\mathbf{X}^{(k)}$ be the design estimate at the kth iteration and $\Delta\mathbf{X}^{(k)}$ be the desired change in design. Instead of using $\Delta\mathbf{X}^{(k)}$ as change in design, usually it is taken as the search direction $\mathbf{d}^{(k)}$ and a step size is calculated along it to determine the new point. We write Taylor's expansion of the cost and constraint functions about the point $\mathbf{X}^{(k)}$ to obtain an approximate the sub-problem as

$$\text{minimize } \bar{f} = \sum_{i=1}^{n} c_i d_i, \quad \text{or} \quad \bar{f} = (\mathbf{c} \cdot \mathbf{d}) \tag{12.4}$$

subject to the linearized equality constraints

$$\sum_{i=1}^{n} \frac{\partial g_j}{\partial X_i}(\mathbf{X}^{(k)}) d_i = e_j, \quad \text{or} \quad (\mathbf{a}^{(j)} \cdot \mathbf{d}) = e_j, \quad j = 1 \text{ to } p \tag{12.5}$$

and the linearized inequality constraints

$$\sum_{i=1}^{n} \frac{\partial g_j}{\partial X_i}(\mathbf{X})^{(k)}) d_i \leq e_j, \quad \text{or} \quad (\mathbf{a}^{(j)} \cdot \mathbf{d}) \leq e_j, \quad j > p, \quad j \in I_k \tag{12.6}$$

where $\mathbf{a}^{(j)} = \nabla g_j(\mathbf{X}^{(k)})$ is the gradient of the jth constraint, $e_j = -g_j(\mathbf{X}^{(k)})$, and $c_i = \partial f(\mathbf{X}^{(k)})/\partial X_i$. Note that a potential constraint strategy for inequality constraints is used in equation (12.6). If it is not to be used, ε can be set to a very large number in defining the index set I_k in equation (12.3).

12.5 METHODS BASED ON LINEAR APPROXIMATIONS

12.5.1 Sequential linear programming

Note that all the functions in equations (12.4)–(12.6) are linear in the variables d_i. Therefore, linear programming methods can be used to solve for d_i. Such procedures are called *sequential linear programming* SLP. Note, however, that the problem defined in equations (12.4)–(12.6) may not have a bounded solution or the changes in design may become too large, thus invalidating the linear approximations. Therefore, limits must be imposed on changes in design. Such constraints are usually called *move limits* in the optimization literature and can be expressed as

$$-\Delta_{il}^{(k)} \leq d_i \leq \Delta_{iu}^{(k)}, \quad i = 1 \text{ to } n \tag{12.7}$$

where $\Delta_{il}^{(k)}$ and $\Delta_{iu}^{(k)}$ are the maximum allowed decrease and increase in the ith design variable, respectively at the kth iteration. The problem is still linear in terms of d_i, so LP methods can be used to solve it. Note that the iteration counter k is used to specify $\Delta_{il}^{(k)}$ and $\Delta_{iu}^{(k)}$, i.e. the move limits may change every iteration. *Selection of these move limits* is of critical importance because it can mean success or failure of the SLP algorithm. It is interesting to note here that the LP problem defined in equations (12.4)–(12.6) can be *transformed to be in the original variables* by substituting $d_i = X_i - X_i^{(k)}$. The move limits on d_i of equation (12.7) can also be in terms of the original variables. This way the solution of the LP problem directly gives the estimate for the next design point.

The sequential linear programming algorithm is a simple and straightforward approach to solve constrained optimization problems. Once a linear approximation for the problem has been developed, existing LP codes can be used to obtain new solution at each iteration. However, there are important *limitations* of the method that should be clearly understood:

1. *The method cannot be used as a black box approach for engineering design problems.* The selection of move limits is a trial and error process and can be best achieved in an interactive mode. The move limits can be too restrictive, resulting in no solution for the LP subproblem, or, they can be too large, resulting in oscillations in the design points.
2. *The method is not convergent* since no descent function is used. Therefore it is difficult to monitor progress towards the minimum point.
3. *The method can cycle between two points* if the optimum solution is not at a vertex (intersection of constraints) of the feasible set S.

Due to the foregoing limitations, the method can be frustrating to use, so *it is not recommended as a technique for general design optimization environment*, although it may be tried to improve specific designs.

12.5.2 Sequential quadratic programming: constrained steepest descent method

QP subproblem definition

As observed previously the SLP algorithm is a simple extension of linear programming to solve constrained optimization problems. However, the method has several drawbacks, the major one being the lack of robustness. To correct the drawbacks, we shall present a method where a *quadratic programming (QP) sub-problem* is solved to find a search direction and a descent function is introduced to determine an appropriate step size. General purpose QP solvers can be used to solve the QP subproblem. This gives a general algorithm that is convergent from an arbitrary starting point.

Performance of the sequential linear programming method depends quite heavily on the selection of proper move limits on design changes. The method cannot be proved to converge to a local minimum from an arbitrary starting design. To overcome these difficulties, other methods have been developed that have superseded SLP. Most of the methods still utilize the linear approximations of equations (12.4)–(12.6) for the non-linear programming problem. However, the linear move limits of equation (12.7) are abandoned in favour of a step-size constraint of the form $\|\mathbf{d}\| \leq \xi$, where $\|\mathbf{d}\|$ is the length of the search direction and ξ is a specified small positive number. We shall see later that the parameter ξ will not have to be specified. Using the definition of the length of a vector and squaring both sides of the foregoing equation, we obtain the following quadratic step-size constraint on $\mathbf{d}$:

$$0.5(\mathbf{d} \cdot \mathbf{d}) \leq \xi^2. \tag{12.8}$$

The factor of 0.5 on the left-hand side is introduced for convenience only to eliminate the factor of 2 during differentiation.

Note that the linearized subproblem defined in equations (12.4)–(12.6) with a step-size constraint of equation (12.8) still has an uncertainty in that the step-size parameter ξ must be specified as an input. It turns out that the direction determined by the foregoing subproblem (equations (12.7)–(12.9) and (12.30)) is the same as determined by the following quadratic programming sub-problem (this can be seen by writing the *Karush–Kuhn–Tucker (KKT) necessary conditions* for the two subproblems (Arora [1]):

Sub-problem QP1   Minimize $\overline{f} = (\mathbf{c} \cdot \mathbf{d}) + 0.5(\mathbf{d} \cdot \mathbf{d})$ (12.9)

subject to the linearized constraints in equations (12.5) and 12.6).

Note that when either there are no constraints or none is active, minimization of the quadratic function in equation (12.9) gives $\mathbf{d} = -\mathbf{c}$ (using the necessary condition, $\partial \overline{f}/\partial \mathbf{d} = \mathbf{0}$). This is just the steepest descent direction for the unconstrained problem. When there are constraints, their effect must be included in calculating the search direction. The search direction must satisfy all the linearized constraints. Since that direction is a modification of the steepest descent direction to satisfy constraints, it is called the *constrained steepest descent direction* [1]. The steps of the resulting constrained steepest descent (CSD) algorithm will be clear once we define a *descent function and the related line search procedure to calculate the step size* along the search direction. This method is a slight modification of the method developed by Pshenichny, a Russian mathematician [34]. He called it the *linearization method*.

Descent function

A descent function for the constrained problems is usually constructed by adding a *penalty for constraint violations* to the current value of the cost function. One of the properties of the descent function is that its value at the optimum point must be the same as that for the cost function. Also, it should be possible to reduce it along the search direction at each iteration. We shall introduce *Pshenichny's descent function* due to its simplicity and success in solving a large number of engineering design problems

[34–36]. Other descent functions shall be discussed later.

Pshenichny's descent function Φ at any point $\mathbf{X}$ is defined as

$$\Phi(\mathbf{X}) = f(\mathbf{X}) + RV(\mathbf{X}) \tag{12.10}$$

where R is a positive number called the *penalty parameter* and $V(\mathbf{X}) \geq 0$ is the *maximum constraint violation* among all the constraints. As an example, the descent function at the point $\mathbf{X}^{(k)}$ during the kth iteration is calculated as

$$\Phi_k = f_k + RV_k \tag{12.11}$$

where f_k, Φ_k and V_k are the values of $f(\mathbf{X})$, $\Phi(\mathbf{X})$ and $V(\mathbf{X})$ at $\mathbf{X}^{(k)}$ as

$$f_k = f(\mathbf{X}^{(k)}), \qquad \Phi_k = \Phi(\mathbf{X}^{(k)}), \qquad V_k = V(\mathbf{X}^{(k)}). \tag{12.12}$$

Note that the penalty parameter may change during the iterative process. Actually, it must be ensured that it satisfies the following necessary condition at the point $\mathbf{X}^{(k)}$:

$$R \geq r_k \tag{12.13}$$

where r_k is a sum of absolute values of all the Lagrange multipliers $u_i^{(k)}$ of the QP subproblem at the kth iteration:

$$r_k = \sum_{i=1}^{m} |u_i^{(k)}|. \tag{12.14}$$

The *parameter* $V_k \geq 0$, *related to the maximum constraint violation* at the kth iteration, is determined using values of the constraint functions at $\mathbf{X}^{(k)}$ as

$$V_k = \max\{0, |g_i|, i = 1 \text{ to } p; g_i, i = p+1 \text{ to } m\}. \tag{12.15}$$

Since the equality constraint is violated if it is different from zero, its absolute value is used in equation (12.15). Note that if all constraints are satisfied at $\mathbf{X}^{(k)}$, then $V_k = 0$.

Step-size determination

Before the constrained steepest descent algorithm can be given, a step size determination procedure is needed. In most practical implementations of any algorithm, only an approximate step size can be determined because determination of an accurate step size can be very time consuming, requiring many analyses of the system. This is done using the so-called *inaccurate* or *inexact line search*. The usual procedure is to start with the trial step size as one. If a certain descent condition (defined later) is not satisfied, the trial step is taken as half of the previous trial. If the descent condition is still not satisfied, the trial step size is bisected again. The procedure is continued, until the descent condition is satisfied.

To implement the foregoing procedure, define a sequence of trial step sizes t_j as follows:

$$t_j = (1/2)^j, \quad j = 0, 1, 2, 3, 4, \ldots \tag{12.16}$$

which gives a sequence of trial step sizes as, 1, 1/2. 1/4, 1/8, 1/16, etc.

Before presenting the descent condition for determining the step size, we shall introduce certain additional notations and variables. We shall use a *second subscript or superscript to indicate values of certain quantities at a trial step size*. For example, let t_j be the trial step size at the kth iteration; then the trial design point is given as

$$\mathbf{X}^{(k+1,j)} = \mathbf{X}^{(k)} + t_j \mathbf{d}^{(k)}. \tag{12.17}$$

The descent function $\Phi_{k+1,j}$ of equation (12.11) evaluated at the trial step size t_j and the corresponding design point $\mathbf{X}^{(k+1,j)}$ is given as

$$\Phi_{k+1,j} \equiv \Phi(\mathbf{X}^{(k+1,j)}) = f_{k+1,j} + RV_{k+1,j} \tag{12.18}$$

with $f_{k+1,j} = f(\mathbf{X}^{(k+1,j)})$ and $V_{k+1,j}$ as the maximum constraint violation at the trial design point calculated using equation (12.15). Note that in evaluating $\Phi_{k+1,j}$ or Φ_k, the most recent value of the penalty parameter R is used. Define a constant β_k using the search direction $\mathbf{d}^{(k)}$ as

$$\beta_k = \gamma \|\mathbf{d}^{(k)}\|^2 \tag{12.19}$$

where γ is a specified constant between 0 and 1. Note that in the kth iteration β_k defined in equation (12.19) is a constant once the search direction $\mathbf{d}^{(k)}$ has been determined.

Now we are ready to define the condition that determines an approximate step size. At the kth iteration, a step size $\alpha_k = t_j$ is acceptable if t_j satisfies the following *descent condition*

$$\Phi_{k+1,j} + t_j \beta_k \leq \Phi_k. \tag{12.20}$$

Thus, the procedure for determining an appropriate step-size is to keep trying the numbers given by the sequence in equation (12.16) until the condition in equation (12.20) is satisfied (note that t_j is the only variable in the inequality (12.20). This procedure is illustrated in Reference [1].

It is important to note the foregoing step-size determination procedure needs only the descent function values and not its derivatives. The descent function of equation (12.10) is not differentiable at certain points in the design space, so any step-size determination procedure that requires more smoothness may not work well with this function. For such methods, differentiable descent functions given later in the chapter should be used [1].

The CSD algorithm

We are now ready to state the constrained steepest descent (CSD) algorithm in a step-by-step form. It has been proved [34] that the solution point of the sequence $\{\mathbf{X}^{(k)}\}$ generated by the following algorithm is a KKT point (i.e. a local minimum point) for the general constrained optimization Problem P. The stopping criterion for the algorithm is that $\|\mathbf{d}\| \leq \varepsilon_2$ for a feasible design point, where ε_2 is a small positive number. The CSD method is now summarized in the form of a *computational algorithm* [1].

Step 1. Set $k = 0$. Estimate initial values for design variables as $\mathbf{X}^{(0)}$. Select an appropriate initial value for the penalty parameter R_0, a constant γ between 0 and 1 ($0 < \gamma < 1$) and two small numbers ε_1 and ε_2 that define the permissible constraint violation and convergence parameter, respectively. Select a small number $\varepsilon > 0$ for determining the potential constraint index set I_k defined in equation (12.3). $R_0 = 1$, $\varepsilon = 0.1$, and $\gamma = 0.2$ is a reasonable selection.

Step 2. At $\mathbf{X}^{(k)}$ compute the cost and constraint functions and their gradients. Calculate the maximum constraint violation V_k as defined in equation (12.15).

Step 3. Using the cost and constraint function values and their gradients define the QP sub-problem given in equations (12.9), (12.5) and (12.6). Solve the QP sub-problem to obtain the search direction $\mathbf{d}^{(k)}$ and the Lagrange multipliers $\mathbf{u}^{(k)}$.

Step 4. Check for the following convergence criteria

$$\|\mathbf{d}^{(k)}\| \leq \varepsilon_2$$

and the maximum constraint violation $V_k \leq \varepsilon_1$. If these termination criteria are satisfied then stop. Otherwise continue.

Step 5. To check the necessary condition of equation (12.13) for the penalty parameter R, calculate the sum r_k of the Lagrange multipliers defined in equation (12.14). Set $R = \max\{R_k, r_k\}$.

Step 6. Set $\mathbf{X}^{(k+1)} = \mathbf{X}^{(k)} + \alpha_k \mathbf{d}^{(k)}$, where α_k is the step size determined to satisfy the descent condition of equation (12.20) as explained earlier.

Step 7. Save the current penalty parameter as $R_{k+1} = R$. Update the iteration counter as $k = k + 1$, and go to Step 2.

The CSD algorithm along with the foregoing step-size determination procedure is convergent provided second derivatives of all the functions are piecewise continuous (this is the so-called Lipschitz condition) and the descent function values for the design points $\mathbf{X}^{(k)}$ are bounded as follows [34]:

$$\Phi(\mathbf{X}^{(k)}) < \Phi(\mathbf{X}^{(0)}), \quad k = 1, 2, 3, \ldots.$$

Observations on the CSD algorithm

Pshenichny's algorithm given above can be looked upon as the constrained analogue of the steepest descent method for unconstrained problems. The following observations are noteworthy for the algorithm:

1. The method does not use any second order derivatives and can treat equality as well as inequality constraints. It converges to a local minimum point starting from an arbitrary point.
2. The *potential constraint strategy* as suggested by Pshenichny and Denilin [34] is not used in the algorithm. Instead, the strategy suggested in equation (12.3) is used. A potential constraint strategy must be used for most engineering design problems [35, 37].
3. The rate of convergence of the CSD algorithm can be further improved by including higher-order information about the problem functions in the QP sub-problem [38]. This will be discussed in the next section.
4. It is important to note that the step size determined using equation (12.16) is not allowed to be greater than one. However, in numerical implementations [38], the algorithm is more efficient for many problems when the step size is allowed to be larger than one. This can be done by allowing the integer j in equation (12.16) to have negative values.
5. It is important to note that although the CSD algorithm is theoretically supposed to converge to a local minimum point starting from any point, its numerical behaviour can be different. The algorithm needs to be implemented in a robust way [39].
6. The starting point can also affect performance of the algorithm. For example, at some points the QP subproblem may not have any solution. This need not mean that the original problem is infeasible. The original problem may be highly non-linear, so the linearized constraints may be inconsistent, giving an infeasible sub-problem. This situation can be handled by either temporarily deleting the inconsistent, constraints or starting from another point. For more discussion on the implementation of the algorithm, References [39] and [40] should be consulted.

Use of potential constraint strategy

As noted in the foregoing, the potential set strategy where only a subset of the constraints is included in defining the QP sub-problem, is incorporated into the CSD algorithm. The procedure is to calculate all the constraint functions for the problem and define a potential set according to equation (12.3). There are several alternative ways to define the potential set, and different procedures can lead to different search directions and the paths to the optimum point.

The main effect of using a potential set strategy in an algorithm is on the efficiency of the entire iterative process. This is particularly true for large and complex applications where the evaluation of gradients of constraints is a computationally expensive proposition. With the potential set strategy, gradients of only the potential constraints are calculated and used in defining the sub-problem. The original problem may have hundreds of constraints, but only a few may be in the potential set. Thus with this strategy, not only the number of gradient evaluations is reduced but also the dimension of the sub-problem is substantially reduced. This can result in additional savings in the computational effort. *Therefore, the potential set strategy is highly beneficial and must be used in practical applications of optimization.*

12.5.3 Other methods

Many other methods and their variations for constrained optimization have been developed and evaluated in the literature [9, 22, 23]. In this section, we shall briefly discuss the basic ideas of three methods that have been used quite successfully for engineering design problems.

Feasible directions method

The method of feasible directions is one of the earliest primal methods for solving constrained optimization problems [41]. The basic idea of the method is to move from one feasible point to an improved feasible point. Thus given a feasible design $X^{(k)}$, an 'improving feasible direction' $d^{(k)}$ is determined such that for a sufficiently small step size $\alpha > 0$, the following two properties are satisfied: (1) the new point, $X^{(k+1)} = X^{(k)} + \alpha d^{(k)}$ is feasible, and (2) the new cost function is smaller than the old one, i.e. $f(X^{(k+1)}) < f(X^{(k)})$. Once $d^{(k)}$ is determined, a line search is performed to determine how far to proceed along $d^{(k)}$ based on minimizing the cost function without violating the constraints. This leads to a new feasible design $X^{(k+1)}$, and the process is repeated from there. Note that the cost function is used as the descent function in this algorithm.

The method is based on the general algorithm described in Section 12.3.1, where the design improvement is decomposed into search direction and step-size determination sub-problems. The direction is determined by defining a linearized sub-problem at the current feasible point, and step size is determined to reduce the cost function as well as maintain feasibility. Since linear approximations are used, it is difficult to maintain feasibility with respect to the equality constraints. Therefore, the method has been developed and applied mostly to inequality constrained problems. Some procedures have been developed to treat equality constraints in these methods [42]. However, we shall describe the method for problems with only inequality constraints.

Now we define a sub-problem that yields an improving feasible direction at the current design point. An *improving feasible direction* is defined as the one that reduces the cost function as well as remains strictly feasible for a small step size. Thus it is a direction of descent for the cost function and has a finite segment passing through the feasible set. The improving feasible direction d satisfies the conditions $(c \cdot d) < 0$ and $(a^{(i)} \cdot d) < 0$ for $i \in I_k$, where I_k is a potential constraint set at the current point as defined in equation (12.3) with $p = 0$. It can be obtained by minimizing the maximum of $(\mathbf{c}\cdot\mathbf{d})$ and $(\mathbf{a}^{(i)}\cdot\mathbf{d})$ for $i\in I_k$. Denoting this maximum by β, the direction-finding sub-problem is defined as

minimize β subject to (12.21)

$(\mathbf{c}\cdot\mathbf{d}) \leq \beta$ (12.22)

$(\mathbf{a}^{(i)}\cdot\mathbf{d}) \leq \beta \quad$ for $i\in I_k$ (12.23)

$-1 \leq d_j \leq 1, \quad j = 1$ to n. (12.24)

The normalization constraint of equation (12.24) has been introduced to obtain a bounded solution. Other forms of normalization constraints can also be used. Let $(\beta, \mathbf{d})$ be an optimum solution for the above problem. If $\beta < 0$, then $\mathbf{d}$ is an improving feasible direction. If $\beta = 0$, then the current design point satisfies the KKT necessary conditions.

There are many different line search algorithms that may be used to determine an appropriate step size along the search direction. Also, to estimate a better feasible direction $\mathbf{d}^{(k)}$, the constraints of equation (12.23) can be expressed as $(\mathbf{a}^{(i)}\cdot\mathbf{d}) \leq \theta_i\beta$, where $\theta_i > 0$ are the 'push-off' factors. The greater value of θ_i, the greater is the direction vector $\mathbf{d}$ pushed into the feasible region. The reason for introducing θ_i is to prevent the iterations from repeatedly hitting the constraint boundary and slowing down the convergence. If θ_i is taken as zero, then the right-hand side of equation (12.23) becomes zero. The direction $\mathbf{d}$ in this case tends to follow the active constraint, i.e. it is tangent to the constraint surface. On the other hand, if θ_i is very large, the direction $\mathbf{d}$ tends to follow the cost function contour. A small value of θ_i will result in a direction which rapidly reduces the cost function. It may, however, rapidly encounter the same constraint surface due to non-linearities. Larger values of θ_i will reduce the risk of re-encountering the same constraint, but will not reduce the cost function as fast. A value of $\theta_i = 1$ yields acceptable results for most problems.

The disadvantages of the method are (1) a feasible starting point is needed — special algorithms must be used to obtain such a point if it is not known — and (2) equality constraints are difficult to impose and require special procedures for their implementation.

Gradient projection method

The gradient projection method was developed by Rosen [43]. Just as in the feasible directions method, the method also uses first order information about the problem at the current point. The feasible directions method requires the solution of an LP at each iteration to find the search direction. In some applications this can be an expensive calculation. Thus, Rosen was motivated to develop a method that does not require the solution of an LP. His idea was to develop a procedure in which the direction vector could be calculated easily, although it may not be as good as the one obtained from the feasible directions approach. Thus, in the gradient projection method, an explicit expression for the search direction is available.

The method starts with an initial estimate for the solution point, as in other methods. If the point is inside the feasible domain, the steepest descent direction for the cost function is used until a constraint boundary is encountered. If the starting point, is infeasible then the constraint correction step is used to reach the feasible set. When the point is on the boundary, a direction that is tangent to the constraint surface is used to change the design. This direction is computed by projecting the steepest descent direction for the cost function on to the tangent hyperplane. This was termed the constrained steepest descent (CSD) direction in Section 12.5.2. A step is executed in the negative projected gradient direction. Since the direction is at a tangent to the constraint surface, the new point will be infeasible. Therefore, a series of correction steps needs to be executed to reach the feasible domain again. Comparing the gradient projection method and the constrained steepest descent method of Section 12.5.2, we observe that at a feasible point where some constraints are active, the two methods have identical directions. The only difference is in the step-size determination.

Philosophically, the idea of the gradient projection method is quite good, i.e. the search direction is easily computable, although it may not be as good as the feasible direction. However, numerically the method has considerable uncertainty. The step size specification is arbitrary; the constraint correction process is quite tedious. A serious drawback is that convergence of the algorithm is tedious to enforce. For example, during the constraint correction steps, it must be ensured that $f(\mathbf{X}^{(k+1)}) < f(\mathbf{X}^{(k)})$ (i.e. the cost function is used as the descent function). If this condition cannot be satisfied or constraints cannot be corrected, then the step size must be reduced and the entire process must be repeated. This can be quite tedious and inefficient. Despite these drawbacks, the method has been applied quite successfully to a wide variety of engineering design problems [5]. In addition, many variations of the method have been investigated [22, 23, 31].

Generalized reduced gradient method

In 1967, Wolfe developed the *reduced gradient method* based on a simple variable elimination technique for problems with equality constraints [44]. The *generalized reduced gradient (GRG) method* is an extension of the reduced gradient method to accommodate non-linear inequality constraints [45]. In this method, a search direction at each iteration is found such that for any small move, the current active constraints remain precisely active (i.e. they are treated as equalities). If some active constraints are not precisely satisfied due to non-linearity of constraint functions, the Newton–Raphson method is used to return to the constraint boundary. Thus, the GRG method can be considered somewhat similar to the gradient projection method in its steps.

Since inequality constraints can always be converted to equalities by adding slack variables, we can form an NLP model with equality constraints. Also, we can employ potential constraint strategy and treat all the constraints in the sub-problem as equalities. The direction-finding sub-problem in the GRG method can be defined in the following way: let the design variable vector be partitioned as $\mathbf{X}^T = [\mathbf{Y}^T, \mathbf{Z}^T]$, where $\mathbf{Y}_{(n-p)}$ and $\mathbf{Z}_{(p)}$ are vectors of independent and dependent design variables, respectively, and p is the number of equalities or active constraints at the current iteration. First-order changes in the cost and active constraint functions (treated as equalities) are given as

$$\delta f = (\mathbf{c}_y \cdot \delta \mathbf{Y}) + (\mathbf{c}_z \cdot \delta \mathbf{Z}) \tag{12.25}$$

$$\delta g_i = ((\partial g_i/\partial \mathbf{Y}) \cdot \partial \mathbf{Y}) + ((\partial g_i/\partial \mathbf{Z}) \cdot \delta \mathbf{Z}) \tag{12.26}$$

where $\mathbf{c}_y = \partial f/\partial \mathbf{Y}$ and $\mathbf{c}_z = \partial f/\partial \mathbf{Z}$. Since we started with a feasible design, any change in the variables must keep the current equalities satisfied at least to first order, i.e. $\delta g_i = 0$. Therefore, equation (12.26) can be written in the matrix form as

$$\mathbf{D}^T \delta \mathbf{Y} + \mathbf{B}^T \delta \mathbf{Z} = \mathbf{0}, \quad \text{or} \quad \delta \mathbf{Z} = -(\mathbf{B}^{-T}\mathbf{D}^T)\delta \mathbf{Y} \tag{12.27}$$

where the ith column of matrix $\mathbf{D}_{(n-p) \times p}$ is given as $\partial g_i/\partial \mathbf{Y}$ and the ith column of $\mathbf{B}_{(p \times p)}$ is given as $\partial g_i/\partial \mathbf{Z}$. Equation (12.27) can be viewed as the one that determines $\delta \mathbf{Z}$ (change in the dependent variable) when $\delta \mathbf{Y}$ (change in the independent variable) is specified. Substituting $\delta \mathbf{Z}$ from equation (12.27) into equation (12.25), we get

$$\delta f = (\mathbf{c}_y^T - \mathbf{c}_z^T \mathbf{B}^{-T} \mathbf{D}^T)\delta \mathbf{Y}.$$

Or, re-writing the equation as $\delta f = ((df/d\mathbf{Y}) \cdot \delta \mathbf{Y})$, the gradient of f with respect to $\mathbf{Y}$ is identified as

$$\frac{df}{d\mathbf{Y}} = \mathbf{c}_y - \mathbf{D}\mathbf{B}^{-1}\mathbf{c}_z. \tag{12.28}$$

This is commonly known as the *generalized reduced gradient* and can be viewed as the gradient of the unconstrained function $f(\mathbf{Y}, \mathbf{Z}(\mathbf{Y}))$.

In the line search, the cost function is treated as the *descent function*. For a trial value of α, the design variables are updated using $\delta \mathbf{Y} = -\alpha \, df/d\mathbf{Y}$ and $\delta \mathbf{Z}$ from equation (12.27). If the trial design is not feasible, then independent design variables $\mathbf{Y}$ are considered to be fixed and dependent variables $\mathbf{Z}$ are changed iteratively by applying the Newton–Raphson (NR) method until we get a feasible design point. If the new feasible design satisfies the descent condition, then line search is terminated; otherwise, the previous trial step size is discarded and the procedure is repeated with a reduced step size. It can be observed that when $df/d\mathbf{Y} = \mathbf{0}$ in equation (12.28), the KKT conditions of optimality are satisfied for the original NLP problem.

The main *computational burden* associated with the GRG algorithm arises from the Newton–Raphson (NR) iterations during line search. Strictly speaking, the gradients of constraints need to be recalculated and the Jacobian matrix $\mathbf{B}$ needs to be inverted at every NR iteration during the line search. This is prohibitively expensive. Towards this end, many efficient numerical schemes have been suggested, e.g. the use of a quasi-Newton formula to update $\mathbf{B}^{-1}$ without recomputing gradients but requiring only constraint function values. This can cause problems if the set of independent variables changes at every iteration. Also selection of independent and dependent variables is not arbitrary; one must ensure that **B** is a non-singular matrix. Another difficulty is to select a feasible starting point. Special algorithms must be used to handle arbitrary starting points, as in the feasible directions method. In spite of these difficulties, the method has been successfully used in the past to solve certain engineering design problems [46].

There is some confusion in the literature on the relative merits and demerits of the GRG method. For example, the method has been declared superior to the gradient projection method, whereas the two methods are considered essentially the same by Sargeant [47]. The confusion arises when studying the GRG method in the context of solving inequality constrained problems; some GRG algorithms convert the inequalities into equalities by adding non-negative slack variables while others adopt a potential constraint energy. It turns out that if a potential constraint strategy is used, the GRG method becomes essentially the same as the gradient projection method [31]. On the other hand, if inequalities are converted to equalities, it behaves quite differently from the gradient projection method. Unfortunately, the inequality constrained problem in most engineering applications must be solved using a potential constraint strategy, as the addition of slack variables to inequalities implies that all constraints are active at every iteration and must, therefore, be differentiated. This is ruled out for large-scale applications due to the enormous computation and storage of information involved. Therefore, as observed by Belegundu and Arora [31], we need not differentiate between gradient projection and reduced gradient methods when solving most engineering optimization problems.

12.6 SEQUENTIAL QUADRATIC PROGRAMMING: QUASI-NEWTON METHODS

12.6.1 Basic idea

Thus far we have used only linear approximation for the cost and constraint functions in defining the search direction determination sub-problem. The rate of convergence of algorithms based on such sub-problems can be slow. It is possible to use quadratic approximations for the cost and constraint functions in defining the sub-problem. This is likely to improve the rate of convergence of the algorithms because curvature information for the functions is used in determining the search direction. However, it turns out that the subproblem defined with quadratic approximations for the functions can be as difficult to solve as the original non-linear optimization problem.

It turns out that the QP sub-problem, defined in the previous section, can be modified slightly to incorporate curvature information about the Lagrangian [48]. The original non-linear constraints are still approximated by linear constraints as in equations (12.5) and (12.6). To define the new QP sub-problem, we need to evaluate the Hessian matrix for the Lagrangian. This procedure still leads to two difficulties:

1. Second-order derivatives of all the constraints and cost function must be evaluated, which is usually a very tedious calculation.
2. Lagrange multiplier estimates for all constraints must be available to calculate

the Hessian of the Lagrangian. These are usually known after the QP sub-problem has been solved.

A great breakthrough in the methods based on the foregoing philosophy occurred when it was realized that the Hessian of the Lagrangian could be approximated using only the first order information [49–52]. The idea is quite similar to that used in unconstrained quasi-Newton methods. We use the gradient of the Lagrangian at the two points and the change in design to update a secant approximation to the Hessian of the Lagrangian. We shall call these *constrained Quasi-Newton methods*. They have been also called *constrained variable metric* (CVM), *sequential quadratic programming* (SQP), or *recursive quadratic programming* (RQP) methods in the literature.

12.6.2 Derivation of quadratic programming sub-problem

There are several ways to derive the quadratic programming (QP) sub-problem that has to be solved at each optimization iteration. Understanding of the detailed derivation of the QP sub-problem is not necessary in using the constrained quasi-Newton methods. However, for better understanding of the basis for the SQP method, we shall derive the QP sub-problem [1].

It is customary to derive the QP sub-problem by considering only the equality constrained model as

Problem PE   Minimize $f(\mathbf{X})$ subject to $g_i(\mathbf{X}) = 0$, $i = 1$ to p. $\qquad(12.29)$

Later on, the inequality constraints are easily incorporated into the sub-problem. The procedure for derivation of the QP sub-problem is to write KKT necessary conditions for the problem defined in equation (12.29) and then attempt to solve them by Newton's method for non-linear equations. Each iteration of Newton's method is then interpreted as being equivalent to the solution of a QP sub-problem.

The Lagrangian for the design optimization problem defined in equation (12.29) is given as

$$L(\mathbf{X}, \mathbf{u}) = f(\mathbf{X}) + \sum_{i=1}^{p} u_i g_i(\mathbf{X}) = f(\mathbf{X} + \mathbf{u} \cdot \mathbf{g}(\mathbf{X})) \qquad(12.30)$$

where u_i is the Lagrange multiplier for the *i*th equality constraint $g_i(\mathbf{X}) = 0$. Note that u_i for an equality constraint is free in sign. The KKT necessary conditions give

$\nabla L(\mathbf{X}) = \mathbf{0}$,   or $\mathbf{c} + \mathbf{A}\mathbf{u} = \mathbf{0}$ $\qquad(12.31)$

$g_i(\mathbf{X}) = 0$, $i = 1$ to p $\qquad(12.32)$

where $\mathbf{A}$ is an $n \times p$ matrix whose *i*th column is the gradient of g_i. Note that equation (12.31) actually represents n equations, since the dimension of the design variable vector is n. These equations along with the p equality constraints in equation

(12.32) give $(n + p)$ equations in $(n + p)$ unknowns (n design variables in $\mathbf{X}$ and p Lagrange multipliers in $\mathbf{u}$). These are non-linear equations, so the Newton–Raphson method can be used to solve them.

Let us write equations (12.31) and (12.32) in a compact notation as

$$\mathbf{F}(\mathbf{y}) = \mathbf{0} \tag{12.33}$$

where $\mathbf{F}$ and $\mathbf{y}$ are identified as

$$\mathbf{F} = \begin{bmatrix} \nabla L \\ \mathbf{g} \end{bmatrix}_{(n+p \times 1)} \quad \text{and} \quad \mathbf{y} = \begin{bmatrix} \mathbf{X} \\ \mathbf{u} \end{bmatrix}_{(n+p \times 1)}. \tag{12.34}$$

Now using the iterative Newton–Raphson procedure, we assume that $\mathbf{y}^{(k)}$ at the kth iteration is known and a change $\Delta \mathbf{y}^{(k)}$ is desired. Using linear Taylor's expansion for equation (12.33), $\Delta \mathbf{y}^{(k)}$ is given as a solution of the following linear system:

$$\nabla \mathbf{F}^T(\mathbf{y}^{(k)}) \Delta \mathbf{y}^{(k)} = - \mathbf{F}(\mathbf{y}^{(k)}) \tag{12.35}$$

where $\nabla \mathbf{F}$ is the $(n + p) \times (n + p)$ Jacobian matrix for the nonlinear equations whose ith column is the gradient of the function $F_i(\mathbf{y})$ with respect to the vector $\mathbf{y}$. Substituting definitions of $\mathbf{F}$ and $\mathbf{y}$ from equation (12.34) into equation (12.35), we obtain

$$\begin{bmatrix} \mathbf{H} & \mathbf{A} \\ \mathbf{A}^T & \mathbf{0} \end{bmatrix}^{(k)} \begin{bmatrix} \Delta \mathbf{X} \\ \Delta \mathbf{u} \end{bmatrix}^{(k)} = - \begin{bmatrix} \nabla L \\ \mathbf{g} \end{bmatrix}^{(k)} \tag{12.36}$$

where the superscript k indicates that the quantities are calculated at the kth iteration, $\mathbf{H}$ is an $n \times n$ symmetric Hessian matrix of the Lagrangian, $\Delta \mathbf{X}^{(k)} = \mathbf{X}^{(k+1)} - \mathbf{X}^{(k)}$, and $\Delta \mathbf{u}^{(k)} = \mathbf{u}^{(k+1)} - \mathbf{u}^{(k)}$. Equation (12.36) can be transformed into a slightly different form by writing the first row as

$$\mathbf{H}^{(k)} \Delta \mathbf{X}^{(k)} + \mathbf{A}^{(k)} \Delta \mathbf{u}^{(k)} = - \nabla L^{(k)}. \tag{12.37}$$

Substituting for $\Delta \mathbf{u}^{(k)} = \mathbf{u}^{(k+1)} - \mathbf{u}^{(k)}$ and $\nabla L^{(k)}$ from equation (12.31) into equation (12.37), we obtain

$$\mathbf{H}^{(k)} \Delta \mathbf{X}^{(k)} + \mathbf{A}^{(k)}(\mathbf{u}^{(k+1)} - \mathbf{u}^{(k)}) = - \mathbf{c}(\mathbf{X}^{(k)}) - \mathbf{A}^{(k)} \mathbf{u}^{(k)}. \tag{12.38}$$

Or, the equation is simplified to

$$\mathbf{H}^{(k)} \Delta \mathbf{X}^{(k)} + \mathbf{A}^{(k)} \mathbf{u}^{(k+1)} = - \mathbf{c}(\mathbf{X}^{(k)}). \tag{12.39}$$

Combining equation (12.39) with the second row of equation (12.36), we obtain

$$\begin{bmatrix} \mathbf{H} & \mathbf{A} \\ \mathbf{A}^T & \mathbf{0} \end{bmatrix}^{(k)} \begin{bmatrix} \Delta \mathbf{X}^{(k)} \\ \mathbf{u}^{(k+1)} \end{bmatrix} = -\begin{bmatrix} \mathbf{c} \\ \mathbf{g} \end{bmatrix}^{(k)}. \tag{12.40}$$

Solution of equation (12.40) gives a change in the design $\Delta \mathbf{X}^{(k)}$ and a new value for the Lagrange multiplier vector $\mathbf{u}^{(k+1)}$. The iterative procedure is continued until convergence criteria are satisfied.

It will now be shown that equation (12.40) is also a solution of the following QP sub-problem defined at the kth iteration as

Sub-problem QP2 Minimize $(\mathbf{c} \cdot \Delta \mathbf{X}) + 0.5(\Delta \mathbf{X} \cdot \mathbf{H} \Delta \mathbf{X})$ (12.41)

subject to linearized equality constraints

$$g_i + (\mathbf{a}^{(i)} \cdot \Delta \mathbf{X}) = 0, \quad i = 1 \text{ to } p. \tag{12.42}$$

The Lagrangian for the problem defined in equations (12.41) and (12.42) is given as

$$\bar{L} = (\mathbf{c} \cdot \Delta \mathbf{X}) + 0.5(\Delta \mathbf{X} \cdot \mathbf{H} \Delta \mathbf{X}) + \sum_{i=1}^{p} u_i(g_i + (\mathbf{a}^{(i)} \cdot \Delta \mathbf{X})). \tag{12.43}$$

The KKT necessary conditions treating $\Delta \mathbf{X}$ as the unknown variable give

$$\mathbf{c} + \mathbf{H} \Delta \mathbf{X} + \mathbf{A} \mathbf{u} = \mathbf{0} \tag{12.44}$$

$$\mathbf{g} + \mathbf{A} \Delta \mathbf{X} = \mathbf{0}. \tag{12.45}$$

It can be seen that if we combine equations (12.44) and (12.45) and write them in a matrix form, we get equation (12.40). Thus, the problem of minimizing $f(\mathbf{X})$ subject to $g_i(\mathbf{X}) = 0$, $i = 1$ to p can be solved by iteratively solving the QP subproblem defined in equations (12.41) and (12.42). This is equivalent to solving the KKT optimality conditions of equations (12.31) and (12.32) for the problem. Note that instead of solving QP2, one can also solve equation (12.40) directly. However, for the problem with inequality constraints, it will be better to solve the QP sub-problem instead of equation (12.40). The reason is that the QP sub-problem has inequality constraints which are difficult to include in equation (12.40).

Just as in Newton's method for unconstrained problems the solution $\Delta \mathbf{X}$ should be treated as a search direction and step size determined by minimizing an appropriate descent function to obtain a convergent algorithm. Defining the search direction $\mathbf{d}$ as $\Delta \mathbf{X}$ and including inequality constraints, the QP sub-problem for the general Problem P is defined as

Sub-problem QP3 Minimize $(\mathbf{c} \cdot \mathbf{d}) + 0.5(\mathbf{d} \cdot \mathbf{H} \mathbf{d})$ (12.46)

subject to contraints of equations (12.5) and (12.6), where $\mathbf{H}$ is the Hessian of the Lagrangian or its approximation.

12.6.3 Quasi-Newton Hessian approximation

Just as for the quasi-Newton methods for unconstrained problems, we can approximate the Hessian of the Lagrangian for the constrained problems. We assume that the approximate Hessian $\mathbf{H}^{(k)}$ at the kth iteration is available and we desire to update it to $\mathbf{H}^{(k+1)}$. The Broyden–Fletcher–Godfarb–Shanno (BFGS) formula for direct updating of the Hessian or other formulae can be used [22, 23]. It is important to note that the updated Hessian should be kept positive definite because, with this property, the QP sub-problem defined in equation (12.46) remains strictly convex. This way a unique search direction is obtained. It turns out that the standard BFGS updating formula can lead to a singular or indefinite Hessian. To overcome this difficulty, Powell [52] suggested a modification to the standard BFGS formula. Although the modification is based on intuition, it has worked well in most applications. We shall give the modified BFGS formula.

Several intermediate scalars and vectors must be calculated before the final formula can be given. We define these as follows:

vector of changes in design: $\mathbf{s}^{(k)} = \alpha_k \mathbf{d}^{(k)}$ (12.47)

a vector: $\mathbf{z}^{(k)} = \mathbf{H}^{(k)} \mathbf{s}^{(k)}$ (12.48)

difference in the gradients of the Lagrangian at two points:

$\mathbf{y}^{(k)} = \nabla L(\mathbf{X}^{(k+1)}, \mathbf{u}^{(k)}) - \nabla L(\mathbf{X}^{(k)}, \mathbf{u}^{(k)})$ (12.49)

scalars: $\xi_1 = (\mathbf{s}^{(k)} \cdot \mathbf{y}^{(k)})$; $\xi_2 = (\mathbf{s}^{(k)} \cdot \mathbf{z}^{(k)})$ (12.50)

a scalar: $\theta = 1$ if $\xi_1 \geq 0.2\xi_2$, otherwise $\theta = \dfrac{0.8\xi_2}{\xi_2 - \xi_1}$, $\xi_1 \neq \xi_2$ (12.51)

a vector: $\mathbf{w}^{(k)} = \theta \mathbf{y}^{(k)} + (1 - \theta) \mathbf{z}^{(k)}$ (12.52)

a scalar: $\xi_3 = (\mathbf{s}^{(k)} \cdot \mathbf{w}^{(k)}) = \theta \xi_1 + (1 - \theta) \xi_2$ (12.53)

an $n \times n$ matrix: $\mathbf{D}^{(k)} = \dfrac{1}{\xi_3} \mathbf{w}^{(k)} \mathbf{w}^{(k)T}$ (12.54)

an $n \times n$ matrix: $\mathbf{E}^{(k)} = \dfrac{1}{\xi_2} \mathbf{z}^{(k)} \mathbf{z}^{(k)T}$. (12.55)

With the preceding definition of matrices $\mathbf{D}^{(k)}$ and $\mathbf{E}^{(k)}$, the approximate Hessian is updated as

$\mathbf{H}^{(k+1)} = \mathbf{H}^{(k)} + \mathbf{D}^{(k)} - \mathbf{E}^{(k)}$. (12.56)

It turns out that if the scalar ξ_1 in equation (12.50) is negative, the original BFGS formula can lead to an indefinite Hessian. The use of the modified vector $\mathbf{w}^{(k)}$ given in equation (12.52) tends to alleviate this difficulty.

Due to the usefulness of incorporating Hessian into an optimization algorithm, several updating procedures have been developed in the recent literature [22]. For example, Cholesky factors of the Hessian can be directly updated. In numerical implementations, it is useful to incorporate such procedures because numerical stability can be guaranteed.

12.6.4 An SQP algorithm

The CSD algorithm of Section 12.5.2 has recently been extended to include Hessian updating where the subproblem QP3 is solved instead of the subproblem QP1 [35, 38, 53]. Such methods have been called sequential quadratic programming (SQP) methods. The original SQP algorithm presented by Han [49, 50], Powell [52] and Schittkowski [54] did not use any potential set strategy. The new algorithm uses a potential set strategy and has been extensively evaluated numerically. Several computational enhancements have been incorporated into it to make it robust as well as efficient [39]. For lack of a better name, the new algorithm that uses a potential set strategy has been called *Pshenichny–Lim–Belegundu–Arora (PLBA) algorithm*:

Step 1. Same as the CSD algorithm, except also set the initial estimate for the approximate Hessian as identity, i.e. $\mathbf{H}^{(0)} = \mathbf{I}$.

Step 2. Calculate the cost and constraint functions at $\mathbf{X}^{(k)}$ and calculate the gradients of the cost function and the potential constraint functions. Calculate the maximum constraint violation V_k as defined in equation (12.15). If $k > 0$, update Hessian of the Lagrangian using equations (12.47)–(12.56). If $k = 0$, skip updating and go to Step 3.

Step 3. Define the QP subproblem of equation (12.46) and solve it for the search direction $\mathbf{d}^{(k)}$ and Lagrange multipliers $\mathbf{u}^{(k)}$.

Steps 4–7. Same as for the CSD algorithm.

We see that the only difference between the two algorithms is in Steps 2 and 3. Schittkowski [54] and Hock and Schittkowski [55] have extensively analysed the methods and evaluated them against several other methods using a set of non-linear programming test problems. Their conclusion is that quasi-Newton methods are far superior to others. Lim and Arora [38], Thanedar *et al.* [37] and Arora and Tseng [56] have evaluated the methods for a class of engineering design problems. Consistent procedures for Hessian updating with the potential constraint strategy and re-starting the algorithm have been developed [38, 39]. Reference [57] has also discussed several enhancements of Pshenichny's algorithm including incorporation of quasi-Newton updates of the Hessian of the Lagrangian. In general, these investigations have shown the quasi-Newton methods to be superior to others.

12.6.5 Descent functions

Besides the non-differentiable function of equation (12.10), there are two other functions that have been used as descent functions. Another *non-differentiable descent function* has been suggested by Han [49, 50] and Powell [52]. We shall denote this as Φ_H and define it as follows at the kth iteration:

$$\Phi_H = f(\mathbf{X}^{(k)}) + \sum_{i=1}^{p} r_i^{(k)} |g_i| + \sum_{i=p+1}^{m} r_i^{(k)} \max\{0, g_i\} \tag{12.57}$$

where $r_i^{(k)} \geq |u_i^{(k)}|$, $i = 1$ to p are the penalty parameters for equality constraints and $r_i^{(k)} \geq u_i^{(k)}$, $i > p$ are the penalty parameters for inequality constraints. The penalty parameters sometimes become very large, so Powell [52] suggested a procedure to adjust them as follows:

First iteration:

$$r_i^{(0)} = |u_i^{(0)}|, \quad i = 1 \text{ to } p; \qquad r_i^{(0)} \geq u_i^{(0)}, \quad i > p. \tag{12.58}$$

Subsequent iterations:

$$\begin{aligned} r_i^{(k)} &= \max\{|u_i^{(k)}|, \tfrac{1}{2}(r_i^{(k-1)} + |u_i^{(k)}|)\}, & i = 1 \text{ to } p \\ r_i^{(k)} &= \max\{u_i^{(k)}, \tfrac{1}{2}(r_i^{(k-1)} + u_i^{(k)})\}, & i > p. \end{aligned} \tag{12.59}$$

Schittkowski [54] has suggested using the following augmented Lagrangian function Φ_A as the descent function:

$$\Phi_A = f(\mathbf{X}) + P_1(\mathbf{u}, \mathbf{g}) + P_2(\mathbf{u}, \mathbf{g}) \tag{12.60}$$

where $P_1(\mathbf{u}, \mathbf{g}) = \sum_{i=1}^{p} (u_i g_i + \tfrac{1}{2} r_i g_i^2) \tag{12.61}$

$$P_2(\mathbf{u}, \mathbf{g}) = \sum_{i=p+1}^{m} \begin{cases} (u_i g_i + \tfrac{1}{2} r_i g_i^2), & \text{if } (g_i + u_i/r_i) \geq 0 \\ \tfrac{1}{2} u_i^2 / r_i, & \text{otherwise} \end{cases} \tag{12.62}$$

where the penalty parameters r_i have been defined previously in equations (12.58) and (12.59). One good feature of Φ_A is that the function and its gradient are continuous.

12.7 NUMERICAL IMPLEMENTATION ASPECTS

Computational algorithms for design optimization must be properly implemented on the computer for use in practical applications. Considerable care, judgement, safeguards and user-friendly features must be designed and incorporated into the software. Numerical calculations must be robustly implemented. Each step of the algorithm must be analysed and proper numerical procedures developed to implement the intent of the step. The software must be properly evaluated for performance by solving many different problems. Transcribing an algorithm into a computer program is an art to some degree. A theoretically convergent algorithm may be implemented badly such that it is neither convergent nor reliable, and vice versa. Robust implementation of an algorithm requires considerable experience and insight, and it is best to leave the task for the specialist. Proper implementation of algorithms also requires a certain amount of heuristics which can be developed only with experience and by solving many problems of varying difficulty. An excellent discussion on this subject is given in References [22] and [39]. A detailed discussion of these aspects from an artificial intelligence point of view is contained in Reference [58]. Each optimization algorithm has certain parameters that must be selected every time a problem is solved. Specification of these parameters can have considerable influence on the performance of the algorithm. This has been amply demonstrated in References [38–40] in a detailed study of the PLBA algorithm. A good optimization algorithm should be relatively insensitive to variations in the parameters. As a minimum requirement, the algorithm must converge with every specification of the parameters within their allowable range. These aspects and the attributes discussed in a previous section can be used in selecting algorithms and the associated software for practical applications.

Good software is essential for design optimization of large complex systems. This has generally been a stumbling block in practical applications of optimization. Recently some general purpose software has become available [12, 17, 26, 56]. Several efforts are also under way to incorporate optimal design capability into general purpose finite element programs. Undoubtedly more effort is needed in this area to make optimization methodology available to designers. This is particularly true for complex systems that require multidisciplinary capabilities. Software system must be highly sophisticated and modular, allowing it to be upgraded and expanded. Modern *object-oriented programming* concepts [59], database design techniques, and database management systems (DBMS) can be quite useful in this regard. Importance of using central databases has been realized and a database design methodology has been developed for structural analysis and design optimization [60]. Design optimization programs that use the DBMS have been designed, implemented and evaluated [58, 61, 62]. It appears that use of a DBMS is of critical importance and its use will increase in the future as more multidisciplinary design problems are treated in an integrated manner.

The steps of the *PLBA algorithm* of the previous section have recently been analysed [39]. Various potential constraint strategies have been incorporated and evaluated. Several descent functions have been investigated. Procedures to resolve inconsistencies in the QP subproblem have been developed and evaluated. As a result of these enhancements and evaluations, a very powerful algorithm for engineering applications and software based on it has become available [56]. Various applications of the algorithm are discussed in the next section. This implementation of the PLBA algorithm can be used as a guideline for analysing and implementing other algorithms.

12.8 APPLICATIONS OF OPTIMIZATION TECHNIQUES

Applications of the optimization techniques abound these days. Several Conferences and Institutes have been organized on the subject in the past. Proceedings of these conferences show a wide variety of engineering applications of optimization techniques [7, 8, 12, 15, 17, 63]. Proceedings of an annual conference called Structures, Structural Dynamics and Materials, organized by the American Institute of Aeronautics and Astronautics (AIAA) also contains latest developments in optimization methods and their applications.

In the past decade, the SQP method has been applied to a wide variety of engineering problems. It has proved to be very reliable and generally applicable. It is now accepted to be the best method for non-linear programming problems. Therefore some of the applications of the method are summarized here.

Since introduction of the basic idea of sequential quadratic programming considerable computational enhancements have taken place and several different implementations of the method have become available. A major enhancement has been to incorporate a potential constraint strategy and a re-start procedure into the algorithm [38, 53, 57, 64, 65]. Several other enhancements and computational aspects are discussed by Tseng and Arora [39, 40]. Four computer codes based on the earlier version of the algorithm that does not use any potential constraint strategy are available [37], and two codes that use this strategy are available [56, 57]. A comparative evaluation of some of these codes has been carried out [37]. Basic conclusions from that study were that (1) for small-scale problems both algorithms performed reasonably well, although the code based on the potential constraint strategy was slightly more robust, and (2) potential constraint strategy was essential for larger-scale problems with regard to efficiency as well as robustness.

In recent years, the SQP algorithm has been used to solve several classes of different optimization problems: dynamic response optimization of mechanical and structural systems [40, 64, 66–8], non-linear response structural optimization problems [69–71], non-linear buckling problems [71], optimal control of structural and mechanical systems [40, 72–4], optimal control of thermal systems [75], non-linear material identification problems [76], biomechanical problems and many other applications. Shape optimization problems have also been treated [77, 78]. In all these applications, a code based on the SQP algorithm with potential constraint strategy has worked extremely well. In most cases, just one execution of the program gave the optimum solution.

12.9 PRACTICAL DESIGN OPTIMIZATION: STRUCTURAL DESIGN WITH FINITE ELEMENTS

12.9.1 Problem formulation

Optimum design formulation of complex engineering systems requires more general tools and procedures for evaluation of functions and their gradients. We shall demonstrate this by considering optimum design of structural systems with finite elements. This is an important class of problems that has a wide range of applications in automotive, aerospace, mechanical and structural engineering. It is chosen here to demonstrate the procedure of problem formulation and explain the treatment of implicit constraints. Other application areas will require similar analyses and procedures.

It is common practice to analyse complex structural systems using the *finite element technique*, which is also available in many commercial software packages. Displacements, stresses and strains at various points, vibration frequencies, and buckling loads for the system can be computed and constraints imposed on them.

Let **X** represent an *n*-component vector containing design parameters for the system. This may contain thicknesses of members, cross-sectional areas, parameters describing the shape of the system, and stiffness and material properties of the elements. Once **X** is specified, a design of the system is known. To analyse the system (calculate stresses, strains and frequencies, buckling load and displacements), the procedure is to first calculate displacements at some key points (called the *grid points* or *nodal points*) of the finite element model. From these displacements, strains (relative displacement of the material particles) and stresses at various points of the system are evaluated. Detailed procedures for these calculations are available in many textbooks [79–81].

Let **U** be a vector having *l* components representing generalized displacements (*state variables*) at key points of the system. The basic equation that determines the displacement vector **U** for a linear elastic system (called the *equilibrium equation in terms of displacements*) is given as

$$\mathbf{K}(\mathbf{X})\mathbf{U} = \mathbf{F}(\mathbf{X}) \tag{12.63}$$

where $\mathbf{K}(\mathbf{X})$ is an $l \times l$ matrix called the *stiffness matrix* and $\mathbf{F}(\mathbf{X})$ is an *effective load vector* having *l* components. The stiffness matrix $\mathbf{K}(\mathbf{X})$ is a property of the structural system which depends explicitly on the design variables, material properties and geometry of the system. Systmatic procedures have been developed to automatically calculate the matrix with different finite elements. The load vector $\mathbf{F}(\mathbf{X})$, in general, can also depend on design variables. We shall not discuss procedures to calculate $\mathbf{K}(\mathbf{X})$ because that is beyond the scope of the present chapter. The objective here is to demonstrate how the design can be optimized once a finite element model for the problem (meaning equation (12.63)) has been developed.

It can be seen that once the design **X** is specified, the displacements **U** can be calculated by solving the linear system of equation (12.63). Note that a different **X** will give, in general, different values for the displacements **U**. Thus **U** is a function of **X**; however, its explicit functional form cannot be written; that is **U** is an implicit function of the design variables **X**. The stress σ_i at the ith point is calculated using the displacements and is an explicit function of **U** and **X** as $\sigma_i(\mathbf{U}, \mathbf{X})$. However, since **U** is an implicit function of **X**, σ_i also becomes an implicit function of the design, variables **X**. The stress and displacement related constraints can be written in a functional form as

$$g_i(\mathbf{X}, \mathbf{U}) \leq 0. \tag{12.64}$$

In many automotive, aerospace, mechanical and structural engineering applications, the amount of material used must be minimized for efficient and cost-effective systems. Thus, the usual cost function for this class of application is the weight, mass or material volume of the system. This is usually an explicit function of the design variables **X**. Implicit cost functions, such as stress, displacement, vibration frequencies, etc., can also be treated by introducing artificial design variables [5].

In summary, a *general formulation for the design problem involving explicit and implicit functions of design variables* is defined as: find an *n*-dimensional vector **X** of design variables to minimize a cost function $f(\mathbf{X})$ satisfying the implicit design constraints of equation (12.64) with **U** satisfying equation (12.63), and other explicit constraints. Note that equality constraints, if present, can be routinely included in the formulation.

12.9.2 Gradient evaluation

To use modern optimization methods, we need to evaluate gradients of constraint functions. When the functions are implicit in terms of design variables, we need to develop and utilize special procedures for gradient evaluation. In this regard substantial work has been done in developing and implementing efficient procedures for calculating derivatives of implicit functions with respect to the design variables [27, 82]. The subject is generally known as *design sensitivity analysis*. For efficiency considerations and proper numerical implementations, References [27] and [82] should be consulted. The procedures have been programmed into general-purpose software for automatic computation of design gradients.

Generally two analytical methods have been used for gradient evaluation in structural optimization: the direct differentiation method (DDM) and the adjoint variable method (AVM). We will present both the methods and compare their computational aspects. The finite difference methods has also been extensively used, though it may be less accurate and has a drawback in terms of uncertainty in the step-size specification.

Direct differentiation method

In this method, we use the chain rule of differentiation to obtain the total derivative of $g_i(\mathbf{X}, \mathbf{U})$ with respect to the *j*th design variable as

$$\frac{dg_i}{dX_j} = \frac{\partial g_i}{\partial X_j} + \left(\frac{\partial g_i}{\partial \mathbf{U}} \cdot \frac{d\mathbf{U}}{dX_j}\right) \quad (12.65)$$

where $\partial g_i/\partial \mathbf{U}$ and $d\mathbf{U}/dX_j$ are given as

$$\frac{\partial g_i}{\partial \mathbf{U}} = \left[\frac{\partial g_i}{\partial U_1} \frac{\partial g_i}{\partial U_2} \cdots \frac{\partial g_i}{\partial U_l}\right]^T \quad \text{and} \quad \frac{d\mathbf{U}}{dX_j} = \left[\frac{\partial U_1}{\partial X_j} \frac{\partial U_2}{\partial X_j} \cdots \frac{\partial U_l}{\partial X_j}\right]^T. \quad (12.66)$$

To calculate the gradient of a constraint using equation (12.65), we need to calculate the partial derivatives $\partial g_i/\partial X_j$ and $\partial g_i/\partial \mathbf{U}$, and the total derivatives $d\mathbf{U}/dX_j$. The partial derivatives $\partial g_i/\partial X_j$ and $\partial g_i/\partial \mathbf{U}$ are quite easy to calculate using the form of the function $g_i(\mathbf{X}, \mathbf{U})$. To calculate $d\mathbf{U}/dX_j$, we differentiate the equilibrium equation (12.63) to obtain

$$\frac{\partial \mathbf{K}(\mathbf{X})}{\partial X_j}\mathbf{U} + \mathbf{K}(\mathbf{X})\frac{d\mathbf{U}}{dX_j} = \frac{\partial \mathbf{F}}{\partial X_j}. \tag{12.67}$$

Or the equation can be rearranged as

$$\mathbf{K}(\mathbf{X})\frac{d\mathbf{U}}{dX_j} = \frac{\partial \mathbf{F}}{\partial X_j} - \frac{\partial \mathbf{K}(\mathbf{X})}{\partial X_j}\mathbf{U}. \tag{12.68}$$

The equation can be used to calculate $d\mathbf{U}/dX_j$. The derivative of the stiffness matrix $\partial \mathbf{K}(\mathbf{X})/\partial X_j$ can easily be calculated since the explicit dependence of $\mathbf{K}$ on $\mathbf{X}$ is known. It is also possible to perform this calculation without explicitly calculating the derivatives of the element stiffness matrices. Such computational aspects and computer implementation details have been discussed recently in Reference [83] and other references cited there. Note that equation (12.68) needs to be solved for each design variable. Once $d\mathbf{U}/dX_j$ are known, the gradient of the constraint is calculated from equation (12.65). The derivative vector in equation (12.65) is often called the *design gradient*.

Adjoint variable method

This method introduces an additional variable (called the *adjoint variable*) for each function needing gradient evaluation. The method is quite general and is applicable to many different clases of problems from various fields [5, 27, 82–5]. Although the method can be derived from several different points of view, we shall derive it using a recently developed *variational principle of design sensitivity analysis* [85]. In this approach, we introduce an adjoint variable vector $\mathbf{U}^a$ (to be determined later) for the equilibrium equation (12.63) and define an augmented functional g_{iA} for the function g_i as

$$g_{iA} = g_i(\mathbf{X}, \mathbf{U}) + (\mathbf{U}^a \cdot [\mathbf{K}(\mathbf{X})\mathbf{U} - \mathbf{F}(\mathbf{X})]). \tag{12.69}$$

According to the foregoing principle, partial deriative of g_{iA} with respect to X_j is the total derivative of g_i with respect to X_j, i.e. $\partial g_{iA}/\partial X_j = dg_i/dX_j$. Therefore,

$$\frac{dg_i}{dX_j} = \frac{\partial g_i}{\partial X_j} + \left(\mathbf{U}^a \cdot \frac{\partial}{\partial X_j}[\mathbf{K}(\mathbf{X})\mathbf{U} - \mathbf{F}(\mathbf{X})]\right). \tag{12.70}$$

Also stationarity of g_{iA} with respect to $\mathbf{U}$ (i.e. $\partial g_{iA}/\partial \mathbf{U} = 0$) give conditions that determine $\mathbf{U}^a$. This lead to the following *adjoint equation*:

$$\mathbf{K}(\mathbf{X})\mathbf{U}^a = -\frac{\partial g_i}{\partial \mathbf{U}}. \tag{12.71}$$

Equation (12.70) can also be derived directly from equations (12.65) and (12.68). Substituting dU/dX_j from equation (12.68) into equation (12.65) and then using equation (12.71), we see that equation (12.70) is obtained.

Comparison of DDM and AVM

Since no approximations have been used in the derivations, both the methods will give the same numerical values for the gradient of g_i. However, the number of calculations needed to obtain the final gradient can be quite different [82]. Comparing various equations for the two methods, we observe that the major difference lies in the right-hand sides of the linear system of equations (12.68) and (12.71). In equation (12.68), the number of right-hand-side vectors is equal to the number of design variables, n. In equation (12.71), there is only one right-hand-side vector; however it depends on the constraint function whose gradient is desired, whereas the right-hand-side of equation (12.68) does not depend on any constraint function. Thus in DDM, the gradient of any function can be easily evaluated once equation (12.68) has been solved; but with the AVM, the functions that need gradient evaluation must be known before equation (12.71) can be defined and solved. Let p be the number of functions that need gradient evaluation. Then usually if $p < n$, the AVM is preferred, otherwise DDM is preferred.

12.10 OPTIMIZATION OF LARGE SYSTEMS

A large-scale non-linear programming problem may be defined as the one that has more than 100 design variables. Also design problems that require excessive computational effort to evaluate the functions and their gradients may be classified as large-scale problems. Such problems pose a special challenge for the optimization algorithms. The number of calculations for the search direction and step size determination can be very large, so the numerical truncation and round-off can become a problem in the optimization process. In addition the number of iterations to reach the solution point can be large. As a matter of fact there is no algorithm for general non-linear programming problems that has been proved to converge to the solution point in a finite number of iterations. It is therefore, obvious that for stability and efficiency of the solution process for large scale problems, the number of calculations within on iteration should be kept to a minimum, and the number of iterations to reach an acceptable solution point needs to be kept reasonable.

Several algorithms have been developed in the past to solve Problem P and many others are under active development. Most of the methods require substantial computational effort for large-scale problems having hundreds of variables and thousands of constraints. The most modern, efficient and reliable methods are based on *sequential quadratic programming* (SQP) procedures. Even though these methods are the best up to now, they have some *drawbacks* for large-scale problems: (1) use of the approximate Hessian results in a larger requirement for the computer memory, and (2) the large number of operations associated with updating and use of this matrix results in inefficiency, and numerical uncertainty and instability. Numerical experiments have shown that the SQP method needs a large number of operations with the Hessian matrix which is the primary cause of its inefficiency for application to large-scale problems.

An algorithm that does not have the foregoing drawbacks is needed for large-scale problems. It is well known in the unconstrained optimization literature that the *conjugate directions method* is a method of choice for large-scale problems [1, 22, 23]. The method turns out to be a very simple modification of the basic *steepest descent method* and is quite easy to implement on the computer; but it performs much better than the steepest descent method. Recently, this concept of *conjugate directions* for unconstrained optimization has been extended to solve general constrained minimization problems [86]. The *constrained steepest descent* (CSD) directions obtained as the solution of a *quadratic programming* (QP) sub-problem are used to generate the *constrained conjugate directions* (CCD). Just as for the unconstrained methods, the CCD method does not generate or require the second-order information resulting in a very simple algorithm. *Finite convergence* and other properties on the algorithm have been proved for convex quadratic programming problems. For general problems, *restart* procedures, *step-size* determination, and *potential constraint strategy* have been developed. A set of 41 structural design *test problems* having 2 to 489 design variables and 7 to 1051 constraints excluding the simple bounds, has been used to evaluate the method and its variations [86]. The new method performs much *better* than the CSD method; however, its performance is *not as good as* the sequential quadratic programming (SQP) method for moderate size problems. This result is expected and is quite similar to the one for the unconstrained counterpart of the method. For a large-scale problem, the new method performs much better than both the CSD and the SQP methods. It has been *concluded* that the basic concept of constrained conjugate directions is a *viable approach* for large-scale optimization problems. However, more work needs to be done on the restart strategies and step-size determination procedures to fully develop the method and realize its potential for large-scale problems.

Besides the optimization algorithms for large-scale problems discussed in the foregoing, several other developments have recently taken place relative to optimum design of large and complex strutural and mechanical systems. Noteworthy among these are the ideas of *substructures, multilevel optimization,* and *use of parallel algorithms.* The idea of dividing a large structure into several smaller substructures has been developed and investigated [5, 87, 88]. This approach can be useful in structural analysis (function evaluation for the optimization problem) and evaluation of gradients of the problem functions. This idea of substructures has been generalized to multilevel optimization of multidisciplinary systems [89]. In this approach, design of the entire system is divided into several levels that interact with each other only through well defined interfaces. Each subsystem is optimized relatively independently and sensitivity of the interacting variables is calculated to integrate their effect into the design of individual subsystems.

Exploitation of parallel computations in the optimization process is another recent development that can help in the treatment of large-scale systems. This has been made possible by the development of parallel computers. There are several steps in the overall process of designing systems where parallel processing can be very useful.

These steps have been recently analysed [90]. Undoubtedly, further studies and developments (in terms of hardware, system software, and parallel algorithms) are needed to fully exploit this new technology for optimum design of large-scale systems.

12.11 CONCLUDING REMARKS

Basic concepts related to design optimization of engineering systems are described and discussed. General concepts of computational algorithms are explained and the definition of a good algorithm is given. Several computational algorithms that use values of the problem functions and their gradients are described. These algorithms have been recently superseded by the methods that generate and use second order information. An algorithm belonging to the latter class, called sequential quadratic programming, is described. This is the most recent method for optimization that has proved to be very reliable in solving highly non-linear problems.

Based on the experience with several algorithms, it is suggested that designers who are not experts in optimization should use only general and reliable methods. Although such methods require more computational effort compared to other algorithms, they are more cost effective in the longer run. Efficiency, although important, can be sacrificed for reliability. A few more CPU cycles of the machine are worth the wait if the algorithm can reliably give optimum solutions. Use of unreliable and *ad hoc* methods and software based on them can actually be more costly and frustrating in terms of designer's time and uncertainty about the solution. There are many numerical difficulties in solving large complex problems. They need not be compounded by the use of unreliable algorithms and programs based on them. A detailed discussion on robustness versus efficiency of methods has been presented recently in Reference [33].

Transcription of an algorithm into good software requires considerable experience, time and resources [39]. This should be a task for optimization experts and software developers. Design engineers need not be involved in algorithm development or its implementation. Their job should be to properly formulate the design optimization problem [1], solve it using available software and interpret the results. They should concentrate on investigating alternative conceptual designs or formulations for their applications. Once a concept is properly formulated, well-developed and robust software can solve the problem very quickly. This way optimization methods and programs can be used as tools in the creative process of designing systems.

The SQP method presented in this chapter uses a potential constraint strategy and re-start procedures. The method has been used successfully to solve a wide range of problems having 3 to 489 designs variables and 6 to 2000+ constraints. Many problems also had equality constraints. One observation about the method is that a the number of design variables increases, the approximate Hessian of the Lagrangian becomes quite large. Updating, storing and performing calculations with it become time consuming. This is a major drawback of the method that hinders its application to large scale problems. Recently, the concept of conjugate gradients for unconstrained problems has been extended to the constrained problems [86]. This concept uses the CSD directions to generate conjugate directions in the constrained space. A method based on this idea has worked very well for large-scale problems. Optimization of large-scale non-linear problems remains an active area of research. Undoubtedly, research will continue to develop efficient and robust methods for such problems.

BIBLIOGRAPHY

[1] Arora, J.S. (1989) *Introduction to Optimum Design*, McGraw-Hill, New York.
[2] Arora, J.S. (1990) Computational design optimization: a review and future directions, *Structural Safety*, **7**, 131–148.
[3] Arora, J.S. (1993) Fundamental concepts of optimum design, and sequential linearization and quadratic programming techniques, Chapters in M. Kamat (ed.), *Structural Optimization: Status and Review*, AIAA Series in Astronautics and Aeronautics, to appear.
[4] Arora, J.S. and Thanedar, P.B. (1986) Computational methods for optimum design of large complex systems, *Computational Mechanics*, **1**, 221–242.
[5] Haug, E.J. and Arora, J.S. (1979) *Applied Optimal Design: Mechanical and Structural Systems*, Wiley Interscience, New York.
[6] Kirsch, U. (1981) *Optimum Structural Design*, McGraw-Hill, New York.
[7] Haug, E.J. and Cea, J. (eds.) (1981) Optimization of distributed parameter structures, Vols. 1 and 2, *Proc. NATO ASI Meeting*, Iowa City (1980), Sijthoff and Noordhoff International Publishers B.V., Alphen aan den Rijn, The Netherlands, Rockville, Maryland.
[8] Morris, A.J. (ed.) (1982) Foundations of structural optimization, *Proc. NATO ASI Meeting*, Liege, Belgium, 1980, Wiley Interscience, New York.
[9] Reklaitis, G.V., Ravindran, A. and Ragsdell, K.M. (1983) *Engineering Optimization*, Wiley Interscience, New York.
[10] Rao, S.S. (1984) *Optimization: Theory and Applications* (2nd edn), Wiley, New York.
[11] Vanderplaats, G.N. (1984) *Numerical Optimization Techniques for Engineering Design*, McGraw-Hill, New York.
[12] Atrek, E., Gallagher, R.H., Ragsdell, K.M. and Zienkiewicz, O.C. (eds.) (1984) New directions in optmum structural design, *Proc. Intern. Symp. Optimum Structural Design*, University of Arizona, Tucson, Arizona, 1981, Wiley, New York.
[13] Sobieski, J. (ed.) (1984) Recent experiences in multidisciplinary analysis and optimization, NASA Conf. Publ. 2327, *Proc. Symp.*, NASA Langley Research Center, Hampton, VA.
[14] Frangopol, D.M. (1985) Structural optimization using reliability concepts in design, *J. Struct. Engg*, ASCE, **111**, 11.
[15] Bennett, J.A. and Botkin, M.E. (eds.) (1986) The optimum shape: automated structural design, *Proc. Intern. Symp.*, GM Research Laboratories, Warren, MI, September 30–October 1, 1985, Plenum Press, New York.
[16] Levy, R. and Lev, O.E. (1987) Recent developments in structural optimization, *J. Struct. Engg*, ASCE, **113**(9), 1939–1962.
[17] Mota Soares, C.A. (ed.) (1987) Computer aided optimal design: structural and mechanical systems, *Proc. NATO Advanced Study Institute*, Troia, Portugal, June 29–July 11, 1986, Series F: Computer and System Sciences, Vol. 27, Springer-Verlag, New York.
[18] Papalambros, P.Y. and Wilde, D.J. (1988) *Principles of Optimal Design: Modeling and Computation*, Cambridge University Press, New York.
[19] Rozvany, G.I.N. (1988) *Structural Design via Optimality Criteria*, Kluwer Academic Publishers, Hingham, Massachusetts.
[20] Haftka, R.T., Kamat, M.P. and Gürdal, Z. (1990) *Elements of Structural Optimization* (2nd edn), Kluwer Academic Publishers, Dordrecht.

[21] Bazaraa, M. and Shetty, C.M. (1979) *Nonlinear Programming: Theory and Algorithms*, Wiley, New York.
[22] Gill, P.E., Murray, W. and Wright, M.H. (1981) *Practical Optimization*, Academic Press, New York.
[23] Luenberger, D.G. (1984) *Linear and Nonlinear Programming*, Addison-Wesley, Reading, Massachusetts.
[24] Osyczka, A. (1984) *Multicriterion Optimization in Engineering*, Halsted Press, Wiley, New York.
[25] Cha, J.Z. and Mayne, R.W. (1989) Optimization with Discrete Variables via Recursive Quadratic Programming, *J. Mech. Des.*, ASME, **111**, 124–136.
[26] Schittkowski, K. (ed.) (1985) Computational Mathematical Programming, *Proc. NATO ASI*, Bad Windsheim, FRG, July 23–August 2, 1984, Springer, Berlin, Heidelberg, New York, Tokyo.
[27] Haftka, R.T. and Adelman, H.M. (1989) Recent developments in structural sensitivity analysis, *Structural Optimization*, **1**(3), 137–152.
[28] Arora, J.S. (1990) Global optimization methods for engineering design, *Proc. Work-in-progress Sessions*, 31st AIAA/ASME/ASCE/AHS/ASC Structures, Structural Dynamics and Materials Conference, Long beach, CA, April.
[29] Arora, J.S. and Belegundu, A.D. (1984) Structural optimization by mathematical programming methods, *AIAA J.*, **22**, 854–856.
[30] Belegundu, A.D. and Arora, J.S. (1984) A computational study of transformation methods for optimal design, *AIAA J.*, **22**, 535–542.
[31] Belegundu, A.D. and Arora, J.S. (1985) A study of mathematical programming methods for structural optimization, Part I: Theory; Part II: Numerical aspects, *Int. J. Num. Meth. Engg*, **21**, 1583–1623.
[32] Arora, J.S., Chahande, A.I. and Paeng, J.K. (1991) Multiplier methods for engineering optimization, *Int. J. Num. Meth. Engg*, **32**(7), 1485–1525.
[33] Thanedar, P.B., Arora, J.S., Li, G.Y. and Lin, T.C. (1990) Robustness, generality and efficiency of optimization algorithms for practical applications, *Structural Optimization*, **2**, 203–212.
[34] Pshenichny, B.N. and Danilin, Y.M. (1978) *Numerical Methods in Extremal Problems*, Mir Publishers, Moscow.
[35] Belegundu, A.D. and Arora, J.S. (1984) A recursive quadratic programming method with active set strategy for optimal design, *Int. J. Num. Meth. Engg*, **20**, 803–816.
[36] Choi, K.K., Haug, E.J., Hou, J.W. and Sohoni, V.N. (1983) Pshenichny's linearization method for mechanical system optimization, *J. Mech. Transm. Autom. Des.*, ASME, **105**, 97–103.
[37] Thanedar, P.B., Arora, J.S., Tseng, C.H., Lim, O.K. and Park, G.J. (1986) Performance of some SQP algorithms on structural design problems, *Int. J. Num. Meth. Engg*, **23**(12), 2187–2203.
[38] Lim, O.K. and Arora, J.S. (1986) An active set RQP algorithm for engineering design optimization, *Comput. Meth. Appl. Mech. Engg*, **57**, 51–65.
[39] Tseng, C.H. and Arora, J.S. (1988) On implementation of computational algorithms for optimum design, *Int. J. Num. Meth. Engg*, **26**(6), 1365–1402.
[40] Tseng, C.H. and Arora, J.S. (1989) Optimum design of systems for dynamics and controls using sequential quadratic programming, *AIAA J.*, **27**(12), 1793–1800.
[41] Zoutendijk, G. (1960) *Methods of Feasible Directions*, Elsevier, Amsterdam.
[42] Vanderplaats, G.N. (1984) An efficient feasible directions algorithm for design synthesis, *AIAA J.*, **22**(11), 1633–1640.
[43] Rosen, J.B. (1961) The gradient projection for nonlinear programming, Part II: Nonlinear constraints, *J. Soc. Ind. Appl. Math.*, **9**, 514–532.

[44] Abadie, J. (ed.) (1970) *Nonlinear Programming*. Chapter 6, North Holland, Amsterdam.
[45] Abadie, J. (1969) Generalization of the Wolfe reduced gradient method to the case of nonlinear constraints, in R. Fletcher (ed.), *Optimization*, Academic Press, New York, pp. 37–47.
[46] Gabriele, G.A. and Ragsdell, K.M. (1980) Large scale nonlinear programming using the generalized reduced gradient method, *ASME J. Mech. Des.*, **102**, 566–573.
[47] Sargeant, R.W.H. (1974) Reduced-gradient and projection methods for nonlinear programming, in P.E. Gill and W. Murray (eds.) *Num. Meth. Constrained Optimization*, Academic Press, New York, pp. 149–174.
[48] Wilson, R.B. (1963) *A Simplical Algorithm for Concave Programming*, Ph.D. Dissertation, Harvard University, Cambridge, Massachusetts.
[49] Han, S.P. (1976) Superlinearly convergent variable metric algorithms for general nonlinear programming, *Math. Program.*, **11**, 263–282.
[50] Han, S.P. (1977) A globally convergent method for nonlinear programming, *J. Optimization Theor Appl.*, **22**, 297–309.
[51] Garcia-Palomares, V.M. and Mangasarian, O.L. (1976) Superlinearly convergent quasi-Newton algorithms for nonlinearly constrained optimization problem, *Math. Program.*, **11**, 1–13.
[52] Powell, M.J.D. (1978) Algorithms for nonlinear functions that use Lagrangian functions, *Math. Program.*, **14**, 224–248.
[53] Thanedar, P.B., Arora, J.S. and Tseng, C.H. (1986) An efficient hybrid optimization method and its role in computer aided design, *Comput. Struct.*, **23**(3), 305–314.
[54] Schittkowski, K. (1981) The nonlinear programming method of Wilson, Han and Powell with an augmented Lagrangian type line search function, Part 1: Convergence analysis: Part 2: An efficient implementation with least squares subproblems, *Num. Math.*, **38**, 83–114; 115–127.
[55] Hock, W. and Schittkowski, K. (1981) Test examples for nonlinear programming codes, in *Lecture Notes in Economics and Mathematical Systems*, Vol. 187, Springer, Berlin, Heidelberg, New York.
[56] Arora, J.S. and Tseng, C.H. (1988) Interactive design optimization, *Engg Optimization*, **13**, 173–188.
[57] Beltracchi, T.J. and Gabriele, G.A. (1987) An investigation of Pshenichny's recursive quadratic programming method for engineering optimization, *J. Mechanisms, Transmissions and Automation in Design*, ASME, **109**(2), 248–253.
[58] Arora, J.S. and Baenziger, G. (1986) Use of artificial intelligence in design optimization, *Comput. Meth. Appl. Engg*, **54**, 303–323.
[59] Lee, H.H. and Arora, J.S. (1991) Object-oriented programming for engineering applications, *Engg Comput.*, **7**, 225–235.
[60] SreekantaMurthy, T. and Arora, J.S. (1986) Database design methodology for structural analysis and design optimization, *Engg Comput.*, **1**, 149–160.
[61] Park, G.J. and Arora, J.S. (1987) Role of database management in design optimization systems, *J. Aircraft*, **24**(1), 745–750.
[62] Al-Saadoun, S.S. and Arora, J.S. (1989) Interactive design optimization of framed structures, *J. Comput. Civil Engg*, ASCE, **3**(1), 60–74.
[63] Rozvany, G.I.N. and Arihaloo, B.L. (eds.) (1988) Structura Optimization *Proc. IUTAM Symp. Structural Optimization*, Melbourne, Australia.
[64] Lim, O.K. and Arora, J.S. (1987) Dynamic response optimization using an active set RQP algorithm, *Int. J. Num. Meth. Engg*, **24**(10), 1827–1840.
[65] Arora, J.S. and Tseng, C.H. (1987) An investigation of Pshenichny's recursive quadratic programming method for engineering optimization: a discussion, *J. Mechanisms, Transmissions and Automation in Design*, ASME, **109**(2), 254–256.

[66] Paeng, J.K. and Arora, J.S. (1989) Dynamic response optimization of mechanical systems with multiplier methods, *J. Mechanisms, Transmissions and Automation in Design*, Transactions of ASME, **111**(1), 73–80.

[67] Hsieh, C.C. and Arora, J.S. (1984) Design sensitivity analysis and optimization of dynamic response, *Comput. Meth. Appl. Mech. Engg*, **43**, 195–219.

[68] Hsieh, C.C. and Arora, J.S. (1985) An efficient method for dynamic response optimization, *AIAA J.*, **23**, 1484–1486.

[69] Ryu, Y.S., Haririan, M., Wu, C.C. and Arora, J.S. (1985) Structural design sensitivity analysis of nonlinear response, *Comput. Struct.*, **21**, 245–255.

[70] Wu, C.C. and Arora, J.S. (1987) Design sensitivity analysis of nonlinear response using incremental procedure, *AIAA J.*, **25**(8) 1118–1125.

[71] Wu, C.C. and Arora, J.S. (1988) Design sensitivity analysis of nonlinear buckling load, *Comput. Mechanics*, **3**(1), 129–140.

[72] Khot, N.S. (1988) Structure/control optimization to improve the dynamic response of space structures, *Comput. Mechanics*, **3**, 179–186.

[73] Grandhi, R.V. (1989) Structural and control optimization of space structures, *Comput. Struct.*, **31**(2), 139–150.

[74] Lin, T.C. and Arora, J.S. (1991) *A Study of Differential Dynamic Programming for Optimal Control*, Technical Report No. ODL-91.2, Optimal Design Laboratory, College of Engineering, The University of Iowa, Iowa City, Iowa.

[75] House, J.M., Smith, T.F. and Arora, J.S. (1991) Optimal control of a thermal system, *Trans. ASHREE*, **97** (Part 2), 991–1001.

[76] Jao, S.Y., Arora, J.S. and Wu, H.C. (1991) An optimization approach for material-constant determination for endochronic constitutive model, *Comput. Mechanics*, **8**(1), 25–41.

[77] Arora, J.S. and Cardoso, J.B. (1989) A design sensitivity analysis principle and its implementation into ADINA, *Comput. Struct.*, **32**(3/4), 691–705.

[78] Jao, S.Y. and Arora, J.S. (1993) Design optimization of nonlinear structures with rate dependent and rate-independent constitutive models, *Int. J. Num. Meth. Engg*, to appear.

[79] Bathe, K.J. (1982) *Finite Element Procedures in Engineering Analysis*, Prentice Hall, Englewood Cliffs, New Jersey.

[80] Huebner, K.H. and Thornton, E.A. (1982) *The Finite Element Method for Engineers*, Wiley, New York.

[81] Chandrupatla, T.R. and Belegundu, A.D. (1991) *Introduction to Finite Elements in Engineering*, Prentice Hall, Englewood Cliffs, New Jersey.

[82] Arora, J.S. and Haug, E.J. (1979) Methods of design sensitivity analysis in structural optimization, *AIAA J.*, **17**(9), 970–974; and *AIAA J.*, **18**(11), 1407.

[83] Arora, J.S., Lee, T.H. and Cardoso, J.B. (1992) Structural shape sensitivity analysis: relationship between material derivative and control volume approaches, *AIAA J.*, **30**(6), 1638–1648.

[84] Cardoso, J.B. and Arora, J.S. (1991) Shape design sensitivity analysis of field problems, *Int. J. Engg Sci.*, **29**(12), 1627–1637.

[85] Arora, J.S. and Cardoso, J.B. (1992) A variational principle for shape design sensitivity analysis, *AIAA J.*, **30**(2), 538–547.

[86] Arora, J.S. and Li, G.-Y. (1993) Constrained conjugate directions method for design optimizations of large systems, *AIAA J.*, **31**(1/2).

[87] Arora, J.S. and Govil, A.K. (1977) An efficient method for optimal structural design by substructuring, *Comput. Struct.*, **7** (4), 507–515.

[88] Arora, J.S. and Haug, E.J. (1979) Sensitivity analysis and optimization of large scale

structures, in C.T. Leondes (ed.), *Control and Dynamic Systems*, Vol. 15, Academic Press, New York, pp. 248–275.
[89] Sobieszczanski-Sobieski, J., James, B.B. and Riley, M.F. (1987) Structural sizing by generalized multilevel optimization, *AIAA J.*, **25**(1), 139.
[90] Lin, T.P. and Arora, J.S. (1991) *Parallel Computations in Structural Optimization*, Technical Report No. ODL-91.22, Optimal Design Laboratory, College of Engineering, The University of Iowa, Iowa City, Iowa.

J.S. Arora
Optimal Design Laboratory
Departments of Civil & Environmental Engineering, and Mechanical Engineering
College of Engineering
University of Iowa
Iowa City
IA 52242
USA

13

Automatic Generation of Finite Element Models

M. S. Shephard

Rensselaer Polytechnic Institute, NY, USA

13.1 INTRODUCTION

The majority of chapters in this volume address techniques for solving the various forms of systems of algebraic equations that arise in the numerical solution of problems in mathematical physics. These systems of algebraic equations arise from a discretization process. A commonly applied process is the finite element discretization of the weak form of the governing equations. The application of this process requires the domain of interest be decomposed, in a highly controlled manner, into a mesh of simple subdomains, each of which represents a finite element. The manual or semi-automatic generation of such meshes is a time-consuming and error-prone process, the cost of which can dominate the analysis of complex three-dimensional domains. The only way to eliminate this bottleneck, as well as to ensure the reliability, is to automate the meshing process. This chapter addresses the basic issues associated with reliable automatic mesh generation, outlines the status of some of the current developments in this area, and discusses a few aspects of the efficient parallel solution to the systems of discrete equations associated with these meshes.

A prerequisite of the reliable application of the finite element technique on general three-dimensional domains is the availability of a mesh generator that can accept a geometry-based definition of the problem and automatically produce a valid finite element discretization of it, without user intervention. The automatic generation of finite element meshes for general three-dimensional geometries has been a topic of active research for nearly a decade. A number of algorithmic approaches have been proposed for performing this task. Reference [1] reviews the algorithmic approaches being considered in 1988. In addition to updating the progress on the development of algorithmic approaches to automatic mesh generation, this chapter considers the basic issues that must be considered in the development of an automatic mesh generation algorithm.

Central to the development of an automatic mesh generator is the definition of what constitutes a valid finite element mesh. Section 13.2 overviews the concept of a geometric triangulation which provides a definition of a valid finite element mesh [2], [3]. Section 13.3 discusses the key issues that must be addressed, in one form or another, in the development of an automatic mesh generator. Section 13.4 discusses octree-based meshing procedures and indicates how the various approaches address the issues critical to the generation of valid finite element meshes. Section 13.5 briefly discusses recent activities on other automatic mesh generation approaches including Delaunay, advancing front (paving), and medial (symmetric) axis transformation mesh generators. Section 13.6 briefly discusses the integration of automatic mesh generators with geometric modelling.

The last section considers the influence the types of meshes generated by automatic mesh generators have on efficient adaptive solution procedures on parallel computers, and indicates some current efforts to address these issues for octree-based automatic mesh generators.

13.2 DEFINITION OF A VALID FINITE ELEMENT MESH

The first step in ensuring a mesh generator will generate valid finite element meshes is to define a valid finite element mesh [2–4].

Since mesh generation is concerned with the decomposition of a geometric domain into a union of simple, non-overlapping geometric entities, the definition of a valid mesh must be in terms of the definition of the geometric domain. The definition of a geometric domain can be considered to consist of two sets of information:

$$G = \{_G S, _G T\} \tag{13.1}$$

where $_G S$ represents the geometric information defining the shape of the entities which define the domain and $_G T$ represents the topological types and associativities of the entities. Since individual finite elements are assumed to be a simple region bounded by simply connected faces, the topological entities associated with the 0- to n-dimensional geometric entities are of interest. For the three-dimensional case ($n = 3$)

$$_G T = \{_G T^0, _G T^1, _G T^2, _G T^3\} \tag{13.2}$$

where $_G T^d$, $d = 0, 1, 2, 3$ are respectively the set of vertices, edges, faces and regions defining the primary topological elements of the geometric domain.

Critical to the definition of a valid finite element mesh are the concepts of mesh classification and mesh compatibility [2–4].

Definition *Classification* [2–4]. The unique association of a topological mesh entity of dimension d_i, ${}_M T_i^{d_i}$, to a topological model entity of dimension d_j, ${}_G T_j^{d_j}$, where $d_i \leq d_j$, is termed classification and is denoted

$$ {}_M T_i^{d_i} \sqsubset {}_G T_j^{d_j} \tag{13.3}$$

where the classification symbol, $\sqsubset$, indicates that the left-hand entity or set is classified on the right-hand entity.

Multiple ${}_M T_i^{d_i}$ can be classified on a ${}_G T_j^{d_j}$.

Definition *Topological compatibility* [2], [3]. Given a non-self-intersecting mesh with all vertices in the vertex set, ${}_M T^0$, classified, and the remaining sets of mesh entities ${}_M T^d$, $1 \leq d \leq n$ with boundary entities sets $\partial({}_M T^d)$, consider a model entity ${}_G T_j^d$ with boundary entities $\partial({}_G T_j^d)$. If each $\partial({}_M T_i^d) \sqsubset {}_G T_j^d$ is used by two ${}_M T_k^d \sqsubset {}_G T_j^d$, and each $\partial({}_M T_i^d) \sqsubset \partial({}_G T_j^d)$ is used by one ${}_M T_k^d \sqsubset {}_G T_j^d$, then the mesh is topologically compatible with the topological entity ${}_G T_j^d$. A mesh is topologically compatible if it is compatible with all topological entities.

Consider Figures 13.1 through 13.3 for a clarification of topological compatibility. Figure 13.1 shows a ${}_G T_j^2$ covered a compatible set of ${}_M T_k^2 \sqsubset {}_G T_j^2$. In this case all the ${}_M T_k^1 \sqsubset {}_G T_j^2$ are used by two ${}_M T_k^2 \sqsubset {}_G T_j^2$ and all ${}_M T_i^1 \sqsubset \partial({}_G T_j^2)$ are used by one ${}_M T_k^2 \sqsubset {}_G T_j^2$. Figure 13.2 contains a topological hole characterized by the fact that the three mesh edges $({}_M T_1^1, {}_M T_2^1, {}_M T_3^1) \sqsubset {}_G T_j^2$ are only used by only one ${}_M T_k^2 \sqsubset {}_G T_j^2$. Figure 13.3 depicts a topological redundancy which is characterized by the four mesh edges $({}_M T_1^1, {}_M T_2^1, {}_M T_3^1, {}_M T_4^1) \sqsubset {}_G T_j^2$ each being used by three ${}_M T_k^2 \sqsubset {}_G T_j^2$.

Starting with these definitions a definition of a valid finite element mesh can be given.

Definition *Geometric triangulation* [2], [3]. Given a set P of M unique points, each classified with respect to the geometry G, an n-dimensional geometric triangulation, ${}_G \Lambda^n$, is a set of N non-degenerate elements $s_i^{d_i}$

$$ {}_G \Lambda^n = \{s_1^{d_1}, s_2^{d_2}, s_3^{d_3}, \ldots, s_N^{d_N}\} \tag{13.4}$$

with $0 \leq d_i \leq n$, satisfying the following properties:

(i) All vertices of each $s_i^{d_i} \in P$
(ii) For each $i \neq j$, interior $(s_i^{d_i}) \cap$ interior $(s_j^{d_j}) = 0$
(iii) ${}_G \Lambda^n$ is topologically compatible with G
(iv) ${}_G \Lambda^n$ is geometrically similar to G.

The simplest explanation of a geometrically similar mesh is one that in the limit of refinement will exactly match the geometry of the domain. This simple definition is not a workable one for the development of algorithms to evaluate and correct mesh validity. In all but complex geometric cases, this requirement is satisfied if

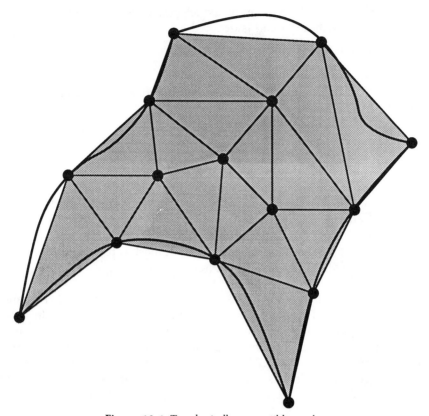

Figure 13.1 Topologically compatible mesh.

topological compatibility is satisfied. However, since there is no *a priori* method to ensure that topological compatibility alone will also ensure geometric similarity, it must be explicitly considered.

One method of ensuring geometric similarity introduced by Schroeder [2] employs the concept of parametric intersection. Any application of this approach requires that each of the geometric entities in the geometric model be uniquely mappable.

Definition *Uniquely mappable* [2]. A geometric entity of dimension d, S^d, is uniquely mappable if for each point $p \in S^d$, there exist a function $f: S^d \to H^d$, that satisfies the following conditions:

1. For each neighbourhood V of $f(p)$, there exist a neighbourhood U of p such that $f(U) \subset V$
2. For each $p \neq \acute{p}$, $f(p) \neq f(\acute{p})$
3. Each S^d is mappable to the hyperplane H^d
4. Each S^d is of finite extent.

The property of unique mappability allows the introduction of a parameterization

13.2 DEFINITION OF A VALID FINITE ELEMENT MESH

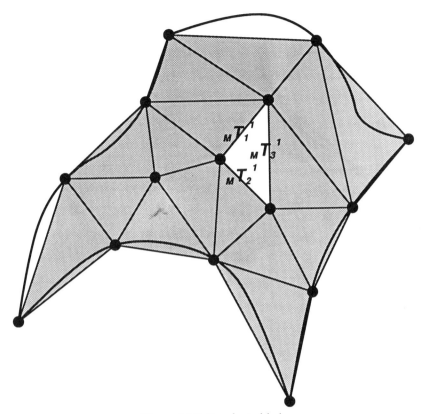

Figure 13.2 Topological hole.

of the individual geometric entities. The development of a practical algorithm does not require the parameterization to be explicitly defined over the entire entity. Instead it can be defined in local neighbourhoods large enough to perform parametric intersections of the mesh entities under consideration.

Definition *Parametric intersection* [2]. Given two mesh entities of order d, $_MT_i^d$ and $_MT_j^d$, classified on a topological model entity of dimension d, $_GT_k^d$, the parametric intersection of $_MT_i^d$ and $_MT_j^d$ is written as

$$_MT_i^d \square^* {}_MT_j^d. \tag{13.5}$$

With the concept of a parametric intersection the conditions of geometric similarity can be given.

Definition *Geometric similarity* [2]. A set of mesh entities of order d, $_MT^d$, is geometrically similar to a topological model entity of order d, $_GT_k^d$, when $_MT^d$ consist of N mesh entities of order d:

436 AUTOMATIC GENERATION OF FINITE ELEMENT MODELS

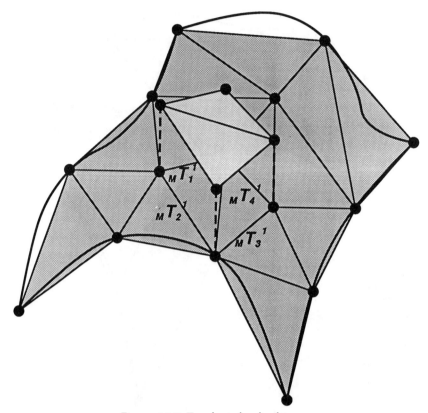

Figure 13.3 Topological redundancy.

$$_M T^d = \bigcup_{i=1}^{N} {_M T_i^d} \tag{13.6}$$

where each mesh entity $_M T_i^d$ is classified on the topological model entity $_G T_k^d$ as

$$_M T_i^d \sqsubset {_G T_k^d}, \quad \forall i = 1, \ldots, N \tag{13.7}$$

and the parametric intersection of any two $_M T_i^d \in {_M T^d}$ is:

$$_M T_i^d \,\square^* {_M T_j^d} = 0. \tag{13.8}$$

Figure 13.4 demonstrates the concept of geometric similarity for a $_G T_j^1$. The mesh entities in 13.4a do not satisfy the geometric similarity conditions because the mesh edge $_M T_1^1$ overlaps mesh edges $_M T_2^1$ and $_M T_3^1$ in the parametric space of the model edge, while $_M T_4^1$ also overlaps $_M T_2^1$ and $_M T_3^1$ in the parametric space of the model edge. The set of mesh edges in Figure 13.4b do satisfy the geometric similarity requirements. In the case of model edges the determination of satisfaction of geometric similarity is straightforward. However, the algorithms required for the

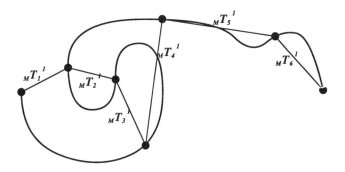

a) violating geometric similarity

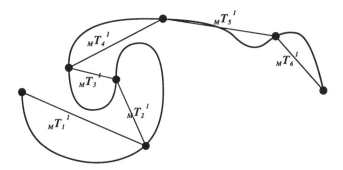

b) satisfying geometric similarity

Figure 13.4 Geometric similarity on a model edge.

determination of geometric similarity for model faces are more complex [2].

13.3 KEY COMPONENTS/ISSUES IN AUTOMATIC MESH GENERATION

The development of an automatic mesh generation algorithm must begin by selecting overall methodologies to address the basic issues involved with the decomposition of a geometric domain into a set of finite elements. Depending on the actual algorithm developed, these issues may be addressed by various combinations by specific algorithmic steps. However, in one manner or another, each of the following issues must be addressed:

1. the geometric representation and its decomposition as used to drive the mesh generation process;

2. the triangulation of the decomposed domain into a set of finite elements;
3. the assurance that the finite element mesh generated represents the geometric domain of interest.

13.3.1 *Geometric representation and decomposition*

Automatic mesh generation is a problem in computational geometry in which the domain being meshed is decomposed into a finite set of simpler (topologically and geometrically) entities. By definition the mesh generator must interact directly with the object's geometric representation. It is common during this process to first decompose the domain of interest into a secondary representation which can be used to effectively drive the process of generating the finite element mesh entities. Often this decomposition introduces additional topological entities by the appropriate subdivision of the domain. In addition, the shape information assigned to the topological entities of the decomposition is an approximation* to that of the original model, $_DS \neq {}_GS$. The form of decomposition used dictates the efforts required to generate the finite element mesh through the triangulation process and to assure that the resulting mesh is a valid geometric triangulation of the original domain.

The interaction of an automatic meshing algorithm with the geometric representation during domain decomposition, triangulation into the finite element entities, element shape improvement, and assurance of a geometric triangulation can take various forms. At a minimum an automatic mesh generation algorithm must interrogate the geometric representation to determine basic interactions. Simple examples would be to determine a set of points along an edge of the domain, or find the intersection of a line with the surface of the domain. In general, the interrogations required during the process of automatic mesh generation require the determination of information not inherently available in the data defining the geometric representations. The required information must be derived through the application of geometric operations acting on the representation [5].

In addition to operators that interrogate a given geometric model to determine specific geometric information, automatic mesh generators could require the application of operations which modify the geometric representation. An example of this type of interaction would be carving a finite element out of the model using the capabilities of the geometric modelling system to update its geometric representation to reflect the removal of that element.

At one limit of interaction, a mesh generator does not employ a domain decomposition that introduces any geometric approximation. In such a case the capabilities of the geometric modelling system would also be used to decompose the object into a finite element mesh. In such an approach it is necessary to update the geometric representation. Although such an approach is possible [6], the computational effort required to perform all meshing operations through geometric modification operators acting directly through the geometric modelling system can be prohibitive.

---

*A common example of such geometric approximation is to assume a triangular face on the surface of the geometric model is planar instead of assigning the actual shape of that portion of the model face.

Most automatic mesh generation algorithms employ a domain decomposition which does introduce geometric approximation into the process. The triangulation process then interacts either exclusively, or primarily, with this approximate geometric representation to define the element mesh. Consideration of this approximation process and its interaction with the original geometry is a critical aspect of ensuring that the resulting mesh is a valid geometric triangulation of the domain.

There are three common domain decompositions employed by automatic mesh generation algorithms. The simplest decomposition is a set P of M unique points, each classified with respect to the geometry G. The symbol $\mathscr{P}$ will be used to denote this decomposition.

The second decomposition, denoted by $D = \mathscr{T}$, approximates to the boundary of the domain, ∂G, with a boundary triangulation, $\partial \Lambda$, which is used as a basis of the triangulation process. In a two-dimensional domain each boundary edge, $_G T^1_i$, is replaced with a series of discrete edges, $_G T^1_i \quad \{_{\mathscr{T}} T^1_j, j = 1, l\}$, where the shape of each discrete edge, $_{\mathscr{T}} S^1_j$, is typically assumed to be a straight line. In a three-dimensional domain each boundary face, $_G T^2_i$, is replaced with a series of discrete faces, $_G T^2_i \quad \{_{\mathscr{T}} T^2_j, j = 1, l\}$, where the shape of each discrete face, $_{\mathscr{T}} S^2_j$, is typically assumed to be a plane when those faces are triangular. As indicated below, the creation of such a surface decomposition is one of the primary steps in the triangulation process. It is important when such decompositions are used to ensure the validity, in the sense of a geometric triangulation, of each of the resulting decompositions.

The $\mathscr{T}$ decomposition has the advantage of forming a discrete closed boundary of the domain for use in driving the triangulation process, while the $\mathscr{P}$ has the advantage of providing the triangulation with information through the domain to help drive the triangulation process. The third form of decomposition attempts to obtain the advantages of both of these while typically adding a structure to the decomposition which improves the efficiency of the triangulation process. These decompositions have been typically based on octree structures originally introduced as a method to speed graphics rendering of geometric objects [7].

In octree decompositions, denoted by $D = \mathscr{O}$, the geometric domain, G, is replaced by an octree representation, $_G T^3_i \quad \{_{\mathscr{O}} T^3_j, j = 1, l\}$, where different forms of geometric approximation of the shape of the individual octant cells, $_{\mathscr{O}} S^3_j$, are used based on the triangulation process used to create the mesh. Further discussion on these approximations is given in Section 13.4.

13.3.2 Triangulation

The goal of the triangulation process is to take the decomposition produced in the previous step and produce a finite element discretization of the domain. Clearly, the algorithmic approach to the triangulation process depends on the domain decomposition provided.

There are a number of algorithmic approaches to triangulate a point set decomposition, $\mathscr{P}$, into a set of non-overlapping mesh entities. Currently the most popular of these approaches employ the Delaunay circumsphere property [8], [9]. If the triangulation process employs no additional information past the classified set of points, it is not possible to ensure the resulting triangulation is a geometric triangulation. In these cases, it is necessary to employ an assurance algorithm which interacts with the geometric model on the resulting triangulation. The basic Delaunay triangulation procedure uses only the point set and therefore requires the application of additional procedures to ensure a geometric triangulation.

One key aspect of mesh generation algorithms that employ point set discretizations is the manner in which they obtain the point set such that the desired mesh gradation and element shape control is maintained.

Triangulation algorithms that operate from a boundary triangulation, $\mathcal{T}$, are typically implemented by triangulating the geometric entities, $_GT^d$, $d = 1, 2, 3$ in increasing dimensional order employing the triangulation of the lower-order entities which bound the current entity as the discretized boundary [10–13]. The triangulation of the model edges, $_GT^1$, is typically a straightforward process of placing mesh vertices, $_MT^0$, distributed as desired along each defining mesh edge, $_MT^1$, between each pair of $_MT^0$. The triangulation of surfaces, $_GT^2$, is typically much more complex. The triangulation algorithm is responsible for defining points on the surface such that it can be connected with other points to create a valid surface triangulation with the desired mesh properties. In the advancing front [10, 11, 14] and paving [12, 13] techniques this process is performed starting from the boundary working into the interior of the surface introducing mesh vertices, mesh edges and mesh faces as the algorithm progresses. The process of triangulating the regions, $_GT^3$, into mesh regions, $_MT^3$, requires the placement of mesh vertices, interior to the region and defining mesh edges, faces and regions using those mesh vertices. Again the typical method of performing this process is to work from the boundary into the interior defining individual mesh vertices and the mesh entities that connect it to the current boundary.

The triangulation of the model faces, $_GT^2$, is the most complex portion of this process when the shape of the faces, $_GS^2$, is not planar. This is because of the geometric complexity introduced by non-planar surfaces. The two primary complexities are placing points on the surface with a distribution to match the desired distribution, and ensuring the triangulation of the surface is compatible and the geometric approximations introduced by the mesh faces does not introduce any intersections. This process is made much easier if a simple parametric mapping exist for the surface. When such parameterization does not exist and/or the surface is trimmed, proper surface triangulation becomes more complex.

Region triangulation is simpler than surface triangulation because the region is nicely bounded by a surface with known simple geometric properties. In this case it is more straightforward to ensure mesh vertices generated are interior to the region since the geometric checks can be performed with respect to mesh entities bounding the region. Surface triangulation can only take advantage of these properties when it is planar or maps to a parametric space which is planar.

Octree decompositions $\mathcal{O}$ are typically performed one octant cell at a time using a cell point triangulation or boundary triangulation as discussed in the next section.

13.3.3 *Ensuring mesh validity*

The process of ensuring a set of mesh entities is a valid geometric triangulation of the domain is a function of the decomposition and triangulation procedures used.

13.3 KEY COMPONENTS

One approach to ensuring mesh validity is to create the mesh by the stepwise discretization of the geometric domain, being sure to maintain the validity of the discretization in each step of the process. Such an approach can be used with $\mathcal{T}$ and specific forms of $\mathcal{O}$ decompositions.

If the mesh is created with no concern for its validity; validity must be regained through the application of a mesh assurance algorithm [2, 3]. The second approach is necessary when a point set discretization, $\mathcal{P}$, is used.

The outline of a general assurance algorithm that can operate from a triangulation of a set of properly classified points which encompasses the convex hull of the domain being meshed is [2]:

1. Initial classification based on necessary conditions. The necessary conditions used to classify a mesh entity are based on the classification of its boundary entities. Given a model entity of dimension n, $_GT_i^n$, the set of mesh entities $_MT_j^m \sqsubset \, _GT_i^n$, $m \leq n$ initially classified on the model entity, H, is given by

$$H = \{_MT_j^m|\}. \tag{13.9}$$

   Initial classification must be done in increasing topological order from mesh edges. An important property of the initial classification process is that all mesh entities which can be classified on an entity are identified as classified on that entity or its boundary. Therefore, later steps in the assurance algorithm will not need to look for additional candidates.

2. Edge compatibility assurance by traversal. Employing the idea of an edge parameter the mesh vertices classified on the model edge can be sorted in order from the start to the end. The mesh is compatible and geometrically similar if there is a single mesh edge connecting each pair of mesh vertices on the mesh edge and these edges do not intersect themselves. Mesh edges that connect to other than consecutive mesh vertices are redundant and are corrected through reclassification. If two consecutive mesh vertices are not connected by a mesh edge, a hole exist. Holes are corrected by either the creation of the correct connection or the insertion of additional points along the model edge between the mesh vertices bounding the hole followed by local retriangulation.

3. Face compatibility by recursive boundary classification. Given the loop(s) of mesh edges bounding a model face mesh, take a mesh face that uses one or more of the bounding mesh edges once and mark it as compatible with the face. For the mesh face under consideration, remove the $_MT_j^1 \sqsubset \partial(_GT_i^2)$ from the loop and insert the $_MT_j^1 \sqsubset \partial(_GT_i^2)$ into the updated boundary. Continue this process until there are no edges remaining in the loop. If the process terminates before all edges are removed from the loop an incompatibility, in terms of either a redundancy or hole, exist. Redundancies are removed by the proper reclassification. Holes are corrected by either the creation of the correct connections or the insertion of points on the face in the area of the hole and local retriangulation. Geometric similarity can be checked during this process through local surface parameterizations.

4. Region compatibility by inheritance. Once the mesh is compatible with the model faces all model regions will be completely bounded by valid sets of mesh faces. Starting with a single mesh region in a region, all its unclassified boundary entities inherit that region classification as does any neighbouring region sharing a mesh face not classified on the boundary of the model region.

The interested reader is referred to [2] for more detail.

Since geometric approximation is typically involved with the various domain decomposition and triangulation processes, it is common in approaches based on $\mathcal{T}$ and $\mathcal{O}$ decompositions to employ the appropriate steps of the assurance algorithm given above at the completion of specific steps of the algorithm.

13.4 OCTREE MESH GENERATORS

13.4.1 *Overview*

Spatial decomposition methods based on octree structures were originally proposed for use as approximate representations of geometric objects [7]. The basic concept of the octree representation consist of placing the object of interest in a parallelepiped, typically a cube, which totally encloses it. This parallelepiped is then subdivided into its eight octants which are then recursively subdivided a number of times based on criteria defined by the application. The number of levels of subdivision typically varies throughout the domain of the representation. As indicated below, the octant refinement requirements for the finite element meshing applications are typically dictated by the size of elements desired and quickness of mesh gradation.

The application of octree and quadtree techniques for three- and two-dimensional mesh generation has a nearly ten year history [2, 15–27]. Although there are major differences in each of these octree-based approaches, there are a number of aspects which are consistent:

1. The octree structure is used to localize many of the meshing processes.

2. The mesh generation process is implemented as a two-step discretization process. The first is the octree and the second is the finite element mesh generated within the octree.

3. Those octants containing portions of the object's boundary receive specific consideration to deal with the actual boundary of the object.

4. The corners of the octants as well as points defined by the interaction of the boundary of the object being meshed with the octants' boundaries (if any) are used as nodes in the mesh.

5. The mesh gradation is controlled by varying the level of the octants within the octree through the domain occupied by the object.

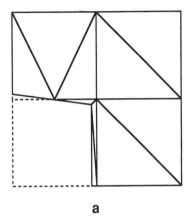 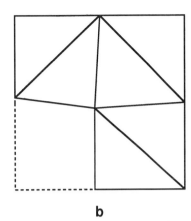

a b

Figure 13.5 Two-dimensional example showing the desire to eliminate disproportionately small pieces of quadrant level geometry: (a) mesh that would be generated if all quadrant/geometry interactions are represented, (b) desired mesh.

13.4.2 Domain decomposition and geometric representation

Before considering the differences in domain decomposition, triangulation techniques and ensuring mesh validity of the current octree-based mesh generators, one aspect of the interactions of an octree representation with the geometry of the artifact being meshed needs to be mentioned. The need to address this issue is a primary reason for the differences in approaches developed. During the recursive subdivision of a bounding parallelepiped it is possible to create octants containing disproportionately small pieces of the object which can lead to poorly shaped elements in the final mesh if they are not eliminated. A simple two-dimensional situation is shown in Figure 13.5, where Figure 13.5(a) shows the mesh that would result if all quadrant/geometry interactions are represented and Figure 13.5(b) shows the desired mesh.

The critical differentiating factor among the various octree-based mesh generators is the specific topological, $_OT$, and geometric, $_OS$, information constructed. The earliest procedure [26] employed unvalidated geometric approximations to simplify the meshing process and deal with limitations of the geometric modelling functionality. Although capable of generating useful meshes, these approximations made it impossible to ensure the resulting mesh was a geometric triangulation using only the resulting octree decomposition.

Current octree-based mesh generators employ three forms of decomposition:

- Pulled octant [16, 17], which employs the octant boundaries which can deform to locally match the model entities plus additional entities defined through allowing specific cuts through the octant;
- Finite octree [24] (also [19, 20] where it is referred to as exact octrees), which employs a full set of topological entities based on the interactions of the model with the octants of the octree;
- Octree–Delaunay [2, 22], which employs a point set defined by the corners of

octants internal to the domain, plus points of intersections of octant boundary entities with model boundary entities.

Since the important differences between these decompositions are concerned with octants containing a portion of the boundary of the domain, consideration will focus on a typical boundary octant.

The pulled octant approach [17] maintains a complete understanding of the portions of the geometric model within the individual octants by representing all topological entities within each of the octants. In the pulled octant approach the octant boundary entities and additional boundary entities introduced by specific cuts through the octant can be used to represent the pieces of the model boundary associated with that octant. Since the geometric model's interactions with the octant can produce entities that are not represented by the limited set of entities, the boundaries of those octants are allowed to deform to the model so that they can be represented. This is an effective approach to eliminate the undesired disproportionately small pieces in an octant. However, such an approach does place a specific limit on the level of topological complexity that can be represented by a single cell. When this does not provide a sufficient number of topological entities, forced subdivision introduces more cells allowing the possibility of more topology being represented. This will often force the mesh to be finer than needed. However, there are situations where no level of refinement will introduce the number of topological entities needed. A simple example is when a single model vertex has more model edges using it than are available at a vertex in the octree including the allowed cuts. With specific care in the definition of method of octant deformation and the introduction of sufficient flexibility in cuts allowed, all but the most complex topological situations can be properly represented.

An alternative method being considered that avoids the use of hexahedral octants with specific cuts, bases the octree on the subdivision of tetrahedra. In this case the additional entities available allow the representation of most topological situations by the deformation of the octree tetrahedra.

The Finite Octree technique [24] also maintains a complete understanding of the portions of the geometric model within the individual octants by representing all topological entities within each of the octants. The topological entities within the octants containing portions of the boundary of the object being meshed is constructed using discrete intersection information based on the interactions of the boundary of the object with the boundary of the octant. The basic intersection information used includes the intersections of model edges with octant faces, and octant edges with model faces. This is the primary information used in a set of procedures which build (1) octant level loops, (2) octant level faces and (3) octant level regions to represent those portions of the model within the octant. Additional discrete geometric information must also be determined during these processes for those cases where the octant discretization defined with the original intersection information is not topologically consistent with and geometrically similar to the original model. Ensuring the topological compatibility and geometric similarity of the octant level topology with that of the geometric model information in that octant at the end of each step is a requirement of this approach.

Within Finite Octree a separate algorithmic step is used to eliminate the

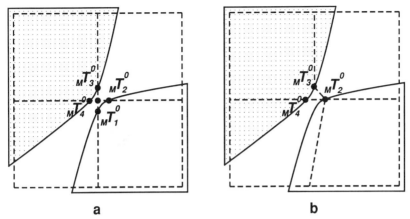

Figure 13.6 Two-dimensional example of valid quadrant deformation to eliminate the disproportionately small geometric features that can be eliminated: (a) before deformation of the quadrant, (b) after deformation of the quadrant.

disproportionately small geometric features by the appropriate octant deformation. The procedure employs a set of topological operators which are required to ensure the resulting deformed octant topology is consistent with the geometric model. Figure 13.6(a) shows a two-dimensional case where there are small geometric features in two quadrants sharing a quadrant corner. Figure 13.6(b) shows a valid quadrant deformation for this situation. Although Finite Octree cannot always eliminate all disproportionately small geometric features, because of cases such as that shown in Figure 13.6, it can validly mesh objects of any topological complexity.

In both pulled octant and Finite Octree approaches it is common to assume that the geometric shape information associated with the octant topological entities is based on straight line connections. That is, the $_OT_i^1$ are straight lines and the $_OT_i^2$ are bounded by straight line edges. This is acceptable so long as these geometric approximations are topologically equivalent to the original model. That is they do not intersect themselves. It should be noted that the use of straight lines to bound $_OT_i^2$ does not ensure they are planar. Therefore, it is not possible at this point to detect all possible intersections introduced by geometric approximations. That must wait until triangulation is complete and the final geometric shape of the mesh entities assigned. This does introduce some complexity into the mesh algorithm that requires special care [24].

The basic domain decomposition in octree–Delaunay mesh algorithms [2], [22] consist of only a set of classified points. As with the finite octree approach, these points are those octant corners interior to the domain, the object's vertices, and the intersections of appropriate octant and model boundary entities. The additional mesh entities are constructed as part of the remainder of the meshing process. The elimination of the undesired small features that can arise is carried out by eliminating points that are too close to each other. The mesh assurance process is performed to ensure the resulting mesh is a geometric triangulation and deals with problems introduced by any inappropriate point elimination. Since the domain decomposition used in octree–Delaunay contains only points, there is no consideration of geometric approximation until a triangulation is generated.

13.4.3 Triangulation process

The availability of a complete octant level representation in the pulled octant or Finite Octree approaches make the application of element removal triangulation procedures on an octant-by-octant level attractive when they cannot be simply meshed with templets. As octants are triangulated, the information on octant face triangulations is made available to the neighbouring octants. Employing the face triangulation for neighbouring octants allows direct generation of compatible meshes. The octant level triangulation procedures typically employ the straight edge geometry used in the generation of the octant level geometric approximations. As the octant faces, $_OT_j^2$, are triangulated into mesh faces, $_MT_j^2$, a faceted approximation to $_OT_j^2$ is defined. This approximation can be used so long as it is topologically compatible with the original geometric model.

Since the domain decomposition of octree–Delaunay approach contains only a set of classified points, the triangulation technique used to get the initial mesh can only employ points in three space. Such a triangulation is easily constructed using one of the point insertion procedures which employ the Delaunay circumsphere property [8], [9]. A drawback of the application of this approach on the entire point set is the computational effort due to the poor computational growth rate. This problem is eliminated by generating Delaunay meshes on an octant-by-octant basis using only the small number of points in that octant. Specific care must be exercised in the application of octant-by-octant triangulation to ensure the matching of mesh faces between octants. This can be done by the proper ordering of the point insertion process. A highly efficient triangulation can be developed by combining this with templets for the appropriate interior octants which satisfy both compatibility and Delaunay requirements [2, 22].

13.4.4 Ensuring mesh validity

Pulled octree and Finite Octree algorithms can be designed to ensure that the resulting mesh is a geometric triangulation by requiring that the octant level geometric information available after each step of the process is valid with respect to the compatibility and geometric similarity conditions. Since the entities that define the resulting mesh entities are defined at various points in the process and only partial checks can be carried out on specific entities during the various steps, it is possible to arrive at situations at later steps that the use of the entities from previously accepted steps will not allow any valid selections in the current step. To resolve these possibilities the finite octree algorithm employ special procedures to improve the local approximation by the introduction of additional entities. These entities may be introduced by specific entity splitting operations or by forcing entire octants to subdivide.

Since the octree–Delaunay technique generates its original triangulation based on a point set only, the general geometric triangulation assurance algorithm outlined above and described in [2] must be applied. Knowing that this procedure must be applied allows the triangulation process used in the octree–Delaunay algorithm to first use the octant-by-octant procedure for the domain of all octants containing any portion of the boundary. The points defined by the intersection of the model and octant boundary entities are then inserted using the appropriate localization procedures. The general mesh assurance algorithm then takes over to convert this triangulation into a geometric triangulation.

13.5 DELAUNAY, ADVANCING FRONT (PAVING), AND MEDIAL (SYMMETRIC) AXIS TRANSFORMATION MESH GENERATORS

13.5.1 *Delaunay mesh generators*

Much of the initial development of Delaunay mesh generation algorithms were built on the basic Delaunay triangulation properties [28–30] and associated computational methods [8, 9, 31]. With these properties one can quickly develop an algorithm that triangulates a set of points in n dimensions into a set of n-dimensional simplices having known geometric properties. Most of the development of finite element mesh generation algorithms based on this type of approach [22, 32–8] have focused most of their attention on the issues of:

- converting or constraining the Delaunay triangulation to produce a valid mesh (geometric triangulation);
- controlling the shape of the elements so that extremely flat elements are avoided;
- creating an appropriate point set to drive the triangulation process.

In the basic application of an n-dimensional Delaunay triangulation process the domain decomposition consist of a point set in n-space. In this case the only reasonable assumption of the n-dimensional simplices is that they are straight sided and have planar faces. The assignment of any higher-order shape information to the mesh entities can only be performed after the mesh entities have been classified with respect to the geometric model of the object being meshed. Once mesh entities have been classified it is possible to check the validity of those assignments and assign higher-order shape information to the mesh entities that are validly classified. Typically higher-order geometric shape information is assigned only to mesh entities classified on the boundary of the domains. The common practice is to place higher-order nodes on the boundary of the domain. However, it is also possible to explicitly assign the shape of the geometric boundary to the mesh entities. When mesh entities on the boundary are assigned higher-order geometric shapes, it is necessary to ensure that they do not intersect other finite element entities on the boundary of the mesh entities they bound. If there are intersections, corrective procedures must be made to eliminate the intersections.

Ensuring the validity of a Delaunay mesh with respect to geometric compatibility and similarity is a critical aspect of the meshing process. In addition to the assurance algorithm outlined above and given in [2], a number of algorithmic approaches have been developed to ensure the Delaunay triangulation process maintains mesh validity by ensuring a valid triangulation of the boundary of the domain [32, 33, 37, 39].

These procedures carefully track the process of the generation of the boundary triangulation. The triangulation of an n-dimensional region into valid n-dimensional mesh entities is straightforward when there is a valid triangulation of the $(n-1)$-dimensional surfaces that bound the region. In some cases these algorithms generate the Delaunay triangulation and perform corrective procedures to obtain a valid triangulation of the region. Others obtain a valid triangulation of the surfaces and constrain the Delaunay process to maintain that triangulation throughout the meshing process.

The Delaunay circumsphere property does provide explicit control of the geometry of the elements of the mesh. In two dimensions this property has been shown to maximize the minimum angles in the mesh [28, 30], which also places some level of control on the maximum angle in the mesh. However, in three dimensions there is not an equivalent property that elements are as well shaped as possible. In particular, satisfaction of the Delaunay circumsphere property in three dimensions does not force the dihedral angle between element faces to be bounded. This property was documented in the work of Cavendish et al. [34] in terms of the 'silver' elements that could be generated.

Most efforts to control this problem have focused on defining point distributions and point insertion and ordering procedures [2, 32] that will yield well-shaped elements. Baker [32] introduced specific geometric checks that can be applied to the sets of points to ensure the resulting mesh has well-shaped elements.

An additional procedure that can be used to create a triangulation with the best geometric properties is to look at local connectivities switches. For example in two dimensions, pairs of neighbouring triangles can be examined to see if switching the diagonal of the quadrilateral defined by their union leads to improved element properties. Typically these procedures consider potential improvements to the resulting angles of the mesh such as maximizing the minimum angle. This concept has been extended to three dimensions where the switching of faces for sets of neighbouring tetrahedron has been applied [40] to locally maximize the minimum solid angle. Although there is no proof that locally maximizing the minimum solid angle leads to a global maximization of the minimum solid angle, Joe [40] has found that the application of iterations of the local maximizing of the minimum solid angle starting from a Delaunay triangulation does lead to substantial improvements in element shapes.

An issue often not given appropriate consideration in the development of Delaunay meshing procedures is the definition of the point sets used in the procedures. In addition to attempting to avoid the generation of poorly shaped elements, the point distributions must be defined to provide the mesh gradations desired. A number of *a priori* procedures have been defined for this process including a recent one based on the use of contours [41]. Other approaches begin with a minimal triangulation based on boundary points spaced to provide the desired mesh gradations and follow by inserting points to obtain the best possible set of element shapes [36, 37].

13.5.2 *Advancing front and paving mesh generators*

Over the past several years element-by-element removal procedures [1] referred to as advancing front [10, 11, 14, 42] and paving [12, 13] techniques have received considerable attention. In these methods the meshing algorithm begins at the boundary and develops elements working into the interior of the domain until it is filled. Two reasons for the popularity of these approaches are the ability to reflect directional mesh gradation information and the fact that the mesh is sensitive to the boundary of the domain.

In most implementations of these approaches the mesh is created by meshing the object's entities following a spatial hierarchy, first meshing the edges, $_GT_i^1$, then the faces, $_GT_i^2$, and finally the regions, $_GT_i^3$ employing the $_MT_j^{k-1} \sqsubset \partial(_GT_i^k)$ as the mesh entities defining the initial front which is used to define the $_MT_l^k \sqsubset {_GT_i^k}$. The meshing process for an entity $_GT_i^k$ begins by defining a front in terms of the $k-1$ order mesh entities that enclose it, $_MT_j^{k-1} \sqsubset \partial(_GT_i^k)$. A particular $_MT_j^{k-1}$ in that front is selected for removal from the front. A $_MT_j^{k-1}$ is removed from the front by defining a mesh entity $_MT_l^k \sqsubset {_GT_i^k}$ going into the region which has that $k-1$ order mesh on its boundary, $_MT_j^{k-1} \in \partial(_MT_l^k)$. The front is then updated to reflect the addition of the mesh entity, $_MT_l^k$. Since this new mesh entity goes into the interior of the entity being meshed, all the $_MT_j^{k-1} \in \partial(_MT_l^k)$ that were part of the front before the new mesh entity was defined are removed from the front and all the new boundary entities created in the definition of the mesh entity are added to the front.

During the process of defining new $_MT_l^k \sqsubset {_GT_i^k}$ decisions must be made to define new and/or to use existing $_MT_j^m \in \partial(_MT_l^k)$, $0 \le m < k$. The procedures used to determine when new $_MT_j^0$ must be defined and where to place them in $_GT_i^k$ are critical to the ability of the meshing algorithm to produce the type of meshes desired. In all cases the node must be placed so that the mesh entities defined are valid with respect to the model entity being meshed. Consideration must also be given to the shape of the resulting elements and to reflecting the requested mesh gradation information including the possibility of directional mesh gradations.

In addition to the basic element creation steps outlined, additional functions can be added to these meshing procedures to improve the quality of meshes generated. As outlined here, these procedures work in a forward direction making decisions considering the entities currently being defined. Such procedures can produce cases where, as the front begins to close, it is difficult to create mesh entities which satisfy the desired mesh gradation requirements. In the development of the paving procedures to generate all quadrilateral meshes [12, 13] specific features were added to the meshing process to correct such situations as they arise.

Since advancing front and paving mesh generators operate working with detailed boundary information, that is they employ a $\mathcal{T}$ domain decomposition, it would be most appropriate to perform the checks to ensure the mesh validity as the meshing process progresses. Ensuring mesh validity as the meshing process proceeds consists of two main steps. As each of the object's entities (edges, faces, regions) are meshed, the new mesh entities which are defined must be introduced such that they are geometrically compatible and similar with the entity being meshed. In addition, as the initial fronts for higher-order object entities are constructed by combining the mesh entities classified on the entity's boundary, care must be taken to see that the front does not violate geometric similarity. Figure 13.7 shows a close-up of a narrow region of a face where mesh entity $_MT_3^1 \sqsubset {_GT_1^1}$ intersects $_MT_9^1$ and $_MT_{10}^1$ which are classified on two straight line edges that come close to $_GT_1^1$. In general, the process of ensuring the initial front is valid requires determining if any of the mesh entities

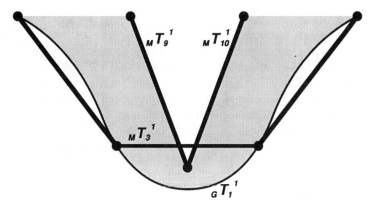

Figure 13.7 Intersection of mesh entities on two edges bounding a planar face.

bounding that front intersect each other. If they do not intersect, it is a valid front and the process can proceed. If there are intersections of mesh entities which are valid with respect to the individual boundary entities, local refinement of the mesh can be used to eliminate the intersections.

13.5.3 Medial (symmetric) axis transformation mesh generators

There have been a number of approaches to automatic mesh generation developed based upon partitioning the domain of interest into sets of subdomains and the subsequent meshing of those subdomains. In some cases specific computational geometry algorithms for domain decomposition employing specific geometric constraints on what constitutes an acceptable partitioning process have been used. Early approaches of this type had some difficulty producing well-controlled meshes in a reliable manner. The other commonly attempted approach was to codify into rule bases the methodologies experts at mesh generation would use to partition domains into regions which can be filled by mapped mesh generators. The two-dimensional implementations of these approaches tended to be slow and did not appear to always produce predictable meshes. To the author's knowledge, no one has yet successfully produced a three-dimensional version that effectively meshes entirely general domains in a controlled manner.

A more recent alternative to this basic approach builds on a more natural partitioning of the domain into subdomains, a decomposition referred to as $\mathscr{A}$. This approach, referred to as the medial [43, 44] or symmetric [45] axis transformation employs the basic concept of a Voronoi diagram to define subdomains associated with each boundary curve (in two dimensions) which is closer to that curve than any other boundary of the domain. This approach has the advantage of reducing the domain of interest into a number of fairly simple subdomains where the new boundaries that are introduced tend to represent a smooth transition between opposite boundaries. This basic transitioning, or blending, is a property common to the mapped mesh generation techniques which give meshes most users find aesthetically superior. Although the subdomains of $\mathscr{A}$ created by this process may not be directly amenable to filling with a mapped mesh generator, the two-dimensional efforts to date indicate that the shapes are simple and can be easily meshed.

A potential drawback of this method is the generation of the medial axes to form $\mathscr{A}$. This appears to be a fairly complex process that requires substantial computational effort in two-dimensions. Algorithms to define medial axes for three-dimensional domains are not yet available and could be somewhat difficult to define.

The most intensive interactions of a medial axis mesh generator with a geometric representation will be during the definition of the various curves and surfaces (in three dimensions) defining the portions of the medial axis separating the domain into subdomains. The shape of these boundary entities are dependent on the shape of the domain boundaries. In the case of simple straight sided planar domains these boundaries are straight lines and arcs. In the case of curved geometry they will be blends between the 'opposite' boundary entities. In addition to the definition of these basic entities, their interaction (intersections) must be terminated so that the appropriate portions can be trimmed off to form the subdomains. This appears to be the most complex and geometrically demanding portion of the process. If the resulting subdomains satisfy the geometric compatibility and similarity requirements, it should be reasonably straightforward to triangulate the subdomains to produce a valid geometric triangulation for the domain.

13.6 INTEGRATION OF AUTOMATIC MESH GENERATION WITH GEOMETRIC MODELLING

By their nature, automatic mesh generators must interact with the geometric representation of the domain being meshed. One approach to provide such information is to provide the meshing algorithm with a data file of the geometric model. Since all automatic mesh generators must at least interrogate the geometric representation to obtain information not inherently in the model's data file, this approach would necessitate the development of those modelling functionalities within the mesh generator. In those cases where these procedures could be developed to reliably perform the required geometric interrogations, this requires the development of those procedures, which is not an easy development. However, a more important issue is the ability of the interrogation procedures to produce answers consistent with that of the geometric modelling system.

All geometric modelling operations are based on some degree of approximation. To maintain consistency during the modelling process geometric modelling systems employ known tolerance information which is consistently applied throughout the process of defining the geometric model. A set of procedures within an automatic mesh generator would need specific information of both the tolerance information and methods used to apply the tolerances if consistent interrogations are to be performed.

An alternative approach is to employ a dynamic interface that directly employs the functionality of the geometric modelling system in the mesh generation process. Such an approach is consistent with the dynamic interface methods of the CAM-I Application Interface specification [46]. The application of such approaches for general finite element applications have been considered in [5].

An example of a mesh generator developed using such an approach is the Finite Octree procedure [24] which interacts with the geometric modelling system through a specific set of 19 geometric operators [47]. The integration of Finite Octree with a new geometric modelling system only requires the creation of the geometric operators. This requires no knowledge of the operation of the mesh generator or the data structures internal to the geometric modelling system. What is required is a knowledge of the basic interrogation operators available from the geometric modelling system which can be used to build the specific Finite Octree geometric operators.

All the operators are keyed via the topological entities of vertex, edge and face. The use of topological entities has the advantage of independence of particular geometric form or method of geometric modelling. Boundary information can be derived for all complete geometric modelling forms, and since most applications are primarily concerned with the boundary of the object being modelled, modelling systems of all forms commonly derive at least basic boundary information.

Approximately one half of the operators request basic topological associativities. Although these operators require the recognition of the basic topological entities and some basic topological adjacencies, there is no requirement that the geometric modelling system support a complete boundary representation*.

The geometric interrogations used are all limited to determining pointwise quantities such as the points of intersections and surface normals. There are two reasons it is desirable to limit these interrogations to pointwise information since requesting higher-order information, such as a curve of intersection between two surfaces. Such interrogations return shape information to the mesh generator, making it dependent on the specific geometric modelling form and geometric modelling system. In addition, such interrogations are computationally expensive to perform and can be avoided during meshing by the use of a limited number of properly defined pointwise intersections. The primary operators used by Finite Octree are line–face intersection, edge–plane intersection, surface normal determination, nearest point determination (can be approximate) and conversion, both ways, between real and parametric coordinates (used only during node point smoothing).

This approach has been successfully applied with multiple geometric modelling, systems with the operators being written either by the developers of the mesh generator or the developers of the geometric modelling system. Most geometric modelling systems provide basic sets of interrogation operators in terms of callable routines. More recent geometric modelling systems are built using a toolkit of such routines, making it easy for applications such as finite element mesh generation to access the needed functionality. Our experience has indicated that the Finite Octree geometric interrogation operators which ask for pointwise geometric information are typically provided by geometric modelling systems. This is critical since there is no easy way to provide these capabilities without getting deep into the geometric modelling system. There have been occasions when not all the topological entities and adjacencies have not been directly available from the geometric modelling system. In these cases it has been possible to use the functionality of the geometric

---

*One of the geometric modellers that Finite Octree has been integrated with is a CSG modeller that is quickly postprocessed to create the basic topological information needed.

13.7 EFFICIENT PARALLEL SOLUTION OF AUTOMATICALLY GENERATED ADAPTIVE MESHES

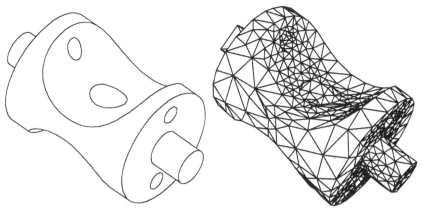

Figure 13.8 CATIA object and its Finite Octree mesh.

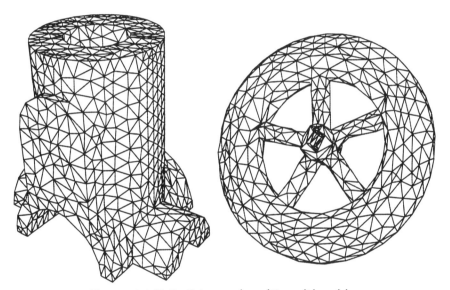

Figure 13.9 Finite Octree meshes of Parasolid models.

modelling system to preprocess the model to provide the additional information required.

One geometric modelling system Finite Octree has been integrated with is CATIA. Figure 13.8 shows a CATIA model and Finite Octree mesh [48] while Figure 13.9 shows meshes of parasolid objects.

13.7 EFFICIENT PARALLEL SOLUTION OF AUTOMATICALLY GENERATED ADAPTIVE MESHES

It can be easily argued that the efficient solution to problems of mathematical physics on general three-dimensional domains requires the use of adaptive techniques on automatically generated meshes. In addition, it is well known that the efficient construction and solution of large systems of algebraic equations requires the effective use of vector and parallel computations. The difficulty that arises is that the systems of algebraic equations that arise from automatically generated, adaptive refined meshes are not ideally suited to the most effective vector and parallel solution techniques.

The reasons that automatically generated, adaptively refined meshes are not well suited for vector and parallel processing are the changing distribution of computational effort during the adaptive solution process and the lack of regular structure for the mesh. As a finite element discretization is enriched as dictated by an adaptive analysis procedure, the distribution of computational effort required for each solution iteration changes. This creates load imbalances in the previous partitioning of the mesh to a set of processors. Although it is possible to re-balance the mesh over the processors, the effort of this re-partitioning, and the communications required to carry it out, can lead to large decreases in speed-up factors. The meshes produced by general domains by automatic mesh generators are unstructured. These unstructured meshes lack the regularity of the structured meshes produced by mapping techniques, which provide for the simple construction of ideal vector lengths for vector computations and allow for simple algorithms to decompose the domain for parallel processing with a minimum of data communication. Put simply, the key to combining automated, adaptive solution techniques on vector and parallel processors is to provide structure to the mesh and, with a minimum of effort, to maintain balanced computations in the solution process.

Efficient parallel solution procedures balance the load between processors, and minimization of interprocessor communications and synchronization steps. Experience indicates that the effectiveness of re-balancing the computational load to a set of processors after each mesh adaptation decreases rapidly with increasing number of processors because of the time required to re-partition the mesh and to communicate the needed information to all the processors. The computational effort required for the solution steps in the adaptive solution of linear elliptic problems tends to be dominated by the last sets of meshes. Therefore, one could save load re-balancing until these meshes are analyzed. Of course this approach requires a good estimate of which are the final meshes. In many cases these estimates can be obtained as part of the adaptive analysis process by estimating how much final refinement is likely to be required. An alternative approach is to use the same information to partition the mesh to the parallel processors at an early stage of the adaptive procedure based on the expected final mesh. This approach has the advantage that the load balance is constantly improving as the mesh is adapted and the problem size grows. Initial efforts at the development of such procedures, employing the elemental error estimates form a coarse discretization, are producing promising results [49].

The most obvious method to provide a general structure to an unstructured mesh is to employ the appropriate adjacencies of the topological entities defining the finite elements and their boundaries. One approach is the full set of topological adjacencies between mesh regions and faces, mesh faces and edges, and mesh edges and vertices [5]. The advantage of this approach is that any desired adjacency can be quickly obtained by, at worst, a local traversal; the disadvantage is the amount of data storage required, especially if full non-manifold representations are to be maintained.

By a careful examination of the specific adjacency information needed the amount of data can be drastically reduced by maintaining just the limited adjacency structures needed to perform load balancing and the solution process.

One effort to develop such relationships on unstructured meshes to support parallel solution procedures on a variety of architectures is the DIME system developed by Williams [50], [51]. The adjacencies used in the two-dimensional version of this structure are the nodes belonging to an element, the elements that use a node and boundary structure which defines the loop of nodes that define the boundary. In this system the mesh is partitioned into subdomain meshes which are then processed on separate processors. The mesh for each subdomain has a complete boundary structure around it and the nodes on the boundaries of the subdomains are repeated on the processors for each subdomain that shares that boundary node.

An alternative method to provide structure to an unstructured mesh is to employ a regular domain decomposition method to partition the domain. The potential advantage of such an approach is the ability to employ the regular nature of the decomposition to control parallel computation. Of course, there is the difficulty of associating the mesh with this structure. One such structure is an octree in three dimensions and a quadtree in two dimensions. These decompositions have been used to support mesh generation [52] and rezoning [53], improving the computational efficiency of the process by localizing searching processes. In both these applications the octree or quadtree is defined by enclosing a given mesh in a cube or square and recursively subdividing it into octants or quadrants until there is only a fixed number of nodes from the mesh in each terminal octant or quadrant.

The use of an octree structure is a natural selection for octree-based mesh generators since the tree structure is defined as part of the meshing process and the finite elements are defined in the terminal octants. When used in adaptive analysis, mesh refinement in octree- and quadtree-based mesh generators is performed by altering the levels of the terminal octants and quadrants [54]. Two specific efforts at employing the tree structure of the mesh generator are storing multiple meshes for multigrid solution procedures, and for developing colouring algorithms for controlling parallel iterative solution procedures.

Multigrid solution procedures are an effective method for the solution of adaptively defined meshes [55]. Key to effective parallel computation of these procedures is an overall structure to support the iteration between the various levels of meshes properly reflecting the adaptively defined local variations. This can be accomplished by storing multiple meshes in a single tree structure without the need to repeat the mesh information common to more than one mesh. A given mesh is associated with a specific tree traversal defined in terms of the terminal octants for that mesh [56]. For a given mesh, any terminal octant points to the set of highest-order mesh entities within that octant. For a given mesh and a given terminal octant, the highest-order mesh entities are entities not connected to higher-order mesh entities within the octant. Figure 13.10 gives an idea of how, given a mesh that is refined twice in the upper left-hand corner, the highest-order entities (here, mesh faces) may be obtained when traversing the tree. Another issue critical to the success of this approach is the storing of mesh topological adjacency relationships. If, for any mesh and any terminal octant, one has access to the set of highest-order mesh entities, downward adjacency is not mesh dependent. However, upward adjacency is mesh dependent and has to

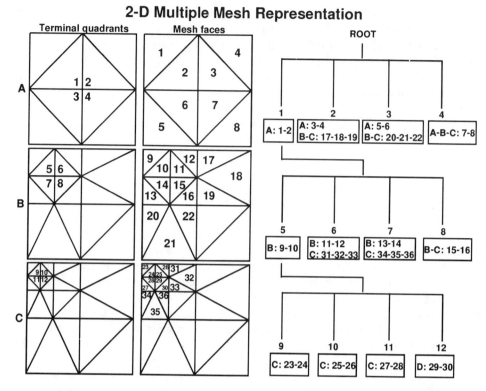

Figure 13.10 Tree–mesh relationships for adaptively defined multiple mesh.

be stored explicitly for each mesh. Figure 13.11 shows the mesh dependency of upward topological adjacency relationships. These relationships are also constructed in a manner that avoids the need to repeat sets of adjacencies by the clever use of pointers [56].

The effective application of iterative equation solvers on shared- or distributed-memory parallel computers requires fast procedures to subdivide the elements of the mesh into groups of elements that can be computed upon independently of the other elements in that group. The typical terminology used is to assign the elements in a group a colour. Therefore, the goal is to develop an algorithm that can quickly colour the mesh using a minimum number of colours since a synchronization step is required each time a new colour is processed.

Benantar et al. [57] have described six- and eight-colour linear-time procedures for colouring the quadrants of quadtree-structured meshes where the one-level difference rule between neighbours sharing at least an edge is enforced. Under this constraint, the maximum level difference for octants sharing a node is two. The eight-colour scheme requires only a simple tree traversal using two sets of four colours. The only specific consideration is to employ the complement colour for the smaller quadrants across a two-level nodal difference. The six-colour algorithm requires a traversal of the quadtree based on a binary directed graph called a 'quasi-binary tree'. In the six-colour scheme three sets of two colours are used with a specific procedure for switching colours in a set and between the sets based on how terminal quadrants are encountered during the traversal of the quasi-binary tree.

2-D Multiple Topological Adjacency Representation

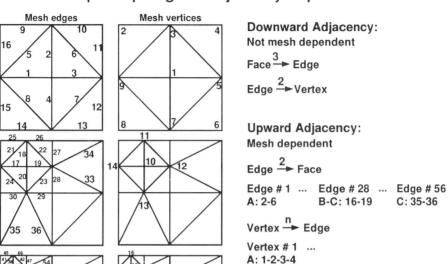

Figure 13.11 Topological adjacency relationships for adaptively defined multiple mesh.

Given a mesh partitioned into a set of independent colours, parallel solutions may be obtained by preconditioned conjugate gradient (PCG) iteration using either an element-by-element (EBE) [58] or a symmetric successive over-relaxation (SSOR) [59] preconditioner. Example computations were done using a piecewise linear finite element basis and PCG iteration with the EBE and SSOR preconditionings on a 16-processor Sequent Balance 21000 shared-memory parallel computer. A mesh having 2594 elements, 1346 vertices, and 1099 quadrants was generated using the Finite Quadtree procedure. The number of elements per quadrant ranged from 2 to 6. Speed-ups using the EBE and SSOR preconditionings on a 15-processor system are 73 and 86% of ideal, respectively. The time to calculate solutions of equal accuracy was three times less with the SSOR preconditioning than with the EBE preconditioning on this example problem. Speed-up results for this model problem demonstrate that the six-colour PCG procedure has a high degree of parallelism on a shared-memory system. Decay in speed-up can be due to processor bus contention, start-up latency, and data granularity. The lower speed-up of the EBE preconditioning relative to the SSOR preconditioning may be due to a lack of global information [60]. For a more detailed description see reference [57].

ACKNOWLEDGEMENTS

The author would like to acknowledge the support of the Industrial Affiliates of the Scientific Computation Research Center for the development of the Finite Octree mesh generator, and the Mathematical and Computer Sciences Division of the Army

Research Office (grant no. p-29167-MA, Dr K. D. Clark, contract monitor) for supporting the development of parallel solution techniques on octree-based meshes.

Thanks go to Mr Hasan Ahmad, Mr M. Benantar, Mr Hugues L. de Cougny, Dr Marcel K. Georges and Dr William J. Schroeder for their contributions discussed in this chapter.

BIBLIOGRAPHY

[1] Shephard, M.S. (1988) Approaches to the automatic generation and control of finite element meshes, *Appl. Mech. Rev.*, **41**(4), 169–185.

[2] Schroeder, W.J. (1991) *Geometric Triangulations: with Application to Fully Automatic 3D Mesh Generation*, PhD thesis, Rensselaer Polytechnic Institute, Scientific Computation Research Center, RPI, Troy, New York.

[3] Schroeder, W.J. and Shephard, M.S. (1989) An O(N) algorithm to automatically generate geometric triangulations satisfying the Delaunay circumsphere criteria, *Engg Comput.*, **5**(3/4), 177–194.

[4] Schroeder, W.J. and Shephard, M.S. (1988) Geometry-based fully automatic mesh generation and the Delaunay triangulation, *Int. J. Num. Meth. Engg*, **26**, 2503–2515.

[5] Shephard, M.S. and Finnigan, P.M. (1989) Toward automatic model generation, in A.K. Noor and J.T. Oden (eds.) *State-of-the-Art Surveys on Computational Mechanics*, pp. 335–366, ASME.

[6] Shephard, M.S. and Lo, J.A. (1991) Automatic generation of coarse three-dimensional meshes using the functionality of a geometric modeler, *Adv. Engg Software*, **13**(5/6), 273–286.

[7] Jackins, C. and Tanimoto, S. (1980) Octrees and their use in the representation of three-dimensional objects, *Comput. Graphics and Image Processing*, **14**, 249–270.

[8] Bowyer, A. (1981) Computing Dirichlet tesselations, *Comput. J.*, **24**(2), 162–166.

[9] Watson, D.F. (1981) Computing the n-dimensional Delaunay tessellation with application to Voronoi polytopes, *Computer J.*, **24**(2).

[10] Jin, H. and Wiberg, N.E. (1990) Two-dimensional mesh generation, adaptive remeshing and refinement, *Int. J. Num. Meth. Engg*, **29**, 1501–1526.

[11] Löhner, R. and Parilch, P. (1988) Three-dimensional grid generation by the advancing front method, *Int. J. Num. Meth. Fluids*, **8**, 1135–1149.

[12] Blacker, T.D. and Stephenson, M.B. (1991) Paving: A new approach to automated quadrilateral mesh generation, *Int. J. Num. Meth. Engg*, **32**(4), 811–847.

[13] Blacker, T.D., Stephenson, M.B. and Canann, S. (1991) Analysis automation with paving: A new quadrilateral meshing technique, *Adv. Engg Software*, **13**(5/6), 332–337.

[14] Zhu, J.Z., Zienkiewicz, O.C., Hinton, E. and Wu, J. (1991) A new approach to the development of automatic quadrilateral mesh generation, *Int. J. Num. Meth. Engg*, **32**(4), 849–866.

[15] Baehmann, P.L., Wittchen, S.L., Shephard, M.S., Grice, K.R. and Yerry, M.A. (1987) Robust, geometrically based, automatic two-dimensional mesh generation, *Int. J. Num. Meth. Engg*, **24**(6), 1043–1078.

[16] Buratynski, E.K. (1988) A three-dimensional unstructured mesh generator for arbitrary internal boundaries, in S. Sengupta, J. Huaser, P.R. Eiseman and J.F. Thompson (eds.), *Numerical Grid Generation in Computational Fluid Mechanics*, pp. 621–631, Pineridge Press.

[17] Buratynski, E.K. (1990) A fully automatic three-dimensional mesh generator for complex geometries, *Int. J. Num. Meth. Engg*, **30**, 931–952.

[18] Kela, A., Perucchio, R. and Voelcker, H.B. (1986) Toward automatic finite element analysis, *Comput. Mech. Engg*, July, 57–71.

[19] Kela, A. (1989) Hierarchical octree approximations for boundary representation-based geometric models, *Computer Aided Design*, **21**, 355–362.
[20] Kela, A. (1989) Exact octree approximations from geometric (csg/brep) models: Their derivation and application in finite element mesh generation, *IEEE Comp. Graphics and Applications*.
[21] Perucchio, R., Saxena, M. and Kela, A. (1989) Automatic mesh generation from solid models based on recursive spatial decompositions, *Int. J. Num. Meth. Engg*, **28**, 2469–2501.
[22] Schroeder, W.J. and Shephard, M.S. (1990) A combined octree/Delaunay method for fully automatic 3-D mesh generation, *Int. J. Num. Meth. Engg*, **29**, 37–55.
[23] Shephard, M.S., Yerry, M.A. and Baehmann, P.L. (1986) Automatic mesh generation allowing for efficient a priori and a posteriori mesh refinements, *Comp. Meth. Appl. Mech. Engg*, **55**, 161–180.
[24] Shephard, M.S. and Georges, M.K. (1991) Automatic three-dimensional mesh generation by the Finite Octree technique, *Int. J. Num. Meth. Engg*, **32**(4), 709–749.
[25] Yerry, M.A. and Shephard, M.S. (1983) Finite element mesh generation based on a modified-quadtree approach, *IEEE Comput. Graphics and Appl.*, **3**(1), 36–46.
[26] Yerry, M.A. and Shephard, M.S. (1984) Automatic three-dimensional mesh generation by the modified-octree technique, *Int. J. Num. Meth. Engg*, **20**, 1965–1990.
[27] Yerry, M.A. and Shephard, M.S. (1985) Automatic three-dimensional mesh generation for three-dimensional solids, *Comput. Struct.*, **20**, 31–39.
[28] Lawson, C.L. (1986) Properties of n-dimensional triangulations, *Computer Aided Geom. Des.*, **3**(4), 231–246.
[29] Preparata, F.P. and Shamos, M.I. (1985) *Computational Geometry, An Introduction*, Springer-Verlag, New York.
[30] Sibson, R. (1978) Locally equiangular triangulations, *Comput. J.*, **21**(3), 243–245.
[31] Green, P.J. and Sibson, R. (1978) Computing Dirichlet tesselation, *Comput. J.*, **2**, 168–173.
[32] Baker, T.J. (1989) Automatic mesh generation for complex three dimensional regions using a constrained Delaunay triangulation, *Engg Comput.*, **5**, 16–175.
[33] Baker, T.J. (1991) Shape reconstruction and volume meshing for complex solids, *Int. J. Num. Meth. Engg*, **32**(4), 665–675.
[34] Cavendish, J.C., Field, D.A. and Frey, W.H. (1985) An approach to automatic three-dimensional mesh generation, *Int. J. Num. Meth. Engg*, **21**, 329–347.
[35] Field, D.A. (1991) A generic Delaunay algorithm for finite element meshes, *Adv. Engg Software*, **13**(5/6), 263–272.
[36] Frey, W.H. (1987) Selective refinement: A new strategy for automatic node placement in graded triangular meshes, *Int. J. Num. Meth. Engg*, **24**, 2183–2200.
[37] George, P.L., Hecht, F. and Vallet, M.G. (1991) Creation of internal points in Voroni's type method: Control and adaptation, *Adv. Engg Software*, **13**(5/6), 303–312.
[38] Lo, S.H. (1989) Delaunay triangulation of non-convex planar domains, *Int. J. Num. Meth. Engg*, **28**, 2695–2707.
[39] George, P.L., Hecht, F. and Saltel, E. (1988) Constraint of the boundary and automatic mesh generation, in *Numerical Grid Generation on Computational Fluid Mechanics '88*, Pineridge Press, Swansea, pp. 589–597.
[40] Joe, B. (1991) Delaunay versus max-min solid angle triangulations for three-dimensional mesh generation, *Int. J. Num. Meth. Engg*, **31**, 987–997.
[41] Lo, S.H. (1991) Automatic mesh generation and adaptation by using contours, *Int. J. Num. Meth. Engg*, **31**, 689–707
[42] Löhner, R. (1989) Adaptive remeshing for transient problems, *Comp. Meth. Appl. Mech. Engg*, **75**, 195–214.
[43] Gursoy, H.N. and Patrikalakis, N.M. (1986) Automatic interrogation and adaptive subdivision of shape using medial axis transform, *Adv. Engg Software*, **1**, 21–42.

[44] Tam, T.K.H. and Armstrong, C.G. (1991) 2D finite element mesh generation by medial axis subdivision, *Adv. Engg Software*, **13**(5/6), 313–324.

[45] Srinivasna, V., Nackman, L.R., Tan, J.-M. and Meshkat, S.N. (1990) *Automatic Mesh Generation Using the Symmetric Axis Transformation of Polygonal Domains*, Technical Report, IBM Research Division, T.J. Watson Research Center, Yorktown Heights, New York. Report RC 16132 (#71695) 9/27/90.

[46] CAM-I (1986) *Applications Interface Specification (Restructured Version)*, Technical Report, CAM-I R-86-GM-01, Arlington, TX, January.

[47] Georges, M.K. (1990) *Geometric Operators for the Finite Octree Mesh Generator*, Technical Report SCOREC, 13-1991, Scientific Computation Research Center, Rensselaer Polytechnic Institute, Troy, New York.

[48] Ahmad, H. (1990) *CATIA/Octree Interface*, Technical Report SCOREC, 34-1990, Scientific Computation Research Center, Rensselaer Polytechnic Institute, Troy, New York.

[49] Flaherty, J.E., Benantar, M., Biswas, R. and Moore, P.K. (1990) *Symbolic and Parallel Adaptive Methods for Partial Differential Equations*, Technical Report, 90–26, Dept. Computer Science, RPI, Troy, New York.

[50] Williams, R.D. (1988) Dime: A programming environment for unstructured triangular meshes on a distributed-memory parallel processor, *Proc. 3rd Int. Conf. on Hypercube Parallel Processors and Applications*; also Caltech Concurrent Computation Project Report C3P-502.

[51] Williams, R.D. (1990) *Performance of a Distributed Unstructured-mesh Code for Transonic Flow*, Technical Report C3P-856, Center for Research in Parallel Computation, California Institute of Technology, 206-249, Caltech, Pasadena, January.

[52] Löhner, R. (1988) Some useful data structures for the generation of unstructured grids, *Commun. Appl. Num. Meth.*, **4**, 123–135.

[53] Niu, Q. and Shephard, M.S. (1989) Transfer of solution variables between finite element meshes, SCOREC Report #4-1990, Scientific Computation Research Center, Rensselaer Polytechnic Institute, Troy, New York.

[54] Baehmann, P.L. and Shephard, M.S. (1989) Adaptive multiple level h-refinement in automated finite element analyses, *Engg Comput.*, **5**(3/4), 235–247.

[55] Schmidt, R. (1991) *Solution of the Euler Equations using Finite Quadtree Grids*, Ph.D. thesis, Department of Mathematical Sciences, Rensselaer Polytechnic Institute, Troy, New York.

[56] de Cougny, H.L., Shephard, M.S. and Georges, M.K. (1992) *An Advanced Octree/Delaunay Mesh Generator (Tentative title)*, Technical Report, Scientific Computation Research Center, Rensselaer Polytechnic Institute, Troy, New York. Masters thesis in progress.

[57] Benantar, M., Biswas, R., Flaherty, J.E. and Shephard, M.S. (1990) Parallel computation with adaptive methods for elliptic and hyperbolic systems, *Comput. Meth. Appl. Mech. Engg*, **82**, 73–93.

[58] Winget, J.M. and Hughes, T.J.R. (1985) Solution algorithms for nonlinear transient heat conduction analysis employing element-by-element strategies, *Comput. Meth. Appl. Mech. Engg*, **52**, 711–815.

[59] Ortega, J. (1988) *Introduction to Parallel and Vector Solution of Linear Systems*, Plenum Press, New York.

[60] Keyes, D.E. and Gropp, W.D. (1987) A comparison of domain decomposition techniques for elliptic partial differential equations and their parallel implementation, in C.W. Gear and R.G. Voigt (eds.), *Selected Papers from the Second Conference on Parallel Processing for Scientific Computing*, pp. s166–s202, SIAM, Philadelphia.

M.S. Shephard
Scientific Computation Research Center
Schools of Engineering and Science
Rensselaer Polytechnic Institute
Troy NY 12180-3590 USA

Author Index

Numbers in **bold type** refer to lists of publications at the ends of chapters

Abadie, J.H.M., 393, 408, 420, 421, **427**
Abdallah, A.A., 226, 231, **255**
Adams, L., 18, 24, **35**, **36**, 136, **155**
Adelman, H.M., 393, 420, 421, **426**
Ahmad, H., 453, **460**
Aithal, R., 98, **122**
Ajiz, M.A., 8, 9, 28, **33**, 136, **155**
Al-Saadoun, S.S., 417, **427**
Alvarez, R., 226, 231, **254**
Alves, J.L.D., 6, 18, 28, **33**, **35**, 226, 231, 243, **255**
Alvin, K.F., 292, **300**
Anagnostopoulos, S.A., 226, 230, **256**
Andersson, B., 71, 82, **87**, **88**
Anderson, R.G., 125, **152**
Angeleri, F., 9, 10, **34**, 100, **123**
Appleyard, J.R., 136, **155**
Argyris, J.H., 28, **37**, 302, 305, 313, 315, 329, 332, 336, 341, 344, **353**, **354**
Arihaloo, B.L., 418, **427**
Arioli, M., 28, **37**
Armstrong, W.H., 277, **299**
Armijo, L., 206, **223**
Armstrong, C.G., 450, **460**
Arnold, R.R., 226, 231, 239, **254**
Arora, J.S., 391, 392, 393, 394, 395, 398, 401, 403, 404, 405, 408, 410, 411, 415, 417, 418, 420, 421, 423, 424, **425**, **426**, **427**, **428**
Ashby, S., 68, 69, 85, **87**
Ashby, S.F., 204, **223**
Ashcraft, C.C., 67, **86**
Atrek, E., 392, 417, 418, **425**
Axelsson, O., 7, 8, 10, 18, 22, 24, 28, **33**, **34**, **35**, **36**, **37**, 127, 136, **154**, **155**, 204, **223**

Babilis, G., 9, 15, 24, 26, **34**
Babuska, I.M., 71, 76, 83, **87**, **88**

Baeckhmann, P.L., 442, 455, **459**, **460**
Balasubramanian, P., 226, 231, **255**
Balestra, M., 359, **388**
Balmer, H., 301, **353**
Balopoulos, V., 193, 194, 208, **220**
Bampton, M.C.C., 125, **151**
Banerjee, P.K., 90, 98, **121**
Bank, R.E., 69, **87**
Barragy, F., 10, **34**
Bassi, P., 125, **152**
Bathe, K.J., 9, 10, **34**, 100, **123**, 125, 126, 138, **152**, **155**, 157, 169, **180**, 189, 190, 201, **219**, **222**, 243, **256**, 419, **428**
Batoz, J.L., 196, **221**
Bayo, E.P., 226, 231, **254**, **255**
Beauwens, R., 22, **35**
Becovier, M., 9, **34**
Beer, G., 98, **121**
Behr, M., 358, 361, 364, **387**
Beliveau, J.G., 126, **152**
Bell, K., 226, 230, **256**
Bellini, P.X., 201, **222**
Belman, R., 281, **300**
Belvin, W.K., 270, 271, 274, 292, **299**, **300**
Belytschko, T., 259, **298**, 304, **354**
Ben Bouzid, M., 22, **35**
Bender, C.F., 127, **153**
Bennet, J.A., 392, 418, **425**
Bernsang, L., 100, **123**, 189, 199, 206, **219**
Bergan, P., 200, 201, 202, **221**, **222**
Berk, A.D., 130, **154**
Bertero, V.V., 245, **257**
Bettes, P., 302, **353**
Bettess, J.A., 99, **122**
Beumgarte, J.W., 264, **298**
Bialecki, R., 98, **121**
Biegler-Konig, F., 207, **223**
Biot, M.A., 130, **154**
Bitoulas, N., 8, 9, 18, 21, 28, **33**, **35**, **37**

Bitzarakis, S., 21, **35**
Bjorstad, P.E., 24, **36**
Blevins, R.D., 375, **389**
Blos, O., 349, **355**
Bohm, G.J., 243, **257**
Borgers, C., 24, **36**
Borggaard, J.T., 98, **122**
Borino, G., 226, 230, 231, **255**
Borja, R.I., 194, **221**
Bostic, S.W., 2, 6, **32**, 158, **180**
Bourgat, J.F., 85, **88**
Boyle, J.M., 126, **152**
Bozek, D.G., 98, **122**
Bradbury, W.W., 127, **153**
Bramble, J.M., 24, **36**, 66, **86**
Brebbia, C.A., 90, 98, **121**
Brezzi, F., 359, **388**
Brinkkemper, S., 22, **35**
Brodlie, K., 190, **220**
Brooks, A.N., 359, **388**
Broyden, C.G., 188, 191, **218**, **219**
Brussino, G., 67, **87**, 100, **123**
Buckley, A., 189, 190, 191, 192, 199, 206, **220**
Bucy, R.S., 271, **299**
Bunch, J.R., 204, **223**
Butterfield, R., 90, 98, **121**

Canann, S., 440, 448, 449, **458**
Cani, I.M., 203, **222**
Cardona, A., 226, 231, 242, 243, **255**, **256**
Cardoso, J.B., 418, 421, **428**
Carey, G.F., 10, **34**, **35**
Carter, W.T., 24, **36**
Cavendish, J.C., 447, 448, **459**
Cea, J., 392, 418, **425**
Cerrolaza, M., 226, 231, **254**
Cervera, M., 194, **221**
Cha, J.Z., 393, **426**
Chahande, A.I., 395, **426**
Chan, A.H.C., 270, **299**, 302, 304, 341, **353**, **354**

Chan, T.F., 24, **36**
Chance, M.S., 125, **151**
Chandrupatla, T.R., 419, **428**
Chang, C.J., 242, 243, 245, 252, **256**
Chang, O.V., 98, **121**
Chargin, M., 226, 231, 239, **254**
Chawdhury, P.C., 226, 231, **255**
Chayapathy, B.K., 83, **88**
Chen, H.C., 226, 231, **254**, **255**
Chen, P., 226, 231, **254**
Cheshire, I.M., 136, **155**
Chiou, J.C., 260, 262, 264, 266, 267, 270, 287, 298, **299**, **300**
Choi, K.K., 401, **426**
Chulya, A., 201, **222**
Ciarelli, D.M., 98, **122**
Ciarelli, K.J., 98, **122**
Cimento, A.P., 189, 190, **219**
Citerley, R.L., 226, 231, 239, **254**
Claret, A.M., 244, **257**
Clarke, M.J., 201, 203, **222**
Clough, R.W., 125, 140, **152**, **155**, 162, 164, **181**, 226, 227, 231, 233, 235, 241, 243, **256**, **257**
Concus, P., 7, 22, **33**, **35**, 67, 68, 69, **87**, 127, **154**
Contro, R., 349, **354**
Cornwell, R.E., 226, 230, 231, **256**
Courlay, A.R., **63**
Coutinho, A.L.G.A., 6, 18, 28, **33**, **35**, 140, **155**, 226, 231, 243, **254**, **255**
Craig, A.W., 76, 83, **88**
Craig, M.M., 125, **151**
Craig, R.R., 158, **180**, 226, 230, 231, **256**
Crisfield, M.A., 24, **36**, 67, **86**, 189, 190, 192, 193, 196, 199, 200, 201, 203, 206, 207, **219**, **220**, **221**, **222**, **223**
Crotty, J.M., 98, **121**
Cullum, J., 84, **88**, 126, 140, **152**, **156**, 158, 162, 164, **180**
Cursoy, H.N., 450, **460**
Curtis, A.R., 192, **222**

Danilin, Y.N., 401, 404, 405, **426**
Das, P.C., 98, **121**
Davidon, W.C., 188, **219**
Davies, T.G., 98, **122**
De Cougny, H.L., 456, **460**
Dembo, R.S., 193, 194, **220**
Demmel, J.W., 28, **37**
Dennis, J.E., 188, 189, 192, **219**
De Roeck, Y.H., 24, **36**, 85, **88**
Dervieux, A., 103, **123**
De Runtz, J.A., 270, 278, **299**
De Sonza, M.M., 344, **354**
Deyo, R.C., 269, **299**
Dhatt, G., 196, **221**
Dickens, J.M., 226, 230, 231, 253, **254**, **255**

Dixon, L.C.W., 206, **223**
Dohmen, D.M., 344, **354**
Dold, A., 136, **154**
Doltsinis, I.St., 301, 302, 305, 313, 327, 328, 332, 336, 340, 341, 344, 345, 347, 349, 350, **353**, **354**, **355**
Donath, W.E., 126, 140, **152**
Donea, J., 302, **353**
Dongara, J.J., 126, **152**
Dowener, J.D., 260, 267, 269, 270, 287, **298**, **299**, **300**
Dracopoulos, M.C., 7, 8, 9, 10, 33
Drake, J.B., 99, **123**
Drija, M., 24, **36**, 69, **87**
Dubois, P.F., 17, **35**
Dubois-Pelerin, Y.B., 278, 281, 282, **299**
Duff, I.S., 28, **37**, 67, **86**, 204, **223**
Dunbar, W.S., 140, **155**
Dungar, R., 243, **256**
Dussault, S., 245, **257**
Dvorkin, E.N., 201, **222**

Ebecken, U.F.F., 6, 18, 28, 33, 140, **155**, 226, 231, 243, **254**, **255**
Eckman, B., 136, **154**
Eijkout, V., 22, **35**
Eisenstat, S.C., 7, 18, **33**, **35**, 99, **123**, 136, **155**
Elghadamsi, F.E., 242, 245, 252, **256**
Elman, H.C., 83, **88**
Elsawaf, A., 187, 192, **218**
Engeli, M., 3, **32**, 184, **218**
Ericsson, T., 126, **152**
Erisman, A.M., 67, **86**
Esfandiari, R.S., 244, **257**
Eskow, E., 204, **223**
Evans, D.J., 6, **33**, 127, 136, **153**, **154**

Fadeev, D.K., 127, **153**
Fadeeva, V.V., 127, **153**
Falk, S., 127, **153**
Falk, U., 71, **87**
Farhat, C., 24, **37**, 67, **87**, 186, **218**, 226, 231, **254**, 260, 262, 278, 281, 282, **298**, **299**, 329, 350, **354**, **355**
Felippa, C.A., 186, 201, **218**, **222**, 259, 270, 278, **298**, **299**, 302, 304, **353**
Fenves, G., 447, **459**
Ferencz, R.M., 6, 10, 13, 28, **33**, **34**, 358, **388**
Ferrari, R.L., 125, **152**
Fezoni, L., 103, **123**
Field, D.A., 447, 448, **459**
Finnigan, P.M., 438, 451, 454, **458**
Fischer, H., 302, 341, 344, **353**
Flaherty, J.E., 454, 456, 458, **460**

Fletcher, R., 127, **153**, 185, 186, 188, 205, **218**, **219**
Florian, P., 132, **154**
Forde, B.W.R., 189, 199, 201, **219**
Foresti, S., 67, **87**
Forsythe, G., 28, **37**
Fox, L., 188, **219**
Fox, R.L., 10, **34**, 42, **63**
Franca, L.P., 358, 359, 363, **387**, **388**, **389**
Frangopol, D.M., 392, **425**
Frey, S.L., 363, **389**
Frey, W.H., 447, 110, **109**
Fried, I., 10, **34**, 42, **63**, 127, 133, **153**, 201, **222**
Friz, H., 336, **354**
Fulton, R.E., 158, **180**

Gabriele, G.A., 410, 418, **427**
Galant, D., 226, 231, 239, **254**
Gallagher, R.H., 98, **122**, 392, 417, 418, **425**
Gambolati, G., 2, 6, 9, **32**, **34**, 125, 127, 128, 131, 132, 134, 136, 145, 146, **152**, **153**, **154**, **155**
Gantes, C.J., 192, 194, 205, **220**, **221**
Gantmacher, F.R., 274, **299**
Garbow, B.S., 126, **152**
Garcia-Palomares, V.M., 411, **427**
Garg, N.K., 194, **220**
Geist, G.A., 99, **123**
Gelin, J.G., 189, **219**
George, A., 40, 51, **63**, 77, **88**, 170, **181**
George, P.L., 447, 448, **460**
Georges, M.K., 442, 443, 444, 445, 452, 456, **459**, **460**
Georgiev, K., 98, **122**
Geradin, M., 127, **153**, 189, 190, **219**
Gershwindner, L.F., 243, **257**
Ghionis, P., 185, 205, **218**
Gibson, W., 205, **223**
Gill, P.E., 188, 191, 192, 194, 204, 206, **219**, **220**, 221, 393, 406, 408, 414, 415, 417, 423, **426**
Gillies, A.G., 242, 244, **256**
Ginsburg, T., 3, 32, 184, **218**
Giuliani, S., **253**, 302
Gloudeman, J.F., 28, **37**
Glowinski, R., 24, **36**, 85, **88**, 359, **388**
Goble, G.G., 205, **223**
Goldstein, A.A., 206, **223**
Golub, G.H., 7, 22, 24, **33**, **35**, **36**, 45, **63**, 67, 68, 69, **87**, 100, **123**, 126, 127, 140, **152**, **154**
Goovaerts, D., 24, **36**
Gossain, D.M., 283, **300**
Gourlay, A., 190, **220**

Govil, A.K., 423, **428**
Gracewski, S., 243, **256**
Grandhi, R.V., 418, **428**
Gray, L.J., 99, **123**
Green, J.M., 125, **151**
Green, P.J., 447, **459**
Greenbaum, A., 17, **35**
Greenstadt, J., 190, **220**
Grice, K.R., 442, **459**
Griffin, O.M., 375, **389**
Grimes, R.G., 67, **86**, 140, **156**
Grimm, R.G., 125, **151**
Grivelli, L., 186, **218**
Gropp, W.D., 23, 24, **35**, 458, **461**
Gruber, R., 125, **151**
Gupta, V.K., 158, 162, **180**
Gurdal, Z., 392, 393, **425**
Guru Prasad, K., 106, **124**
Gustafsson, A.I., 8, 10, 24, 28, **33**, **35**, **36**, **37**, 136, **154**, **155**

Habibullah, A., 253, **257**
Hadjikov, L., 98, **122**
Haftka, R.T., 392, 393, 420, 421, **425**, **426**
Haisler, W.E., 243, **257**
Hall, J.F., 24, **37**
Halleux, J.P., 302, **353**
Hallquist, J.O., 10, 13, **34**
Han, S.P., 411, 415, 416, **427**
Hancock, G.J., 201, 203, **222**
Hanna, M.N., 243, **256**
Hansbo, P., 358, **388**
Hansteen, O.E., 226, 230, **256**
Harrian, M., 418, **428**
Harrold, A.J., 28, **37**
Hassanzadeh, S., 67, **87**
Hatjikonstantinou, K., 21, **35**
Haug, E.J., 262, 269, **298**, **299**, 391, 392, 401, 408, 418, 420, 421, 423, **425**, **428**
Hayes, L., 7, 10, **33**, **34**, 259, **298**
Heath, M.T., 99, **123**
Hecht, F., 447, 448, **460**
Henkel, C.S., 99, **123**
Hestenes, M.R., 3, **32**, 39, **62**, 127, **153**
Hibbeler, R.C., 158, **180**
Hinton, E., 189, 190, 194, 208, **219**, **221**, **223**, 302, **353**, 440, 448, **458**
Hock, W., 415, **427**
Hockney, R.W., 166, **181**
Hodous, M.F., 98, **122**
Hofmeister, L.D., 243, **257**
Hogge, M., 189, 190, **219**
Horrigmoe, G., 201, **222**
Hosteny, R.P., 127, **153**
Hou, J.W., 401, **426**
House, J.M., 418, **428**
Hsieh, C.C., 418, **428**
Huckelbridge, A.A., 226, 231, **255**

Huebner, K.H., 158, **180**, 419, **428**
Hughes, T.J.R., 10, 13, **34**, 41, 50, **63**, 103, **123**, 138, 140, **155**, 304, **354**, 358, 359, 363, **387**, **388**, **389**, 457, **461**
Hulbert, G.M., 358, **387**
Humar, J.L., 227, **256**
Hunter, J.A., 283, **300**
Hussani, M.Y., 136, **155**
Hvidsten, A., 24, **36**

Ibrahimbegovic, A., 226, 231, 243, 244, **254**
Idelsohn, S., 189, 190, 201, 203, **219**, **222**
Idelsohn, S.R., 226, 231, 242, 243, **255**, **256**
Ilin, V.P., 22, **35**
Irons, B.M., 125, **152**, 187, 192, **218**

Jackins, C., 439, 442, **458**
Jackson, C.P., 136, **155**
Jacobs, D.A.H., 136, **155**
Jacucci, G., 349, **354**
Jagadeesh, J.G., 226, 231, **255**
James, B.B., 423, **429**
Jankowski, M., 28, **37**
Jao, S.Y., 418, **428**
Jennings, A., 8, 9, 22, 28, **33**, 136, **154**, **155**, 157, 169, **180**, 315, 318, **354**
Jensen, P.S., 157, **180**
Jesshope, C.R., 166, **181**
Jeusette, J.P., 201, 203, **222**
Jin, H., 98, 122, 440, 448, **458**
Joe, B., 448, **460**
Johan, Z., 103, **123**
Johnsen, Th.L., 28, **37**
Johnson, C.P., 226, 230, 231, **256**
Johnson, J.L., 125, **151**
Johnson, O.G., 17, 35, 136, **155**
Jones, M.T., 158, **180**
Joo, K.J., 226, 231, 233, **254**
Joubert, W., 68, **87**, **124**
Juliani, F., 349, **354**

Kalabis, H.P., 344, **354**
Kalathas, N., 201, **223**
Kalman, R.E., 271, **299**
Kamat, M.P., 192, 199, **220**, **221**, 392, 393, **425**
Kane, J.H., 92, 94, 98, 99, 102, 103, 106, 118, **122**, **123**, **124**
Karasudhi, P., 243, **257**
Karush, W., 127, **153**
Kasai, K., 242, **256**
Katnik, R.B., 98, **122**
Katz, I.N., 71, **87**
Kela, A., 442, 443, **459**
Kershaw, D.S., 8, 33, 41, 51, **63**, 127, 136, **154**

Keyes, D.E., 22, 24, **35**, 103, **123**, 458, **461**
Khot, N.S., 418, **428**
Kincaid, D.R., **124**
King, R., 371, 375, **389**
Kirsch, U., 392, **425**
Kline, K.A., 226, 231, **254**
Knight, N., 243, **256**
Knight, N.F., 172, **181**
Komzsik, L., 158, 162, **180**
Konig, M., 301, **353**
Koopmann, G.H., 375, **389**
Kopp, R., 344, **354**
Kouhia, R., 201, 203, **222**
Koyamada, Y., 125, **152**
Krakeland, B., 201, **222**
Kumar, B.L.K., 92, 94, 98, 99, **122**, **124**
Kwarkernaak, H., 270, **299**

Lachat, J.C., 98, **121**
Lachet, G., 201, 203, **222**
Landau, L., 6, 28, **33**, 140, **155**, 226, 231, 243, **254**, **255**
Lanczos, C., 4, **32**, 39, **62**, 126, 136, **152**, 162, **180**
Law, K.H., 24, **36**
Lawson, C.L., 447, 448, **459**
Lee, H.H., 417, **427**
Lee, S.H., 189, 192, 206, **219**, **223**
Leger, P., 226, 230, 231, 233, 234, 235, 236, 241, 242, 245, **254**, **255**, **257**
Lemieux, P., 126, **152**
LeNir, A., 189, 190, 191, 199, 206, **220**
Le Tallec, P., 85, **88**
Lett, G.S., 76, **88**
Leung, A.Y.T., 226, 231, **255**
Lev, O.E., 392, **425**
Levit, I., 10, **34**, 41, **63**, 358, **388**
Levy, R., 392, **425**
Levy, S., 28, **37**
Lewis, J.G., 67, **86**, **87**, 140, **156**
Lewis, R.W., 302, **353**
Li, G., 423, **428**
Li, G.Y., 395, 425, **426**
Li, M.R., 52, 53, **63**
Li, Y., 226, 231, **254**
Lim, O.K., 405, 415, 417, 418, 426, **427**
Lima, E.C.P., 6, 28, **33**, 140, **155**, 226, 227, 231, 233, 235, 241, 243, **254**, **255**
Lin, T.C., 395, 418, **426**, **428**
Lin, T.P., 424, **429**
Lindskog, G., 8, 10, **34**, **35**
Liou, J., 10, **34**, 358, 359, 361, 364, 368, 369, **387**, **388**, **389**
Liu, J.W., 40, 51, **63**, 77, **88**, 170, **181**
Liu, Y.C., 194, **221**

Lively, R.K., 39, **63**
Lo, J.A., 438, **458**
Lo, S.H., 447, 448, **460**
Lohner, R., 440, 448, 455, **458**, **460**
Longsine, D.E., 127, **153**
Luginsland, J., 336, 340, 344, 347, 350, **354**
Luenberger, D.G., 69, **87**, 204, 205, 206, **223**, 271, **299**, 393, 406, 408, 414, 423, **426**
Lukkunapracit, P., 243, **257**
Lundren, T., 360, **389**
Lyzenga, G., 24, **36**, **64**

Ma, S., 158, **180**
Mahajan, U., 158, **180**
Maison, B.F., 242, **256**
Makimoto, T., 125, **152**
Malik, G.M., 8, 9, 22, **33**, 136, **154**
Mallet, M., 358, **388**
Mandel, J., 24, **36**, 65, 66, 73, 75, 76, 82, **86**, **88**
Mangasarian, O.L., 411, **427**
Manini, L., 349, **354**
Mansour, N.N., 360, **389**
Manteufell, T.A., 8, 10, **33**, 67, 68, 69, 85, **87**, 127, 136, **154**, 204, **223**
Margenov, S., 98, **122**
Martin, C.W., 28, **37**
Marvil, E.S., 192, **220**
Matthies, H., 188, 189, 190, 191, 206, **219**
Matzenmiller, A., 189, 199, **219**
Mayne, R.W., 393, **426**
McConnel, R.E., 203, **222**
McCormick, S.F., 67, 69, **87**, 127, **153**
McGuire, W., 201, **222**
Meek, J.L., 208, **223**
Meijerink, J.A., 8, **33**, 67, **87**, 127, 136, **154**
Meshkat, S.N., 450, **460**
Meurant, G., 22, 24, **35**, **36**
Meyers, D., 24, **36**
Michelli, C.A., 17, **35**, 136, **155**
Miehe, C., 203, **222**, 338, 341, **354**
Mikkola, M., 201, 203, **222**
Mittal, S., 358, 359, 360, 361, 363, 364, 371, **387**, **388**
Mohraz, B., 242, 243, 245, 252, **256**
Mojtahedi, S., 226, 230, **256**
Molina, J., 226, 231, **254**
Moler, C.B., 28, 37, 126, **152**
Molnar, A.J., 243, **257**
Moore, P.K., 454, **460**
More, J.J., 188, 189, 192, **219**
Morris, A.J., 392, 418, **425**
Morris, N.F., 243, **257**
Moscarello, R., 199, 201, **221**
Mota-Soares, C.A., 392, 417, 418, **425**

Munksgaard, N., 8, **33**, **34**, 204, **223**
Murray, W., 188, 191, 192, 194, 204, 206, **219**, **220**, **221**, 393, 406, 408, 414, 415, 417, 423, **426**
Muscolino, G., 226, 230, 231, 243, **255**, **256**
Mustoe, G.G.W., 98, **122**

Nachtigal, N.M., 101, **123**
Nackman, L.R., 450, **460**
Nahavandi, A.N., 243, **257**
Nahlik, R., 98, **121**
Nash, S.G., 194, 204, 206, **221**
Natori, M.C., 287, **300**
Navarra, A., 103, **123**
Nazareth, L., 189, 191, 192, **219**
Nesbet, R.K., 127, **153**
Newell, J.F., 158, 162, **180**
Newman, M., 162, **180**
Nguyen, T., 10, **34**, 359, 364, 368, 369, **388**
Nickell, R.E., 242, **256**, 277, **299**
Nielsen, H.B., 100, **124**, 204, **223**
Nin, Q., 455, **460**
Nisbet, R.M., 127, **153**
Nocedal, J., 189, 190, 191, 199, **220**
Nomikos, N., 194, **221**
Noor, A.K., 227, **256**
Notling, S., 302, 327, 328, 336, 340, 344, 345, 347, 349, 350, **354**, **355**
Nour-Omid, B., 6, 10, 24, 28, **33**, **34**, **36**, 41, 45, 49, 52, 53, **63**, **64**, 128, 136, 140, **154**, 162, 164, **181**, 193, 194, **220**, 226, 231, 233, **254**, **255**
Novakova, M., 98, **122**

Oden, J.T., 277, **299**, 315, **354**
Oettli, W., 28, **37**
Ojalvo, I.U., 162, **180**
O'Leary, D.P., 7, 22, **33**, 67, 68, 69, **87**, 127, **154**, 194, 204, 206, **221**
Ong, E.G., 18, 24, **35**, **36**
Oppe, T.C., **124**
Ordiz, M., 10, **34**, 41, 53, **63**
Ortega, J.M., 18, **35**, 100, **123**, 315, 318, **354**, 457, **461**
Osyczka, A., 393, **426**
Overmann, A.L., 100, **123**, 169, 171, **181**
Owen, D.J.R., 208, **223**

Padovan, J., 199, 201, **221**, **222**
Paeng, J.K., 395, 418, **426**, **428**
Paige, C.C., 5, **32**, 126, 138, 140, **152**, **156**, 162, 164, **180**, 204, **223**
Pantazopoulos, G., 190, 198, **220**

Parilch, P., 440, 448, **458**
Papadrakakis, M., 2, 3, 5, 6, 7, 8, 9, 10, 15, 18, 21, 24, 26, 28, **32**, **33**, **34**, **35**, **37**, 127, 128, **153**, **154**, 184, 185, 186, 190, 192, 193, 194, 198, 199, 200, 203, 204, 205, 206, 207, 208, **218**, **220**, **221**, **222**, **223**
Papalambros, P.Y., 392, **425**
Park, G.J., 405, 417, 418, **426**, **427**
Park, K.C., **64**, 200, 201, **221**, 259, 260, 262, 264, 266, 267, 270, 271, 274, 277, 278, 279, 280, 281, 282, 287, 292, **298**, **299**, **300**, 302, 304, **353**
Parlett, B.N., 6, 10, **33**, **34**, 41, 45, 48, 49, 52, 53, **63**, 126, 128, 140, **152**, **154**, **156**, 157, 158, 162, 164, **180**, 193, 194, 204, **220**, **222**, **223**
Parsons, I.D., 24, **37**
Pasciak, J.E., 24, **36**, 66, **86**
Patera, A.T., 82, **88**
Patrick, M.L., 158, **180**
Patrikalakis, N.M., 450, **460**
Paul, D.K., 270, **299**
Paul, G., 17, **35**, 136, **155**
Pegoretti, A., 349, **354**
Penzien, J., 125, **152**, 227, **256**
Perdon, A.M., 127, 132, 136, 145, 146, **154**
Perego, U., 226, 231, **254**
Periaux, J., 24, **36**, 103, **123**
Perry, A., 192, **220**
Perucchio, R., 442, **459**
Peyton, B.W., 67, **86**
Pica, A., 189, 190, **219**
Picart, P., 189, **219**
Pierce, D.J., 67, **87**
Pimenta, P.M., 315, **354**
Pini, G., 9, **34**, 125, 127, 131, 136, **152**, **154**, **155**
Pinsky, P.M., 10, **34**, 41, **63**
Pipano, A., 127, **153**
Pitkaranta, J., 76, **88**, 359, **388**
Polak, E., 185, **218**
Polman, B., 22, **35**
Poole, E.L., 100, **123**, 169, 171, **181**
Poole, S., 10, **34**, 359, 364, 368, 369, **388**
Powell, M.J.D., 188, 192, 193, **219**, **220**, 411, 414, 415, 416, **427**
Prager, W., 28, **37**
Prepareta, E.P., 447, **459**
Psenichny, B.N., 401, 404, 405, **426**
Przemieniecki, J.S., 21, **35**

Raefsky, A., 10, 24, **34**, **36**, **64**
Ragsdell, K.M., 392, 406, 410, 417, 418, **425**, **427**
Ramamourti, V., 226, 231, **255**

AUTHOR INDEX

Ramaswany, S., 126, **152**
Ramm, E., 189, 190, 199, 201, 207, **219**, **222**
Rao, S.S., 392, **425**
Ravindran, A., 392, 406, **425**
Ray, S.E., 359, 360, 363, 371, **388**
Reaser, M.H., 199, **221**
Reddy, S.C., 110, **123**
Reeves, C.M., 185, 205, **218**
Reid, J.K., 3, **32**, 40, **63**, 67, **86**, 192, 204, **220**, **222**
Reinboldt, W.C., 315, 318, **354**
Reklaitis, G.V., 392, 406, **425**
Remseth, S.N., 242, **256**
Resasco, D.C., 24, **36**
Ribiere, G., 185, **218**
Rice, R.C., 206, **223**
Ricles, J.M., 226, 231, 233, 239, 240, **254**, **255**
Riks, E., 200, 201, 203, **221**, **222**
Riley, M.F., 423, **429**
Robayo, L.J., 226, 231, 233, **255**
Robinson, P.C., 136, **155**
Romine, C.H., 99, **123**
Ronquist, E.M., 82, **88**
Rosanoff, R.A., 28, **37**
Rosen, J.B., 407, **426**
Rosenthal, A., 9, **34**
Rozvany, G.I.N., 392, 418, **425**, **427**
Roy, J.R., 28, **37**
Ruge, J.W., 69, **87**
Ruhe, A., 126, 127, 132, **152**, **153**
Runesson, K., 98, **122**, 189, 199, 206, **219**
Rutishauser, H., 3, **32**, 184, **218**
Ryn, Y.S., 418, **428**

Saad, Y., 17, 18, 24, **35**, 100, 103, 104, **123**, 136, **155**, 204, **223**, 358, 365, **388**
Sackman, J.L., 277, **299**
Saigal, S., 92, 94, 98, 99, **122**, **123**
Saltel, E., 447, **460**
Sameh, A.H., 126, 127, **152**
Samuelsson, A., 98, 100, **122**, **123**, 189, 199, 206, **219**
Sanehchi, J., 127, **153**
Sargeant, R.W.H., 410, **427**
Sarpkaya, T., 371, 375, **389**
Sartoretto, F., 125, 127, 128, 131, 132, **152**, **153**, **154**
Saunders, M.A., 5, **32**, 203, **223**
Saxena, M., 442, **459**
Say, C.W., 243, **257**
Saylor, P.E., 69, 85, **87**, 204, **223**
Schatz, A.H., 24, **36**, 66, **86**
Schittkowski, K., 393, 415, 416, 417, **426**, **427**
Schleicher, E., 7, **33**
Schmidt, R., 455, **460**
Schnabel, R., 188, 204, **219**, **223**

Schroeder, W.J., 432, 433, 434, 435, 437, 441, 443, 445, 446, 447, 448, **458**, **459**
Schultz, M.H., 100, 103, 104, **123**, 358, 365, **388**
Schwarz, H.R., 127, **153**
Schweizerhof, K., 189, 190, 199, 201, 207, **219**, **222**
Scott, D.S., 6, **33**, 140, **156**, 162, 164, **181**
See, T., 203, **222**
Sehmi, N.S., 225, **254**
Shakib, F., 6, **33**, 103, **123**, 358, 363, **387**
Sham, T.L., 24, **36**
Shamos, M.I., 447, **459**
Shanno, D.F., 192, **220**
Shah, V.N., 243, **257**
Shavitt, I., 127, **153**
Shephard, M.S., 431, 433, 438, 441, 442, 443, 444, 445, 447, 451, 452, 454, 455, 456, 458, **458**, **459**, **460**
Shepherd, R., 242, 244, **256**
Sherman, A.H., 194, **220**
Shetty, C.M., 393, **426**
Shih, R., 359, 360, 363, 371, **388**
Shugar, T.A., 186, **218**
Sibson, R., 447, 448, **459**
Silvester, P.P., 125, **152**
Simo, J.C., 203, **222**, 268, **299**, 341, **354**
Simon, H.D., 6, **33**, 41, 48, 55, 57, **63**, 67, **86**, 139, 140, **155**, **156**, 162, 164, 179, **181**
Sivan, R., 270, **299**
Skeel, R.D., 28, **37**
Skeie, G., 201, **222**
Smith, B.F., 24, **36**
Smith, B.T., 126, **152**
Smith, L.A., 205, **223**
Smith, T.F., 418, **428**
Smerou, S., 5, 6, 28, **33**
Sobh, N., 24, **37**
Sobieski, J., 392, **425**
Sobieszczanski-Sobieski, J., 423, **429**
Sohoni, V.M., 401, **426**
Sonnad, V., 9, 10, **34**, 67, **87**, 100, **123**
Soreide, T.H., 201, **222**
Soriano, L.H., 226, 230, **255**
Sotolino, E.D., 53, **63**
Soucy, Y., 126, **152**
Sreekanta Murthy, T., 417, **427**
Srinivasan, A.V., **124**
Srinivasna, V., 450, **460**
Stanton, E., 188, **219**
Stanton, E.L., 10, **34**, 42, **63**, 205, **223**
Stegmuller, H., 189, **219**
Steihaug, T., 193, 194, **220**
Stephenson, M.P., 440, 448, 449, **458**

Steve, H., 103, **123**
Stiefel, E., 3, **32**, 39, **62**, 127, **153**, 184, **218**
Stiemer, S.F., 189, 199, 201, **219**
Storaasli, O., 158, **180**
Stoufflet, B., 103, **123**
Strang, G., 188, 189, 190, 191, 206, **219**
Stricklin, J.A., 243, **257**
Stroud, J.W., 172, **181**
Stuben, K., 69, **87**
Suhas, H.K., 226, 231, **255**
Suri, M., 71, **87**
Szabo, B.A., 71, 83, **87**, **88**
Szepessy, A., 358, **388**
Szyld, D.B., 84, **88**

Tam, T.K.H., 450, **460**
Tan, J.M., 450, **460**
Tan, L.H., 9, 10, **34**
Tan, H.S., 208, **223**
Tanimoto, S., 439, 442, **458**
Tapia, R.A., 194, **220**
Taylor, R.L., 10, **34**, 128, 140, **154**, 193, 194, **220**, 226, 231, **254**, **255**, 347, **354**
Telles, J.C.F., 98, **121**
Tewarson, R.P., 192, **220**
Tezduyar, T.E., 10, **34**, 358, 359, 360, 361, 363, 364, 368, 369, 371, **387**, **388**, **389**
Thanedar, P.B., 391, 392, 395, 405, 415, 418, 424, **425**, **426**, **427**
Theoharis, A.P., 200, 206, 207, **221**
Thorton, E.A., 158, **180**, 419, **428**
Toint, P.L., 192, **220**
Tomlin, G.R., 98, **121**
Tovichakchaikul, S., 201, **222**
Trail Nash, R.W., 226, 231, **256**
Trefethen, L.N., 101, **123**
Troina, L.M., 18, **35**
Tsao, N.K., 98, **122**
Tseng, C.H., 405, 415, 417, 418, **426**, **427**
Tuff, A.D., 8, **33**, 136, **154**

Uang, C.M., 245, **257**
Udwadia, F.E., 244, **257**
Underwood, P.G., 186, **218**
Underwood, R.R., 22, **35**, 126, 140, **152**
Ussher, T.H., 283, **300**

Vallet, M.G., 447, 448, **460**
Vanden Brink, D.J., 192, **220**
Van der Houven, P.J., 304, **354**
Vanderplaats, G.N., 392, 406, **425**, **426**
Van der Sluis, A., 69, **87**
Van der Vorst, H.A., 8, **33**, 67, 69, **87**, 127, 136, 139, 149, **154**
Van Kats, J.M., 139, 149, **155**

Van Loan, C.F., 45, **63**, 69, **87**, 100, **123**
Vashi, K.M., 243, **257**
Vassilevski, P.S., 24, **37**
Vaugham, C.T., 18, **35**
Venancio Filho, F., 226, 230, 231, 243, 244, **255**
Vidrascu, M., 85, **88**
Voelcker, H.B., 442, **459**
Vogelius, M., 83, **88**

Waburton, G.B., 227, **256**
Wachspress, E.L., 49, **63**
Wagner, W., 203, **222**
Walker, S., 90, 98, **121**
Wang, H., 99, **122**, **123**
Wappi, A., 226, 231, **254**
Watson, D.F., 439, 446, 447, **458**
Watson, J.O., 98, **121**
Watson, L.T., 192, 199, **220**, **221**
Wempner, G.A., 200, **221**
Wehage, R.A., 262, **298**
Wiberg, N.E., 100, **123**, 440, 448, **458**
Widartawan, S., 243, **257**
Widlund, O.B., 24, **36**, 69, **87**
Wigton, L.B., 103, **123**
Wilkinson, J.H., 28, **37**
Willoughby, R.A., 84, **88**, 126, 140, **152**, **156**, 158, 162, 164, **180**

Winget, J., 10, **34**, 41, 50, **63**, 358, **388**
Wilde, D.J., 392, **425**
Wilson, E.L., 9, **34**, 125, 126, **152**, 157, **180**, 226, 227, 230, 231, 233, 235, 241, 243, 244, 253, **254**, **255**, **257**, 277, **299**, 329, **354**
Williams, R.D., 455, **460**
Wills, J., 199, 203, **221**, **222**
Wilson, R.B., **124**, 410, **427**
Winget, J.M., 457, **461**
Wisniewski, J.A., 126, 127, **152**
Wittchen, S.L., 442, **459**
Wolfe, M.A., 188, **219**
Wong, K.K., 268, **299**
Wong, Y.S., 136, **155**
Woodbury, A.D., 140, **155**
Wozniakowski, M., 28, **37**
Wriggers, P., 201, 203, 207, **222**
Wright, M.H., 188, 206, **219**, 393, 406, 408, 414, 415, 417, 423, **426**
Wrobel, L.C., 98, **121**, 226, 231, **254**
Wu, C.C., 418, **428**
Wu, J., 440, 448, **458**
Wu, H.C., 418, **428**
Wustenberg, H., 302, 315, 341, 344, 353, **354**

Xiong, S., 226, 231, **254**

Yakoumidakis, M., 128, **154**, 203, **222**
Yamamoto, S., 125, **152**
Yan, M.W., 226, 231, 253, **254**
Yang, Y.B., 201, **222**
Yerry, M.A., 442, 443, **459**
Yin, Y.C., 226, 231, **255**
Young, D.M., 7, **33**
Young, D.P., 103, **123**
Yserentant, H., 24, **37**
Yu, N.J., 103, **123**
Yuan, M., 226, 231, **254**

Zang, T.A., 136, **155**
Zhang, Y., 192, **220**
Zhu, J.Z., 440, 448, **458**
Ziegler, H., 341, **354**
Zienkiewicz, O.C., 39, **63**, 125, **152**, 270, **299**, 302, 304, 341, 347, **353**, **354**, 392, **425**, 440, 448, **458**
Zilli, G., 9, **34**, 136, **155**
Zlatev, Z., 100, **124**
Zoboli, M., 125, **152**
Zutendijk, G., 406, **426**
Zyvoloski, G., 136, **155**

Subject Index

Adaptive finite element formulation, 24
Adaptive iterative solver, 70
   numerical tests, 76–81
Advancing front techniques, 400
   mesh generators, 450–451
Angular momentum conservation law, 268
Angular velocity, 266
Arc length method, 200–201
Augmented Lagrangian, 395

BFGS inverse update, 189–190
   constrained, 198–199
Block-type factorization, 22
Boolean connectivity matrix, 42
Broyden's inverse update, 189

Capacitance matrix, 22; see also Schur complement matrix
CG for eigenvalue problems, 126, 131–135
   comparison to Lanczos method, 140–150
   preconditioned, 131–135
   storage requirements, 149
Chebyshev polynomials, 17
Chebyshev semi-iterative method, 184
Central difference algorithm, 266
Clustered EBE, 358, 369–370, 386–389
   comparison to direct methods, 371
   Crout factorization, 370
   numerical tests, 370–371
Coefficient matrix, 53
   block arrow structure, 53
Colouring algorithms, 456
Computation design optimization, 394
Computation labour, 1
Condensed zone matrix, 93
Condition number, 7, 41, 68
Conjugate gradient method (CG), 3–4, 39, 45–47
   algorithmic form, 48, 68
   asymptotic convergence, 134–135
   conjugacy condition of, 46
   domain decomposition algorithmic form, 23
   EBE algorithmic form, 16
   in indefinite problems, 47, 204
   instability of iterative process, 47
   non-linear, 185
   relation to Lanczos method, 6

   special, 192
   special constrained, 199
   unsymmetric, 100, 103
Conjugate gradient normal (CGN), 100–101
   preconditioned, 101–103
Conjugate-Newton method, 187
Constrained conjugate direction method, 423–424
Constrained quasi-Newton, 411
   Hessian approximation of, 414
   modified BFGS formula, 414
Constraint move limits, 399
Constraint normalization, 396
Constraint set, 392
   estimation of, 68
Constrained steepest descent method, 400–401
   computational algorithm, 404
   Newton technique, 314, 343
   non-linear iterations, 312–314
Control-structure interaction (CSI) problems, 260–269
   efficiency of, 293
   equation of, 270–271
   parallel implementation, 292–293
   solution algorithms, 269–277
Cost function, 392; see also Objective function
Coupled-field direction integration, 303–312
   stability analysis of, 305–312
Coupled-field problems, 260, 270, 301–355
   finite element modelling, 337
   formulation of, 303–312
   numerical tests, 282–297
   parallel processing of, 326–335
   solution of, 312–340
Coupled-field solution with subdomains, 315–326
   amplification matrix, 316, 318
   comparison of solution procedures, 319–323
   Gauss–Seidel iteration, 318, 326–336
   Jacobi iteration, 315, 326–336, 339
   parallel processing of, 326–335
Coupled-field solution taken as a whole, 312–315
Convexity, 393
Critical points, 195
   bifurcation, 202
   snap-back, 202
Crout EBE preconditioner, 13
Crout factorization, 367

468 SUBJECT INDEX

Current stiffness parameter, 202

Davidon's inverse update, 189
Deflation in eigenvalue problems, 133, 141
  with PCG, 141, 144–145, 150
Deforming-spatial domain/space–time
  (DSD/ST) procedure, 358, 363
Delaunay mesh generators, 447–448
Delaunay triangulation, 440
Descent function, 396, 416; see also Merit function
Design optimization model, 392
  basic algorithm, 394
  convergence of an algorithm, 397
  definition of a good algorithm, 398
  numerical algorithm, 394–398
  numerical implementation aspects, 417
  primal methods, 395
Design sensitivity analysis, 420
Design variables, 392
Deviatoric viscocity coefficient, 347
DFP inverse update, 189
Diagonal scaling, 9, 10, 19, 31; see also
  Diagonal preconditioning
Direct methods, 1, 2, 7, 23, 39, 40, 65
  comparison to iterative solvers, 11–13, 66–67,
    76–81, 86, 110, 113, 115–116, 371
  iterative improvement, 28
Direct sparse solvers, 65
  comparison to GMRES in BEA, 110–111,
    113–116, 119
  in boundary element analysis BEA, 98–99
  multi-frontal, 77
Direct time integration methods, 259, 357,
  361–364
  in coupled-field problems, 303–312
Discretization error, 24
Divide and conquer, 269
Domain decomposition, 21, 350, 439, 443
  algebraic level, 21, 25
  element level, 350
  multi-block algorithm, 350
Domain decomposition-based preconditioning,
  21–26
  numerical tests, 24–26
  partitioned matrix implementation, 22
Duality, 393
Dynamic relaxation method, 185–186

ε active constraints, 392
EBE, 10, 11
EBE splitting, 42
Edge compatibility, 441
Eigen-dependent vector bases, 226
Eigenvalues in engineering, 128–131, 158–162, 225
  in buckling analysis, 129, 161
  in dynamic analysis, 128–129
  in optical waveguides analysis, 129
  in vibration analysis, 129, 159
Element-by-element (EBE) preconditioners, 10–21,
  31, 49–51, 358, 369–370, 386–389
  additive form, 15–21, 32, 53
  Cholesky, 50

  computer implementation, 51
  LU split, 50
  product form, 13–15, 50
Element grouping algorithm for arbitrary meshes,
  366–367
Energy norm, 69
Error in energy balance, 245
Errors in finite element computation, 26
  discretization, 18
  round-off, 26, 27, 28, 47
  truncation, 26, 27, 29
Exact arithmetic, 6, 39, 69
Explicit integration, 265, 304, 343–347, 351
Explicit methods, 184

Face compatibility, 441
Factorization, 1, 8
  block-triangular unsymmetric, 97–98
  block-type, 22
  Cholesky, 42, 49
  Crout, 367
  Gauss–Seidel, 368
  incomplete, 8, 9
  LU, 5
  symbolic, 22
Feasible set, 392; see also Constraint set
Fill-in, 2, 40, 66, 73, 91
  block in BEA, 114
Finite precision arithmetic, 6, 26–30, 47
  in Lanczos vectors, 163

Galerkin approximation, 43–44
Galerkin/least-squares (GLS)
  stabilization, 357, 363, 386
Gauss–Seidel method, 15, 318, 326–336, 339,
  343, 368
  convergence behaviour, 335
  parallel processing of, 326–335
Gauss–Seidel non-symmetric EBE preconditioner,
  368
Gauss–Seidel symmetrized EBE preconditioner, 15
Generalized eigenvalue problem, 125–159
  numerical tests, 140–151
Generalized minimum residual method (GMRES),
  103–106
  comparison to direct solvers in BEA, 110–111,
    113, 115–116, 119
  non-restarted, 105
  preconditioned algorithm, 105
  preconditioners of, 105–106, 114
  restarted, 105
Generalized reduced gradient method, 408–410
Geometric modelling, 451–453
  multiple, 452
Geometric similarity, 435
Geometric triangulation, 433
Global connectivity matrix, 12
Global preconditioners, 7–10, 31
  numerical tests, 9
Gradient evaluation, 420
  adjoint variable method, 421
  direct differentiation method, 420

SUBJECT INDEX

Gradient projection method, 407–408
Graph colouring, 53
Gram–Schmidt orthogonalization, 4, 43, 104, 132, 231
Grouped EBE, 359, 386–389
   Crout factorization, 367
   Gauss–Seidel factorization, 368
   GMRES, 365, 369
   two-pass preconditioner, 368
   vectorization/parallelization aspects, 364–369

Hierarchical basis, 24
Hierarchical serendipity elements, 70
Hierarchy of discretization, 67
High performance computers, 164–165
Heat conductivity matrix, 342
Heat capacity, 342
   matrix, 342
Hessenberg matrix, 104
Hessian matrix, 412
Hybrid solution methods, 5, 7, 52, 62

Ill-conditioning, 6, 9, 10, 28, 31, 41
   effect of, 58–61
   geometric, 55
   in highly anisotropic materials, 83
   in incompressible materials, 83
   material, 55
   with respect to element aspect ratio, 18, 76, 83
Implicit integration, 304, 343, 347, 351
Implicit methods, 184, 186–194
   algorithmic form, 190–191
   constrained, 197–200
   numerical tests, 208–217
Incomplete factorization preconditioning, 8, 9, 22, 31
   algorithmic form, 8
   diagonal modification, 10
   by magnitude, 8, 9
   numerical tests, 9, 11–14
   by position, 8, 9, 25, 114, 136
   for RQ optimization, 136
   stability of, 10
Incompressible flows, 357
Inertia of a matrix, 203
I/O operations, 6, 49, 81, 99
Iterative methods, 2, 7, 65
   comparison to direct solvers, 11–13, 66–67, 76–81, 86, 110, 113, 115–116, 371
   effect of ill-conditioning, 58–61
Iterative methods under finite precision arithmetic, 26–30
   numerical tests, 28–30
Iterative solution techniques in BEA, 99–100
   accuracy considerations, 112–113
   advantages of preconditioning, 99
   unsymmetric preconditioning, 100

Jacobi method, 315, 326–336
   convergence behaviour, 335

Krylov subspace, 4, 43, 103, 136, 365

Lagrange multipliers, 261–262, 264
   Hessian matrix of, 410, 412
Lagrangian for design optimization, 411
Lanczos method, 4, 5, 6, 28, 39, 43–45
   accumulated, 5
   algorithmic form, 5, 45
   effect of preconditioning, 56
   orthogonality property, 48
   partial reorthogonalization, 28
   preconditioned algorithmic form, 5
   relation to CG, 6
   reorthogonalization, 6
   semi reorthogonalization, 6
   the three-term of, 4, 43
Lanczos method for eigenvalue problems, 126, 136–140, 157, 162
   accuracy of, 148
   block, 158, 179
   in buckling analysis, 160
   comparison of PCG, 140–141
   convergence of, 147
   implementation aspects, 168–171
   numerical tests, 172–179
   orthogonalization aspects, 139, 149, 151, 178–179
   parallel implementation, 176
   storage requirements, 149
   in vibration analysis, 160
   vectorized implementation, 172–176
Lanczos vector matrix, 137
Lanczos vectors, 43–44, 163
Least squares, 17
Line search, 204–207
   algorithmic form, 205–206
   with the arc length, 207
   curve fitting technique, 205
   with fixed loading, 204, 206
   regula falsi technique, 205
   stability test, 206
   in unstable equilibrium, 206–207
   with variable loading, 206–207
Linear programming (LP), 392
Load dependent transformation method (LDM), 231–234, 244
   generation of vectors, 231–232
   block vector algorithmic form, 234
   numerical tests, 234–242
   single vector algorithmic form, 234

Macro element matrix, 12
Matrix-vector multiplication, 171–173
Mechanical work, 342
   hydrostatic, 342
   deviatoric, 342
Merit function, 396; *see also* Descent function
Mesh classification, 432
Mesh generation, 336, 345, 432
   advancing front, 450–451
   automatic, 437
   Delaunay, 447–448
   geometric representation and decomposition, 438

key issues, 437–442
medial axis transformation, 450–451
mesh validity, 440–442
paving, 448–450
solution procedures for adaptively defined meshes, 455
triangulation, 439–440
Metal forming, 341, 352
Mid-point implicit rule, 266
Mixed precision arithmetic, 29, 32
Modal summation methods, 229–231
mode displacement summation method (MDM), 230, 244
numerical tests, 234–242
Mode superposition methods, 226–231; see also Vector superposition methods
algorithmic form, 228
for non-linear problems, 242–253
Modified central difference algorithm, 261
Modified Newton–Raphson method, 187
constrained, 197
Multi-element group partitioning, 32; see also Domain decomposition; Substructures
Multibody dynamic (MBD) analysis, 260–269
constraint degrees of freedom, 260
explicit integration, 265–266
parallell implementation, 262–264
Schur complement of, 263
techniques for handling constraint conditions, 262–265
time integration, 265–269
Multi-grid methods, 66–67, 69, 82
Multiplier methods, 395; see also Augmented Lagrangian
Multi-zone BEA, 89–97
numerical tests, 107–118
performance of, 107–118
with substructures, 95–97, 111

Navier–Stokes equations, 360
Non-linear programming (NLP), 392
model, 393
standard NLP model, 392
Non-linear transient finite element equations, 40
Neumann series, 17
Newton method, 186, 314, 343
Newton–Raphson method, 186
constrained, 197
conventional, 186–188
modified, 187
step length of, 187

Objective function, 392; see also Cost function
Octree mesh generators, 442
domain decomposition, 443
finite octree, 443–444
geometric representation, 443
mesh validity, 446
pulled octant, 443–444
Optimality conditions, 393, 400
Orthogonality aspects, 47–49, 62, 139–149
effect of, 55

Parallel Cholesky decomposition, 326–329, 333
performance of, 334
Parallel processing, 166–168
granularity, 167
dedicated versus batch mode, 168
Parallel processing of coupled-field problems, 326–335
numerical tests, 332–336
physical decomposition, 330–332
spatial decomposition for, 327
Parallel solution of automatically generated adaptive meshes, 453–458
Parallel substructuring, 329–330, 333, 349
performance of, 333
Parametric intersection, 435
Partial orthogonalization, 73
Partial reorthogonalization, 28; see also Semi-reorthogonalization
Partitioned solution for CSI, 269, 272–277
stabilization of, 272–277
Paving techniques, 440
Powell's restart criterion, 191
Penalty approach, 341
Penalty parameter, 55–56, 402
Polynomial preconditioning, 15, 17–21
numerical tests, 18–21
Potential constraint strategy, 396, 405
Preconditioned conjugate gradient method (PCG), 3, 71–75
algorithmic form, 3
Preconditioning matrix, 3, 5, 7, 22, 40
adaptive, 73–75
band decomposition, 102, 110–111
block, 7
block diagonal, 65, 73, 102, 106, 109, 110–111, 114, 118
block incomplete factorization, 22
Cholesky decomposition, 75
clustered EBE, 358, 369–370, 386–389
diagonal, 26, 41, 102, 106, 109, 110–111, 117
domain decomposition, 7, 21–26, 66, 69, 85
global, 7–10
grouped EBE, 359, 386–389
EBE, 7, 10–21
for eigenvalue problems, 135–136, 141, 143, 145
Jacobi, 365
incomplete LU decomposition, 84, 114
incomplete symmetric elimination, 65, 72
indefinite, 8, 205
LU decomposition, 5
parallel implementation aspects, 18, 83
polynomial, 15, 17, 18
for the Schur complement, 24
SSOR, 7, 457
substructure-by-substructure, 51
wrap-around, 6
Pressure-stabilizing/Petrov–Galerkin (PSPG) stabilization, 359, 363
Profile solver, 62

Pseudo-force method (PFM),
    242–245
  algorithmic form, 246–247
Pshenichny's descent function,
    401–402, 404–405
p-version adaptive finite
    element formulation, 24, 65, 70–86

Quadratic programming (QP), 392
Quasi-Newton methods, 188–193, 410–416;
    see also Variable metric methods
  constrained, 197–200, 411
  limited memory, 191–193
  sparsity preserving update, 188
  special, 191
  rank-one update, 188
  rank-two update, 188
  an SQP algorithm, 415

Rayleigh quotient (RQ), 126
    the Hessian of, 133, 135
    optimization of RQ by accelerated CG, 131–135
Recursive quadratic programming (RQP), 411
Region compatibility, 441
Regula-falsi technique, 205
Rejection parameter, 7, 8, 9, 17, 19, 29
Reorthogonalization of Lanczos vectors, 163
Residual vector, 3
  of CG, 28
  defining formula of, 28
  of GMRES, 104
  of Lanczos method, 28
  recursive expression of, 28
Rough non-linearities, 186, 201
Round-off unit, 6, 48

Schur complement matrix, 22, 263; see also
    Capacitance matrix
Search direction, 3
Secant-Newton techniques, 192
Second order Richardson process, 184
Selective reorthorgonalization, 49, 140
Semi-reorthogonalization, 6, 49, 55, 139
  the cost of, 57
Sequential linear programming (SLP), 399–400
Sequential quadratic programming (SQP),
    400–406, 410–411, 422
Shuttle remote manipulator system, 283–284
Similarity transformation, 203
Simultaneous solution for CSI, 269, 271–272
Skyline storage, 9, 11
Smooth non-linearities, 186–201
Sparse matrix, 1, 25, 40
Sparse storage schemes, 9, 11, 51
Sparcity pattern, 9
  irregular, 9
  regular, 9
Spurious eigenvalues, 163
SSOR preconditioning, 7, 9, 26, 31
Stabilization of staggered solution procedures,
    278–282
  analysis of accuracy, 280

analysis of stability, 280
computational sequence, 282
Stabilized space–time finite element formulation,
    361
  direct time integration of,
    361–364
Staggered implicit–explicit solution
    procedures, 278
  stabilization of, 278–282
Staggered MBD solution procedure, 260
  implicit, 261
Standard eigenvalue problem, 158
Step length parameter, 3, 187, 394
Step size, 394
  determination of, 402–403
Stokes flow, 358
Storage reduction schemes, 179
Streamline-upwind/Petrov–Galerkin (SUPG)
    stabilization, 359, 363
Subspace iteration, 126, 157, 237
  comparison to LDM, 237–240
Substructures, 21, 329–330, 333, 349, 423;
    see also Superelements
  boundary element, 92–97
Successive over-relaxation method (SSOR), 3, 194
Superelements, 17, 32, 51; see also Substructures
Sylvester theorem, 204
Symbolic factorization, 22

Tangent spectrum method (TSM), 242–245
  algorithmic form, 246–247
Thermal-structure interaction problems, 260
  semidiscrete equations, 277
  solution of, 277–282
  staggered solution procedures, 278
Thermomechanical coupling, 323–326, 352
  energy balance, 342
  numerical tests, 341–353
Three-term recursive expression, 3, 184
Topological compatibility, 433
Tracing non-linear equilibrium paths, 194–204
  algorithmic form, 197
  constraint relationship, 200
  incremental control along the equilibrium path,
    201–203
  incremental-iterative formulation, 194–197
  iterative control inside an increment, 200–221
Transformation matrix, 71
Truncated Newton-like methods, 193
  forcing sequence of, 193

Uniquely mappable, 434
Unsteady incompressible flows, 360
  geometric equations of, 360–361
  numerical tests, 370–386
  past a circular cylinder 370
Unsymmetric conjugate gradient method, 100
  algorithmic form, 103

Valid finite element mesh, 432–437
Variable band Cholesky solver in parallel–vector
    computers, 169–171
  column storage versus skyline storage, 170–171

reordering of nodes, 170
storage schemes, 169
Variable metric methods, 188; *see also* Quasi-Newton methods
  constrained, 411
Vector processing, 165–166
  compiler directives, 166
  local memory
  loop unrolling, 165
Vector superposition methods, 226–231; *see also* Mode superposition methods
  formation and solution of, 227–229
Vector superposition methods in non-linear problems, 242–252
  comparison of, 243, 249–250
  solution of, 244–245
Viscosity matrix, 342
Volumetric viscosity coefficient, 347

Zone condensation and expansion algorithm, 95